Air Quality

4th Edition

Thad Godish

LEWIS PUBLISHERS

A CRC Press Company
Boca Raton   London   New York   Washington, D.C.

Cover image provided by the SeaWiFS Project, NASA / Goddard Space Flight Center and ORBIMAGE.

Library of Congress Cataloging-in-Publication Data

Godish, Thad.
    Air quality / Thad Godish. -- 4th ed.
        p.   cm.
    Includes bibliographical references and index.
    ISBN 1-56670-586-X
    1. Air--Pollution. 2. Air quality management. I. Title.

TD883.G57 2003
363.739′2—dc21   2003047412

Visit the CRC Press Web site at www.crcpress.com

© 2004 by CRC Press LLC
Lewis Publishers is an imprint of CRC Press LLC

No claim to original U.S. Government works
International Standard Book Number 1-56670-586-X
Library of Congress Card Number 2003047412
Printed in the United States of America   5 6 7 8 9 0
Printed on acid-free paper

Dedication

To the memory of Ed Lewis, publisher and friend.

This book is your continuing legacy.

Preface

This fourth edition of *Air Quality* has been written (as have previous editions) to provide readers (students, instructors, consultants, government personnel, and others) with a comprehensive overview of air quality, the science that continues to provide a better understanding of atmospheric chemistry and effects on public health and the environment and the regulatory and technological management practices employed in achieving air quality goals.

It has been nearly two decades since I began to pen the words that comprised the first edition. Much has changed. The Internet, that enormous library of information that was so much a part of the writing of this edition, was virtually unknown outside the military and academe (and even there it was still a foundling). The word processor had not quite replaced the typewriter (when "cutting and pasting" was literally that). How long ago technologically the early to mid-1980s now seem.

In the U.S. we are now well into our fourth decade of significant regulatory efforts to "protect and enhance" the quality of the nation's air. We accepted the challenge of expanding our concerns to protect the Earth's atmosphere from the effects of ozone-destroying chemicals that have threatened to destroy the ozone layer. We are, at present, timidly moving toward accepting the challenges that the planet faces from greenhouse gas-associated increases in global surface temperatures and the environmental changes predicted to occur and already occurring.

In the 1970s and 1980s, acidic deposition was identified as a major environmental concern in North America, and we responded by enacting and then implementing major acidic deposition control measures. In conducting research studies on acidic deposition, it became increasingly apparent that it was only a part of a much larger environmental concern, atmospheric deposition, which includes mercury, nitrate nitrogen, and organochlorine compounds such as pesticides, polychlorinated biphenyls (PCBs), dioxins, and furans.

The health protection issues that dominate air quality management in the U.S. continue to evolve as more powerful statistical procedures increasingly demonstrate that pollutant exposures at levels previously considered safe cause adverse health effects, with a resultant need for more stringent regulatory requirements. This has been particularly the case for ozone and $PM_{2.5}$ (particulate matter with an aerodynamic diameter of ≤ 2.5 μm).

Though we Americans and other citizens of the planet want clean air, regulatory efforts impose real costs on the regulated community. Not surprisingly, American industry is reluctant to bear the costs of new regulatory initiatives and is increasingly challenging the historically better and better science that is driving the need for more stringent environmental regulation.

This edition represents a significant revision of the previous one, including updated content and changes to the way in which subject matter is presented. Chapters 2, 4, 5, 6, and 10 reflect the first category, and Chapters 3, 6, 7, 9, 11, and 12 the second. This edition also includes an extensive glossary.

Air Quality continues to be a very readable text for advanced-level undergraduate and beginning-level graduate students in environmental science; environmental management technical programs at the junior college level; and programs serving public health, industrial hygiene, and engineering needs. It is useful as a supplement to engineering curricula where the primary focus is the design and operation of pollution control equipment. It is also written for a variety of nonuniversity and noncollege readers who have a professional or personal interest in the field.

Thad Godish
Muncie, Indiana

Acknowledgments

I am indebted to my late colleague, Clyde Hibbs, professor of natural resources at Ball State University, who encouraged me to write an air pollution book to serve the needs of our environmental science majors. I am also indebted to the late Ed Lewis, who visited my office one day in 1984 and invited me to be an author in his fledgling Lewis Publishers, Inc. Flattered, I nevertheless felt overwhelmed by my increasingly expanding research, consulting, and public service activities in the then-emerging field of indoor air quality. Ed's entreaties continued, and the first edition of *Air Quality* was the outcome. Thus began a professional and personal relationship that significantly shaped my career.

In this writing effort I was provided the invaluable assistance of my spouse, Diana, whose editorial skills considerably improved the readability and ultimate quality of this edition.

As always, I am indebted to the many colleagues in colleges and universities in the U.S. and other parts of the world who have elected to use *Air Quality* in their courses. You have allowed me to expand my classroom to a much wider domain.

Contents

Chapter 1   The Atmosphere
1.1   Chemical Composition 1
    1.1.1 General Composition 1
    1.1.2 Water 4
1.2   Physical Phenomena 6
    1.2.1 Solar Radiation 6
    1.2.2 Terrestrial Thermal Radiation 7
    1.2.3 Energy Balance 7
    1.2.4 Surface Temperatures...9
    1.2.5 Vertical Temperature Gradients 9
    1.2.6 Gravity 11
    1.2.7 Atmospheric Density 11
    1.2.8 Atmospheric Pressure 12
1.3   Atmospheric Motion 12
    1.3.1 Winds 13
    1.3.2 Pressure Gradient Force 13
    1.3.3 Coriolis Effect 14
    1.3.4 Friction..14
    1.3.5 Cyclones and Anticyclones 15
    1.3.6 General Circulation of the Atmosphere 17
1.4   Evolution of the Atmosphere...19
Readings 21
Questions 21

Chapter 2   Atmospheric Pollution and Pollutants
2.1   Natural Air Pollution 23
2.2   Anthropogenic Air Pollution 24
    2.2.1   Smog 24
    2.2.2   Haze 26
    2.2.3   Nontraditional Air Pollutants 26
    2.2.4   Gases and Particles 26
        2.2.4.1 Sources 27
        2.2.4.2 National Emission Estimates................................27
        2.2.4.3 The National Emissions Picture 29
2.3   Gas-Phase Pollutants 31
    2.3.1   Carbon Oxides 31
        2.3.1.1   Carbon Dioxide 31
        2.3.1.2   Carbon Monoxide 33
    2.3.2   Sulfur Compounds 36
        2.3.2.1   Sulfur Oxides 36
        2.3.2.2   Reduced Sulfur Compounds 38
    2.3.3   Nitrogen Compounds 38
        2.3.3.1   Nitrous Oxide 39

        2.3.3.2   Nitric Oxide 39
        2.3.3.3   Nitrogen Dioxide 40
        2.3.3.4   Nitrogen Oxides 41
        2.3.3.5   Ammonia 42
        2.3.3.6   Hydrogen Cyanide 42
        2.3.3.7   Organic Nitrate Compounds 42
    2.3.4   Hydrocarbon Compounds 43
        2.3.4.1   Hydrocarbon Chemistry 43
        2.3.4.2   Oxyhydrocarbons 45
        2.3.4.3   Nonmethane Hydrocarbons 46
        2.3.4.4   Methane 49
        2.3.4.5   Halogenated Hydrocarbons 50
    2.3.5   Photochemical Oxidants 51
        2.3.5.1   Tropospheric Ozone Concentrations 54
        2.3.5.2   Ozone Sinks 55
2.4   Atmospheric Particles 55
    2.4.1   Classes and Sources 56
    2.4.2   Size 56
    2.4.3   Particle Shape 60
    2.4.4   Chemical Composition 60
    2.4.5   Source Apportionment 63
    2.4.6   Behavior 63
    2.4.7   Concentrations 65
    2.4.8   Long-Range Transport from Sources Outside the U.S. 66
Readings 67
Questions 68

Chapter 3   Atmospheric Dispersion, Transport, and Deposition
3.1   Dispersion and Transport 71
    3.1.1   Wind 73
    3.1.2   Turbulence 74
    3.1.3   Atmospheric Stability 75
    3.1.4   Inversions 76
        3.1.4.1   Radiational Inversions 77
        3.1.4.2   Subsidence Inversions 78
    3.1.5   Mixing Height 80
    3.1.6   Topography 81
    3.1.7   Pollutant Dispersion from Point Sources 82
        3.1.7.1   Pollutants and Diffusion...82
        3.1.7.2   Plume Rise and Transport 82
        3.1.7.3   Plume Characteristics 83
    3.1.8   Large-Scale Transport and Dispersion 85
        3.1.8.1   Long-Range Transport 85
        3.1.8.2   Urban Plume 85
        3.1.8.3   Planetary Transport 85
    3.1.9   Stratosphere–Troposphere Exchange 86

  3.1.10 Stratospheric Circulation 86
3.2 Atmospheric Removal and Deposition Processes 87
  3.2.1 Atmospheric Lifetimes 87
  3.2.2 Deposition Processes 88
3.3 Meteorological Applications: Air Pollution Control 89
  3.3.1 Episode Planning 90
  3.3.2 Air Quality Modeling 90
Readings 90
Questions...91

Chapter 4 Atmospheric Effects
4.1 Visibility 93
  4.1.1 Natural Visibility Reduction 94
  4.1.2 Visibility Concerns Associated with Human Activities 94
    4.1.2.1 Smog 95
    4.1.2.2 Regional Haze 95
    4.1.2.3 Pristine Air 95
    4.1.2.4 Arctic Haze...95
    4.1.2.5 South Asian Haze 96
  4.1.3 Factors Affecting Visibility Reduction 97
  4.1.4 Meteorological Effects 99
  4.1.5 Visibility Measurements 100
  4.1.6 Visibility Patterns and Trends 101
4.2 Turbidity 103
4.3 Thermal Air Pollution 104
4.4 Effects on Precipitation and Precipitation Processes 105
4.5 Atmospheric Deposition 106
  4.5.1 Acidic Deposition 106
    4.5.1.1 Characteristics of the Phenomenon 106
    4.5.1.2 Environmental Impacts 108
    4.5.1.3 Trends in Sulfate and Acidic Deposition 110
  4.5.2 Nitrogen Deposition 110
  4.5.3 Mercury Deposition 111
  4.5.4 Deposition of Pesticides, PCBs, Dioxins, and Furans ...113
4.6 Stratospheric Ozone Depletion 113
  4.6.1 Ozone Depletion Concerns 113
  4.6.2 Ozone Layer Dynamics 114
  4.6.3 Role of Nitrogen Oxides 115
  4.6.4 Role of Chlorofluorocarbons and Other Halogenated
     Hydrocarbons 116
  4.6.5 The Antarctic Ozone Hole 118
  4.6.6 Ozone Depletion in the Arctic 121
  4.6.7 Trends 121
  4.6.8 Changes in Surface UV Light Radiation 122
  4.6.9 Ozone Depletion Effects on Human Health 123
  4.6.10 Recovery of the Ozone Layer 124

4.7 Global Warming 124
    4.7.1 Climate and Global Climate 124
        4.7.1.1 Orbital Changes and Climatic Variation 125
        4.7.1.2 Variations in Solar Radiation 126
    4.7.2 The Greenhouse Effect and Global Warming 128
    4.7.3 Radiative Forcing 128
    4.7.4 Greenhouse Gases 129
    4.7.5 Evidence of Climate Change 131
    4.7.6 Linking Climate Change to Anthropogenic Influences ... 133
    4.7.7 Projections of the Earth's Future Climate 134
    4.7.8 Global Warming Uncertainties 135
    4.7.9 Environmental Impacts 136
    4.7.10 Stratospheric Gases and Atmospheric Temperature
            Changes 137
Readings 137
Questions 139

Chapter 5   Health Effects
5.1 Air Pollution Episodes 141
5.2 Pollutant Exposures and Cause–Effect Relationships 142
    5.2.1 Epidemiological Studies 142
        5.2.1.1 Confounding Factors 143
        5.2.1.2 Interaction Effects 143
        5.2.1.3 Exposure Assessment 145
        5.2.1.4 Population Susceptibility 145
    5.2.2 Toxicological Studies 147
        5.2.2.1 Human Studies 147
        5.2.2.2 Animal Studies 148
    5.2.3 Occupational Exposure Studies 148
5.3 Impact of Pollutants on the Human Body 149
    5.3.1 Eye Irritation 149
    5.3.2 Effects on the Cardiovascular System 149
    5.3.3 Effects on the Respiratory System 150
        5.3.3.1 Units of the Respiratory System 150
        5.3.3.2 Defense Mechanisms of the Respiratory System 152
        5.3.3.3 Air Pollution and Respiratory Disease 153
5.4 Health Effects of Regulated Air Pollutants 156
    5.4.1 Carbon Monoxide 156
    5.4.2 Sulfur Oxides 159
        5.4.2.1 Sulfur Dioxide 159
        5.4.2.2 Acid Sulfate 161
    5.4.3 Particulate Matter 162
    5.4.4 Hydrocarbons 166
    5.4.5 Nitrogen Oxides 167
    5.4.6 Ozone 169
    5.4.7 Lead.. 170

    5.4.8   Hazardous Air Pollutants 174
5.5   Personal Air Pollution 175
5.6   Risk Assessment and Management 177
Readings 179
Questions 181

Chapter 6   Welfare Effects
6.1   Effects on Agricultural Crops, Ornamental Plants, and Trees 183
    6.1.1   Plant Injury 185
          6.1.1.1   Sulfur Dioxide 185
          6.1.1.2   Ozone 186
          6.1.1.3   Peroxyacyl Nitrate 187
          6.1.1.4   Fluorides 188
          6.1.1.5   Ethylene 189
          6.1.1.6   Acidic Deposition 189
          6.1.1.7   Interaction Effects 190
    6.1.2   Economic Losses 190
    6.1.3   Forest Declines 191
          6.1.3.1   California Pines 191
          6.1.3.2   Eastern White Pine 192
          6.1.3.3   Red Spruce and Fraser Fir 192
          6.1.3.4   Pines in the Southeast 192
          6.1.3.5   Hardwoods in the Northeast............................... 193
          6.1.3.6   Declines in Central Europe 193
6.2   Effects on Domesticated Animals 194
    6.2.1   Fluoride 194
    6.2.2   Lead 195
    6.2.3   Other Toxic Exposures 195
6.3   Ecological Effects 195
    6.3.1   Phytotoxic Pollutants 196
    6.3.2   Atmospheric Deposition 197
          6.3.2.1   Acidic Deposition.. 197
          6.3.2.2   Nitrogen 199
          6.3.2.3   Mercury and Other Trace Metals 201
          6.3.2.4   Organochlorine Compounds 201
    6.3.3   Stratospheric Ozone Depletion 203
    6.3.4   Global Warming 203
          6.3.4.1   Polar Regions 204
          6.3.4.2   Coastal Zones 204
          6.3.4.3   Plant and Animal Communities 204
6.4   Effects on Materials... 205
    6.4.1   Metal Corrosion 206
    6.4.2   Building Materials 207
    6.4.3   Paints 208
    6.4.4   Textiles and Textile Dyes 209
    6.4.5   Paper and Leather 210

    6.4.6   Rubber 210
    6.4.7   Glass 211
    6.4.8   Economic Losses 211
6.5   Odor Pollution 211
    6.5.1   Odor Measurement 212
    6.5.2   Odor Problems 213
Readings 213
Questions 215

Chapter 7   Air Quality and Emissions Assessment
7.1   Air Quality Monitoring 217
    7.1.1   Monitoring Considerations 218
        7.1.1.1   Sampling 218
        7.1.1.2   Averaging Times 219
        7.1.1.3   Sampling Techniques 219
    7.1.2   Reporting Pollutant Concentrations 224
        7.1.2.1   Accuracy, Precision, Bias 226
        7.1.2.2   Calibration 228
    7.1.3   Federal Monitoring Requirements 229
        7.1.3.1   Air Quality Monitoring Networks 230
        7.1.3.2   Federal Reference Methods 231
        7.1.3.3   Quality Assurance Programs 235
        7.1.3.4   Data Summarization and Characterization 235
        7.1.3.5   Air Quality Index 239
7.2   Emissions Assessment 240
    7.2.1   Source Sampling and Monitoring 240
    7.2.2   Opacity 242
    7.2.3   Emission Factors 242
7.3   Air Quality Modeling 245
    7.3.1   Gaussian Models 246
    7.3.2   Model Options 252
Readings..253
Online Sources 254
Questions...254

Chapter 8   Regulation and Public Policy
8.1   Nonregulatory Legal Remedies 257
    8.1.1   Nuisance..258
    8.1.2   Trespass 258
8.2   Regulatory Strategies and Tactics 259
    8.2.1   Strategies 259
        8.2.1.1   Air Quality Management 259
        8.2.1.2   Emission Standards 260
        8.2.1.3   Economics-Based Approaches....................261
    8.2.2   Tactics 264

8.3   Federal Legislative History 265
        8.3.1   1955 Clean Air Legislation 265
        8.3.2   1963 Clean Air Act 265
        8.3.3   Motor Vehicle Air Pollution Control Act 265
        8.3.4   The Air Quality Act of 1967 266
        8.3.5   1970 Clean Air Act Amendments 266
        8.3.6   1977 Clean Air Act Amendments 267
        8.3.7   1990 Clean Air Act Amendments 267
        8.3.8   Federalization 267
8.4   Air Pollution Control under the 1970, 1977, and 1990 CAA
        Amendments 268
        8.4.1   Air Quality Standards 268
        8.4.2   Air Quality Criteria 268
        8.4.3   Air Quality Control Regions 269
        8.4.4   State Implementation Plans 269
                8.4.4.1   Episode Plans 270
                8.4.4.2   Legal Authority 270
                8.4.4.3   State Implementation Plan Approval...............270
        8.4.5   Nonattainment of Air Quality Standards 270
        8.4.6   New Source Performance Standards 272
        8.4.7   Prevention of Significant Deterioration 273
        8.4.8   Hazardous and Toxic Air Pollutants 275
        8.4.9   Stratospheric Ozone Depletion 276
        8.4.10 Acidic Deposition 277
        8.4.11 Motor Vehicle Emission Control 278
                8.4.11.1   Vehicle Emissions 278
                8.4.11.2   Regulatory History 278
                8.4.11.3   Exhaust Gas Standards 279
                8.4.11.4   Light-Duty Motor Vehicle Emission Control
                          Requirements..281
                8.4.11.5   Fuel Additives and Cleaner Fuels 282
                8.4.11.6   Emission Test Cycles 282
                8.4.11.7   Inspection and Maintenance 283
                8.4.11.8   Tampering Prohibitions 283
                8.4.11.9   Onboard Vapor Recovery Systems 283
                8.4.11.10 Onboard Diagnostic Systems 284
                8.4.11.11 Emission Requirements for Trucks and Other
                          Vehicles 284
                8.4.11.12 Clean-Fuel Vehicles 284
                8.4.11.13 Clean Fleets 284
        8.4.12 Enforcement...284
        8.4.13 Permit Requirements 285
        8.4.14 Citizen Suits 286
        8.4.15 Pollution Prevention...286
8.5   State and Local Air Pollution Control Functions 287
        8.5.1   Boards and Agencies 288

      8.5.2   Effectiveness of State and Local Programs 288

8.6   Public Policy Issues 289
      8.6.1   Global Warming and the Kyoto Protocol 289
      8.6.2   Revised Ozone and Particulate Matter NAAQSs 290
      8.6.3   New Source Review 291
      8.6.4   Motor Vehicle Emissions 292
Readings 294
Questions 295

Chapter 9   Control of Motor Vehicle Emissions
9.1   Motor Vehicle Engines 299
      9.1.1   Otto Cycle Spark Ignition Engines 300
             9.1.1.1   Engine Characteristics 300
             9.1.1.2   Emissions 301
             9.1.1.3   Formation of Combustion Pollutants 301
             9.1.1.4   Emission Control Techniques and Systems 303
      9.1.2   Compression Ignition Engines 310
9.2   Automotive Fuels 312
      9.2.1   Gasolines 312
             9.2.1.1   Octane Rating 312
             9.2.1.2   Gasoline Composition and Emissions 313
             9.2.1.3   Reformulated Gasolines and Oxygenated Additives 313
      9.2.2   Alternative Fuels 314
             9.2.2.1   Alcohol Fuels 314
             9.2.2.2   Compressed and Liquified Gases 315
9.3   Low Emission and Zero Emission Vehicles 316
      9.3.1   Low Emission Vehicles 316
      9.3.2   Hybrid Electric Vehicles 316
      9.3.3   Electric Vehicles 318
Readings 318
Questions 319

Chapter 10   Control of Emissions from Stationary Sources
10.1 Control Practices 321
      10.1.1 Tall Stacks 321
      10.1.2 Fuel Use Policies 322
      10.1.3 Fugitive Emissions 322
      10.1.4 Pollution Prevention 323
             10.1.4.1 Substitution 323
             10.1.4.2 Process Equipment Changes 324
             10.1.4.3 Plant Operating Practices 324
             10.1.4.4 Maintenance 324
             10.1.4.5 Process Changes 324
             10.1.4.6 Energy Conservation 324
  10.2 Gas Cleaning Technology 325
      10.2.1 System Performance 325

10.2.2 Particle Collection Efficiency 326
10.2.3 Gas-Phase Contaminant Collection Efficiency 327
10.3 Control of Particulate-Phase Pollutants 328
10.3.1 Cyclonic Collectors 328
10.3.2 Electrostatic Precipitation 329
10.3.3 Filtration 331
10.3.3.1 Fabric Filters 331
10.3.3.2 Wet Scrubbers 332
10.4 Control of Gas-Phase Pollutants 334
10.4.1 Thermal Oxidation 335
10.4.1.1 Flare 335
10.4.1.2 Conventional Oxidizers 336
10.4.1.3 Catalytic Oxidation....................................... 336
10.4.1.4 Operating Considerations 337
10.4.2 Adsorption 337
10.4.3 Absorption 339
10.4.4 Control of Sulfur Oxides 341
10.4.4.1 Coal Beneficiation 342
10.4.4.2 Solvent Refining 342
10.4.4.3 Coal Gasification and Liquefaction 342
10.4.4.4 Flue Gas Desulfurization 342
10.4.5 Fluidized Bed Combustion 345
10.4.6 Biological Treatment 346
10.4.7 Control of Nitrogen Oxides 347
Readings 347
Questions 349

Chapter 11 Indoor Air Quality
11.1 Personal, Indoor, and Outdoor Pollutant Relationships 352
11.1.1 Personal Pollutant Exposures 352
11.1.2 Indoor and Outdoor Pollutant Concentrations 352
11.2 Indoor Air Quality Concerns 353
11.3 Major Indoor Pollutants 354
11.3.1 Asbestos 354
11.3.2 Radon 357
11.3.2.1 Concentrations and Sources 357
11.3.2.2 Health Effects and Risks 358
11.3.3 Combustion By-Products 359
11.3.3.1 Unvented Combustion Systems 359
11.3.3.2 Vented Combustion Systems 362
11.3.3.3 Miscellaneous Sources 363
11.3.4 Aldehydes 368
11.3.4.1 Formaldehyde 368
11.3.4.2 Acrolein 370
11.3.4.3 Glutaraldehyde 371
11.3.5 Volatile and Semivolatile Organic Compounds 371

        11.3.5.1 Volatile Organic Compounds 371
        11.3.5.2 Semivolatile Organic Compounds 372
    11.3.6 Biological Contaminants...375
        11.3.6.1 Illness Syndromes 376
        11.3.6.2 Other Health Concerns 378
11.4 Problem Buildings 379
    11.4.1 Building-Related Illness 380
    11.4.2 Sick Building Syndrome 380
    11.4.3 Field Investigations and Systematic Building Studies 380
    11.4.4 Risk Factors 382
11.5 Investigating Indoor Air Quality Problems 383
11.6 Measurement of Indoor Contaminants 385
11.7 Prevention and Control Measures 385
    11.7.1 Source Management 385
        11.7.1.1 Exclusion 385
        11.7.1.2 Source Removal 386
        11.7.1.3 Source Treatment 386
    11.7.2 Contaminant Control 386
        11.7.2.1 Ventilation 386
        11.7.2.2 Air Cleaning 389
11.8 Indoor Air Quality and Public Policy 390
Readings 391
Questions 392

Chapter 12   Environmental Noise
12.1 Sound and Its Measurement 395
    12.1.1 Sound Energy 396
    12.1.2 Sound Measurement...398
        12.1.2.1 Weighting Scales 399
        12.1.2.2 Meter Response and Instrument Accuracy 401
        12.1.2.3 Measurement of Impulse Sound 401
        12.1.2.4 Spectrum Analysis 401
12.2 Sound Exposure Descriptors..402
    12.2.1 A-Weighted Sound Level 402
    12.2.2 A-Weighted Sound Exposure Level 402
    12.2.3 Equivalent Sound Level 403
    12.2.4 Day–Night Sound Level 404
12.3 Loudness...405
12.4 Effects of Environmental Noise Exposure 406
    12.4.1 Hearing Impairment...406
    12.4.2 Speech Communication 409
    12.4.3 Sleep Disturbance 411
    12.4.4 Psychophysiological Effects 411
    12.4.5 Performance 412
    12.4.6 Annoyance and Social Behavior..412
12.5 Nonregulatory and Regulatory Control Approaches 412

12.5.1 Community Exposure Guidelines 412
12.5.2 Community Regulatory Efforts.. 413
12.5.3 Federal Noise Control Programs 414
12.6 Control Measures 415
Readings 415
Questions 416

Glossary 419

Index...441

CHAPTER 1

The Atmosphere

1.1 CHEMICAL COMPOSITION

1.1.1 General Composition

The Earth's atmosphere is a mixture of gases and particulate-phase substances. The most abundant of these, nitrogen (N_2) and oxygen (O_2), comprise approximately 78 and 21%, respectively, of atmospheric mass and volume. A number of trace gases make up the remaining 1%. Average concentrations (with the exception of stratospheric ozone, (O_3) are reported in Table 1.1. These include gases present in essentially constant concentrations: N_2, O_2, argon (Ar), neon (Ne), helium (He), krypton (Kr), hydrogen (H_2), and xenon (Xe). Others vary temporally and spatially. These include water vapor (H_2O), carbon dioxide (CO_2), carbon monoxide (CO), methane (CH_4), O_3, the nitrogen oxides (nitrous oxide (N_2O), nitric oxide (NO), and nitrogen dioxide (NO_2)), ammonia (NH_3), formaldehyde (HCHO), sulfur dioxide (SO_2), a number of reduced sulfur compounds (dimethyl sulfide (($CH_3)_2S$), carbon disulfide (CS_2), carbonyl sulfide (COS), and hydrogen sulfide (H_2S)), and odd hydrogen species (hydroxyl radical (OH·), hydroperoxyl radical (HO_2·), and hydrogen peroxide (H_2O_2)). In addition to these gas-phase substances, the atmosphere contains trace quantities of particulate nitrate (NO_3^-), ammonium (NH_4^+), and sulfate (SO_4^{-2}).

Though N_2 is the most abundant constituent of the atmosphere, it has a relatively limited direct role in atmospheric and life processes. It serves as a precursor molecule for the formation of NO_3^-, from which plant processes synthesize amino acids, proteins, chlorophyll, and nucleic acids (organic molecules that are directly or indirectly essential to all living things). The conversion of N_2 to NO_3^- occurs as a result of atmospheric and symbiotic biological processes.

Nitrogen reacts with O_2 to produce nitrogen oxides (NO_x) that include N_2O, NO, NO_2, gas-phase nitric acid (HNO_3), and short-lived substances such as dinitrogen pentoxide (N_2O_5) and nitrate radical (NO_3·). Concentrations of these compounds or substances, unlike their precursors (N_2 and O_2), vary significantly in time and space. Nitrous oxide, a relatively inert gas, was, until several decades ago, thought to be

Table 1.1 Atmospheric Gases

Chemical Species	Symbol	Concentration (%, ppmv, ppbv, pptv)
Nitrogen	N_2	78.084%
Oxygen	O_2	20.948
Argon	Ar	0.934
Water vapor	H_2O	0.1–30,000 ppmv
Carbon dioxide	CO_2	~370
Neon	Ne	18.18
Ozone (stratosphere)	O_3	0.5–10
Helium	He	5.24
Methane	CH_4	~1.72
Krypton	Kr	1.14
Hydrogen	H_2	0.50
Xenon	Xe	0.09
Nitrous oxide	N_2O	~0.33
Carbon monoxide	CO	110 ppbv (50–200)
Ozone	O_3	20
Ammonia	NH_3	4
Formaldehyde	HCOH	0.1–1
Sulfur dioxide	SO_2	~1
Nitrogen dioxide	NO_2	~1
Carbonyl sulfide	COS	500 pptv
Carbon disulfide	CS_2	1–300
Dimethyl sulfide	$(CH_3)_2S$	10–100
Hydrogen sulfide	H_2S	~50
Nitric oxide	NO	~50
Hydroxyl radical	OH·	0.1–10
Hydroperoxyl radical	$HO_2·$	0.1–10

present in the atmosphere at constant levels. It is one of a number of substances whose concentrations are increasing as a result of human activities.

The evolution of free atmospheric O_2 at elevated concentrations set the stage for the evolution of oxidative metabolism, the series of energy-transferring chemical reactions that sustain most life forms. Oxygen, as a consequence, is vital to almost all living things. It is a precursor for the production of O_3 and development of the O_3 layer, the stratospheric region that absorbs high-energy ultraviolet (UV) light radiation streaming into the Earth's atmosphere from the sun. By absorbing most of the UV radiation incident on the Earth and its atmosphere, the O_3 layer shields most organic materials and living things from UV's destructive energy.

On average, background O_3 levels in the troposphere are approximately 20 parts per billion by volume (ppbv). Ozone concentrations vary, and there is evidence that average levels are increasing as a result of human activities. At heights of 10 to 50 km (6.2 to 31 mi), O_3 concentrations increase dramatically, with peak mixing ratio (parts per million by volume (ppmv)) concentrations (defined in Chapter 7) observed at 35 km (21.7 mi) (Figure 1.1); concentrations based on partial pressure or molecular density peak at 25 km (15.5 mi). At about 15 km (9.3 mi), O_3 concentrations are ~0.1 to 0.5 ppmv; at 20 km (12.4 mi), 3 ppmv; at 35 km, 8 to 10 ppmv; and at 50 km, 2 ppmv.

In addition to these altitude-based differences, stratospheric O_3 concentrations vary from day to day, from one season to another, and latitudinally. There are also

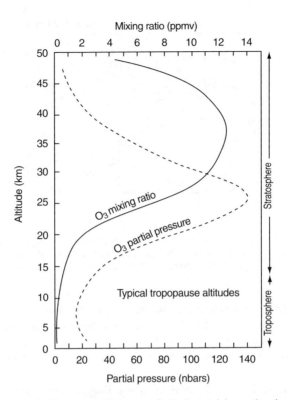

Figure 1.1 Atmospheric O_3 concentrations expressed as mixing ratios (ppmv) and partial pressure (nbars) as a function of height. (From National Research Council, *Rethinking the Ozone Problem in Urban and Regional Air Pollution,* National Academy Press, Washington, D.C., 1991. With permission.)

significant variations that result from quasi-biennial oscillations (air movement in the stratosphere), the 11-year sunspot cycle, and volcanic eruptions.

Ozone concentrations are also measured and reported as total column O_3 (vertical sum, reported as Dobson units). Mean monthly column O_3 concentrations are illustrated as a function of latitude and time of year in Figure 1.2. Column O_3 concentrations are highest at latitudes where O_3 production is relatively low. Highest O_3 values are found at high latitudes in winter and early spring; the lowest values are found in the tropics. Most O_3 production occurs near the equator where high levels of solar radiation are received; the observed stratospheric O_3 distribution reflects a strong poleward transport during winter.

In contrast to N_2 and O_2, the atmospheric concentration of CO_2 is relatively low, about 0.037%, or 370 ppmv. Carbon dioxide is enormously important. It is one of the two principal raw materials from which green plants (in the process of photosynthesis) make food molecules on which most living things depend. Life is carbon based, and CO_2 is the source of that carbon. Carbon dioxide is also a major greenhouse gas and, because of its thermal absorptivity, is responsible in good measure for maintaining a favorable global heat balance.

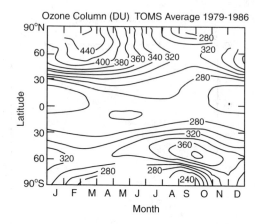

Figure 1.2 Total column O_3 (Dobson units) averaged over the period 1979–1986 reflecting latitudinal and seasonal differences. (From WMO, *Report of the International Ozone Trends Panel: 1988*, World Meteorological Organization, Geneva, 1990.)

Water vapor is the atmospheric constituent with the highest degree of variability (0.1 to 30,000 ppmv). Like CO_2, it is a major greenhouse gas, absorbing thermal energy radiated from the Earth's surface. Water vapor is significant in the atmosphere because it readily changes phase. On cooling, it condenses to form tiny droplets of liquid H_2O.

Helium, Ar, Ne, Kr, and Xe are noble gases. They are inert and, as a consequence, do not appear to have any major effect on, or role in, the atmosphere.

Many of the variable trace gases are produced either bio- or geogenically. Most important among these are NH_3, CH_4, H_2S, CO, and SO_2. Ammonia, CH_4, and H_2S are produced primarily by biological decomposition. Methane is a thermal absorber and serves as a greenhouse gas.

Other gas-phase species may be found in the atmosphere (in addition to those listed in Table 1.1). These include organic substances released by plants, microorganisms, and, to a lesser extent, animals. During the growing season, plants emit large quantities of volatile substances such as isoprenes and pinenes. These biogenic substances are rapidly removed by sink processes. Therefore, they are not described as being "normal" constituents of the atmosphere.

1.1.2 Water

Water in its liquid phase covers approximately 70% of the Earth's surface; in its solid phase it forms the polar ice caps, glaciers, episodic snow and ice of our seasons, and ice crystals of cold and very cold clouds. In its vapor phase it is a significant component of the atmosphere, varying from trace levels up to ~3%.

Water is a unique constituent of the atmosphere in that it occurs under ambient conditions as a solid, liquid, and gas. In some circumstances, all three phases are present at the same time. Phase changes are significant in atmospheric processes that produce "weather."

In its vapor phase, H_2O represents a variable component of the mixture of gases that comprise the atmosphere. The combined pressure of these gases (Dalton's law) describes atmospheric pressure. The partial pressure of H_2O vapor in an atmosphere is described as vapor pressure. Vapor pressure increases and decreases with increasing or decreasing concentrations of H_2O vapor. Water vapor concentrations may be reported as absolute or relative humidity. Absolute humidity may be expressed as density (mass/volume) or vapor pressure (millimeters of mercury).

Relative humidity is expressed in percent; it is the percent H_2O vapor a volume of air holds at a given temperature. Since air can hold more H_2O vapor at higher temperatures, relative humidity values decrease as temperature increases. As air is cooled, relative humidity increases until it reaches 100%, or the saturation value (saturation vapor pressure). If cooled below the saturation value, condensation occurs. This is described as the dew point, and the temperature at which the dew point occurs is called the dew point temperature.

Relationships between temperature and relative humidity, relative humidity and dew point temperature, and relative and absolute humidity can be seen in the psychrometric chart in Figure 1.3.

As warm, moist air rises, it cools below the dew point. Subsequent condensation produces clouds with droplet sizes one millionth the size of a rain droplet. Clouds are masses of air with enormous numbers of condensed H_2O droplets or ice particles that initially are too small to appreciably fall through the atmosphere. As these droplets grow by a millionfold, precipitation occurs. Clouds have very important

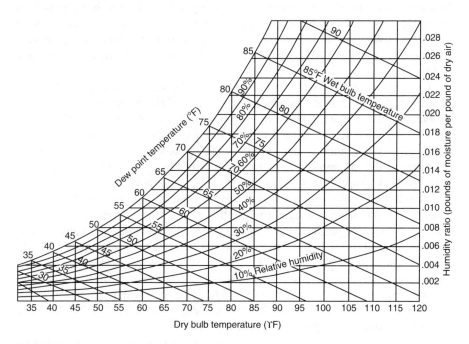

Figure 1.3 Psychrometric chart.

physical effects. They reflect sunlight back to space and are major contributors to the Earth's albedo, i.e., its ability to reflect sunlight. They also absorb heat, retarding its flow back to space.

1.2 PHYSICAL PHENOMENA

The atmosphere is characterized by its constituent gas- and particulate-phase substances and the physical forces and phenomena that act upon and within it. These include solar radiation, thermal energy, gravity, air density and associated pressure, and the movement of individual air molecules and the atmosphere itself.

1.2.1 Solar Radiation

The sun radiating energy approximately as a blackbody (perfect absorber and emitter of radiation at all wavelengths) at an effective temperature of 6000 K "showers" the atmosphere and the Earth's surfaces with an enormous quantity of electromagnetic energy. The total amount of energy emitted by the sun and received at the extremity of the Earth's atmosphere is constant, $1370 \text{ W/m}^2/\text{sec}$. That received per unit area of the Earth's surface is $343 \text{ W/m}^2/\text{sec}$.

Spectral characteristics of solar radiation, both external to the Earth's atmosphere and at the ground, can be seen in Figure 1.4. Over 99% of the energy flux from the sun is in the spectral region of 0.15 to 4 μm, with approximately 50% in the visible light region of 0.4 to 0.7 μm. The solar spectrum peaks at 0.49 μm, the green portion of visible light. Because of the dominance of visible light in the solar spectrum, it is not surprising that life processes such as photosynthesis and photoperiodism (responses to changing day length) are dependent on specific visible wavelength bands.

Incoming solar radiation is absorbed by atmospheric gases such as O_2, O_3, CO_2, and H_2O vapor. UV light at wavelengths of <0.18 μm (180 nm) are strongly absorbed by O_2 at altitudes above 100 km (62 mi). Ozone below 60 km (37.2 mi) absorbs most UV between 0.2 and 0.3 μm (200 to 300 nm). Atmospheric absorption above 40 km (24.8 mi) results in attenuation of ~5% of incoming solar radiation. Under clear-sky conditions, another 10 to 15% is absorbed by the lower atmosphere or is scattered back to space, with 80 to 85% reaching the ground.

The Earth, however, has variable but significant cloud cover so that under average conditions of cloudiness, a significant portion of the sun's energy incident on the atmosphere does not reach the Earth's surface. Cloud droplets and other atmospheric aerosols, as well as the Earth's surfaces, scatter sunlight back to space. This back-scatter as seen from the moon makes the Earth a relatively bright planet. This phenomenon describes the Earth's albedo. The albedo varies considerably, with an average value estimated to be ~30 to 35%; 30% is the most widely referenced value. Approximately 55% of the Earth's albedo is due to clouds, 23% to cloud-free atmosphere, and about 22% to the Earth's surface.

As can be seen in Figure 1.4, a significant portion of solar radiation incident on the atmosphere is not received at the ground. In addition to reflection by clouds and

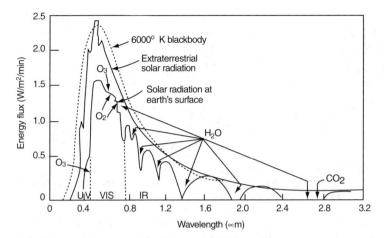

Figure 1.4 Solar spectra. (From Krondratjev, K.Y., *Radiation in the Atmosphere*, Academic Press, New York, 1969. With permission.)

atmospheric aerosols, considerable attenuation of solar energy results from absorption of different portions of the solar spectrum by atmospheric constituents such as H_2O vapor, CO_2, O_3, and O_2. Carbon dioxide and H_2O vapor primarily absorb in the infrared region, O_3 in the UV and portions of the visible region, and O_2 in portions of the visible and infrared regions.

1.2.2 Terrestrial Thermal Radiation

Since the Earth's temperature is less than that of the sun, it absorbs and reradiates energy at longer wavelengths than it receives. The Earth and its atmosphere radiate energy nearly as a blackbody, with an effective temperature of 290 K. This results in an emission spectrum primarily in the infrared region of 1 to 30 μm, with a peak at ~11 μm. A nighttime thermal emission spectrum is illustrated in Figure 1.5 (note the significant absorption due to CO_2 and H_2O vapor). The atmosphere is particularly transparent to wavelengths in the infrared spectral region of 7 to 13 μm (described as the atmospheric window). Nearly 80% of thermal energy emitted by the Earth's atmospheric window escapes to space. It is notable that most non-CO_2/H_2O vapor greenhouse gases (i.e., O_3, CH_4, N_2O, and chlorofluorocarbons (CFCs)) have strong absorption bands in this spectral region. Water vapor is the most significant of all greenhouse gases, absorbing five times the amount of thermal radiation as all other greenhouse gases combined.

1.2.3 Energy Balance

The Earth's overall energy balance is illustrated in Figure 1.6. Incoming solar radiation produces an energy input of 343 W/m²/sec. Of this, about 103 W/m²/sec is reflected back to space. The fraction absorbed by the Earth–atmosphere system is about 240 W/m²/sec. This corresponds to the net outgoing radiation flux of 240

Figure 1.5 Nighttime Earth atmospheric emission spectrum. (From Gates, D.M., *Energy Exchange in the Atmosphere*, Harper and Row Monographs, New York, 1962. With permission.)

$W/m^2/sec$ (Figure 1.6). The radiative flux emitted at the Earth's surface is 390 $W/m^2/sec$, which substantially exceeds the outgoing flux of 240 $W/m^2/sec$. The difference reflects the significant effect that clouds, H_2O vapor, and greenhouse gases have in absorbing and emitting thermal energy. Absorption of thermal energy by the atmosphere provides the Earth with a relatively moderate average global surface temperature of ~15°C (59°F).

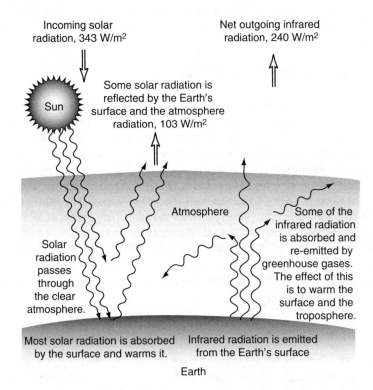

Figure 1.6 Earth's energy balance. (From Seinfeld, J.H. and Pandis, S.N., *Atmospheric Chemistry and Physics: Air Pollution to Climate Change*, John Wiley & Sons, New York, 1998. With permission.)

1.2.4   Surface Temperatures

The quantity of solar radiation received at any location on the Earth's surface is a direct function of the sun's angle of inclination. As a consequence, considerably more energy is received and absorbed in equatorial than polar regions. In the tropics, average temperatures exceed 22°C (71.6°F). Average temperature differences between the tropics and poles are approximately 35°C (95°F).

As a result of these differences, thermal energy is transported poleward by air and ocean currents. Warm currents move toward the poles; cold currents move toward the equator. Considerable energy transfer also occurs by evaporation and condensation.

The unequal distribution of solar energy on the Earth's surface and heat transfer mechanisms result in observed differences in regional climate. Changes in the Earth's orientation to the sun that occur over the course of the year result in the temporal and climatic temperature differences associated with the seasons, most notably in middle and high latitudes.

1.2.5   Vertical Temperature Gradients

The atmosphere is characterized by significant temperature gradients in the vertical dimension. These temperature gradients define the various layers or zones of the atmosphere (Figure 1.7).

The troposphere, the lowest layer of the atmosphere, extends upward to the tropopause, which begins at altitudes of 8 to 18 km (4.96 to 11.2 mi) depending on the latitude and time of year. The troposphere is at its maximum height over the equator and its lowest height over the poles. Temperature decreases steadily from an average of ~15°C (59°F) to –60°C (–76°F) at 15 km (9.3 mi). On average, temperature decreases at a rate of –6.5°C/km. This progressive temperature decline is due to the increasing distance from the sun-warmed Earth.

Because of the intense movement of thermal energy associated with heating of the Earth's surface and subsequent transfer to the atmosphere, the troposphere is relatively unstable. This instability results in changes in day-to-day temperature, cloud cover, and precipitation, a phenomenon we call weather. The troposphere is where we humans live and all biological activity occurs.

The troposphere can be divided into two regions, i.e., the planetary boundary layer (PBL), which extends upward to 1 km (0.62 mi) or more, and the free troposphere, which extends up to the tropopause.

The tropopause, like other pauses, is an atmospheric zone where temperatures are isothermal; i.e., no temperature change occurs with height. It is a relatively stable region forming a boundary between the troposphere, where temperature decreases with height, and the stratosphere, where temperature increases with height. The height of the tropopause is often represented as extending from a few kilometers to ~10 km (~1 to 6 mi) (Figure 1.7).

Above the tropopause, temperature increases to an altitude of 45 to 55 km (27.9 to 34.1 mi) (where temperature is ~ –2°C (28.4°F). As such, stratospheric temperatures are inverted relative to the troposphere. Consequently, the stratosphere is

Figure 1.7 Vertical temperature profile and associated atmospheric zones.

relatively stable, with little vertical convective mixing. The increase of temperature with height in the stratosphere results from the absorption of UV by O_3. The highest temperatures occur at the top of the stratosphere because of high-efficiency UV absorption.

Few clouds (though stratospheric cloudiness is increasing) are present in the stratosphere, and air movement is primarily horizontal. There is little manifestation of weather-type processes.

The stratosphere and the atmospheric layers above it make up the upper atmosphere, the troposphere being the lower atmosphere. Above the stratopause, temperature decreases with altitude to –90°C (–130°F) at 85 km (52.7 mi), the coldest location in the atmosphere. This region, the mesosphere, is characterized by rapid vertical mixing.

The thermosphere begins above the mesopause altitudes of 90 km (55.8 mi) and extends outward to the atmosphere's extremity (~1000 km, 620 mi). The high temperatures that characterize it are produced as a result of absorption of short-wavelength radiation by molecules of N_2 and O_2. Temperatures reach ~1200°C (2128°F) at 500 km (310 mi). These are, however, thermodynamic temperatures, which are a measure of the energy of molecules. Because relatively few molecules are present, standard thermometer temperatures are lower since fewer air molecules collide with the thermometer.

Molecules of N_2 and O_2 are photoionized in the thermosphere region of 90 to 400 km (55.8 to 248 mi). These molecules and atoms lose one or more electrons and thus become positively charged ions. Electrons are set free and travel as electrical currents. The principal chemical species found in the thermosphere are N_2, O_2, N, O, and the ions N_2^+, O_2^+, N^+, and O^+.

This ionized layer of the thermosphere is described as the ionosphere. It is responsible for formation of the aurora borealis (northern lights) and aurora australis (southern lights). Auroras are closely related in time with solar flare activity and occur at the Earth's magnetic poles. As masses of protons and electrons approach the Earth from the sun, they are captured by its magnetic field, which moves them toward the poles. As these ions move, O_2 and N_2 are energized, emitting light characteristic of auroral displays.

The electrical nature of the ionosphere varies as a result of changes in ion density. Since ion production requires direct solar radiation, the concentration of charged particles decreases at night. The ionosphere reflects AM radio signals back to Earth during nighttime hours.

1.2.6   Gravity

Atmospheric gases are held close to the Earth's surface by gravity. According to Newton's law of universal gravitation, all bodies in the universe attract each other with a force equal to

$$F = Gm_1m_2/r^2 \tag{1.1}$$

where m_1 and m_2 are the masses of two bodies, G is the universal gravity constant (6.67×10^{-11} Nm^2/kg^2), and r is the distance between the two bodies. It can be seen from Equation 1.1 that the force (F) of gravity decreases as an inverse square of the distance between two bodies.

In addition to holding air molecules close to the Earth's surface, gravity plays a significant role in atmospheric processes. As air cools, it becomes more dense and sinks. On warming, it becomes less dense and rises. These mass/density changes result in vertical air motions and contribute to planetary air circulation. Gravity also affects the atmospheric residence time of atmospheric aerosols.

1.2.7   Atmospheric Density

Though mixing ratios of atmospheric gases remain relatively constant up to an altitude of 80 km (49.6 mi), the absolute concentration or numbers of molecules decreases exponentially with height. As molecules become fewer, their density (mass per unit volume) decreases (Figure 1.8). Approximately 80 to 90% of atmospheric mass is below an altitude of 12 km (7.4 mi); 99% is below 33 km (20.5 mi). The total mass of the atmosphere is estimated to be ~5×10^{18} kg (11×10^{18} lb). The molar mass or molecular weight of the lower atmosphere is 28.96 g/mol.

1.2.8 Atmospheric Pressure

Pressure, defined as force per unit area, is produced when constantly moving air molecules strike an object, rebound, and transfer momentum to it. Momentum transferred is a function of the average kinetic energy associated with air molecules and is proportional to the absolute temperature. At higher temperatures, molecules have more kinetic energy, and thus atmospheric pressure increases.

Though molecules in the atmosphere are in constant motion, they are, nevertheless, pulled toward the Earth's surface by gravity. As such, the atmosphere exerts pressure on all surfaces.

Objects (including humans) support a column of air directly above them. This column exerts a force on such objects equal to its weight or density. This force is, on average, maximal at sea level, where atmospheric pressure is equal to a standard atmosphere, 760 mmHg or 760 torr. In the international system of units (SI), the unit of pressure is Newtons per square meter, called a pascal (Pa). The pressure of the standard atmosphere in SI units is 1.01325×10^5 Pa. Many meteorologists describe atmospheric pressure in millibars; 1 mbar = 100 Pa. At sea level, atmospheric pressure is 1013.25 mbars. In U.S. units it is 29.92 in Hg and 14.7 lb/in.^{2} (PSI).

As atmospheric density decreases with height, atmospheric pressure decreases as well. Since pressure is influenced by temperature, this decrease is not quite the smooth exponential curve observed for atmospheric density in Figure 1.8.

The mean surface pressure at sea level, as indicated, is 1013 mbars. The average global surface pressure, which includes pressure values over oceans and land, is estimated to be 984 mbars (29 in Hg).

Small horizontal and vertical pressure differences occur in the troposphere as a result of thermal conditions. As air warms atmospheric pressure increases, it expands and becomes less dense. As it cools, it becomes more dense. In the first case, atmospheric pressure decreases; in the second, it increases. Thermally-related pressure differences characterize large-scale air movements (low- and high-pressure systems are described later in this chapter).

1.3 ATMOSPHERIC MOTION

The atmosphere is characterized by a continuous exchange of energy, momentum, and moisture, with the ocean and land beneath it. Though the amount of energy entering and leaving the Earth's atmosphere is essentially equal, it is not equal over every part of the Earth. This unequal distribution of solar energy resulting from differences in latitude, cloud cover, and absorptivity of the Earth's surface is responsible for local, regional, and large-scale air motions.

Density (Kg/m³)

Pressure (N/m²)

Figure 1.8 Changes in atmospheric density and pressure with altitude. (From National Atmospheric and Oceanographic Administration, *U.S. Standard Atmosphere*, NOAA S/T 76-1562, Washington, D.C., 1976.)

1.3.1 Winds

Wind is the term commonly used to describe air movement in the horizontal dimension. Winds result from differences in air pressure that are caused by unequal heating of the Earth's surface. In the absence of friction and the Earth's rotation, air would flow from areas of high to low pressure. The direction of airflow is controlled by a combination of the pressure gradient force, Coriolis effect, and friction.

1.3.2 Pressure Gradient Force

Pressure differences in the lower atmosphere occur along a gradient in the horizontal dimension. The force associated with these differences is called the pressure gradient force (PGF). The nature of PGFs can be seen from weather maps on which lines of equal pressure (isobars) are plotted (Figure 1.9). The spacing of isobars describes pressure gradients. Isobars relatively close together indicate a strong or steep pressure gradient with high associated winds. If these isobars were

Figure 1.9   Surface weather map over North America.

contour intervals, they would describe steep or mountainous terrain. Isobars relatively far apart are characterized by small pressure gradients and light winds. The
PGF is directed perpendicular to isobars from high to low pressure.

1.3.3   Coriolis Effect

As air flows from either the tropics or the high (55°) middle latitudes, it appears
to be deflected from its expected path north- or southward. This apparent deflection
to the east (in the northern hemisphere) in poleward flow and to the west in flows
toward the equator is due to the Earth's rotation. As the Earth moves from west to
east, air appears to lag behind the Earth. This effect of the Earth's rotation is described
as the Coriolis effect (CE). Because of the CE, wind does not move perpendicular
across isobars as it would if only the PGF were acting on it. The observed apparent
deflection depends on latitude and wind velocity. The CE is greatest at high latitudes
and is, for the most part, nonexistent at the equator. Deflection associated with the
CE increases with wind speed, since faster winds cover a greater distance than slower
winds in the same period of time.

1.3.4   Friction

Below a height of several kilometers, friction opposes PGF and thus moderates
it. Friction reaches its maximum potential near the Earth's surface. It reduces wind
velocity and, as a consequence, "deflection" associated with the CE. Surface winds

Figure 1.10   Surface day–night winds speeds as a function of elevation. (From Turner, D.B., EPA Publication AP-26, Figure 6-19, EPA, Washington, D.C., 1969.)

turn toward low pressure until friction and the CE balance the PGF. Friction turns the wind in the direction of low pressure, decreasing upward.

The extent to which the gradient wind is reduced and its direction changed depends on roughness of the terrain (surface roughness and temperature gradients in the frictional layer). As friction decreases with increasing height, wind speeds increase. As convective airflows diminish at night, frictional effects are less. This can be seen in the day–night differences in wind speed as a function of height (Figure 1.10).

Above the PBL (~1 km thick) the effects of friction are relatively small. The CE causes wind to turn so that when the PGF and CE are in balance, wind flows parallel to isobars at constant speed. Under these idealized conditions, airflow is said to be in geostropic (turned by the Earth) balance. Geostropic winds flow parallel to isobars with velocities proportional to PGFs. Strong winds are associated with steep pressure gradients, light winds with weak gradients. The geostropic model approximates airflows aloft, although they are not purely geostropic.

1.3.5   Cyclones and Anticyclones

On weather maps, isobars form circular or roughly circular cells of high or low pressure. As such, airflow in these cells follows a curved path that tends to be parallel to isobars. These winds blowing at a constant speed are called gradient winds. They are produced as a result of the combined effects of the PGF and CE.

In the northern hemisphere, the CE deflects airflow to the right; the combined effects of CE and PGF create deflection to the left in low-pressure systems. Consequently, airflow is counterclockwise (cyclonic). Air flowing out from the center of the cell (directed by the PGF) in a high-pressure system is opposed by the CE directed inward; therefore, air flows clockwise.

Because airflow in low-pressure systems is cyclonic, centers of low pressure are called cyclones. Cyclonic flow is counterclockwise in the northern hemisphere and clockwise in the southern hemisphere. Airflow in a high-pressure system is anticyclonic (opposite the Earth's rotation). It is clockwise in the northern hemisphere and counterclockwise in the southern hemisphere.

When the force of friction is added to the PGF and CE, air moves across the isobars at various angles depending on roughness of the terrain, but always moving from higher to lower pressure.

In low-pressure systems, pressure decreases inward and friction causes net airflow (convergence) toward their centers. In high-pressure systems, pressure decreases outward and friction causes air to flow away (divergence) from their centers.

Both low- and high-pressure systems are characterized by airflows in the vertical dimension. Near the Earth's surface, air spirals inward in low-pressure systems, reducing the area occupied by the air mass. As this convergence occurs, air must flow upward to compensate. As such, surface convergence results in divergence aloft (Figure 1.11). As rising air results in cloud formation and precipitation, low-pressure systems are characterized by unstable atmospheric conditions and stormy weather.

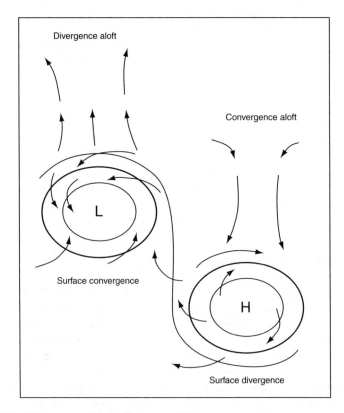

Figure 1.11   Airflows associated with low- and high-pressure systems.

In anticyclones, the outflow of air near the Earth's surface is accompanied by convergence from aloft and subsidence (sinking) of air toward the surface (Figure 1.11). As air descends, it compresses and warms air beneath it (see Chapter 3). As a consequence, cloud formation and precipitation generally do not occur. High-pressure systems are characterized by fair, sunny weather during warm months and cold, sunny weather in cold months.

Most cyclones and anticyclones have cross-sectional diameters of 100 to 1000 km (62 to 620 mi). They are typically migratory, moving from one area of the Earth to another. Cyclones of very large scale are called hurricanes or typhoons.

Many migratory cyclones and anticyclones form in temperate regions as a consequence of movements of tropical and polar air. They move from west to east at ~800 km/day (500 mi/day) and typically have a life span of a few weeks.

1.3.6 General Circulation of the Atmosphere

In the 18th century, Sir George Hadley attempted to explain global air circulation using a simple model in which sensible heat would be transported from the equatorial region to the poles by two large, simple convective cells. Hadley theorized that intense solar heating of land and water surfaces at the equator would produce warm air that would rise to the top of the troposphere where it would lose its thermal energy, sink to the Earth at the poles, and return along the surface to the equator where it would be reheated. Because of factors such as the CE and friction, Hadley's model does not adequately describe global air circulation.

Global air circulation is better described by the three-zone model illustrated in Figure 1.12. Hadley-type circulation cells are formed between the equator and 30°N and 30°S latitude. Thermal air circulation on both sides of the equator is characterized by the northward or southward flow of warm tropical air aloft and cool southerly or northerly flows along the surface toward the equator. Airflows to the south (in the northern hemisphere) are easterly; i.e., they appear to be moving from an easterly direction. These are the trade winds. Their flow is determined in part by the CE.

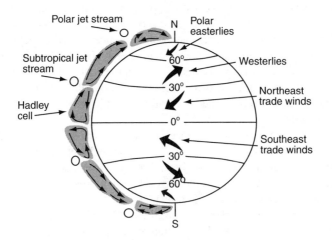

Figure 1.12 Three-zone model describing planetary airflow.

Hadley-type airflows also occur in polar regions. Warm air from temperate regions moves poleward at relatively high tropospheric altitudes, cools by radiation, and descends at the poles. Cool or cold polar air then moves southward (in the northern hemisphere). These winds appear to be coming from the east and are described as the polar easterlies.

Due to the intrusion of air from the tropics and polar regions, atmospheric circulation in temperate regions (40 to 55° latitude) is highly variable. Consequently, airflow patterns are less of the Hadley-type characteristic of tropical and polar zones. In temperate regions of the northern hemisphere, surface winds generally flow from the southwest.

Areas of relative calm occur at approximately 30°N and 30°S latitude. These areas are characterized by the subsidence of relatively cool air, commonly associated with semipermanent marine high-pressure systems. They have sunny skies, low precipitation, and low winds. These systems are centered over oceans, with their positions shifting only slightly in summer and winter. Deserts are common on the west coasts of continents where such semipermanent marine high-pressure systems occur.

In the upper troposphere (7.5 to 12 km, 4.7 to 7.4 mi) west to east airflows occur in a wavelike fashion (in contrast to the relatively closed high- and low-pressure systems that occur near the Earth's surface). These streams or rivers of westerly flowing air (jet streams) meander for thousands of kilometers around the Earth. They have widths that vary from 100 to 500 km (62 to 310 mi) and are usually only a few kilometers deep. Wind speeds are relatively high, with peak speeds of >200 km/h (124 mi/h).

Jet streams form along zones called fronts (in the discontinuities associated with circulation cells in the three-zone model). A polar jet stream in middle latitudes is associated with a polar front that is situated between the cool polar easterlies and the warm westerlies. These relatively fast-moving upper air winds are produced as a result of strong temperature differences at the surface that produce strong pressure gradients aloft.

As the polar front moves, so does the polar jet stream. As such, it meanders and sometimes splits. It travels at average speeds of ~125 km/h (77.5 mi/h) in the winter and ~60 km/h (37.2 mi/h) in the summer; the difference is a consequence of seasonal differences in temperatures and pressure gradients. During winter, the northern hemispheric polar jet stream may extend as far south as 30°N. In summer, its position moves much farther north, with an average position of 50°N in midsummer. It typically migrates between 30 and 70°N latitude and is often referred to as the mid-latitude jet stream.

A semipermanent jet stream also occurs over the tropics in winter. This subtropical jet stream is somewhat slower than the mid-latitude jet stream and is centered at 25°N latitude at an altitude of about 13 km (8.1 mi). It sometimes merges with the mid-latitude jet stream. The generalized flow patterns of these northern hemispheric jet streams can be seen in Figure 1.13.

Jet streams significantly affect surface airflows. When they accelerate, divergence of air occurs at the altitude of the jet stream; this promotes convergence near the surface and induction of cyclonic motion. Jet streams supply energy to surface storms

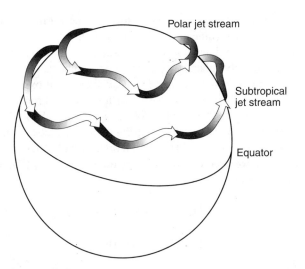

Figure 1.13 Generalized flow patterns of northern hemispheric jet streams.

and direct their path. Jet streams also cause a convergence of air aloft and subsidence near the surface, resulting in intensification of high-pressure systems. As such, jet streams are often described as weather makers.

1.4 EVOLUTION OF THE ATMOSPHERE

Our solar system (including the Earth) is widely held by Earth scientists to have evolved from condensation of an interstellar cloud of gases and dust about 4.6 billion years ago. The Earth's core consisted of iron and iron alloys; its mantle and crust were made up of oxides and silicates of metals.

The very early or primeval atmosphere is believed to have been a mixture of CO_2, CO, N_2, and H_2. While large amounts of gases were lost to space, other gases were brought to the Earth's surface through volcanism. These gases (CO_2, CO, N_2, and H_2) are emitted into the present day's atmosphere by volcanoes.

As volcanoes brought gases to the surface, chemical reactions made possible a variety of new gas-phase species and increases in the levels of others. Water and CO were formed by reactions catalyzed by metal oxides in the Earth's crust.

$$3 H_2 + CO_2 \rightarrow CH_4 + H_2O \qquad (1.2)$$

$$H_2 + CO_2 \rightarrow CO + H_2O \qquad (1.3)$$

As an enormous amount of H_2O vapor was produced over time, the Earth's atmosphere underwent significant changes. As H_2O vapor condensed, the world's oceans were formed. Because of the solubility of CO_2 in H_2O, large quantities of CO_2 dissolved in the newly formed oceans. Dissolved CO_2 was precipitated (first by chemical and then biological processes) to form the Earth's large deposits of sedimentary carbonate rock (limestone, dolomite, etc.).

Because N_2 is relatively inert, insoluble in water, and noncondensable, it accumulated to become the most abundant gas in the atmosphere.

Though O_2 was present in combined forms such as metal oxides, silicates, and CO_2, there was no free O_2 present in early Earth's history. Production of free O_2 required the evolution of life-forms, which began approximately 4 billion years ago.

The evolution of life was preceded by formation of molecular species such as hydrogen cyanide (HCN), NH_3, and H_2S. Small organic molecules such as methanol (CH_3OH) and HCHO were produced from reactions with CO_2, CO, and H_2 in the presence of UV light and catalysts.

$$CO_2 + 2\ H_2 \rightarrow CH_3OH \tag{1.4}$$

$$CO + H_2 \rightarrow HCHO \tag{1.5}$$

These small molecules served as precursors for more complex molecules such as amino acids and simple peptides.

Very early life-forms, which likely evolved in the Earth's oceans, lived in an environment without free O_2. Approximately 3.5 billion years ago, some organisms developed the ability to photosynthetically produce organic molecules and free O_2.

$$CO_2 + H_2O \rightarrow CH_2O + O_2 \tag{1.6}$$

Most of this early O_2 reacted with crustal materials and was removed from the evolving atmosphere. About 2 billion years ago, free O_2 began to accumulate, reaching current levels about 400 million years ago. Present-day steady-state O_2 concentrations are maintained by photosynthesis and biological removal processes.

The abundance of free O_2 led to the formation of O_3 and development of the O_3 layer, which shielded the Earth from biologically destructive UV light and made possible the development of terrestrial life.

As life evolved, CO_2 continued to be removed from the atmosphere by sedimentary activity and formation of large deposits of coal, oil, and natural gas. As a consequence, CO_2 became a minor or trace gas. In the last half to 1 billion years, average levels of the major atmospheric constituents have remained relatively constant.

Noble gases are characterized by their lack of chemical reactivity. Noble gases such as Ar and He were present in the Earth's early atmosphere and continue to be there. Both originated as products of crustal radioactive decay and have increased in concentration over time. However, because of its heavier atomic mass, Ar tends to be retained in the atmosphere, whereas He (because of its low mass) tends to be slowly lost to space.

Hydrogen also has a low mass and, as such, has a greater potential to be lost to space.

Both He and H_2 were once believed to have been left over from the interstellar cloud that formed the Earth. It appears, however, that present-day He and H_2 levels are maintained by geogenic processes. As indicated previously, the atmosphere consists of a variety of trace gases that comprise less than 1% of the atmosphere. Concentrations of some of these trace gases (e.g., CO_2, CH_4, and N_2O) have increased significantly over the last two centuries. Increases have occurred in tropospheric O_3 and decreases in stratospheric O_3; the atmospheric chlorine (Cl) level has increased fivefold.

The atmosphere has evolved considerably over the past 4.6 billion years. For most of that time, its evolution was shaped entirely by natural forces. For the past two centuries, atmospheric evolution has come under the influence of industrial, agricultural, and cultural activities of humans. The atmosphere continues to evolve. Humans now have a significant influence on the nature and pace of this evolution. As such, it is a new day in the Earth's history.

READINGS

Boubel, R.W., Fox, D.L., Turner, D.B., and Stern, A.C., *Fundamentals of Air Pollution*, 3rd ed., Academic Press, Orlando, FL, 1994.

Brasseur, G.P., Orlando, J.J., and Tyndall, G.S., Eds., *Atmospheric Chemistry and Global Change*, Oxford University Press, Oxford, 1999.

Bristlecombe, P., *Air Composition and Chemistry*, 2nd ed., Cambridge University Press, Cambridge, U.K., 1996.

Finlayson-Pitts, B.J. and Pitts, J.N., Jr., *Chemistry of the Upper and Lower Atmosphere: Theory, Experiments and Applications*, Academic Press, Orlando, FL, 2000.

Liou, K.N., *Radiation and Cloud Processes in the Atmosphere*, Oxford University Press, Oxford, 1992.

Lutgens, F.K. and Tarbuck, E.G., *The Atmosphere*, 7th ed., Prentice Hall, Saddlebrook, NJ, 1998.

National Oceanic and Atmospheric Administration, *U.S. Standard Atmosphere*, NOAA S/T 76-1562, Washington, D.C., 1976.

Seinfied, J.H. and Pandis, S.N., *Atmospheric Chemistry and Physics: From Air Pollution to Climate Change*, John Wiley & Sons, New York, 1998.

Turkekian, K.L., *Global Environmental Change: Past, Present and Future*, Prentice Hall, Saddlebrook, NJ, 1996.

Warneck, P., *Chemistry of the Natural Atmosphere*, Academic Press, Orlando, FL, 1988.

QUESTIONS

1. What are the major nonvarying constituents of the atmosphere?
2. What is the significance of atmospheric O_2?
3. Why are concentrations of water vapor and CO_2 so variable?
4. What gases absorb incoming solar radiation?
5. What gases absorb outgoing thermal radiation?
6. Why don't atmospheric gases simply flow away from the Earth's surface and be lost to space?

7. With one exception, the relative concentrations of atmospheric gases do not change with height. Indicate this exception, its concentration patterns, and reasons why this is the case.

8. There is a considerable difference between solar energy received at the extremity of the Earth's atmosphere and that received at the surface. Describe these differences and the factors responsible.

9. Significant temperature changes occur with height out to a distance of 500 km. Describe these differences and the factors that contribute to them.

10. Thermal emissions from the Earth and its atmosphere are characterized by electromagnetic energy in which spectral range?

11. What is the atmospheric window? What is its significance?

12. What is atmospheric pressure? What causes it?

13. Why is water vapor such an important atmospheric constituent?

14. The atmosphere is in constant motion. Why?

15. Why does air flow counterclockwise in a low-pressure system in the northern hemisphere?

16. Air flowing from the equator in the middle latitudes in the northern hemisphere appears to flow from the southwest to the northeast. Why?

17. The three-zone model is used to describe what type of air motion?

18. Why does airflow aloft differ from that near the surface?

19. What causes the Earth to have an albedo? What is the significance of the Earth's albedo?

20. How, where, and why are jet streams formed?

21. How did O_2 originate in the atmosphere?

22. How have atmospheric CO_2 levels changed over time?

23. What is the origin of H_2 and He in the atmosphere?

24. The composition of the Earth's atmosphere has changed in the past 200 years. What were/are these changes?

CHAPTER 2

Atmospheric Pollution and Pollutants

As seen in Chapter 1, the atmosphere is a mixture of gaseous substances produced over the Earth's long history by biogenic, geogenic, and atmospheric processes. As humans evolved, the atmosphere began to be affected by human activities. Initially, these effects were small and insignificant. With the enormous growth in human numbers, resource use, and technological advancements, the impact of humans on the atmosphere has been and continues to be significant.

We generally view (in the absence of human activity) the atmosphere and other environmental media, such as water and land, to be "clean." On contamination with our "wastes," they become "polluted." The concept of pollution includes a sense of degradation, a loss of quality, a departure from purity, and adverse environmental effects. Air becomes polluted when it is changed by the introduction of gas- or particulate-phase substances or energy forms (heat, noise, radioactivity) so that the locally, regionally, or globally altered atmosphere poses harm to humans, biological systems, materials, or the atmosphere itself.

Air pollution concerns, historically, have focused on the freely moving air of the outdoor environment described as ambient air. It is pollution of the ambient air that is the primary focus of regulatory programs, as well as this book. However, it is but one of several air pollution or air quality concerns. Significant air pollutant exposures occur in industrial workplaces and the built environments of our homes, offices, and industrial buildings, and from personal habits such as smoking.

2.1 NATURAL AIR POLLUTION

Contamination or pollution of the atmosphere occurs as a consequence of natural processes as well as human activity (the latter are called anthropogenic). Though regulatory programs focus exclusively on anthropogenic air pollution, it is important to understand that nature contributes to atmospheric pollution and, in some instances, causes significant air quality problems. As evidenced by volcanic activity in the 19th century (e.g., Tambora and Krakatoa) and the more recent eruptions of Mt. St. Helens in Washington State, El Chinchon in Mexico, and Mt. Pinatubo in the Philippines, as well as the 1988 forest fires in Yellowstone National Park and the western U.S., pollutants produced naturally can have significant effects on regional and global air quality. In addition to volcanoes and forest fires, natural air pollution occurs from a variety of sources: plant and animal decomposition, pollen and spores, volatile hydrocarbons (HCs) emitted by vegetation, ocean spray, soil erosion and mineral weathering by wind, emission of gas-phase substances from soil and water surfaces, ozone (O_3) and nitrogen oxides (NO_x) from electrical storms, and O_3 from stratospheric intrusion and photochemical reactions. Emissions to the atmosphere associated with geochemical processes are described as geogenic; those associated with biological processes are biogenic.

Natural pollutants pose air quality concerns when generated in significant quantities near human settlements. With the exception of major events such as dust storms, forest fires, and volcanoes, natural air pollution has not been a major societal concern. On a mass basis, nature pollutes more than humans do, as vocal opponents of contemporary air pollution control efforts maintain. However, such pollution has relatively low significance in causing health and welfare effects since (1) levels of contaminants associated with natural air pollution are typically very low; (2) large distances often separate sources of natural pollution and large human populations; and (3) major sources of natural pollution, such as forest fires, dust storms, and volcanoes, are episodic and transient.

A major potential exception to this relegation of natural air pollution to a status of relative insignificance is biogenic emission of photochemically-active nonmethane hydrocarbons (NMHCs) such as isoprene and α-pinene from vegetation. There is increasing evidence that natural NMHCs play a significant role in photochemical oxidant production in some nonurban and urban areas.

2.2 ANTHROPOGENIC AIR POLLUTION

Anthropogenic air pollution has been and continues to be viewed as a serious environmental and public health problem. Its seriousness lies in the fact that elevated pollutant levels are produced in environments where harm to human health and welfare is more likely. It is this potential that makes anthropogenic air pollution the significant concern that it is.

In a historical sense, air pollution became a serious problem when humans discovered the utility of fire. Smoke generated in the incomplete combustion of wood plagued early man and generations to come. For civilizations in western Europe, the decimation of forests and a switch to use of soft coal as a fuel source in the 14th century added a new dimension to air pollution problems. Emissions of pollutants from coal combustion were intense, and coal smoke became a major environmental concern. The industrial revolution that began in the early 19th century and the technological developments of the 19th and 20th centuries (e.g., motor vehicles) are what have given rise to the atmospheric pollution characteristic of our times.

2.2.1 Smog

In London and other British industrial cities, smoke from coal combustion episodically mixed with fog from the North Sea. These occurrences were marked by an extreme reduction in visibility, the stench of sulfurous emissions, and, in severe episodes, illness and death. Such pollution fogs were experienced in London for more than a century. The term *smog*, used to describe severe ambient air pollution conditions, was derived from the words *smoke* and *fog*. Such smog was gray in color and referred to as a "gray smog" or "London-type" smog.

The term *smog* as it is used today is broadly applied to atmospheric pollution conditions characterized by a significant reduction in visibility. The term is applied without consideration of pollutant types, sources, or smog-forming processes.

The best-known smog problem in the U.S. is associated with the Los Angeles Basin of southern California (Figure 2.1). Although this Los Angeles-type smog is characterized by a significant reduction in visibility, conditions that produce smog in southern California are very different from those that produced smog in London and other northern European cities. The climate of the Los Angeles Basin is semiarid, with extended periods of sunny weather. Additionally, Los Angeles is situated at approximately 30°N latitude, where the subsidence of air from a tropical Hadley cell (see Chapter 1) produces semipermanent marine high-pressure systems with cell centers over Hawaii. This subsidence of air produces elevated inversions that are closest to the surface at the eastern edge, or continental side, of the marine high, i.e., over the southern coast of California. These relatively low subsidence inversions significantly reduce pollution dispersion in the vertical dimension. In addition, mountains to the north and east serve as barriers, inhibiting air movement in northerly and easterly directions and further reducing dispersion.

Figure 2.1 Smog over Los Angeles. (Courtesy of USEPA.)

Poor dispersion conditions in the Los Angeles Basin contribute to smog formation. Other factors also play a role. The major source of pollutants is light-duty motor vehicles (an estimated 5+ million) that emit exhaust and evaporative gases into the atmosphere. Motor vehicle-generated pollutants, abundant sunlight, poor dispersion conditions resulting from inversions (see Chapter 3), and topographical barriers contribute to the development of chemically complex smog. Because of the presence of nitrogen dioxide (NO_2), Los Angeles-type smog is brown.

Smog produced over the southern coast of California is a photochemical smog. The visibility-reducing particulate pollutants, as well as hundreds of gas-phase substances, are produced as a consequence of atmospheric photochemistry.

Los Angeles and London smogs reflect relatively unique conditions of smog development. Smogs that occasionally form over large cities in the eastern U.S. represent sources and atmospheric conditions somewhat intermediate between the two. These cities have, or have had, significant emissions of pollutants from both automobiles and coal combustion. They do not have the climatic conditions that have been characteristic of the smogs of either Los Angeles or London. (Note: Because of pollution control efforts and changes in local climate, London smogs experienced for a century or more no longer occur.)

2.2.2   Haze

The term *haze* is often used in the nomenclature of air quality. Although haze and smog are closely related, both representing atmospheric conditions characterized by visibility reduction, they differ in both intensity and geography. Haze typically refers to the large-scale, low-level, pollution-caused visibility reduction observed during the summer months over much of the Midwest, Northeast, and Southeast. Because of the relative success of pollution control efforts, it is haze and not smog that is characteristic of many cities in North America.

2.2.3   Nontraditional Air Pollutants

In traditional treatments of the subject of air pollution, pollutants are characterized as being particulate- or gaseous-phase substances. Air quality can, however, be degraded by other contaminants. These include noise, heat, ionizing radiation, and electromagnetic fields associated with the transmission and use of electricity. With the exception of noise, discussed extensively in Chapter 12, nontraditional air pollutants are considered to be relatively minor environmental concerns, and thus their treatment in this book is limited.

2.2.4   Gases and Particles

Because of effects on human health, the atmosphere, vegetation, and materials, pollutant gases and particles have received considerable research and regulatory attention. Though the terms *gases* and *particles* are used, pollutants exist in all three phases of matter. Particles include both solid and liquid substances. When dispersed in the atmosphere, solid and liquid particles form aerosols. Aerosols may reduce visibility, soil materials, and affect human health.

Fume aerosols are produced by condensation of metal vapors, dust aerosols from fragmentation of matter, mists from atomization of liquids or condensation of vapors, and smokes from incomplete combustion of organic materials. The photochemical aerosol of Los Angeles-type smog, as well as haze, is produced by condensation of vapors resulting from photochemical reactions in the atmosphere. Acid aerosol, which contributes to the formation of haze and acidification of precipitation, is, for the most part, produced photochemically.

2.2.4.1 Sources

Gaseous pollutants are emitted from a variety of identifiable sources, including those associated with transportation, stationary fuel combustion, industrial processes, waste disposal, etc. They are also produced in the atmosphere as a result of chemical reactions. The former are classified as primary pollutants; the latter, secondary. Sources of primary pollutants may be classified as mobile or stationary, combustion or noncombustion, area or point, direct or indirect. The use of these classifications reflects both regulatory and administrative approaches to implementing air pollution control programs at federal, state, and local levels.

Mobile sources include automobiles, trains, airplanes, etc.; others are stationary. A point source is a stationary source whose emissions degrade air quality. An area source is comprised of a number of small stationary and mobile sources that individually have only a small effect on air quality. Viewed collectively over an area, they may have an enormous impact. Area sources include emissions from motor vehicles, aircraft, trains, open burning, and a variety of small miscellaneous sources. These include indirect sources such as shopping centers and athletic stadiums that contribute to elevated pollutant levels as a result of attracting motor vehicle traffic.

2.2.4.2 National Emission Estimates

National emission estimates for five regulated primary pollutants and major source categories are presented in Table 2.1 and Table 2.2. Emission estimates in Table 2.1 are for 1970, when the U.S. began a major regulatory effort to control ambient air pollution. Emission estimates reflecting the status of this nation relative to the same pollutants and source categories more than two decades later (1997) are summarized in Table 2.2.

Transportation sources include motor vehicles, aircraft, trains, ships, boats, and a variety of off-road vehicles. Major stationary fuel combustion sources include fossil fuel-fired electrical generating plants, industrial and institutional boilers, and home space heaters. Industrial process losses include pollutants produced in a broad range of industrial activities, including mineral ore smelting, petroleum refining, oil and gas production and marketing, chemical production, paint application, industrial organic solvent use, food processing, mineral rock crushing, etc. Emissions from solid waste disposal result from onsite and municipal incineration and open burning. Miscellaneous sources include forest fires, agricultural burning, coal refuse and

Table 2.1   Estimates of National Emissions of Five Primary Pollutants (10^6 t/year) for the U.S. in 1970

Source	CO	PM_{10}[a]	NMHCs	NO_x	SO_x	Total	%
Transportation	98.64	0.46	14.51	9.02	0.43	123.06	54.7
Stationary source fuel combustion	4.63	2.87	0.72	11.69	23.46	43.37	19.3
Industrial processes	9.84	7.67	12.33	0.91	7.09	37.84	16.8
Solid waste disposal	7.06	0.99	1.98	0.08	0.01	10.25	4.5
Miscellaneous	7.91	0.84	1.10	0.30	0.10	10.25	4.5
Total	128.08	12.84	30.65	22.00	32.09	224.66	
%	57.00	5.7	13.6	9.8	13.8		100.0

[a]Does not include fugitive emissions.
Source: From USEPA, EPA/454-R-94-027, EPA, Washington, D.C., 1997.

Table 2.2   National Emission Estimates of Five Primary Pollutants (10^6 t/year) for the U.S. in 1997

Source	CO	PM_{10}[a]	NMHCs	NO_x	SO_x	Total	%
Transportation	67.01	0.73	7.66	11.60	1.38	88.38	55.1
Stationary source fuel combustion	4.81	1.10	0.86	10.72	17.26	34.75	21.7
Industrial processes	6.05	1.28	9.84	0.92	1.72	19.81	12.3
Miscellaneous	9.56	6.70	0.86	0.35	0.01	17.48	10.9
Total	87.43	9.81	19.22	23.58	20.37	160.41	
%	54.5	6.1	12.0	14.7	12.7		100.0

[a]Does not include fugitive emissions.
Source: From USEPA, EPA/454-R-94-027, EPA, Washington, D.C., 1997.

structural fires, and a variety of organic solvent uses. In addition to the five pollutant categories indicated in Table 2.1 and Table 2.2, the U.S. has major regulatory programs for lead (Pb) and O_3. Ozone is not inventoried, as it is a secondary pollutant produced as a result of atmospheric reactions. Emissions of Pb to the atmosphere in 1970 and 1997 are summarized in Table 2.3.

National emission estimates are published periodically by the U.S. Environmental Protection Agency (USEPA). Since 1970, the process has been modified several times. Nonmethane hydrocarbon emissions are now reported as volatile organic compound (VOC) emissions, and particulate matter emissions as PM_{10} (particulate matter with an aerodynamic diameter of ≤ 10 µm); particulate matter emissions now include fugitive emissions (which dominate PM_{10} emissions). Solid waste disposal, because of its relative insignificance at this time, is no longer included as a distinct major source. Emissions estimates presented in Table 2.1 and Table 2.2 do not include fugitive particulate matter emissions (so temporal comparisons can be made). Though USEPA now reports HCs as VOCs, the author has chosen to maintain use of the historical term, NMHCs.

2.2.4.3 The National Emissions Picture

Table 2.1 provides a variety of insights into the nature of the air quality problem the U.S. faced in 1970 as it began to enact and implement tough air pollution control legislation and subsequent regulation. As can be seen, (1) transportation was responsible for over 50% of emissions of five primary pollutants, (2) transportation was the major source of carbon monoxide (CO) and NMHCs and second-largest source of NO_x, (3) stationary source fuel combustion was the major source of sulfur oxides (SO_x) and NO_x, and (4) industrial processes were the largest source of PM_{10} and second-largest source of SO_x.

The impact of transportation on the nation's air quality, as evidenced in Table 2.1, has been considerable. In the transportation category, motor vehicles, which include light-duty cars and trucks, accounted for 75% of all emissions. Motor vehicles were also the major source of atmospheric Pb. Of the seven pollutant categories for which major control programs have been initiated, transportation, particularly motor vehicles, had been the major direct or indirect source for CO, HCs, Pb, and O_3, and the second-largest source of NO_x. Consequently, motor vehicles have justifiably received enormous regulatory attention and investment in pollution control systems.

Stationary source fuel combustion is the largest source of SO_x and NO_x emissions. Sulfur oxides emitted by fossil fuel (especially coal)-fired electrical power generating stations have been of particular concern, and since 1990, these facilities have been required to significantly reduce emissions of SO_x and NO_x for purposes of controlling acidic deposition (see Chapter 4 and Chapter 8).

The relative effectiveness of pollution control efforts over the past three decades can be seen when Table 2.1 and Table 2.2 are compared. A relatively modest 29% reduction has been achieved in emissions of the five primary pollutants. Emission reductions associated with individual source categories include transportation, 28%; stationary source fuel combustion, 19%; and industrial process losses, 48%.

Emission reductions from industrial processes and solid waste disposal (though the USEPA no longer quantifies the latter category) have been significant. These reflect the regulatory attention given to these source categories. Though transportation has been intensively regulated, emission reductions appear to be relatively modest. This has been due, in part, to (1) an increase in the motor vehicle population, (2) a significant increase in miles traveled, and (3) a relatively small population (~10%) of primarily older motor vehicles that disproportionately contributes to motor vehicle emissions. The relatively modest reductions in emissions associated with stationary source fuel combustion reflect an increased number of sources, regulatory policies that have focused primarily on ground-level concentrations rather than emissions, and implementation of only the first phase of emission reductions required under the acidic deposition control program.

On a pollutant basis, emission reductions have included a 33% reduction of CO, 24% reduction of PM_{10}, 38% reduction of NMHCs, and 35% reduction of SO_x. Because of changes in the USEPA's emission estimation procedures, PM_{10} estimates from miscellaneous sources for 1997 were significantly increased. If we factor out these apparent increases from miscellaneous sources, reductions in PM_{10} emissions were on the order of 70+%.

Significant emissions reductions have been achieved for PM_{10} from stationary source fuel combustion and industrial processes, with more modest reductions for CO, NMHCs, and SO_x, and a slight increase in NO_x. The modest reductions in NMHC emissions reflect emission issues associated with motor vehicles and slowly evolving regulatory efforts to control NMHC emissions from industrial and commercial sources. The apparent lack of progress associated with NO_x emissions reflects a variety of factors. These include (1) a belated implementation of NO_x controls on motor vehicles, (2) policies that focused on ground-level concentrations, and (3) an O_3 control strategy that focused exclusively on reducing NMHC emissions from motor vehicles.

Decreases in CO and NMHC emissions are a result of the application of catalytic converters and NMHC control technology on light-duty motor vehicles. Decreases in PM_{10} can be attributed to the installation of pollution control equipment on industrial sources, changes in fuel use practices in the direction of cleaner-burning fuels, the banning of open burning of solid wastes, and limitations on combustion practices used for solid waste incineration. The observed decrease in SO_x emissions has been due primarily to use of lower-sulfur (S) fuels, installation of flue gas desulfurization systems on coal-fired plants, and implementation of the first phase of acidic deposition control requirements.

Transportation has replaced stationary source fuel combustion as the major source of NO_x. This is primarily due to NO_x emission reduction requirements on fossil fuel-burning power plants. Consequently, transportation is now the major source of three of the six primary pollutants for which emission estimates are reported.

Reductions in Pb emissions to the atmosphere (Table 2.3) have been considerable (~98% overall). Emission reductions have been particularly significant in the transportation sector with the phaseout of Pb use in gasoline.

While the U.S. has made significant progress in reducing emissions of Pb and PM_{10}, and modest overall progress in reducing emissions of other primary pollutants and enhancing the quality of ambient air, it is apparent that continuing efforts to limit emissions are required to achieve air quality goals. Many air quality control regions in the country still do not have acceptable air quality. Significant reduction

Table 2.3 National Lead Emission Estimates for the U.S. in 1970 and 1997 (10^3 t/year)

Source	1970	1997	% Reduction
Transportation	180.30	0.52	99.7
Stationary source fuel combustion	10.62	0.49	95.4
Industrial processes	26.40	2.90	89.0
Solid waste disposal	2.20	—	—
Total	219.47	3.91	98.2

Source: From USEPA, EPA/454-R-94-027, EPA, Washington, D.C., 1997.

in emissions of atmospheric pollutants can be expected as (1) the population of motor vehicles with stricter emission limits increases, and (2) toxic pollutant (most of which are NMHCs) reductions continue as a result of new regulatory requirements. A significant reduction of SO_2 from 1997 levels occurred with implementation of Phase II of the acidic deposition reduction requirements.

The emission estimates in Table 2.1, Table 2.2, and Table 2.3 only include primary pollutants for which air quality standards have been promulgated. They do not include O_3-destroying chemicals or greenhouse gases such as carbon dioxide (CO_2) and methane (CH_4). Carbon dioxide emissions are very high and continue to increase.

2.3 GAS-PHASE POLLUTANTS

A large variety of gas-phase substances are emitted to the atmosphere from both anthropogenic and natural sources. Hundreds of different chemicals may be present in the polluted air of the many metropolitan areas that exist on the planet. Over 400 different gas-phase species have been identified in automobile exhaust alone, and many additional substances are produced as a result of atmospheric chemistry.

Only a relatively small number of the many hundreds of gaseous substances found in the atmosphere are either toxic enough or present in sufficient concentrations to directly or indirectly cause harm to humans or their environment. Major gas-phase substances identified as having potentially significant effects on humans or our environment are discussed in the following pages in the context of their nature, sources, background and urban–suburban levels, atmospheric chemistry, and sink processes.

2.3.1 Carbon Oxides

Billions of metric tons of the carbon oxides, CO and CO_2, are emitted by natural and anthropogenic sources. Because of its mammalian toxicity, CO has been a major pollutant and exposure concern for decades. Carbon dioxide, on the other hand, is relatively nontoxic. Though many atmospheric scientists began to express concern about increasing CO_2 levels and their potential effect on global climate decades ago, it has been only recently (since the late 20th century) that CO_2 emissions and global warming have become major public policy issues.

2.3.1.1 Carbon Dioxide

Carbon dioxide is a relatively abundant and variable constituent of the atmosphere. It is produced and emitted naturally to the atmosphere by aerobic biological processes, combustion, and weathering of carbonates in rock and soil. It is taken up by plants and used in the process of photosynthesis. Anthropogenic sources are fossil fuel combustion and land use conversion. The latter case includes clearing of forest lands with subsequent biomass burning and use for agricultural production.

Enormous changes in atmospheric CO_2 levels have occurred over the course of the Earth's history. This has included removal of carbon (C) from its storage (in vast quantities) in the Earth's crust as coal, oil, and natural gas. Because of their availability and ease of use, these fossil energy sources have made the industrial revolution and our modern technological world possible. As these are consumed at a prodigious rate, it is not surprising that global CO_2 levels have risen from ~280 ppmv in preindustrial times to ~370 ppmv at the turn of the millennium. Current estimates indicate that global CO_2 levels are increasing at approximately 1.5 to 1.7 ppmv per year. The pattern of this increase can be seen from data collected at Mauna Loa observatory in Hawaii in the period 1958–1994 (Figure 2.2). Note the strong upward trend, as well as pronounced seasonal variations. The latter (with variability of ~7 ppmv) is associated with photosynthetic consumption during spring and summer, with an excess of plant respiratory emissions during autumn and winter. The seasonal cycle varies from 1.6 ppmv at the South Pole to 15 ppmv at Point Barrow, AK.

The cyclic movement of C and other substances between organisms and their environment, as well as various components of the environment, is addressed in most biological, ecological, and environmental texts. As CO_2 becomes an increasing environmental concern, the traditional C cycle is being modified to include reservoirs and C fluxes associated with anthropogenic emissions. Such a global C cycle is illustrated in Figure 2.3. The values indicate C in gigatons (Gt) (10^9 t) and fluxes (Gt C/year).

Prior to the industrial revolution, the global C cycle was ostensibly in balance, with ~98% of potentially exchangeable CO_2 (carbonate rocks in the Earth's crust are not included) in the oceans and only 2% in the atmosphere. Of the CO_2 emitted to the Earth's atmosphere as a result of human activities, only about 42% has dissolved in the oceans.

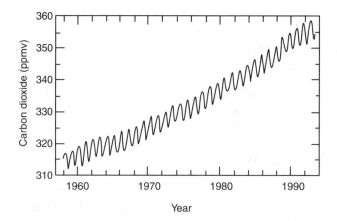

Figure 2.2   Annual trends and seasonal variation in CO_2 levels measured at Mauna Loa Observatory, HA, from 1958 to 1994. (From National Oceanic and Atmospheric Administration Climate Monitoring and Diagnostics Laboratory, http://www.cmdl.noaa.gov/ccgg/flask/index.html.)

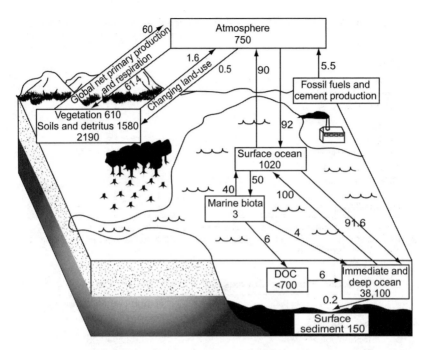

Figure 2.3   Global carbon cycle indicating reservoirs and fluxes. (From Houghton, J.T., Meira-Filho, L.G., Callender, B.A., Harris, N., Kallenberg, A., and Maskell, K., *The Science of Climate Change*, IPCC, Geneva, 1995. With permission.)
DOC = dissolved organic.

Though CO_2 is relatively soluble in water (H_2O) and is taken up by plants, the timescale for C cycling in the terrestrial and ocean spheres is on the order of decades to thousands of years. It has been estimated that for every 100 CO_2 molecules released to the atmosphere, only about 30% dissolve in the oceans in the span of a decade, increasing to 60% over six decades. Consequently, CO_2 has a relatively long atmospheric lifetime (~100 years).

A global CO_2 budget presented in Table 2.4 attempts to quantify CO_2 emissions from human activities and sink processes that remove them. The total anthropogenic flux of 7.1 Gt C/year is small compared to fluxes from the ocean (90 Gt C/year) and terrestrial ecosystems (60 Gt C/year). Although sink processes are relatively slow, they are nevertheless significant. Sinks include oceans, northern hemispheric forests, and the atmosphere. An unknown sink apparently exists since there is an imbalance of 1.4 ± 1.5 Gt C/year between anthropogenic sources and sink fluxes.

2.3.1.2 Carbon Monoxide

Carbon monoxide is a colorless, odorless, tasteless gas. It is emitted from a number of anthropogenic and natural sources and produced in the atmosphere from oxidation of CH_4 and other NMHCs (Table 2.5). Major sources of anthropogenic

Table 2.4   Global Carbon Dioxide Budget Associated with Human Activities

	CO_2 Flux (Gt C/year)
Sources	
Fossil fuel combustion/production	5.5 ± 0.5
Deforestation/land use changes	1.6 ± 1.0
Total sources	7.1 ± 1.1
Sinks	
Oceans	2.0 ± 0.8
Northern hemispheric forest	0.5 ± 0.5
The atmosphere	3.2 ± 0.2
Total sink losses	5.7 ± 1.0
Net imbalance	$+1.4 \pm 1.5$

Source: From Houghton, J.T., Meira-Filho, L.G., Callender, B.A., Harris, N., Kallenberg, A., and Maskell, K., *The Science of Climate Change*, IPCC, Geneva, 1995. With permission.

Table 2.5   Estimated Sources and Sinks of Carbon Monoxide

	Range (10^8 t CO/year)
Sources	
Technological	3.30–6.06
Biomass burning	3.30–7.71
Biogenic	0.07–1.76
Oceans	0.22–2.20
Methane oxidation	4.40–11.01
NMHC oxidation	2.20–6.60
Total sources	19.80–29.73
Sinks	
Reaction with OH·	15.42–28.63
Soil uptake	2.75–7.05
Stratospheric loss	~1.10
Total sink losses	23.13–33.04

Source: From Seinfeld, J.H. and Pandis, S.N., *Atmospheric Chemistry and Physics: Air Pollution to Climate Change*, John Wiley & Sons, New York, 1998. With permission.

emissions are technological processes (combustion and industrial) and biomass burning.

Carbon monoxide is an intermediate product in combustion oxidation of fuels and biomass:

$$2 \, C + O_2 \rightarrow 2 \, CO \tag{2.1}$$

$$2 \, CO + O_2 \rightarrow 2 \, CO_2 \tag{2.2}$$

If insufficient O_2 is present, combustion is incomplete and significant quantities of CO are produced.

Direct anthropogenic emissions from combustion and industrial processes and photolytic oxidation of anthropogenic NMHCs account for approximately 25 to 30% of CO emissions and production in the northern hemisphere. Carbon monoxide emissions from transportation sources are particularly significant in North America and Europe.

2.3.1.2.1 Concentrations

Background CO levels vary spatially and temporally, with a concentration range of 40 to 200 ppbv and an average of 120 ppbv at 45°N latitude. Background concentrations have been increasing (because of emissions and production associated with human activity) at a rate of ~1% per year, with most of the increase occurring in the northern middle latitudes. Highest concentrations are observed in winter when sink processes are less efficient.

Ambient CO concentrations in metropolitan areas are orders of magnitude higher than background. These levels are primarily associated with transportation emissions and are closely related to traffic density and meteorological conditions. Highest concentrations occur along major traffic arteries during morning and evening rush hours and decrease relatively rapidly with distance from roadways. Peak levels averaged over 1 h vary from 5–10 to >35 ppmv (the 1-h National Ambient Air Quality Standard (NAAQS)) in some cities. Elevated CO levels tend to plague cities that have high traffic densities and are located at relatively high altitudes (e.g., Denver, CO). Overall average urban–suburban CO levels in North America show a strong downward trend as regulatory requirements continue to result in decreased emissions to the atmosphere. The opposite trend is being observed in developing countries such as Mexico (notably high-altitude Mexico City), China, and other South Asian countries.

2.3.1.2.2 Sink Processes

When sink processes are considered, CO becomes significantly more important than regulatory concerns would indicate. Carbon monoxide affects tropospheric concentrations of both hydroxyl radical (OH·) and O_3 and, as a consequence, the oxidizing potential of the atmosphere. Because CO and CH_4 (a major greenhouse gas) compete for OH·, CO indirectly affects tropospheric concentrations of CH_4 and stratospheric H_2O vapor (derived from CH_4 oxidation).

Major sink processes include OH·–related photochemistry and uptake by soil. In Table 2.5, these account for an estimated 14.02 to 26.03 $\times 10^8$ and 2.5 to 6.4 $\times 10^8$ metric t/year (15.42 to 28.63 $\times 10^8$ and 2.75 to 7.05 $\times 10^8$ t/year) CO, respectively. Carbon monoxide is oxidized in the atmosphere to CO_2 on reaction with OH·:

$$CO + OH· \rightarrow CO_2 + H· \tag{2.3}$$

Subsequent reactions of H· and oxygen (O_2) produce the highly reactive hydroperoxyl radical, $HO_2·$:

$$H\cdot + O_2 + M \rightarrow HO_2\cdot + M \tag{2.4}$$

where M is an energy-absorbing molecule, e.g., nitrogen (N_2) or O_2. Summarizing,

$$CO + H\cdot + O_2 \rightarrow CO_2 + HO_2\cdot \tag{2.5}$$

Nitrogen dioxide is produced on reaction with $HO_2\cdot$ in the presence of significant concentrations of nitric oxide (NO):

$$HO_2\cdot + NO \rightarrow NO_2 + OH\cdot \tag{2.6}$$

Nitrogen dioxide is subsequently photolyzed to produce O_3:

$$NO_2 + h\nu \rightarrow NO + O(^3P) \tag{2.7}$$

$$O(^3P) + O_2 + M \rightarrow O_3 + M \tag{2.8}$$

$$CO + O_2 + h\nu \rightarrow CO_2 + O_3 \tag{2.9}$$

where $O(^3P)$ is ground-state atomic oxygen and $h\nu$ is a photon of light energy.

As a result of these reactions, CO is converted to CO_2, with one molecule of O_3 produced for each CO molecule oxidized. As such, CO oxidation in the troposphere is an important (but low-level) source of O_3.

As indicated previously, CO competes with CH_4 for $OH\cdot$. Methane oxidation, as seen in Table 2.5, is also a major source of CO, accounting for about 15 to 20% of its background atmospheric concentration. The atmospheric residence time of CO is ~1 month in the tropics to 4 months at mid-latitudes in the spring and autumn.

2.3.2 Sulfur Compounds

Sulfur compounds are emitted from a variety of natural and anthropogenic sources and produced as a result of atmospheric chemical processes. Major atmospheric S compounds include sulfur dioxide (SO_2) and reduced S compounds such as hydrogen sulfide (H_2S), dimethyl sulfide ((CH_3)$_2$S), carbon disulfide (CS_2), and carbonyl sulfide (COS).

2.3.2.1 Sulfur Oxides

Sulfur oxides are emitted to the atmosphere by volcanoes and a number of anthropogenic sources. They are also formed in the atmosphere as a result of oxidation of naturally- or anthropogenically-produced reduced S compounds.

The sulfur oxides, which include sulfur trioxide (SO_3) and SO_2, are produced when fuels and biomass are combusted and S-containing metal ores smelted. Because of its high affinity for H_2O vapor, the atmospheric lifetime of SO_3 is relatively short (seconds). Formation of SO_2, SO_3, and sulfuric acid (H_2SO_4) by relatively slow, direct oxidation processes is described by the following equations:

$$S + O_2 \rightarrow SO_2 \qquad (2.10)$$

$$2\ SO_2 + O_2 \rightarrow 2\ SO_3 \qquad (2.11)$$

$$SO_3 + H_2O + M \rightarrow H_2SO_4 + M \qquad (2.12)$$

Sulfur dioxide is a colorless gas with a sulfury odor that can be easily detected at concentrations of 0.38 to 1.15 ppmv. It has a pungent, irritating odor above 3.0 ppmv.

2.3.2.1.1 Concentrations

Background levels of SO_2 vary from ~20 pptv over marine surfaces to 160 pptv over relatively clean areas of the North American continent. Average annual concentrations in urban areas range from 1.5 ppbv in Auckland, New Zealand, to 90+ ppbv in cities in China. Shorter-term, 24-h concentrations may vary from 5 ppbv in Auckland to 300 to 400 ppbv in Chinese cities. Maximum 1-h concentrations of several hundred parts per billion by volume (ppbv) were once common in U.S. cities such as Chicago and Pittsburgh, and are still common in industrialized cities in developing countries. Highest maximum 1-h concentrations of several parts per million by volume (ppmv) have been reported downwind of nonferrous metal smelters.

2.3.2.1.2 Sink Processes

After emission to the atmosphere, SO_2 is oxidized by various gas- and aqueous-phase reactions. Oxidation may be direct (Equation 2.11 and Equation 2.12), photochemical, or catalytic. Gas-phase oxidation may occur as a result of reactions with OH·, O_3, O(^{3}P), and peroxy radicals (HO$_2$·, RO$_2$·). The dominant sink mechanism for SO_2 is reaction with OH·:

$$SO_2 + OH· \rightarrow HOSO_2 \qquad (2.13)$$

$$HOSO_2 + O_2 \rightarrow HO_2^· + SO_3 \qquad (2.14)$$

$$SO_3 + H_2O \rightarrow H_2SO_4 \qquad (2.15)$$

The reaction of SO_3 with H_2O vapor produces H_2SO_4, which forms an aerosol by nucleation and condensation processes. Sulfuric acid may react with ammonia (NH_3) to form ammonium salts. The lifetime of SO_2 (based on reaction with OH·) is ~1 week.

Sulfur dioxide can also be oxidized to form dilute sulfurous acid (H_2SO_3) in fog, cloud, and rain droplets, as well as hygroscopic aerosols:

$$SO_2 + H_2O \rightarrow SO_2 \cdot H_2O \qquad (2.16)$$

$$SO_2 \cdot H_2O \rightarrow H_2SO_3 \qquad (2.17)$$

In the aqueous phase, SO_2 can be oxidized to H_2SO_4 by reaction with nitrous acid (HNO_2), O_3, hydrogen peroxide (H_2O_2), organic peroxides, and catalysis by iron (Fe) and manganese (Mn).

Both SO_2 and its oxidation products are removed from the atmosphere by wet and dry deposition processes. Dry deposition to plant and other surfaces is relatively efficient. At a dry deposition velocity of 1 cm/sec, the atmospheric lifetime in the 1-km-thick planetary boundary layer has been estimated to be ~1 day. This removal is enhanced when clouds are present.

2.3.2.2 Reduced Sulfur Compounds

Carbonyl sulfide is the S gas present at highest levels in the background global atmosphere, with an average concentration of 500 pptv. It has low chemical reactivity, with an atmospheric lifetime of ~44 years. Carbon disulfide is produced from sources similar to COS. However, it is more photochemically reactive, with a lifetime of ~12 days. Global background levels are variable, ranging from 15 to 190 pptv. Dimethyl sulfide is released from the oceans in large quantities. It has a short atmospheric lifetime (0.6 days) and is rapidly converted to SO_2.

Hydrogen sulfide is the only reduced S compound that poses environmental concerns. Though it is a relatively toxic gas, atmospheric concentrations are usually too low to pose a threat to human health. It has a characteristic rotten egg odor, detectable by humans at concentrations as low as 500 pptv. It is an air quality concern because of the malodor it produces. Other reduced S malodorants include methyl and ethyl mercaptans, which smell like rotting cabbage.

Anthropogenic sources of H_2S include oil and gas extraction, petroleum refining, coke ovens, and kraft paper mills. Though anthropogenic sources produce less than 5% of global emissions, highest ambient concentrations (and, as a consequence, malodor problems) occur near anthropogenic emission sources. Concentrations in industrial and surrounding ambient environments are usually above the odor threshold of 500 pptv. Background concentrations are in the 30- to 100-pptv range.

2.3.3 Nitrogen Compounds

A number of gas- and particulate-phase nitrogen (N) compounds are found in the atmosphere. These include N_2, nitrous oxide (N_2O), NO, NO_2, nitrate radical ($NO_3 \cdot$), dinitrogen pentoxide (N_2O_5), HNO_2, nitric acid (HNO_3), organic nitrates such as peroxyacyl nitrate ($CH_3COO_2NO_2$), i.e., PAN, and reduced N compounds such as NH_3 and hydrogen cyanide (HCN). The ionic species, NO_2^-, NO_3^-, and NH_4^+, are found in the aqueous phase.

Nitrogen is the most abundant gas in the atmosphere. Under ambient conditions, it is relatively unreactive and plays no direct role in atmospheric chemistry. However, as a result of combustion reactions and biological processes, it serves as a precursor molecule for production of NO and NO_2, which play significant roles in tropospheric and stratospheric chemistry. Because they are readily converted from one to the other, NO and NO_2 concentrations are expressed as NO_x.

Reactive nitrogen (NO_y) includes NO_x compounds as well as their atmospheric oxidation products, e.g., HNO_2, HNO_3, $NO_3\cdot$, N_2O_5, peroxynitric acid (HNO_4), PAN and its homologues, alkyl nitrates ($RONO_2$), and peroxyalkyl nitrates ($ROONO_2$). Total NO_y is a measure of the total oxidized N content of the atmosphere.

Despite relatively low concentrations, NO, NO_2, organic nitrates, gas- and particulate-phase N acids, and NH_3 play significant roles in atmospheric chemistry (particularly in polluted environments). Nitric oxide and NO_2 are of major environmental concern since atmospheric concentrations increase significantly as a result of human activities, and they serve as precursor molecules for many important atmospheric reactions.

2.3.3.1 Nitrous Oxide

Nitrous oxide is a colorless, slightly sweet, relatively nontoxic gas. It is widely used as an anesthetic in medicine and dentistry. Human exposure to elevated concentrations produces a kind of hysteria, and as such, it is often referred to as laughing gas.

Nitrous oxide concentrations in the atmosphere have been increasing steadily since preindustrial times, from ~280 to ~320 ppbv today. Estimated annual emissions to the atmosphere are 13.8 Tg (tetragrams) N/year, with ~70% being produced by nitrification and denitrification processes in undisturbed terrestrial environments and the world's oceans. About 3 Tg N/year, or ~8%, is associated with agricultural tillage, fertilizer use, and animal wastes.

Nitrous oxide has no known tropospheric sink. As a result, it has a very long atmospheric lifetime (~150 years). Its photolysis and subsequent oxidation by singlet oxygen ($O(^1D)$) in the stratosphere is the only known sink process. As such, N_2O is the major natural source of NO_x in the stratosphere, where it plays a significant role in stratospheric O_3 chemistry.

Increased atmospheric N_2O levels pose two major environmental concerns: stratospheric O_3 depletion and, because of its thermal absorptivity, global warming (see Chapter 4).

2.3.3.2 Nitric Oxide

Nitric oxide is a colorless, odorless, tasteless, relatively nontoxic gas. It is produced naturally in soil through biological nitrification and denitrification processes and as a result of biomass burning, lightning, and oxidation of NH_3 by photochemical processes. It is also transported from the stratosphere into the troposphere.

Nitric oxide is produced anthropogenically in high-temperature combustion processes:

$$N_2 + O_2 \rightarrow 2\ NO \tag{2.18}$$

As this reaction is endothermic, equilibrium moves to the right at high temperatures and to the left at lower temperatures. If cooling is rapid, equilibrium is not maintained and significant NO emissions occur. High combustion temperatures, rapid cooling associated with gas flows from combustion chambers, and instantaneous dilution contribute to high NO emissions.

Major anthropogenic sources include gasoline- and diesel-powered vehicles, fossil fuel-fired electric generating stations, industrial boilers, municipal incinerators, and home space heating.

Nitric oxide concentrations are not generally reported separately from those of NO_2. However, early studies indicate that NO concentrations peak during the early-morning rush hour (6 to 9 A.M.) and then are rapidly depleted by oxidation to NO_2 and various scavenging processes (Figure 2.4).

2.3.3.3 Nitrogen Dioxide

Nitrogen dioxide is a colored gas that varies from yellow to brown depending on its concentration in the atmosphere. It has a pungent, irritating odor and, because of its high oxidation rate, is relatively toxic and corrosive.

Nitrogen dioxide can be produced in a number of atmospheric processes. These include the relatively slow, direct oxidation reaction,

$$2\ NO + O_2 \rightarrow 2\ NO_2 \tag{2.19}$$

and more rapid photochemical reactions involving O_3, RO_2, and odd hydrogen species ($OH\cdot$, $HO_2\cdot$, H_2O_2, etc.):

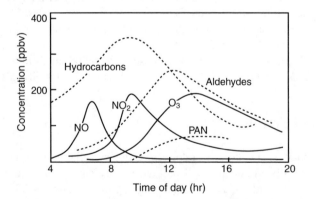

Figure 2.4   Time-dependent variation of NO, NO_2, NMHC, and O_3 in an urban atmosphere on a smoggy day. (From Goody, R., *Principles of Atmospheric Physics and Chemistry*, Oxford University Press, Oxford, 1995. With permission.)

$$NO + O_3 \rightarrow NO_2 + O_2 \qquad (2.20)$$

$$RO_2 \cdot + NO \rightarrow NO_2 + RO \cdot \qquad (2.21)$$

$$HO_2 \cdot + NO \rightarrow NO_2 + OH \cdot \qquad (2.22)$$

As it is primarily produced by photochemical oxidation of NO, peak NO_2 levels in urban–suburban areas typically occur in midmorning (Figure 2.4).

2.3.3.4 Nitrogen Oxides

2.3.3.4.1 Concentrations

Average NO_x concentrations (based on the sum of NO and NO_2) measured in remote locations have been reported to range from 0.02 to 0.04 ppbv in marine environments and 0.02 to 10 ppbv in tropical forests. In rural U.S. locations, NO_x concentrations are reported to range from 0.02 to 10 ppbv; in urban–suburban areas, from 10 to 1000 ppbv. Maximum 3-h average concentrations of 200 ppbv may occur during early-morning hours (6 to 9 A.M.), with average concentrations of <100 ppbv. Yearly average concentrations in urban areas are typically well below the NAAQS of 50 ppbv.

2.3.3.4.2 Sink Processes

Sink processes for NO and NO_2 involve chemical reactions that convert NO to NO_2 and NO_2 to HNO_3. Reaction of NO_2 with OH· is the major sink process:

$$NO_2 + OH \cdot + M \rightarrow HNO_3 + M \qquad (2.23)$$

Nitrogen dioxide is also converted to HNO_3 by nighttime reactions involving O_3:

$$NO_2 + O_3 \rightarrow NO_3 \cdot + O_2 \qquad (2.24)$$

$$NO_2 + NO_3 \cdot \rightarrow N_2O_5 \qquad (2.25)$$

$$N_2O_5 + H_2O \rightarrow 2 \, HNO_3 \qquad (2.26)$$

Reactions between O_3 and NO_2 produce $NO_3 \cdot$ that subsequently reacts with NO_2 to produce N_2O_5 (which reacts rapidly with H_2O vapor to produce HNO_3). Nitric acid is also produced in reactions involving $NO_3 \cdot$ and formaldehyde (HCHO) or hydrocarbon radicals (RH·):

$$NO_3 \cdot + HCHO \rightarrow HNO_3 + CHO \cdot \qquad (2.27)$$

$$NO_3 \cdot + RH \cdot \rightarrow HNO_3 + R \cdot \qquad (2.28)$$

Nitrate radical plays a significant role in nighttime oxidation of biogenically pro-duced volatile organic compounds (VOCs) such as isoprene and the pinenes — compounds that have a major role in tropospheric O_3 chemistry. Nitrate radical is rapidly photolyzed at sunrise by a shift in the equilibrium between NO_2, $NO_3·$, and N_2O_5 toward NO_2.

Nitric acid remains in the gas phase until it reacts with NH_3 to form ammonium salts:

$$NH_3 + HNO_3 \rightarrow NH_4NO_3 \tag{2.29}$$

Nitric acid and its salts are removed from the atmosphere by wet and dry deposition processes. In the lower troposphere the atmospheric lifetime of nitrates is ~1 week.

Because of their low solubility, NO and NO_2 are readily transported to the upper troposphere, with the subsequent production of HNO_3 on reaction with OH·. Nitric acid produced in the upper troposphere is not subject to dry deposition, and generally not to wet deposition. It can be transformed to NO_2 by photolysis,

$$HNO_3 + hv \rightarrow NO_2 + OH· \tag{2.30}$$

and, to a lesser extent, oxidation by OH·. Because these reactions are relatively slow, HNO_3 serves as a reservoir species and carrier for NO_x in atmospheric transport.

2.3.3.5 Ammonia

Ammonia is the most abundant basic chemical substance in the atmosphere and the third most abundant N compound. Major sources include anaerobic decompo-sition of organic matter, animals and their wastes, biomass burning, soil humus formation, and application of anhydrous NH_3 to cropland. Other sources include coal combustion and industrial emissions. Background concentrations have been reported to vary from 100 pptv over remote oceans to 1 to 10 ppbv over continents. Because it is readily absorbed by H_2O and soil surfaces, NH_3's atmospheric lifetime is relatively short (~10 days).

Ammonia reacts rapidly with strong acids (e.g., H_2SO_4 and HNO_3) to produce ammonium salts. As a consequence, NH_3 plays an important role in removing SO_x and NO_x from the atmosphere. Ammonia can also be oxidized on reacting with OH·.

2.3.3.6 Hydrogen Cyanide

Hydrogen cyanide occurs in the atmosphere in low background levels (~160 pptv). Though it reacts relatively slowly with OH·, little is known about its atmo-spheric chemistry.

2.3.3.7 Organic Nitrate Compounds

Organic nitrate compounds are produced as a result of photochemical reactions involving NO_x and NMHCs. Peroxy radicals are formed from the photochemical oxidation of short-chained NMHCs such as acetaldehyde (CH_3CHO). Acetaldehyde, on reaction with OH·, produces an acetyl radical that on reaction with O_2 produces a peroxyacetyl radical. Subsequent reaction with NO_2 produces PAN:

$$CH_3CHO + OH· \rightarrow CH_3CO + H_2O \qquad (2.31)$$

$$CH_3CO + O_2 \rightarrow CH_3COO_2· \qquad (2.32)$$

$$CH_3COO_2· + NO_2 + M \leftrightarrow CH_3COO_2NO_2 + M \qquad (2.33)$$

Reactions involving propionylaldehyde and butrylaldehyde produce peroxypropionyl nitrate (PPN) and peroxybutryl nitrate (PBN).

Peroxy nitrates were first identified as components of smog in the Los Angeles Basin in the 1950s. For most of the past 40+ years, PAN was thought to be associated only with polluted urban atmospheres with high photochemical activity. It is now known to be ubiquitously present in the atmosphere, being found in urban, rural, and remote environments. It is a potent eye irritant and the cause of "smog injury" on sensitive agricultural crops.

Peroxyacetyl nitrate and its homologues are relatively inert photochemically and insoluble in water. As such, they have relatively long atmospheric lifetimes. The principal sink process is thermal degradation to acetylperoxy radicals and NO_2. In the upper troposphere, PAN is quite stable and transported over long distances. It is an important reservoir and carrier of NO_x (similar to HNO_3, described previously). Because its decomposition increases with temperature, it can be formed in cooler regions, transported, and then decompose to release NO_2 in warmer regions. Peak concentrations of PAN in urban–suburban areas have been reported in the range of 10 to 50 ppbv.

2.3.4 Hydrocarbon Compounds

2.3.4.1 Hydrocarbon Chemistry

Hydrocarbon compounds include a large number of chemicals that exist under ambient environmental conditions as gases, volatile liquids, semivolatile substances, and solids. They include compounds that contain only hydrogen (H) and C (HCs) and HC derivatives that contain substances such as O, N, S, or various halogens.

All HC species are characterized by covalent chemical bonds between C and H, C and C, or C and other atomic species such as O. The covalently bonded atoms share electron pairs to attain chemical stability. A single covalent bond is formed when an electron pair is shared. In the simplest HC, CH_4, each H is bonded to C by a single bond. Hydrocarbon compounds that contain only single bonds are called alkanes or paraffins. Larger molecules contain two or more carbons covalently bonded to each other. The paraffins include compounds that are straight chained (e.g., CH_4 and ethane (C_2H_6)), branched (e.g., isobutane), and cyclic (e.g., cyclopropane). The alkanes or paraffins are relatively unreactive and therefore chemically stable; e.g., n-butane has a relatively long atmospheric lifetime (~5 days).

Carbon atoms can share one to three electron pairs. A double bond is formed when two pairs of electrons are shared. Hydrocarbons that contain one or more double bonds are called alkenes or olefins; the simplest is ethylene (C_2H_4). The olefins also include a number of longer-chained compounds (e.g., propylene and butylene). Because double bonds are somewhat unstable, olefinic compounds are very reactive. Olefins and other compounds with double bonds serve as primary reactants in photochemical processes that produce haze and smog. Their atmospheric lifetimes are on the order of several hours.

If two H atoms were removed from ethylene, the two carbons would share three electron pairs, forming acetylene (C_2H_2), which has a triple bond. Unlike the olefins that have atmospheric lifetimes of hours, alkynes such as C_2H_2 are long-lived, with an estimated lifetime of 14 days.

Hydrocarbons described above, i.e., the paraffins, olefins, and alkynes, are classified as aliphatic HCs because of their basic chain structure. Hydrocarbons whose structure is based on the ring structure of benzene (C_6H_6) are called aromatic HCs. Aromatic HCs other than C_6H_6 include (among others) ethylbenzene, toluene, and the xylenes. (This group of four aromatic compounds is formulated to produce the fuel additive, BTEX, which increases the octane rating of unleaded gasolines.)

Because of their widespread use as solvents and fuel additives, aromatic HCs are commonly found in ambient air. Though they contain at least three double bonds, aromatic HCs are generally less reactive than olefins. They vary in their reactivity and therefore estimated atmospheric lifetimes (C_6H_6, ~12 days; toluene, ~2 days; m-xylene, ~7.4 h). Some reactive HCs (e.g., m-xylene) are likely to be quickly consumed in urban photochemistry, whereas others (e.g., C_6H_6) are even more stable than some of the paraffins.

Aromatic HCs described above are simple compounds based on a single C_6H_6 ring. Benzene rings also serve as the basic structural unit of larger, more complex compounds, including napthalene (two C_6H_6 rings) and a family of polycyclic aromatic hydrocarbons (PAHs) that are solids under ambient atmospheric conditions. The best-known compound in this group is benzo-α-pyrene, which contains five C_6H_6 rings. The PAHs are produced in the combustion chemistry of organic fuel and materials and are commonly found in the elemental C (EC) fraction of atmospheric aerosol. Most PAHs are potential human carcinogens.

Hydrocarbons with only single bonds are described as being saturated; those with one or more double or triple bonds are unsaturated. Compounds that contain 1 to 4 C atoms are gases under ambient conditions; those with 5 to 12 C atoms, volatile liquids. Those with higher C numbers tend to be semivolatile liquids or solids, or nonvolatile solids.

Hydrocarbon compounds and their derivatives are classified on the basis of their boiling points and volatility. Very volatile organic compounds (VVOCs) have boiling points that range from 0 to 50–100°C (32 to 122–212°F); VOCs, from 50–100 to 240–260°C (464 to 500°F); semivolatile organic compounds (SVOCs), from 240–260 to 380–400°C (716 to 752°F); and solid organic compounds, >400°C.

Hydrocarbon emissions are regulated by the USEPA. Because of the low reactivity of CH_4, HC concentrations in the atmosphere and national emissions estimates are reported as NMHCs. The USEPA has recently begun to report emission estimates as VOCs (which include both VOCs and VVOCs).

A variety of substances are derived from the basic HC structure. These include, among others, oxygenated and halogenated HCs, both of which represent major atmospheric pollution concerns. Because they are subject to unique regulatory requirements, halogenated HCs are discussed apart from other NMHCs in this chapter.

2.3.4.2 Oxyhydrocarbons

A variety of HC compounds can be formed as a result of reaction with O_2 or O-containing compounds. These include alcohols, aldehydes, ketones, ethers, organic acids, and esters.

Alcohols, e.g., methanol (CH_3OH) and ethanol (C_2H_5OH), are characterized by having an –OH moiety attached to one of their carbon atoms. The substitution of an –OH in the structure of CH_4 produces CH_3OH. Short-chain alcohols are liquid at ambient conditions and commonly used as solvents. Methanol is used as an automotive fuel and C_2H_5OH as a fuel additive. Alcohols with two or more –OH moieties are called polyols. Low-volatility polyols such as ethylene glycol are used as motor vehicle coolants and in water-based paint and varnish products. Because of their widespread use, alcohols and polyols are frequently emitted to the atmosphere.

Aldehydes are characterized by a carbonyl moiety (C=O) on a terminal C atom. The simplest aldehyde is HCHO. It is widely used in industry as a chemical feedstock for production of adhesives and other polymers. It is also a common by-product of combustion processes, as are the aldehydes CH_3CHO and acrolein. Significant aldehyde emissions are associated with light-duty motor vehicle operation. Aldehydes are also produced in photochemical reactions, with peak atmospheric levels occurring at solar noon. Both HCHO and CH_3CHO are important reactants in a variety of photochemical processes. Acetaldehyde has an estimated atmospheric lifetime of 11 h and HCHO, ~1.2 days.

The ketones contain a carbonyl attached to a nonterminal C. Ketones are widely used as solvents and, because of their high volatility, quickly released into the atmosphere.

Acetone, one of the most widely used ketones, enters the atmosphere from both biogenic and anthropogenic sources and oxidation of paraffinic and olefinic compounds. Because of its long atmospheric lifetime (~100 days), acetone is found in the free troposphere at concentrations of 1 ppbv.

Ethers are VVOCs used as solvents. They are characterized by a moiety in which O is bonded to two C atoms. They are not significant atmospheric pollutants. However, several ethers are used as a source of O in reformulated automotive fuels. These include methyl-t-butyl ether (MTBE), ethyl-t-butyl ether (ETBE), and tertiary-amyl-methyl ether (TAME).

Organic acids are characterized by one or more carboxyl groups (–COOH). Organic acids commonly used in industry include acetic and formic acids. Though emissions are limited, organic acids are common in the atmosphere in the gas phase.

Acetic acid is found in urban areas in the low ppbv range. A variety of dicarboxylic acids (two –COOH groups) have been observed in urban areas. These include oxalic, succinic, and malonic acids. They are produced from the photochemical oxidation of organic compounds in the atmosphere. Because of their low vapor pressures, dicarboxylic acids and other organic acids are predominantly found in or on aerosol particles. Organic acids are removed from the atmosphere by reaction with OH· and by wet and dry deposition processes.

Esters are organic compounds produced by reaction of alcohols with organic acids. They are widely used as solvents and industrial feedstock. Emissions to the atmosphere are relatively limited and esters have negligible atmospheric significance.

2.3.4.3 Nonmethane Hydrocarbons

2.3.4.3.1 Sources and Emissions

Because of their role in atmospheric photochemistry, emissions of NMHCs and VOCs are regulated in many developed countries. Major anthropogenic sources include mobile and stationary source fuel usage and combustion; petroleum refining; petrochemical manufacturing; industrial, commercial, and individual solvent use; gas and oil production; and biomass burning. Emissions of NMHCs have been of particular concern in urban areas. A source apportionment study for Sydney, Australia, conducted in 1983 estimated that 36% of NMHC emissions were due to motor vehicle exhausts, 16% to gasoline spillage, 16% to gasoline evaporation, 23% to solvent evaporation, 4% to commercial use of natural gas, and 5% to industrial processes. In this and an earlier source apportionment study in Los Angeles, motor vehicle exhaust and gasoline usage were responsible for ~70% of NMHC emissions in urban areas. In a study of Mexico City, the major source of NMHCs appeared to be emissions of unburned liquefied petroleum gas (LPG) used for heating and cooking, not motor vehicles and gasoline.

Emissions associated with gasoline usage and petroleum-related industries include a variety of paraffinic, olefinic, and aromatic HCs; with natural gas, light paraffins; and with solvents, higher paraffinic and aromatic HCs. Total global emissions from these anthropogenic sources are estimated to be 8.1×10^7 metric t/year (8.9×10^7 t/year). A comparatively similar quantity is emitted as a result of biomass burning in the tropics. Biomass burning emissions consist primarily of light paraffins and olefins.

Emissions from biogenic sources, which include trees and grasslands, and emissions from soils and oceans are approximately an order of magnitude higher on a global basis than anthropogenic emissions. Emissions from trees consist primarily of isoprene and monoterpenes, with some paraffins and olefins; from grasslands, light paraffins and higher HCs; from soils, mainly C_2H_6; and from oceans, light paraffins, olefins, and C_9 to C_{28} paraffins.

The USEPA has estimated that 40% of NMHC anthropogenic emissions over the U.S. result from transportation, with light-duty cars and trucks making up the largest fraction; 32% result from solvent usage and 28% from industrial manufacturing activities and fuel combustion.

2.3.4.3.2   Identification

Many investigators have conducted studies to identify the many NMHC species and their derivatives that may be present in the atmosphere. Because of the complexities and expense associated with NMHC sampling and analysis, the atmosphere has not been well characterized, particularly in rural and remote areas. Many NMHC species produced as a result of photochemical reactions are short-lived or present in subtrace quantities and, in many cases, below the limit of detection of sampling systems utilized. Consequently, the presence of many NMHC compounds must be inferred from smog chamber studies and analysis of motor vehicle exhaust. In the latter case, over 400 NMHCs and oxyhydrocarbon derivatives have been identified. Oxyhydrocarbons, which include aldehydes, ketones, organic acids, alcohols, ethers, esters, and phenol, have been reported to comprise 5 to 10% of the total NMHC concentration in auto exhaust. The aldehyde fraction is dominated by HCHO and CH_3CHO.

A variety of NMHC species have been found, or are believed to be present, in the atmosphere. These include straight, branch-chained, and cyclic paraffinic compounds; olefinic compounds with one or two double bonds; acetylene-type compounds; and C_6H_6 and its derivatives. Compounds with multiple C_6H_6 rings (PAHs) are found condensed on aerosol particles.

2.3.4.3.3   Concentrations

Most concentration data are based on measurement of total NMHCs, usually averaged from 6 to 9 A.M. (corresponding to early-morning commuting traffic in urban areas). Concentrations are determined from flame ionization detector measurements and reported in CH_4 or C equivalents. The air quality standard for this 3-h period is 0.24 ppmv, with most urban concentrations at levels less than this. Substantially higher levels were reported in the Los Angeles Basin in the 1950s and 1960s, in the period prior to emission control requirements on motor vehicles.

Concentrations of individual NMHC species based on C equivalents (ppbc) are presented for 35 major HC species identified in samples collected over Atlanta, GA (Table 2.6). These include a large number of short- and long-chained paraffins, a relatively limited number of olefins, and a variety of aromatic HCs. Para-cymene and isoprene are biogenic in origin. Acetylene is an indicator of motor vehicle emissions.

2.3.4.3.4   Sink Processes

The primary sink process for NMHCs and their oxygenated derivatives is the oxidative production of alkylperoxy radicals (ROO·) on reaction with OH· or O_3. In the presence of NO, ROO· is converted to alkoxy radical (RO·), which reacts with O_2 to produce aldehydes or, in the case of longer-chained NMHCs, butane, aldehydes, and ketones. Oxidation of C_2H_6, a relatively unreactive substance, is summarized in the following series of equations:

Table 2.6　Major Hydrocarbon Compounds (Reported in C Equivalent Concentrations) Measured in Atlanta, GA, in 1981

Species	Concentration (ppbC)	Species	Concentration (ppbC)
i-Pentane	19.8	3-Methylpentane	3.4
n-Butane	16.9	Ethylene	3.0
Toluene	14.7	3-Methylcyclopentane	2.9
p-Cymene	11.0	Ethylbenzene	2.8
n-Pentane	9.4	o-Xylene	2.8
Benzene	8.8	3-Methylhexane	2.6
m- and p-Xylene	7.6	2,3-Dimethylpentane	2.5
2-Methylpentane	5.9	1,4-Diethylbenzene	2.4
Cyclohexane	5.4	Isobutene	2.2
2-Methylhexane	5.2	2,2,4-Trimethylpentane	2.2
Ethane	5.0	1,2,4-Trimethylpentane	2.2
Undecane	4.9	i-Butylbenzene	1.8
Propane	4.8	2-Methyl-2-butene	1.8
i-Butane	4.8	1,3,5-Trimethylbenzene	1.8
Isoprene	4.6	Cyclopentane	1.6
Acetylene	4.3	Propene	1.5
n-Hexane	3.8	i-Propylbenzene	1.5
m- and p-Ethyltoluene	3.6		
		Total ppbC	197

Source: From National Research Council, *Rethinking the Ozone Problem in Urban and Regional Air Pollution*, National Academy Press, Washington, D.C., 1991. With permission.

$$C_2H_6 + OH\cdot \rightarrow H_2O + C_2H_5\cdot \tag{2.34}$$

$$C_2H_5\cdot + O_2 \rightarrow C_2H_5OO\cdot \tag{2.35}$$

$$C_2H_5OO\cdot + NO \rightarrow C_2H_5O\cdot + NO_2 \tag{2.36}$$

$$C_2H_5O\cdot + O_2 \rightarrow CH_3CHO \tag{2.37}$$

Ethane oxidation produces CH_3CHO, which, like other aldehydes, reacts more readily with $OH\cdot$ than its NMHC precursor. As a consequence, CH_3CHO reacts with $OH\cdot$ and, by a series of reactions, produces HCHO, which undergoes further oxidation or photodecomposition.

Formaldehyde, CH_3CHO, and acetone can photodecompose on absorption of ultraviolet (UV) light in the wavelength range of 330 to 350 nm. Photodecomposition of HCHO proceeds via either of two pathways. In both cases, photodecomposition produces CO. These reactions are summarized below:

$$HCHO + h\nu \rightarrow HCO + H \tag{2.38}$$

$$H + O_2 + M \rightarrow HO_2\cdot + M \tag{2.39}$$

$$HCO + O_2 \rightarrow HO_2\cdot + CO \tag{2.40}$$

$$HO_2\cdot + NO \rightarrow NO_2 + OH\cdot \tag{2.41}$$

In the first pathway, photodecomposition of HCHO leads to production of $HO_2\cdot$, which, on reaction with NO, generates OH·, which then becomes available for NMHC oxidation. In another pathway (Equation 2.42), H + HCO are rapidly converted to H_2 and CO:

$$H + HCO \rightarrow H_2 + CO \tag{2.42}$$

Aldehydes and ketones produced from NMHC oxidation are removed from the atmosphere by scavenging processes involving dry and wet deposition. Oxidation of longer-chain aliphatic and aromatic HCs may lead to formation of condensable products such as dicarboxylic acids, which enter the particulate phase and are removed by wet and dry deposition processes.

All NMHCs react relatively rapidly with OH·. However, NMHCs in the atmosphere vary considerably in their OH· reactivity, with unsaturated HCs, including olefinic and aromatic HCs, being the most reactive. Many unsaturated NMHCs react with O_3 at rates that are competitive with OH·. As indicated previously, the lifetime of olefins and biogenically-produced terpenes is on the order of hours, whereas for less reactive paraffinic species, the atmospheric lifetime may be days. Since the most reactive compounds are removed at a faster rate, the abundance of NMHCs changes in the direction of less reactive species such as C_2H_6 in urban areas as well as those downwind.

2.3.4.3.5 Photochemical Precursors

Oxidation products produced in NMHC sink processes, e.g., ROO·, RO·, $HO_2\cdot$, and CO, serve as major reactants in the production of photochemical smog and associated elevated tropospheric levels of O_3 and other oxidants (the formation of which will be discussed in a following section). In addition to oxidants, photochemical reactions produce a large variety of species that may comprise 95% of atmospheric NMHCs during smog episodes.

2.3.4.4 Methane

Methane has low reactivity with OH· relative to olefins, aromatics, and even other paraffins, and as a consequence, it has limited significance in photochemical reactions responsible for producing elevated O_3 levels in urban areas. Methane and other low-reactivity paraffins such as C_2H_6 become important when polluted air masses (depleted of reactive NMHCs) travel downwind of urban sources. In addition, methane is a thermal absorber and, as a result of its increasing atmospheric levels, an important greenhouse gas.

Global CH_4 levels in 1995 averaged ~1.72 ppmv. In the 1980s and 1990s, they were increasing at a rate of ~20 ppbv/year. Increases of CH_4 over the last 300 years, as measured in air trapped in Antarctic and Greenland ice as well as recent atmosphere measurements, can be seen in Figure 2.5. Note the significant inflection of the curve in the past 100 years.

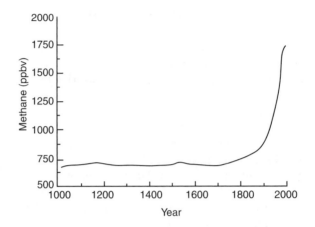

Figure 2.5   Trends in atmospheric methane levels. (From IPCC. Climate Change 2001: The
Scientific Basis, http://www.grida.no/climate/ipcc_tar/wgl/images/fig4–1.gif.)

Emission of CH_4 to the atmosphere occurs from natural sources, natural sources
influenced by human activity, and anthropogenic sources. Biogenic emissions result
from anaerobic decomposition of organic matter in the sediments of swamps, lakes,
rice paddies, and sewage wastes, and from digestion of cellulose by livestock (and
other ruminants) and termites. Methane emissions also occur from coal and lignite
mining, oil and natural gas extraction, petroleum refining, leakage from natural gas
transmission lines, and automobile exhaust; CH_4 accounts for 15% of all motor
vehicle exhaust HCs.

The major sink process for CH_4 is reaction with OH·:

$$OH· + CH_4 \rightarrow CH_3· + H_2O \tag{2.43}$$

Subsequent reactions produce HCHO, CO, and ultimately CO_2. The stratosphere is
another important sink, with CH_4 photodecomposition the major source of stratospheric H_2O vapor. The OH· sink removes approximately 10 times more CH_4 than
the stratosphere. One of the major factors contributing to increased atmospheric CH_4
levels may be CO depletion of OH· (Equation 2.3). The atmospheric lifetime of CH_4
is estimated to be 10 years.

2.3.4.5 Halogenated Hydrocarbons

A variety of halogenated HCs are emitted to the atmosphere from natural and
anthropogenic sources. These include both volatile and semivolatile substances that
contain one or more atoms of chlorine (Cl), bromine (Br), and fluorine (F). Due to
their persistence, halogenated compounds are relatively unique atmospheric contaminants. They include volatile substances such as methyl chloride (CH_3Cl), methyl
bromide (CH_3Br), methyl chloroform (CH_3CCl_3), trichloroethylene (CH_2CCl_3), perchloroethylene (C_2Cl_4), and carbon tetrachloride (CCl_4). Methyl chloride and CH_3Br
are produced naturally; other volatile halogenated HCs have been used industrially

and commercially. A variety of semivolatile halogenated HCs are land, air, water, and biological contaminants. These include chlorinated HCs such as DDT, chlordane, dieldrin, aldrin, etc.; polychlorinated biphenyls (PCBs) used as solvents and electrical transformer insulators; polybrominated biphenyls (PBBs) used as fire retardants; and dioxins and furans that are by-products of pesticide manufacture, bleaching of wood fiber to make paper, and incineration of paper and plastics.

Halogenated HCs that include Cl and F in their chemical structure are called chlorofluorocarbons (CFCs). The most commonly used CFCs have been trichloro-fluoromethane (CCl_3F), difluorodichloromethane (CF_2Cl_2), and trichlorotrifluoro-ethane ($C_2Cl_3F_3$). These are generally described as CFC-11, CFC-12, and CFC-13, respectively. As a result of their important industrial and commercial properties (including low reactivity, low toxicity, thermal absorption, and solvent properties), CFCs have been used as aerosol propellants, refrigerants, degreasing solvents, and foaming agents.

Because of their low chemical reactivity, halogenated HCs break down slowly, and consequently are characterized by long atmospheric lifetimes, estimated at ~1.5 years for CH_3Cl, 6 years for CH_3CCl_3, ~40 years for CCl_4, ~50 years for $CFCl_3$, and ~102 years for CF_2Cl_2.

Though halogenated compounds such as CH_3Cl and CH_3CCl_3 have tropospheric sinks, CFCs do not. As a consequence, concentrations in both the troposphere and stratosphere had been increasing until the early 1990s. Tropospheric concentrations of CFCs and other halocarbons are reported in the pptv range. Average approximate global concentrations reported for 1995 were CF_2Cl_2, 503 pptv; CH_3Cl, 600 pptv; CCl_3F, 268 pptv; CH_3CCl_3, 160 pptv; CCl_4, 132 pptv; $C_2Cl_3F_3$, 82 pptv; and CH_3Br, 12 pptv. Concentrations of CFCs and other halogenated HCs vary spatially over the surface of the Earth, with highest concentrations observed in source regions over the northern hemisphere.

Total Cl concentrations in the troposphere peaked around 1994, with an average concentration of ~3.7 ppbv, slowly decreasing thereafter. Tropospheric Br concentrations, however, were still increasing. Both stratospheric Cl and Br peaked around the year 2000.

Brominated HCs are of environmental interest since Br atoms (as do Cl atoms) catalytically destroy O_3. Methyl bromide is the most abundant Br compound in the atmosphere, with average atmospheric concentrations in the range of 9 to 13 pptv. Its major sources include marine phytoplankton, agricultural pesticide use, and biomass burning. Other atmospheric Br sources include CF_3Br (Halon 1301) and CF_2BrCl (Halon 1211), used as fire control agents.

Various halocarbons can be characterized by their O_3-depleting potential. These include $CFCl_3$ (1.0), CF_2Cl_2 (0.82), $C_2Cl_3F_3$ (0.90), CCl_4 (1.2), CH_3CCl_3 (0.12), CH_3Br (0.64), CF_3Br (12.0), and CH_3Cl (0.02). Thus, CF_2Cl_2 is about 18% less effective in destroying O_3 than $CFCl_3$, and CCl_4 is 20% more effective than $CFCl_3$.

2.3.5 Photochemical Oxidants

Photochemical oxidants are produced in the atmosphere as a consequence of chemical reactions involving sunlight, NO_x, O_2, and a variety of NMHCs.

Photochemical oxidants produced from these reactions include O_3, NO_2, PAN, odd hydrogen compounds, and RO_2·. Because of the significant environmental effects of elevated levels associated with human activities, tropospheric O_3 has received major scientific and regulatory attention.

As indicated in Chapter 1, O_3 is a normal constituent of the atmosphere, with peak concentrations in the middle stratosphere. As a consequence of anthropogenic activities, O_3 represents two major and distinct environmental concerns. In the stratosphere, it is a depletion issue (Chapter 4); in the troposphere, it is one of the levels elevated significantly above background. Tropospheric O_3 production is the focus of this discussion.

Ozone is formed in the atmosphere when molecular O_2 reacts with $O(^3P)$:

$$O_2 + O(^3P) + M \rightarrow O_3 + M \qquad (2.44)$$

In the troposphere, photodissociation of NO_2 at wavelengths of 280 to 430 nm is the primary source of $O(^3P)$:

$$NO_2 + hv \rightarrow NO + O(^3P) \qquad (2.45)$$

The reaction of $O(^3P)$ with O_2 produces O_3, which reacts rapidly with NO to regenerate NO_2:

$$NO + O_3 \rightarrow NO_2 + O_2 \qquad (2.46)$$

This O_3-producing process is represented diagrammatically in Figure 2.6.

Reactions shown in Equation 2.44 to Equation 2.46 proceed rapidly, producing a small steady-state concentration of ~20 ppbv under solar noon conditions in mid-latitudes at an atmospheric $NO_2:NO$ concentration ratio of 1.

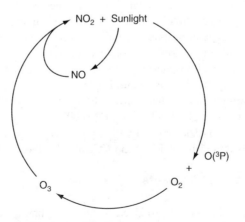

Figure 2.6   Photolysis of NO_2 to produce O_3.

Ozone concentrations in both urban and nonurban atmospheres are often higher than those associated with NO_2 photolysis. Elevated O_3 levels occur as a result of chemical reactions that convert NO to NO_2 without consuming O_3. In very polluted and even lightly polluted atmospheres, shifts in O_3 chemistry occur in the presence of $RO_2\cdot$, produced by oxidation of NMHCs (Equation 2.34 and Equation 2.35). Peroxy radicals then react with NO:

$$RO_2\cdot + NO \rightarrow NO_2 + RO\cdot \tag{2.47}$$

$$NO_2 + hv \rightarrow NO + O(^3P) \tag{2.48}$$

$$O(^3P) + O_2 + M \rightarrow O_3 + M \tag{2.49}$$

$$\text{Net: } RO_2\cdot + O_2 + hv \rightarrow RO\cdot + O_3 \tag{2.50}$$

These reactions are represented diagrammatically in Figure 2.7.

The rate of O_3 formation is closely related to the concentration of $RO_2\cdot$. Peroxy radicals are produced when OH· and other odd hydrogen species react with vapor-phase NMHCs. Odd hydrogen species such as OH· are produced by reactions involving the photodissociation of O_3, aldehydes, and HNO_2. Ozone photodissociation and formation of OH· are shown in Equation 2.51 and Equation 2.52:

$$O_3 + hv \rightarrow O(^1D) + O_2 \tag{2.51}$$

$$O(^1D) + H_2O \rightarrow 2\ OH\cdot \tag{2.52}$$

In very polluted atmospheres, O_3 concentrations at any particular time and place depend on the intensity of sunlight; NO_2:NO ratios; reactive NMHCs and their

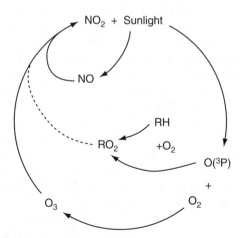

Figure 2.7 Photochemical reactions between NO_x and hydrocarbon radicals resulting in the production of O_3.

concentrations; and other pollutants, e.g., aldehydes and CO, that react photochemically to produce RO_2·. The change in NO_2:NO ratios caused by atmospheric reactions involving RO_2· may result in elevated tropospheric O_3 levels.

2.3.5.1 Tropospheric Ozone Concentrations

In the relatively clean atmospheres of remote sites, ground-level O_3 concentrations are estimated to range from 20 to 50 ppbv during warmer months of the year. Major sources of background O_3 are photochemical processes and the movement of stratospheric O_3 in the region of so-called tropopause folds into the lower atmosphere, particularly during spring months. There is general agreement among atmospheric scientists that the former is a major source of background O_3. Depending on location, O_3 associated with processes uninfluenced by human activities may account for 20 to 50% of observed rural monthly average concentrations, particularly during warmer months.

Ozone concentrations over large, economically developed landmasses such as North America, Europe, eastern Australia, and Southeast Asia are often significantly higher than those found in remote and rural areas. These relatively high O_3 levels are associated with anthropogenic emissions of NMHCs and NO_x, biogenic emissions of reactive NMHCs, and subsequent photochemical reactions summarized in Equation 2.47 to Equation 2.50.

Our understanding of atmospheric chemistry associated with the formation of elevated tropospheric O_3 levels has evolved considerably from the 1950s and early 1960s, when it was understood to be an urban problem associated with cities such as Los Angeles. At that time, peak levels were observed at solar noon, with a subsequent rapid destruction associated with NO_x-scavenging processes (Figure 2.4). In the 1970s, elevated O_3 levels were observed in many rural areas far from urban centers that were the focus of O_3 concerns. Rural O_3 levels elevated above background were observed to be associated with (1) long-range transport of O_3 isolated from normal scavenging processes by ground-based inversions, and (2) transport of O_3 precursors such as NO_x and less-reactive NMHCs. In the mountain areas of the eastern U.S., elevated O_3 is observed to both be persistent and occur at night, contradicting the previous paradigm of O_3 production and destruction that was characteristic of urban areas with large motor vehicle populations.

Our understanding of O_3 chemistry continues to evolve as we come to comprehend the significant role that NMHCs emitted by trees play in producing elevated O_3 levels on a regional basis.

In the U.S., noontime peak concentrations have been at their highest in the precursor- and sunlight-rich Los Angeles Basin, with 1-h average concentrations ranging as high as 0.40 ppmv during severe smog episodes and concentrations of 0.15 to 0.30 ppmv not uncommonly reported. Daily second-highest maximum concentrations (1-h averaging time) are summarized for selected U.S. cities in Table 2.7 for the year 1993. All levels were above the then-ambient air quality standard of 0.12 ppmv (1-h average). The number of U.S. cities exceeding the 0.12-ppmv air quality standard varied from year to year, with a hundred cities or more exceeding the standard in very hot summers.

Table 2.7   Ozone Levels Measured in Selected U.S.
Metropolitan Areas in 1993

Metropolitan Area	O_3 (ppmv)[a]
Baltimore, MD	0.15
Birmingham, AL	0.13
Bridgeport, CT	0.17
Dallas, TX	0.14
Danbury, CT	0.14
Los Angeles, CA	0.25
Louisville, KY	0.14
New London, CT	0.13
Oakland, CA	0.13
Orange County, CA	0.17
Philadelphia, PA	0.14
Phoenix, AZ	0.13
Riverside, CA	0.23
St. Louis, MO	0.13

[a]Second-highest daily maximum 1-h concentration.
Source: From USEPA, EPA/600/AP-95/001c, EPA, Washington, D.C., 1995.

Though elevated urban O_3 levels have been a concern in the U.S. for decades, the problem is not limited to the U.S. In Mexico City, O_3 levels exceeded the Mexican 1-h standard of 0.11 ppmv on 71 and 98% of the days in 1991 and 1992, respectively, with values in the 0.20- to 0.30-ppmv range common and values as high as 0.48 ppmv recorded. During episodes of elevated O_3 in the more solar radiation-limited countries of northern Europe, concentrations of 0.10 ppmv are common, with a severe episode in southern England that exceeded 0.25 ppmv.

2.3.5.2   Ozone Sinks

The two major sink processes for O_3 are surface destruction or deposition and photochemical reactions. Surface deposition includes reaction with plants, bare land, ice and snow, and man-made structures. Deposition of O_3 is at its greatest over forests and croplands during daylight hours.

The primary sink for O_3 is its photodissociation on absorption of UV light and subsequent formation of OH· (Equation 2.51 and Equation 2.52). In polluted atmospheres, O_3 reacts with NO to produce $NO_2 + O_2$ (Equation 2.46). During nighttime hours, it reacts with NO_2 to produce HNO_3 (Equation 2.24 to Equation 2.26).

2.4 ATMOSPHERIC PARTICLES

Atmospheric particles include both solid- and liquid-phase substances that vary in size and density. As such, larger and denser particles settle out in a matter of minutes, whereas smaller, less dense particles remain suspended in the atmosphere for periods of days to weeks. Individually, particles differ in size, mass, density, morphology, chemical composition, and various chemical and physical properties.

Particles may be emitted from natural and anthropogenic sources. They may also be produced secondarily as a result of chemical reactions involving gas-phase pollutants.

Of special concern are particles of <20 μm in aerodynamic diameter, since they may remain suspended in the atmosphere and, depending on their size, settle out slowly. Such suspended particles characterize atmospheric aerosol.

The direct emission of particles from anthropogenic sources (and, to a lesser degree, natural sources) and formation of secondary particles as a result of atmospheric chemistry pose major environmental concerns. These include (1) toxic effects of particle exposures on humans; (2) visibility reduction due to the light-scattering ability of particles; (3) local, regional, and potential global climatic effects associated with the ability of particles to scatter light back to space and absorb incoming solar radiation and outgoing thermal radiation; and (4) nuisance effects associated with the soiling potential of deposited particles.

2.4.1   Classes and Sources

Atmospheric particles can be described as being primary or secondary based on their origin and processes by which they form. Primary particles are emitted directly into the atmosphere from a variety of natural and anthropogenic sources. In the former case, these include volcanoes, forest fires, ocean spray, biologic sources (mold, pollen, bacteria, plant parts), and meteoric debris; the latter includes transportation, fuel combustion in stationary sources, a variety of industrial processes, solid waste disposal, and miscellaneous sources such as agricultural activities and fugitive emissions from roadways.

Secondary particles are produced from both naturally emitted gas-phase substances and anthropogenic sources as a result of chemical processes involving gases, aerosol particles, and H_2O and H_2O vapor. Secondary aerosol particles include sulfates, nitrates, and oxyhydrocarbons produced by direct, catalytic, and photochemical oxidation of S, N, and volatile NMHCs. Approximately 30% of atmospheric aerosol in the particle size range of <5 μm may be produced as a consequence of chemical reactions involving gases and vapors from natural sources.

2.4.2   Size

Particles that comprise atmospheric aerosol range in size over many orders of magnitude, from <0.005 nm to several hundred micrometers. Size ranges for a variety of particles, as well as their physical properties, are summarized in Figure 2.8.

Particle sizes (Figure 2.8) are based on what are described as aerodynamic equivalent diameters (AEDs). The AED refers to a spherical particle of unit density (1 g/cm^3) that settles at a standard velocity. It can also be expressed as the Stokes equivalent diameter (defined as the diameter of a sphere that has the same density and settling velocity as the particle). Stokes diameters are based on settling velocities, while the AED includes a standardized density of 1.

Size is an important property of particles since it affects their atmospheric history, which includes growth, accumulation of substances on their surface, concentrations,

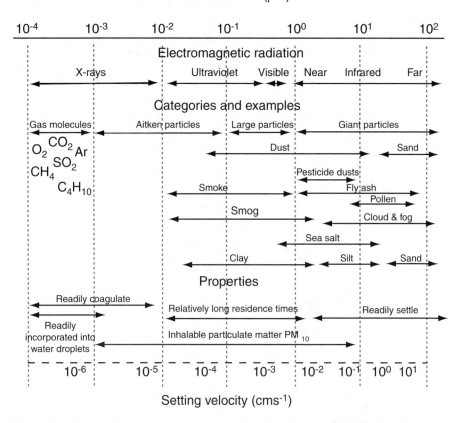

Figure 2.8 Size distribution and physical properties of common aerosol particles. (From Van Loon, G.W. and Duffy, S.J., *Environmental Chemistry: A Global Perspective*, Oxford University Press, Oxford, 2000. With permission.)

atmospheric lifetime, light scattering, and deposition in the human respiratory tract. Most atmospheric particles are very small (<0.1 μm), whereas most aerosol mass is associated with particles >0.1 μm.

The size distribution of aerosol particles can be expressed in several different ways. These include particle number, mass, volume, and surface area in different size ranges (plotted logarithmically). Important characteristics of urban aerosols emerge when urban particle data are plotted as number, volume, surface, or mass distributions. In particle number plots, a large peak is observed at ~0.02 μm and a slight peak around 0.1 μm. Particle concentration distributions based on volume show two peaks, one in the 0.1- to 1.0-μm range and the second in the 1- to 2-μm range. In surface area distribution plots, a major peak is observed near 0.1 μm, with smaller peaks between 0.01 and 0.1 μm, and between 1 and 10 μm. These particle distribution plots indicate that urban aerosol concentrations are multimodal. Until recently, it was widely accepted that atmospheric aerosols consisted of three distinct particle modes. With increasing scientific attention being given to ultrafine particles (UFPs) (diameters of <0.01 μm), atmospheric particles are now being described in terms of the four particle modes illustrated in Figure 2.9. Though no vertical-axis label is shown, it could represent particle distribution based on number, mass, surface area, or volume distribution. Figure 2.9 also shows major sources and particle removal processes for each mode. The UFP mode and the two peaks sometimes observed in the accumulation mode are illustrated by dashed lines.

As can be seen (Figure 2.9), atmospheric aerosols are divided into two major fractions based on their size, origin, and physical properties. Coarse particles have diameters of >2.5 μm, fine particles of <2.5 μm. These particle modes or fractions are distinct relative to (1) their origin, (2) how they are affected by atmospheric

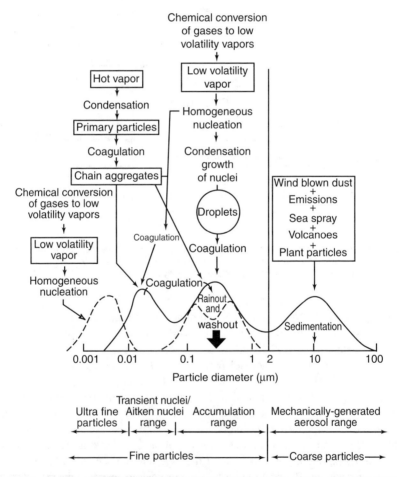

Figure 2.9   Particle size characteristics of atmospheric aerosols. (From Finlayson-Pitts, B.J. and Pitts, J.N., Jr., *Chemistry of the Upper and Lower Atmosphere: Theory, Experiments and Applications*, Academic Press, Orlando, FL, 2000. With permission.)

processes, (3) chemical composition, (4) removal processes, (5) optical properties and light-scattering ability, and (6) deposition in the human respiratory tract.

The fine particle fraction includes three of the four modes illustrated; these are the ultrafine, Aitken nuclei, and accumulation modes. Particle diameter ranges of these modes are <0.01, 0.01 to 0.08, and ~0.10 to 1.2 μm, respectively. Most atmospheric particles are in the fine fraction. It comprises one third of the particulate mass found in nonurban areas and one half in urban areas.

Identification and definition of the UFP mode has been a relatively new development. It is sometimes referred to as the nucleation mode. UFPs are generated by poorly understood gas-to-particle conversion processes. Though they contribute little to particulate mass, they are often present in large numbers. Because of their human respiratory tract depositional and toxicological properties, UFPs are increasingly becoming a major public health concern (see Chapter 5).

Aitken nuclei are produced in the ambient atmosphere by gas-to-particle conversion, as well as by combustion processes in which hot, supersaturated vapors are formed and undergo condensation. They also serve as nuclei for condensation of low-vapor-pressure gas-phase substances. Because of their small size, they account for only a few percent of the mass of airborne particles. Their numbers rapidly diminish as they coagulate with each other to form larger particles.

Accumulation mode particles form from the coagulation of Aitken nuclei particles and condensation of vapors on existing particles. Their rate of growth depends on the number of particles present and their velocity and surface area. Because of growth-limiting physical factors, they do not grow into coarse particles. The term *accumulation mode* is used since removal processes are inefficient in this size range, with particles "accumulating" as a result. Though representing only a small fraction of particle numbers (~5%), they are a significant portion of aerosol mass (~50%). The accumulation mode appears to be bimodal in relatively polluted urban areas, with peaks at ~0.2 and 0.7 μm.

Coarse particles are usually produced by mechanical processes (fragmentation of matter and atomization of liquids). Most coarse particles are in the 2.5- to 10-μm size range, but they may be as large as 30 μm or more. Both fine and coarse particles up to a size range of 10 μm tend to remain suspended in the atmosphere for varying time periods and are therefore described as suspended particles. Particles with diameters of >100 μm tend to settle out rapidly and are described as settleable particles.

In urban areas, particulate aerosols are generally evenly divided (on a mass basis) between fine and coarse particles. Under calm atmospheric conditions, the fine particle mass begins to exceed the coarse mass with increasing downwind distance from urban areas. Because of low settling velocities, fine particles are transported relatively long distances from source regions.

Particle size classes have been identified that reflect particle penetration and deposition in the human respiratory system, and sampling procedures have been designed to reflect the health-affecting potential of atmospheric aerosols. Size ranges of health significance are described and illustrated in Chapter 5.

A variety of sampling devices have been used to measure particle concentrations to determine compliance with air quality standards. These have included the high-volume sampler (Hi-Vol) and PM_{10} and $PM_{2.5}$ samplers. These reflect size-selective sampling considerations. They include the collection of particles below or within a specified aerodynamic size range defined by the upper 50% cut-point size (where 50% of the particles at the specified diameter are collected). The upper cutoff size for the Hi-Vol (used for regulatory purposes until 1987) varied from 25 to 40 μm, depending on the wind speed and direction. Hi-Vol concentrations were reported as total suspended particulates (TSPs). PM_{10} samplers have an upper 50% cutoff diameter of 10 μm, $PM_{2.5}$ samplers of 2.5 μm. PM_{10} includes most particles less than 10 μm and some particles greater than 10 μm. $PM_{2.5}$ includes most particles less than 2.5 μm and some particles greater than 2.5 μm.

PM_{10} samplers were designed to collect particles that can enter the thoracic region of the human respiratory system (see Chapter 5); $PM_{2.5}$ samplers were designed to collect particles that have a high probability of being deposited in lung tissue. However, since $PM_{2.5}$ samples include a fraction of coarse mode particles, $PM_{2.5}$ is equivalent to neither the fine particle mode nor the respirable particle mode described in Chapter 5.

2.4.3 Particle Shape

Particles vary in shape from simple spheres or crystalline cubes to the many that are irregular in their morphology. Some are flat and layered; others are globular. Shape is often determined by a particle's chemical composition and the processes by which it was formed.

Particles occur singly or form aggregates that may appear as single compact particles. Carbon particles produced in combustion processes often form aggregates with irregular shapes. Some are characterized as small clumps, while others are chain-like. Particle surfaces vary as well. Some are smooth, while others are uneven and porous.

2.4.4 Chemical Composition

Particles in atmospheric aerosol vary in their chemical composition. This reflects their origin, how they were initially produced, and their subsequent atmospheric history. Because of the large number of sources of primary particles, formation of secondary particles by atmospheric chemistry, growth of some atmospheric particles, and sorption of gas-phase substances by others, particles suspended in the atmosphere contain hundreds of different chemical species. Notable because of their relatively high concentrations in PM samples are the following chemical groups: sulfates, nitrates, elemental C (EC), condensed organic compounds, organic carbon (OC), and metals.

Chemical composition varies as a function of particle size. Chemical substances are not equally distributed across the particle size spectrum. They tend to be in specific size ranges characteristic of sources and particle development. The smallest UFPs are produced by homogeneous nucleation and, as a result, contain secondary species such as sulfate and organic compounds. Those in the Aitken range are produced by combustion, coagulation of smaller particles, and condensation of products of gas-phase reactions. Such particles, as well as accumulation mode particles, tend to contain EC, metals characteristic of combustion, sulfates, nitrates, and polar organic compounds. Coarse particles tend to contain chemical species characteristic of materials that were mechanically fragmented or atomized. These include dusts from minerals processing and soil erosion. The latter contain common crustal (derived from the Earth's crust) elements such as O, silicon (Si), aluminum (Al), magnesium (Mg), calcium (Ca), etc. Nitrates and sulfates in coarse particles come primarily from the reaction of these gas-phase substances with preexisting coarse particles or deposition on them.

Sulfate compounds comprise a major portion of the fine particle mass collected in ambient air in the U.S.: 44% of the fine particle mass at mixed eastern U.S. rural, suburban, and urban sites, and 56% at rural sites in the eastern U.S. (Table 2.8). At rural western U.S. sites, particulate sulfate levels are considerably lower, representing about one third of the composition of the relatively low-mass $PM_{2.5}$ concentrations.

Sulfates in $PM_{2.5}$ samples include ammonium sulfate ((NH_4)$_2SO_4$), ammonium acid sulfate (NH_4HSO_4), H_2SO_4, a variety of metal salts, and calcium sulfate ($CaSO_4$). Of these, (NH_4)$_2SO_4$ is the most common and abundant sulfate species present. Sulfuric acid may also comprise a significant fraction of atmospheric sulfate. In the global context, sulfate aerosol concentrations are in the range of 1 to 2 $\mu g/m^3$ in remote areas, <10 $\mu g/m^3$ in nonurban continental areas, and >10 $\mu g/m^3$ in areas under urban and anthropogenic influence.

EC, on average, represents a much smaller fraction of U.S. $PM_{2.5}$ mass, with 5% average composition at rural and mixed eastern sites, escalating to as high as 14% in mixed western sites. The major source of EC is fuel combustion. Particles containing EC are relatively small, with average aerodynamic diameters of 0.01 μm. EC can, however, form larger aggregates. Concentrations of EC, as a percentage of collected $PM_{2.5}$, as high as 25% have been reported for Phoenix, AZ.

The OC fraction comprises one fourth to one third of $PM_{2.5}$ mass in the U.S., with the highest percentage levels in western states at all site locations (Table 2.8). In cities such as Los Angeles, the OC fraction may be as high as 45%. $PM_{2.5}$ OC is

Table 2.8 Chemical Composition of $PM_{2.5}$ Particles[a]

Constituent	Rural Measurements (1999) Eastern U.S. % Contribution	Western U.S.	1996 AQCD Eastern U.S. % Contribution	Western U.S.
SO$_4$	56	33	44	11
EC	5	6	5	14
OC	27	36	27	38
NO$_3$	5	8	1	15
Crustal	7	17	6	14
$PM_{2.5}$ concentration ($\mu g/m^3$)	11.0	3.9	31.0	37.3

[a]Obtained in (1)1999 rural studies of the eastern and western U.S. and (2) mixed rural, suburban, and rural areas summarized in the 1996 Air Quality Criteria Document (AQCD) for particulate matter.

Source: From USEPA, Third External Review Draft of Air Quality Criteria for Particulate Matter (April 2002), Vol. 1, http://www.epa.gov/ncea/pdfs/partmatt/Vol_1_AQCD_PM_3rd_Review_pdf.

derived from biogenic as well as anthropogenic sources. The composition of this fraction is complex even in remote areas. Nonurban aerosols contain a wide variety of organic compounds, including paraffins, olefins, aromatic HCs, fatty acids, alcohols, and organic bases. Paraffins in the C_{15} to C_{35} range are common.

Organic compounds found in or on particles associated with human activities and in "aged" air masses are more complex than in remote regions since there is more opportunity for oxidation reactions to take place. As a consequence, such particles contain, in addition to organic compounds from biogenic emissions, complex compounds emitted directly from anthropogenic sources or formed as a result of atmospheric chemical reactions. For example, direct emissions from automobiles and heavy-duty diesel trucks are known to be sources of particulate-phase n-alkanes, n-alkanoic acids, aromatic aldehydes and acids, PAHs and their oxidized derivatives, steranes, pentacyclic triterpones, and azanapthalenes. In secondary aerosols, difunction substitute derivatives of the form $X–(CH_2)_n–X$ (with n = 1 to 5) have been reported in the Los Angeles area. These include dicarboxylic acids, dialdehydes, glycols, esters, and organic nitrite and nitrate compounds. In urban areas of central Japan, dicarboxylic acids are reported to comprise 30 to 50% of total OCs in PM samples.

Inorganic nitrates are a major component of the fine particle aerosol mass, varying from 5 to 8% of $PM_{2.5}$ collected at rural sites to as high as 15% at mixed, western U.S. sites. Under severe smog conditions, inorganic nitrate levels as high as 22.6 $\mu g/m^3$ have been reported in Los Angeles, accounting for 25% of $PM_{2.5}$ mass. Ammonium nitrate is the major nitrate species in urban aerosol.

The crustal fraction represents mineral fragments from the Earth's crust. These become airborne as a result of soil erosion, mineral weathering processes, and dust storms. The crustal component of atmospheric aerosol is considerably higher in both rural and mixed sites in western states in the U.S. than in eastern states.

Metals as a group are also major constituents of atmospheric aerosol. Indeed, atmospheric aerosol is a major transport mechanism for a variety of metals around the planet, in some cases (e.g., mercury, Hg) resulting in ecosystem effects on a global scale.

Metals in crustal dust include iron (Fe), manganese (Mn), zinc (Zn), Pb, vanadium (V), chromium (Cr), nickel (Ni), copper (Cu), cobalt (Co), Hg, and cadmium (Cd). Biogenic particle emissions include low concentrations of Fe, Mn, Zn, Pb, V, Cr, Ni, Cu, Co, Hg, Cd, arsenic (Ar), and antimony (Sb).

Metals are commonly associated with aerosol particles of anthropogenic origin. Major sources include power plants, smelters, incinerators, cement kilns, home heating systems, and motor vehicles. The major source of metals from coal- and oil-fueled power plants is fly ash, the inorganic residue remaining after combustion. Fly ash is characterized by high concentrations of Fe, Al, silicon oxides, and carbonaceous compounds that have undergone various degrees of oxidation. In addition, fly ash from coal-fired power plants contains high concentrations of Mn, Zn, Pb, V, Cr, Ni, Cu, Co, As, and Sb. Oil fly ash can contain not only significant quantities of V and Ni, but also abundant quantities of Mn, Zn, Pb, Cr, Cu, Co, As, and Sb. Metals in oil are often unique to particular oil fields, and as such, they may be used as tracers to determine the origin of atmospheric aerosols of concern (e.g., those in

Arctic haze). Incineration of municipal waste is increasingly becoming an important source of atmospheric metals, including Hg.

Metals in atmospheric aerosol have been assumed to be present as insoluble oxides. However, in the case of oil fly ash, most metals may exist in soluble forms. Significant metal solubilization in atmospheric H_2O occurs by photoreduction processes that involve sulfate. Metals also condense on the surface of fly ash particles, thereby affecting their mobility and bioavailability in the environment.

Metal pollution is increasingly becoming a major environmental and public health concern. Metals released even in what has been believed to be small quantities can accumulate over time and biomagnify in food chains. There is increasing evidence that Hg emissions from coal-fired power plants, and mercury's subsequent accumulation and mobility in the environment, pose a new threat to aquatic and terrestrial ecosystems as well as to humans. Increasing attention is being focused on the metal components of atmospheric aerosol because of their catalytic properties and effects on biological functions. These potential effects of metals on the environment are described in greater detail in Chapter 5 and Chapter 6.

2.4.5 Source Apportionment

Since some sources emit particles with distinct chemical profiles, one should be able to calculate source contributions to PM samples. Such source apportionment can be done using models based on chemical mass balance, factor analysis, principal component analyses, multiple linear regression, and Lagrangian techniques. Source apportionment modeling of PM samples (using a chemical mass balance model) from sampling conducted in downtown Los Angeles (1982) indicated that diesel vehicle emissions, gasoline vehicle emissions, road dust, burning vegetation, secondary sulfate, and secondary nitrate accounted for approximately 18.8, 5.7, 12.4, 9.6, 20.9, and 7.4% of collected PM mass, respectively. In Philadelphia, diesel emissions, road dust, and secondary sulfate accounted for 8.5, 5.1, and 88.5% of collected PM mass, respectively.

2.4.6 Behavior

Aerosol particles have their own unique atmospheric history. It may begin with emission of primary particles from various natural or anthropogenic sources or by nucleation of condensable vapors produced as a consequence of chemical reactions between low-molecular-weight gas-phase chemical species. Depending on various chemical and physical characteristics, as well as the ambient environment, particles may change size, sorb and release vapor-phase molecules, collide, coalesce or adhere to each other, sorb and condense H_2O vapor, undergo deliquescence (rapid uptake of H_2O), undergo changes in electrical condition, and be removed from the atmosphere by depositional processes.

Fine particles produced by gas-to-particle conversion processes commonly undergo changes in size. Homogeneous gas-phase chemical reactions produce condensable species such as H_2SO_4 that can nucleate to form a new particle (nucleation) or condense on the surface of an existing particle (condensation). The relative amount of nucleation compared to condensation depends on the rate of formation of the condensable species and surface area of existing particles. In urban areas, new particle formation appears to occur only near major sources of nuclei such as freeways. In many cases, the available surface area associated with aerosol particles rapidly scavenges newly formed condensable species.

Particle growth may occur as a result of condensation of homogeneous-phase reaction products on particle surfaces, condensation of H_2O vapor, collisions between particles and subsequent coalescence or adherence, surface adsorption, and heterogeneous chemical reactions. In the last case, both gas- and particulate-phase substances participate in chemical reactions (as when SO_2 is oxidized to sulfite in an aqueous droplet). The rate of particle growth by this mechanism is limited by chemical reactions at the surface of or within the particle. Collisions with other particles that result in coalescence or adherence also result in particle growth. Associated with this phenomenon is a decrease in particle number and total surface area, while the average volume per particle increases. Coagulation produces chain agglomerates in soot and some metal-based particles.

Though fine particles are usually thought of as having the potential to increase in size as a result of the processes described above, in theory they also have the potential to decrease in size. Aerosol species such as PAHs are semivolatile and thus can exist in the gas and particulate phases, depending on vapor pressure, particle surface area and composition, and atmospheric temperature. Daily temperature fluctuations can result in their desorption and absorption, with subsequent changes in particle mass as well as chemical composition.

Water is a very important component of many atmospheric particles. The behavior of atmospheric particles in association with changes in ambient humidity is of considerable importance in the global hydrologic cycle and energy budget, and in atmospheric chemistry and optical phenomena. Atmospheric sulfates are both hygroscopic (take up H_2O) and H_2O soluble. Their hygroscopicity results in physical and chemical changes that affect their size, shape, pH, reactivity, and refractive index. Crystalline solid particles such as $(NH_4)_2SO_4$ undergo a phase change with increasing relative humidity (RH) to become aqueous solution particles. This phase transition is abrupt, with a sudden uptake of H_2O taking place at an RH above the deliquescence point (usually around 80% RH).

Particles may undergo a change in their electrical state as they change size. A particle's charge depends on the state of its surface, the value of its dielectric constant, and its size. In general, particles greater than 3 μm carry a negative charge, and particles less than 0.01 μm carry a positive charge. Electrical charges can affect coagulation and the rate of dry deposition.

Particle behavior also includes their ability to scatter visible light and absorb and reradiate thermal energy. This behavior is affected by particle size, geometry, and chemical composition (see Chapter 4).

Like other components of the atmosphere, particles move in both horizontal and vertical dimensions. They are carried downwind by advective airflows and transported both upward and downward by convective air motion. Depending on their size, particles in the atmosphere are affected to different degrees by gravity. In the absence of significant convection, particles (particularly large particles) are drawn toward the surface of the Earth by gravity. Gravitational forces are opposed by a frictional force that is a function of air viscosity, particle velocity, and particle diameter. As a result, larger particles "settle" more rapidly than smaller ones (see Chapter 3). As particle size decreases, the effect of gravity is reduced and particles settle out slowly. Small particles (<1 μm) are subject to random movement-induced collisions with air molecules. This random movement of both air molecules and small particles is called Brownian motion or diffusion. Such movement results in migration of particles from areas of high to low concentration. In a still atmosphere, particles of <0.1 μm are primarily transported by Brownian motion. It is largely responsible for collisions between Aitken nuclei to form larger accumulation mode particles. It is also responsible for collisions that result in adsorption of gas-phase substances onto particle surfaces.

2.4.7 Concentrations

Particulate matter measurements have been routinely made in the U.S. at 1000+ sites since the early 1970s. Initially, these were based on use of a Hi-Vol and reported as TSP. After the NAAQS was changed to a PM_{10} standard (50 μg/m^3 annual average, 150 μg/m^3 maximum 24-h average) in 1987, measurements were made with a sampler with a 50% cutoff diameter of 10 μm. In addition, measurements of fine particles ($PM_{2.5}$) have been made in anticipation of implementation of a $PM_{2.5}$ NAAQS (promulgated in 1997 with an anticipated compliance date of 2004). As such, considerable air quality data have been collected on TSP, PM_{10}, and $PM_{2.5}$ concentrations at various locations in the U.S.

Average annual background concentrations of PM_{10} are estimated to range from 4 to 8 μg/m^3 in the western U.S. to 5 to 11 μg/m^3 in the East. PM_{10} levels in selected urban areas in the U.S. for 1993 are summarized in Table 2.9. These include both annual average and 24-h maximum concentrations. As can be seen, some cities exceeded either the annual average or 24-h maximum standards.

Recent analyses (1999) of PM_{10} data indicate that with few exceptions, sampling locations in the U.S. report PM_{10} concentrations that are below the NAAQS. Indeed, since promulgation and implementation of the PM_{10} NAAQS, annual average concentrations have decreased by about 20% in the East and about 38% in the Northwest. Decreases in PM_{10} levels of 10% or more have been observed in most locations in the U.S.

Though a $PM_{2.5}$ standard (15 μg/m^3 annual average, 65 μg/m^3 maximum 24-h concentration) was promulgated in 1997, its implementation has been postponed until 2004 to allow for completion of baseline monitoring of approximately 1000 locations in the U.S. As this baseline monitoring is being conducted at sampling stations co-located with PM_{10} samplers, considerable data are now available for comparison purposes.

Based on 1999 data, many areas of the country were above the 15 μg/m^3 annual mean $PM_{2.5}$ NAAQS concentration, with some urban areas exceeding 20 μg/m^3. Ninety-eight percent of sample sites had 24-h concentrations of <65 μg/m^3. Most locations exceeding the NAAQS value were in California. Decreases in PM_{10} concentrations observed throughout the U.S. have resulted in decreases in $PM_{2.5}$,

Table 2.9 PM_{10} Levels ($\mu g/m^3$) in Selected U.S. Urban Areas in 1993

Urban Area	Annual Average	Second-Highest 24-h Concentration
Ashville, NC	22	58
Baltimore, MD	35	70
Birmingham, AL	36	85
Casper, WY	18	41
Chicago, IL	47	147
Dallas, TX	30	74
Danbury, CT	19	46
Denver, CO	41	142
Gary, IN	34	122
Honolulu, HI	24	58
Los Angeles, CA	47	101
Louisville, KY	37	43
New Haven, CT	52	178
New York, NY	47	86
Philadelphia, PA	34	531
Phoenix, AZ	44	92
Provo, UT	40	209
Santa Fe, NM	15	35
San Francisco, CA	29	72
St. Louis, MO	44	101
Steubenville, OH	40	177

Source: From USEPA, EPA/600/AP-95/001c, EPA, Washington, D.C., 1995.

particularly where measurements of $PM_{2.5}$ have had a longer history. Annual average $PM_{2.5}$ concentrations decreased by 50% on the southern coast of California and 30% in the San Francisco Bay area during the period 1990–1995.

The availability of PM_{10} and $PM_{2.5}$ concentration data has made other assessments possible. By subtracting concentrations of $PM_{2.5}$ from PM_{10}, an estimate of coarse particle concentrations can be made. The nationwide median annual concentration for coarse particles ($PM_{10-2.5}$) is ~10 $\mu g/m^3$, with values as high as 54 $\mu g/m^3$. Highest coarse fraction concentrations are reported for western states, especially California. $PM_{10}:PM_{2.5}$ mean ratios of 0.70 have been reported for the Northeast, Southeast, and the industrial Midwest; lower ratios have been reported for the Northwest (0.50), Southwest (0.38), and southern California (0.47).

Ambient PM_{10} and $PM_{2.5}$ vary seasonally, with highest concentrations typically reported during summer months during high-pressure, hazy days. Diurnal variations in $PM_{2.5}$ concentrations have also been reported, with maximum concentrations occurring during the morning and evening.

2.4.8   Long-Range Transport from Sources Outside the U.S.

It has been known for decades that both precursor gas-phase molecules and, to a lesser extent, particulate substances (e.g., acid sulfates) can be transported hundreds of kilometers or more. Because of the availability of data from intensive monitoring studies of particulate aerosols and analysis of their chemical constituents, as well as satellite imaging, there is substantial evidence that particles are transported into the U.S. from very distant sources (thousands of kilometers). This includes transport from biomass burning in southern Mexico and Central America, which occasionally causes anomalously high PM levels in southern Texas and generally elevated levels in central and southeastern states; dust from Africa crossing the Atlantic and affecting air quality in Florida (African dust is the dominant constituent of ambient aerosol during the summer in Florida); and dust from the Gobi desert being transported to Washington, Oregon, California, Idaho, and British Columbia, resulting in elevated ambient concentrations. Dust from Asian deserts can be seen in the satellite image over China shown in Figure 2.10.

Figure 2.10 Asian dust clouds formed over Chinese deserts. (From National Aeronautics and Space Administration satellite photo.)

READINGS

Atkinson, R., Gas-phase tropospheric chemistry of organic compounds: a review, *Atmos. Environ.*, 24A, 1, 1990.

Brasseur, G.P., Orlando, J.J., and Tyndall, G.S., *Atmospheric Chemistry and Global Change,* Oxford University Press, Oxford, 1999.

Bridgman, H.A., *Global Air Pollution*, John Wiley & Sons, New York, 1994.

Brimblecombe, P., *Air Composition and Chemistry,* 2nd ed., Press Syndicate, Cambridge University Press, Cambridge, U.K., 1996.

Elsom, D.M., *Atmospheric Pollution: A Global Problem*, 2nd ed., Blackwell Publishers, Oxford, 1992.

Finlayson-Pitts, B.J. and Pitts, J.N., Jr., *Chemistry of the Upper and Lower Atmosphere: Theory, Experiments and Applications*, Academic Press, Orlando, FL, 2000.

Ghio, A.J. and Samet, J.M., Metals and air pollution particles, in *Air Pollution and Health,* Holgate, S.T., Samet, J.M., Koren, H.S., and Maynard, R.L., Eds., Academic Press, San Diego, 1999, p. 635.

Giant, L.D., Shoaf, C.R., and Davis, J.M., United States and international approaches to establishing air standards and guidelines, in *Air Pollution and Health,* Holgate, S.T., Samet, J.M., Koren, H.S., and Maynard, R.L., Eds., Academic Press, San Diego, 1999, p. 947.

Harrison, R.M., Ed., *Pollution: Causes, Effects and Control,* Royal Society of Chemistry, Thomas Graham House, Cambridge, U.K., 1990.

Hidy, G.M., *Aerosols: An Industrial and Environmental Science,* Academic Press, Orlando, FL, 1984.

Houghton, J.T., Meira-Filho, L.G., Callender, B.A., Harris, N., Kallenberg, A., and Maskell, K., *The Science of Climate Change,* Cambridge University Press, Cambridge, U.K., 1995.

Isidorov, V.A., *Organic Chemistry of the Earth's Atmosphere,* Springer-Verlag, Berlin, 1990.

Jacob, D.J., *Introduction to Atmospheric Chemistry,* Princeton University Press, Princeton, NJ, 1999.

Lippmann, M. and Maynard, R.L., Air quality guidelines and standards, in *Air Pollution and Health,* Holgate, S.T., Samet, J.M., Koren, H.S., and Maynard, R.L., Eds., Academic Press, San Diego, 1999, p. 983.

National Research Council, *Global Tropospheric Chemistry: A Plan for Action,* National Academy Press, Washington, D.C., 1984.

National Research Council, *Ozone Depletion, Greenhouse Gases and Climate Change,* National Academy Press, Washington, D.C., 1989.

National Research Council, *Rethinking the Ozone Problem in Urban and Regional Air Pollution,* National Academy Press, Washington, D.C., 1991.

Pooley, F.D. and Milne, M., Composition of air pollution particles, in *Air Pollution and Health,* Holgate, S.T., Samet, J.M., Koren, H.S., and Maynard, R.L., Eds., Academic Press, San Diego, 1999, p. 619.

Rowland, F.S., Chlorofluorocarbons and the depletion of stratospheric ozone, *Am. Sci.,* 77, 36, 1989.

Rowland, F.S. and Isaken, I.S.A., *The Changing Atmosphere,* John Wiley & Sons, Chichester, U.K., 1987.

Seinfeld, J.H. and Pandis, S.N., *Atmospheric Chemistry and Physics: From Air Pollution to Climatic Change,* John Wiley & Sons, New York, 1998.

Singh, H.B., Ed., *Composition, Chemistry and Climate of the Atmosphere,* Van Nostrand Reinhold, New York, 1995.

USEPA, *Air Quality Criteria for Particulate Matter,* Vol. 1, EPA 1600/AP-95/001, EPA, Washington, D.C., 1996.

Van Loon, G.W. and Duffy, S.J., *Environmental Chemistry: A Global Perspective,* Oxford University Press, Oxford, 2000.

Warneck, P., *Chemistry of the Natural Atmosphere,* Academic Press, Orlando, FL, 1988.

Watson, A.Y., Bates, R.R., and Kennedy, D., Eds., *Air Pollution: The Automobile and Public Health,* Health Effects Institute, National Academy Press, Washington, D.C., 1988.

QUESTIONS

1. Describe sources of natural air pollution and their air quality significance.
2. How does ambient air pollution differ from other forms of air pollution?
3. What are the unique characteristics of Los Angeles-type smog?
4. How are smog and haze related?
5. What are the sources of primary and secondary pollutants?

6. Describe the significance of primary and secondary pollutants in the context of air pollution control.
7. Transportation is currently the major source of which of the primary pollutants?
8. Describe the relative success of regulatory programs in reducing emissions of CO, NO_x, sulfur oxides, and lead over the past three decades in the U.S.
9. Why are CO and CO_2 air quality concerns?
10. How are CO and CO_2 removed from the atmosphere?
11. Describe the relative significance of emissions of SO_x and reduced sulfur compounds to the atmosphere.
12. Three major nitrogen oxides are naturally present in the atmosphere and are released to it by natural and anthropogenic sources. These are N_2O, NO, and NO_2. Describe the environmental significance of each.
13. What roles do paraffinic, olefinic, and aromatic hydrocarbons play in atmospheric chemistry?
14. How do peroxy radicals affect tropospheric O_3 levels?
15. What role does ammonia play in atmospheric chemistry?
16. How do aldehydes and organic acids affect air quality?
17. Why are biogenically produced hydrocarbons important in the atmospheric chemistry of tropospheric O_3?
18. What are photochemical oxidants?
19. Describe the role that OH· plays in the atmospheric sink processes involving CO, CH_4, and SO_2.
20. Describe types, uses, emissions, and environmental concerns associated with halogenated hydrocarbons.
21. What factor or factors determine the lifetime of a substance in the atmosphere?
22. Sink processes involve the conversion of substances to more chemically stable substances. What is the final chemical form to which the following pollutants are converted: SO_2, H_2S, O_3, and NO_2?
23. Particles differ in size and shape. Despite these differences, they are often described in the context of their aerodynamic diameter. Explain.
24. How do coarse and fine particles differ relative to their size, mechanisms of formation, and chemical composition?
25. Distinguish between ultrafine, Aitken nuclei, and accumulation mode particles.
26. Atmospheric particles vary in their chemical composition. Why is this the case?
27. Describe how physical and chemical properties of particles affect their atmospheric behavior.
28. Describe the relative significance of metals found in particulate matter.

CHAPTER 3

Atmospheric Dispersion, Transport, and Deposition

The atmosphere has served as a sink for emissions of volcanoes and a variety of geological processes, forest and grassland fires, and decomposition and other biological processes for hundreds of millions (if not billions) of years. It has also served as a sink for pollutants generated by human activities, proceeding from man's first use of fire to the smelting of metal ores and use of fossil fuels such as coal, oil, and natural gas to motor vehicle and other emissions from our very industrialized and technologically advanced modern times.

Despite its vastness, the atmosphere (at least in the short term) is not a perfect sink. Its ability to carry away (transport), dilute (disperse), and ultimately remove (deposition) waste products released to it is limited by various atmospheric motion phenomena. Pollutant concentrations may reach unacceptable levels as a result of local or regional overloading of the near-surface atmosphere, topographical barriers, and micro-, meso-, and macroscale air motion phenomena.

The atmosphere serves as a medium for atmospheric chemical reactions that ultimately serve to remove contaminants. These reactions may produce pollutants that may themselves pose significant environmental concerns. Levels of long-lived pollutants such as methane (CH_4), nitrous oxide (N_2O), and carbon dioxide (CO_2) may increase, causing global warming and, in the case of halogenated hydrocarbons, stratospheric ozone (O_3) depletion.

Because of technological and economic limitations, we have little choice but to use the atmosphere for the disposal of airborne wastes. Like other natural resources, our use of the atmosphere has to be a wise one, recognizing its limitations and using it in a sustainable way.

3.1 DISPERSION AND TRANSPORT

Pollutants released from ground-level and elevated sources (smokestacks) are immediately subject to atmospheric processes, with dispersion in ever-increasing volumes of air by both vertical and horizontal transport. Transport is the process by which air motions carry gas- and particulate-phase species from one region of the atmosphere to another. Transport enhances dispersion and provides an opportunity for pollutants from different sources to interact.

Pollutant transport and dispersion are affected by atmospheric dynamics, fluid physical phenomena that occur in the atmosphere, and physical laws that govern them. These may facilitate or constrain transport and dispersal.

Pollutants are initially released into the planetary boundary layer (PBL), that portion of the atmosphere most directly affected by the Earth's surface. The PBL is subject to fluxes of heat and water (H_2O) vapor from the surface, and other physical forces. Its depth, on average, ranges from a few hundred meters to 1 to 2 km. Above the PBL is a relatively stable layer of air that separates it from the free troposphere above.

Pollutant transport and dispersion are affected by different scales of motion. These are the microscale, mesoscale, synoptic scale, and macro-, or planetary, scale (Table 3.1). Microscale refers to air motion in the near vicinity of a source and includes phenomena that affect plume behavior; mesoscale, atmospheric motions on the order of tens to hundreds of kilometers and such phenomena as fronts, airflows in river valleys, and coastal airflows; synoptic scale, systems on the order of 10^6 km^2 or more such as high- and low-pressure systems responsible for day-to-day weather variations; and planetary scale, atmospheric motions on the order of continents or larger. These scales of air motion are useful in describing atmospheric phenomena. It is important to note that the atmosphere is one continuous flowing fluid and all motions are a part of this larger flow.

In the context of a few months, the PBL is relatively well mixed. However, on shorter timescales and near the Earth's surface (where pollutants are emitted), transport and dispersion are often limited by atmospheric conditions. Some atmospheric conditions result in elevated ground-level pollutant concentrations that may potentially harm humans and our environment. Consequently, they are discussed in detail below. Particular attention is given to horizontal wind (speed and direction), turbulence, topography, atmospheric stability, and inversions.

Table 3.1 Meteorological Scales of Air Motion

Scale	Geographical Area (km^2)	Period	Phenomena
Microscale	2–15	Minutes	Plume behavior Downwash
Mesoscale	15–160+	Hour to days	Sea, lake, and land breezes Mountain valley winds
Synoptic scale	>10^6	Days	Migratory high- and low-pressure systems Cold and warm fronts Semipermanent high-pressure systems
Planetary scale		Weeks to months	Hadley cell flows Tropical storms Jet stream meanders Cold and warm fronts

3.1.1 Wind

Horizontal winds are characterized by both velocity (wind speed) and direction. As seen in Chapter 1, wind speed is affected by horizontal pressure and temperature gradients (the higher the pressure gradient, the higher the wind speed) and friction, which is proportional to the roughness of the Earth's surfaces (surface roughness). Relationships between surface roughness and wind speed for urban, suburban, and rural areas can be seen in Figure 3.1. The maximum height of each wind profile indicates where surface effects end and the gradient wind (wind affected by pressure differentials and the Coriolis effect) begins. For the urban area depicted, this occurs at ~500 m (1650 ft); for the suburban area, ~300 m (990 ft); and for the rural area, ~250 m (825 ft).

For continuously emitting stack sources, dilution begins at the point of release. This plume dilution is inversely proportional to wind speed; i.e., by doubling wind speed, pollutant concentration is decreased by 50% of its initial value. The effect of wind speed is to increase the volume of air available for pollutant dispersal. As seen in Figure 3.1, urban areas are characterized by relatively high surface roughness and, as a consequence, diminished wind speeds. This is ironic in the sense that urban areas, because of their relatively high pollutant emissions, are in greater, not lesser, need of being ventilated by the wind.

Winds have directional aspects. These include the prevailing northeasterly flows in the subtropics, southwesterly flows in the middle latitudes, and easterly flows at high latitudes in the northern hemisphere. They also include the cyclonic (clockwise) and anticyclonic flows associated with migrating low- and high-pressure systems. Because flows are somewhat circular, wind direction will depend on one's position

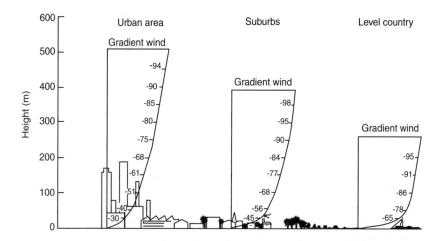

Figure 3.1 Effect of surface roughness on wind speed as a function of height over urban, suburban, and rural areas. (Adapted from Turner, D.B., *Workbook for Atmospheric Dispersion Emissions*, EPA Publication AP–26, EPA, Washington, D.C., 1969.)

in the circulating pressure cell. It also depends on local topography. At night in river valleys, airflows are downslope and downriver; they are upslope during daylight hours. Along sea and lake coasts, winds during clear weather flow inland during the day and waterward at night.

Wind direction is quite variable, with large changes often occurring over relatively short periods of time. A change in wind direction of 30° or more in 1 h is not uncommon. Over a period of 24 h it may shift by 180°. Seasonal factors may cause wind direction variations of 360°.

Wind direction and variability can have significant effects on air quality. Areas downwind of point sources where winds are relatively persistent may experience relatively high ground-level concentrations compared to other areas at similar distances. If the wind is more variable, pollutants will be dispersed in a larger volume of air and be more equally distributed around the source; ground-level concentrations are therefore likely to be lower.

Wind direction is particularly important in the transport and dispersion of pollutants over large geographical areas. It is southwesterly airflows that carry acid precursors from the U.S. Midwest to the northeastern states and southeastern Canada. Similar flows have transported pollutants from countries of southeastern Asia to the West Coast of the U.S.

3.1.2 Turbulence

Airflows within the PBL are influenced by prevailing high-altitude air motion, frictional drag of the Earth's surface, and vertical airflows. Turbulence is characterized by circular eddies that may be vertical, horizontal, and various other orientations. These eddies represent air movements over shorter timescales than those that determine mean wind speeds. Turbulent eddies are produced by both mechanical and thermal forces.

Mechanical turbulence is induced by wind moving over and around structures and vegetation. It increases with wind speed and surface roughness. It is also produced by the shearing effect of fast-moving air aloft as it flows over air slowed by friction.

Thermal turbulence results from the heating or cooling of air near the Earth's surface. On clear days, solar heating of ground surfaces transfers heat to air above it. Convection cells of rising warm air and descending cooler air develop. Under intensive surface heating, convective eddies are generated that extend vertically on the order of 1000 to 1500 m (~3600 to 5000 ft).

For the most part, mechanical and thermal turbulence are daytime phenomena. Both are dampened by nighttime radiative cooling of the ground and air adjacent to it.

The effect of both mechanical and thermal turbulence is to enhance atmospheric mixing and pollutant dispersion. As a consequence, pollutant concentrations are significantly decreased. An exception is downwash phenomena that cause plumes to be brought to the ground near smokestacks (by mechanical turbulence). In most cases, turbulence has a positive effect on air quality.

Downwash results in high pollutant concentrations in the turbulent wake downwind of a source. Downwash can also occur as a consequence of the shearing effect of high-velocity winds (>70 km/h, ~ 40 mi/h).

3.1.3 Atmospheric Stability

The atmosphere, particularly in the PBL, is characterized by highly variable horizontal and vertical air movements. In turbulent flows (described previously), the atmosphere is unstable and pollutants are rapidly dispersed. Turbulent flows associated with heating of the Earth's surface are dampened on cloudy days. Under such conditions, the atmosphere is more stable and pollutants are less rapidly dispersed.

When an entity is undergoing rapid change or has the immediate potential to do so, it is said to be unstable. When it is undergoing little or no change, and is even resistant to change, it is said to be stable. Stability and instability represent opposite ends in a continuum of possibilities. This continuum is implicit in the phrase *atmospheric stability.*

Vertical air motion is significantly affected by temperature gradients. The rate of temperature change with height is described as the lapse rate. Tropospheric temperatures, on average, decrease with height (Figure 1.7). This decrease, or normal lapse rate, is $-6.5°C/km$ ($18.9°F$) or $-0.65°C/100$ m.

The normal lapse rate differs from what would be expected if a parcel of warm, dry air were released into a dry atmosphere. In this theoretical case, the buoyant parcel would rise and expand adiabatically (i.e., no energy is transferred to surrounding air). As it rises, its temperature decreases at a constant rate. This theoretical change of temperature with height is called the adiabatic lapse rate. It has values of $-10°C/km$ ($-29°F$) or $-1°C/100$ m.

Because parcels of air released into the lower atmosphere contain H_2O vapor, the adiabatic lapse rate is used to describe how air cools when it rises in a dry atmosphere. Under real-world conditions, air contains a significant amount of H_2O vapor that cools as the air parcel rises. When air reaches its saturation vapor pressure, heat is released as H_2O vapor condenses (heat of vaporization is released). Air is warmed, resulting in a somewhat smaller decrease of temperature with height than that predicted for adiabatic conditions.

The normal lapse rate represents a summation and averaging of many different lapse rate conditions that vary from more negative (than the adiabatic lapse rate) to positive values. Individual lapse rates (environmental lapse rates) are determined from vertical temperature profile measurements. Though environmental lapse rates are reported as a single value, they represent a summation and averaging of temperature variations with height. Because they represent temperature changes with height, environmental lapse rates are used as indicators of atmospheric stability and the dispersion potential of pollutants.

The relationship between environmental lapse rates and stability can be seen in Figure 3.2. Line A indicates a slight decrease of temperature with height. Since in this case sources or sinks of thermal energy are present (there is little or no heating or cooling of the ground and adjacent air), air cools as it expands and its pressure decreases. This temperature change is close to the adiabatic rate. As a consequence, atmospheric conditions are described as neutral. A neutral lapse rate occurs in response to (1) cloudy conditions that inhibit incoming solar radiation and outgoing thermal radiation, (2) windy conditions that rapidly mix heated or cooled air near the Earth's surface, and (3) transitional circumstances near sunrise and sunset when changes in stability occur. Under such neutral conditions, dispersion is relatively good.

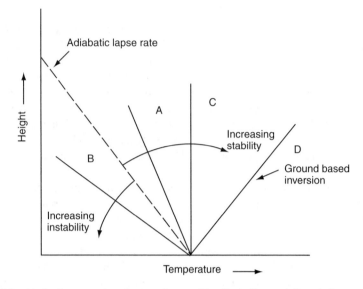

Figure 3.2   Vertical near-surface temperature profiles illustrating variations in lapse rate conditions and increasing–decreasing atmospheric stability.

The lapse rate characterized by line B shows a temperature decrease that is greater (–2°C/10 m) than those under neutral conditions (near the adiabatic lapse rate). A parcel of polluted air in such an environment will rise rapidly. The lapse rate is described as superadiabatic. Atmospheric conditions are very unstable, with strong vertical air motion. Such instability occurs on clear days with light winds at midday. Pollutant dispersion, as expected, is excellent.

Line C represents an isothermal lapse rate (i.e., temperature does not change with height). If a parcel of warm polluted air were released into this somewhat stable environment, it would rise slowly and soon cool to the temperature of its surroundings. Because the atmosphere is more stable, dispersion is more limited. Dispersion potential under these lapse rate conditions can be characterized as moderate.

Line D indicates lapse rate conditions in which temperature increases with height. As such, the temperature change is inverted from "normal." Under such conditions the atmosphere is very stable. Because of warmer temperatures above the ground, a warm parcel of polluted air will quickly come into equilibrium with the temperature of its surroundings. Vertical air motion is suppressed and the dispersion potential of emitted pollutants is poor. Such stable ground-level inverted lapse rate conditions occur at night under clear skies with calm to light winds.

Atmospheric stability, described above, represents changes in near-surface air temperatures that occur over the course of a single day. They do not include larger-scale meteorological conditions associated with high-pressure systems.

3.1.4   Inversions

As indicated, when lapse rate temperatures increase with height, they are inverted from the normal. Such atmospheric conditions are described as inversions. They can be surface based (occur near the ground) or elevated (occur aloft). Atmospheric processes may produce frontal, advective, radiational, and subsidence inversions.

In a frontal inversion, air from a warm front flows over cold air in an adjoining air mass. Inverted temperatures occur at the interface of the two fronts. Because of the movement of these air masses and the interaction between them, frontal inversions have only limited effects on air quality.

An advective inversion forms when warm air flows over a cold surface or cold air. They are commonly associated with land and sea breezes and may be surface-based or elevated.

Radiational and subsidence inversions pose significant air quality concerns since they suppress vertical mixing over industrialized river valleys and urban areas, resulting in elevated pollutant levels.

3.1.4.1 Radiational Inversions

Radiational inversions are produced as a result of the radiational cooling of the ground. Since they form at night, they are also called nocturnal inversions. Radiational inversions are, in most cases, ground-based.

Radiational inversions only occur on clear nights. Surface-based inversions begin to form as the sun sets and intensify throughout the night until sunrise. As the Earth is a net radiator of heat at night, it begins a cooling process that subsequently cools air immediately above it. Relatively warm air overlays an increasingly deepening layer of cool air beneath. This inversion layer may be only 10 to 20 m (33 to 66 ft) deep over flat terrain. With the exception of ground-based sources, such inversions have only a limited impact on air quality.

Radiational inversions in river valleys are of major environmental importance because of their historically heavy industrialization and pollutant emissions and the fact that such inversions are intensified as a result of the effects of topography.

River valleys were formed by the erosive force of water as it moved from higher elevations to the sea. Valleys serve as conduits for water as well as cool, dense air flowing downslope from the radiative cooling of ridges bordering the valley. On reaching the valley floor, this cool, dense air runs under warmer air and forces it aloft. The flow is downslope and downriver. Because of its density and volume, cool air floods the valley floor. This river of cool air deepens over the nighttime hours and reaches its maximum depth just before sunrise. In some mountain valleys the top of the inversion layer may be 100 to 200 m (330 to 660 ft) above the valley floor. A radiational inversion in a mountain valley and its vertical temperature profile are illustrated in Figure 3.3.

The height of the inversion layer has a significant effect on how well or how poorly pollutants are dispersed. In river valleys, emissions from sources having stack heights up to 100 m may be trapped in the inversion layer. As such, dispersion is very poor and pollutant concentrations near the top of the inversion layer are high.

Polluted air emitted from a source will rise to an altitude where its temperature is the same as its surroundings. This occurs near the top of the inversion layer. As a consequence, a layer of intensely smoky or hazy air forms at this height. This is most evident in the early morning hours in a river valley before inversion breakup.

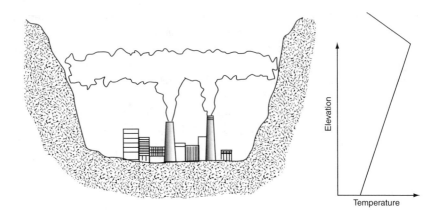

Figure 3.3   Nocturnal, radiational inversion in an industrialized river valley and associated
temperature profile.

During the winter when days are short, radiational inversions may persist up to
16 to 18 h in northern latitudes; they are typically considerably shorter during the
summer. Pollutant levels, as can be expected, are higher under more persistent
inversion conditions.

Radiational inversions begin to break up as the sun starts to warm the ground
and the air above it. Increasingly large convection cells of turbulent air are formed,
which cause complete inversion breakup several hours after sunrise. At that time the
heavily polluted air mass near the top of the inversion is brought to the ground. This
phenomenon is described as fumigation because of the high ground-level concentrations produced.

Despite radiative cooling of the ground, surface-based inversions generally do
not form in urban areas located on flat terrain. This is due to the fact that urban
surfaces emit considerable quantities of heat that produce a well-mixed layer of air
above them. Emitted heat, however, can be absorbed by the polluted air mass (often
described as a dust dome) that forms above many cities. As heat is absorbed by
pollutants, a layer or two of warm air forms aloft (Figure 3.4). Though these are
elevated inversions, they are, nevertheless, produced by radiational cooling.

3.1.4.2 Subsidence Inversions

Subsidence inversions are formed over large geographical areas as the result of
the subsidence of air in high-pressure systems. As air subsides (sinks) to lower
altitudes, it compresses air beneath it, causing temperatures to rise. Since turbulence
almost always occurs near the ground, air in this part of the atmosphere is relatively
unaffected by subsidence occurring above. As a consequence, an inversion layer
(that may be 50 m (165 ft) thick) forms between the subsiding air and the relatively
turbulent air below it (Figure 3.5).

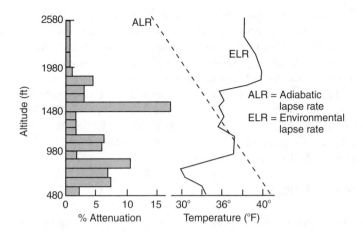

Figure 3.4 Effects of dust levels on environmental lapse rate over the Cincinnati, OH, metropolitan area. (From Bach, W., *Geogr. Rev.*, 61, 573, 1971. With permission.)

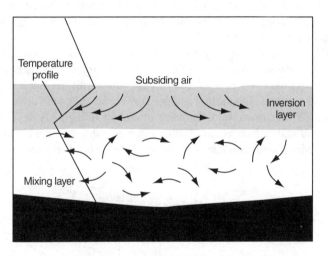

Figure 3.5 Formation of a subsidence inversion and its associated temperature profile.

The height of the inversion layer varies. It is highest near the center of the cell and lowest near the cell's periphery. As a result of turbulence, elevated inversion layers do not reach the ground. Though commonly associated with high-pressure systems, subsidence inversions only have significant effects on air quality when the inversion layer is relatively close to the ground (e.g., 300 to 400 m, ~990 to 1300 ft) and persistent (3 to 5 days).

Most high-pressure systems are migratory; i.e., they move over large expanses of the Earth's surface after they form. These migrating systems contribute to the hazy summer conditions over the American Midwest, Southeast, and Northeast. Occasionally, they have inversion layers that are relatively low and more persistent than normal. Such stagnating systems may result in production of elevated pollutant levels near the ground. These occurrences are called episodes. Fortunately, severe pollution episodes associated with migrating high-pressure systems are relatively rare in North America and northern Europe. The most severe and persistent inversions occur in middle latitudes in autumn. Subsidence inversions of particular note are those associated with semipermanent marine high-pressure systems (see Chapter 1). The inversion layer comes closest to the ground on the easterly or continental side of such systems. As a consequence, west coasts of continents have relatively low and persistent inversion conditions. Inversions below 800 m (2600 ft) occur over the southern coast of California approximately 90% of the time during the summer. Such inversions are primarily responsible for the smoggy and hazy atmospheric conditions over the Los Angeles Basin.

3.1.5 Mixing Height

The mixing height (MH) is the height of the vertical volume of air above the Earth's surface where relatively vigorous mixing and pollutant dispersion occurs. A definable MH is assumed to occur under unstable and neutral conditions. It cannot be defined when the air mass above the surface is stable. Elevated inversions frequently place a cap on the MH. Average MHs for selected U.S. cities are indicated on the map of the U.S. in Figure 3.6.

The MH varies both diurnally and seasonally. It is markedly affected by topography and high-pressure systems. During the day, minimum MHs occur just before sunrise. The MH increases progressively as the sun warms the Earth and the Earth warms the air above it. Increasingly larger convective cells are formed so that the MH reaches its maximum value in early afternoon (commonly several thousand meters). Maximum values occur during summer, with minimum values in late autumn and winter in the middle latitudes of the northern hemisphere.

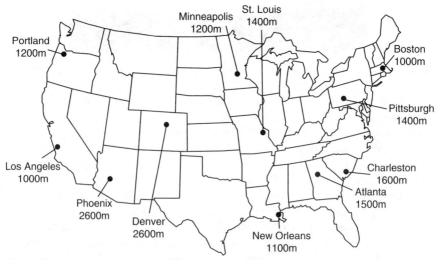

Figure 3.6   Average summertime MHs for selected U.S. cities.

Topographical features such as water surfaces, with their large heat-absorbing capacities, have lower maximum and average MHs than land surfaces with little vegetative cover (e.g., deserts). Not surprisingly, Phoenix, AZ, has a relatively high summertime MH, with coastal cities among the lowest (Figure 3.6).

MHs are affected by semipermanent marine high-pressure systems (note Los Angeles in Figure 3.6) and migrating high-pressure systems. Subsidence of air and formation of inversions cause MHs to decrease. MHs are an important variable used in air quality modeling (see Chapter 7).

3.1.6 Topography

Both micro- and mesoscale air motions are affected by nearby topographical features. Topography can have significant effects on both air movement and pollutant levels. These include differential vertical airflows associated with forests, plowed agricultural fields, parking lots, etc. Such flows affect the behavior of smokestack plumes.

In river valleys, downslope airflows at night intensify (deepen) surface-based inversions, and valley winds during the day help move pollutants upslope and out of the valley. Mountains also serve as barriers to air movement. In the Los Angeles Basin, the San Bernardino Mountains retard airflow in northerly and easterly directions, further intensifying smog and haze conditions. Mountains also increase surface roughness, thereby decreasing wind speeds.

The smog problem over the Los Angeles Basin is also affected by the adjacent cool waters of the Pacific. When sea breezes bring cool air in from the ocean, warmer air is pushed aloft, further intensifying the elevated inversion.

Mesoscale airflow patterns occur on relatively calm days in coastal areas as a result of differential heating and cooling of land and water surfaces. During summer, when skies are clear and prevailing winds light, land surfaces warm more rapidly than sea and lake water. The subsequently warmed air flows up and waterward. As a consequence of temperature and pressure differences, air flows landward at the surface from the water, forming a sea or lake breeze. Air moving from the land cools and descends to form a weak circulation cell. At night, rapid radiational cooling of the land results in surface airflows toward water, forming a land breeze. These land breezes are generally lighter than lake and sea breezes.

Land, sea, and lake breezes, and the circulation patterns that form with them, occur only when prevailing winds are light. They are overridden when winds are strong. In the case of the south coast of California, sea breezes intensify subsidence inversions. They may also cause advective inversions, which commonly occur in late spring when large bodies of water are still cold relative to adjacent land areas. As water-cooled air moves inland, it warms; the inversion is broken up and replaced by superadiabatic lapse rate conditions. The weak circulation cells associated with land, lake, and sea breezes may allow pollutants to be recirculated to some degree and carried over from one day to the next.

3.1.7 Pollutant Dispersion from Point Sources

Point sources may occur at ground level, or as is often the case, pollutants are emitted from smokestacks that vary in height. The subsequent history of plumes formed depends on (1) the physical and chemical nature of pollutants, (2) meteorological factors such as wind speed and atmospheric stability, (3) location of the source relative to physical obstructions, and (4) topographical factors that affect air movement. As these affect plume rise, its spread horizontally and vertically, and its transport, they also affect maximum ground-level concentrations (MGLCs) and the distance of MGLCs from the source.

3.1.7.1 Pollutants and Diffusion

In many cases, point source plumes are a mixture of gas- and particulate-phase substances. Particles with aerodynamic diameters of ≥20 μm have appreciable settling velocities, and as a consequence, deposition occurs relatively close to their sources. Smaller particles, particularly those with aerodynamic diameters of ≤1μm, have very low settling velocities and dispersion behavior similar to that of gases and vapors. The gaseous nature of plumes, given sufficient time, may allow for dispersion by simple diffusion, whereby the random motion of molecules results in pollutants migrating from areas of high concentration (the center of the plume) to areas of low concentration (the plume's periphery). Diffusion causes plumes to spread both horizontally and vertically. As a consequence, the effect of diffusion can be seen to increase with downwind distance.

3.1.7.2 Plume Rise and Transport

Dispersion from a smokestack source is significantly affected by its physical height as well as plume rise. In Figure 3.7, the plume is seen to rise to a maximum height and then level off. The distance from the top of the stack to the center of the plume is described as plume rise. The distance from the ground to the center of the plume (including the stack) is the effective stack height.

Figure 3.7   Plume rise and effective stack height.

Dispersion is enhanced with increasing stack height and plume rise. Ground-level concentrations will be lower, and at constant wind speed, the distance at which MGLCs occur will be increased as effective stack height increases.

The height of the plume at the point it levels off depends on the exit temperature of stack gases, cross-sectional diameter of the stack, emission velocity, horizontal wind speed, and atmospheric stability (as indicated by the vertical temperature gradient).

The effective stack height for a source can be increased by building taller stacks. Tall-stack technology has been used since the 1960s by electrical utilities operating large coal-fired power plants. Such stacks are commonly 250 to 300 m (850 to 1000 ft) high, with some stacks ~400 m (1300 ft). They were designed and are operated on the principle that pollutants could be dispersed from such facilities without causing unacceptable downwind ground-level concentrations.

Wind speed in the horizontal dimension significantly affects both plume rise and ground-level concentrations. Higher wind speeds decrease effective stack height. However, due to the increased volume of air associated with increasing wind speeds, ground-level concentrations are usually reduced. Higher wind speeds decrease the distance at which MGLCs occur. At very high speeds (~80 km/h, 50 mi/h), plume rise may be negligible; the plume may be brought to the ground immediately downwind of the source.

Tall stacks are designed to take advantage of the higher wind speeds that occur aloft as frictional drag of the Earth's surface is diminished. Pollutant dispersion is enhanced as a result of these higher wind speeds.

Plume rise, as indicated, is subject to atmospheric stability. Under unstable conditions, significant plume rise occurs; under stable conditions, plume rise is markedly reduced. In the latter case, dispersion is decreased and higher ground-level concentrations can be expected.

3.1.7.3 Plume Characteristics

As a plume moves downwind of a source, it expands by diffusion in both horizontal and vertical dimensions. Plumes take forms and behavior patterns that reflect stability conditions in the atmosphere. Major plume types are illustrated in Figure 3.8. In the first case (Figure 3.8(a)), the lapse rate is superadiabatic with relatively calm winds. There is significant initial plume rise and a subsequent "looping" motion. This motion results from portions of the plume being buoyed up by convective rising, with subsequent descending of air. Upward and downward air motion is considerable. Since eddies that produce this motion are oriented in all directions, significant horizontal dispersion takes place as well.

A coning plume is illustrated in Figure 3.8(b). Coning plumes form when lapse rates are neutral to isothermal. As such, the atmosphere is slightly unstable to slightly stable. Such lapse rates occur on cloudy or windy days or at night. Atmospheric turbulence is primarily mechanical; turbulent eddies may have different orientations that result in dispersion in a relatively symmetrical pattern.

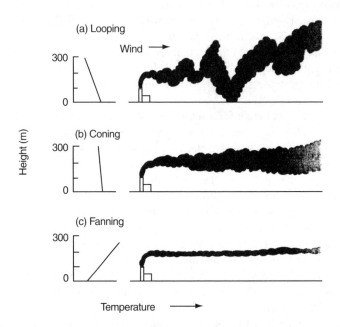

Figure 3.8 Plume form and behavior associated with different atmospheric stability conditions.

A fanning plume is illustrated in Figure 3.8(c). It is produced under stable conditions in which the top of a ground-based inversion is well above the stack. The plume rises slightly, there being little vertical air movement and dispersion. Horizontal motions, however, are not inhibited. As a consequence, the plume may be characterized by varying degrees of horizontal spreading.

Other plume forms and behaviors (not illustrated here) occur. Lofting, fumigating, and trapping plumes are associated with inversions. A lofting plume may be produced when the atmosphere above a surface inversion is unstable, with the stack above the surface-based inversion. The plume rises upward as its downward movement is restricted by the inversion beneath it. Lofting plumes are produced at sunset on clear nights, over open terrain. A fumigating plume is produced when a surface-based inversion breaks up after sunrise. Pollutants are brought to the ground by the downward movement of convective cells. When inversions occur both above and below a smokestack, the plume is trapped; in appearance it is somewhat similar to, but deeper than, a fanning plume.

As plumes move downwind, they generally mix with air that is less polluted. Such mixing occurs as a result of diffusion, advection, displacement, convection, and mechanical turbulence. Consequently, concentrations decrease with downwind distance. Depending on their height and atmospheric stability, plumes may reach the ground within a few kilometers of a source or may remain airborne for extended distances. Plumes can be seen because of the presence of light-scattering particles. Plumes became less and less visible as particle concentrations decrease as a result of dilution.

3.1.8 Large-Scale Transport and Dispersion

3.1.8.1 Long-Range Transport

Historically, pollutant concerns have focused on dispersion of pollutants in urban areas and those downwind of large point sources. It was once assumed that when pollutants were diluted to acceptable levels or plumes were no longer visible, they were not a problem. In the early 1970s, the phenomenon of long-range transport was identified as the cause of elevated nocturnal rural O_3 levels. Both O_3 and its precursors were transported hundreds of kilometers from urban areas (urban plume). The phenomenon of acidic deposition and its attendant ecological effects and associations with the long-range transport of acid precursors in North America became known in the mid- to late 1970s. This was followed by the recognition that Arctic haze over Barrow, AK (in the northern hemispheric spring), was associated with long-range transport of pollutants over the North Pole from northern Europe and Asia. There was, in the late 1980s, increasing evidence that tropospheric pollutants were being transported into the stratosphere. In more recent times, we have come to better understand processes by which long-lived substances such as carbon dioxide (CO_2), CH_4, halogenated hydrocarbons, and particles move around the planet.

3.1.8.2 Urban Plume

Large urban centers include numerous individual point and mobile sources. These contribute collectively to large polluted air masses that affect air quality for tens to hundreds of kilometers downwind. This air mass is described as an urban plume. The transport and dispersion of pollutants in an urban plume occur over larger geographical areas and timescales than those typically associated with individual sources. Of particular importance are airflows associated with migrating high- and low-pressure systems.

3.1.8.3 Planetary Transport

A stable layer of air at the top of the PBL retards vertical mixing and isolates it from the free troposphere above. Because upward air movement is somewhat impeded, timescales on the order of a few hours to a few days are needed for transport of pollutants out of the PBL. As a consequence of convective energy flows, baroclinic (pressure-related) instability, and heat release from condensation of water vapor, pollutants are transported to the top of the troposphere, with uniform mixing occurring in about a week.

Air in the troposphere is continuously stirred by convective air movement and other atmospheric phenomena. As a consequence, substances with lifetimes of several months or more are well mixed within the troposphere. However, significant concentration differences exist between northern and southern hemispheres.

Atmospheric phenomena at the equator retard airflows from one hemisphere to another. As a consequence, cross-equatorial mixing time is approximately 1 year.

3.1.9   Stratosphere–Troposphere Exchange

The troposphere and stratosphere are characterized, respectively, by decreasing and increasing temperatures with height. They are separated by an isothermal layer of air (the tropopause). The increasing temperatures with height in the stratosphere serve to limit the upward and downward movement of atmospheric gases between the troposphere and stratosphere. Trace gas measurements have shown, however, that such exchange takes place, albeit relatively slowly.

As a consequence of stratosphere–troposphere exchange processes, long-lived chemical species such as chlorofluorocarbons (CFCs), CH_4, and nitrous oxide (N_2O) are transported into the stratosphere, and chemical species originating in the stratosphere are transported into the troposphere. These include O_3, nitric oxides (NO_x), and substances such as chlorine nitrate ($ClONO_2$) and hydrogen chloride (HCl) from the photodestruction of CFCs.

The time required to exchange the mass of the entire troposphere with the stratosphere is estimated to be 18 years. Because of differences in mass, it takes about 2 years for the entire stratosphere to mix with the troposphere.

A number of potential pathways and mechanisms have been proposed to explain stratosphere–troposphere exchange. The simplest of these consists of a single Hadley-type cell in each hemisphere with a uniform rising motion across the tropical tropopause, poleward movement in the stratosphere, and return flow into the troposphere outside of the tropics. This is consistent with low H_2O vapor levels in the tropical stratosphere and high O_3 levels in the lower polar stratosphere. This circulation cell is described as a wave-driven "extratropical pump."

This and a second pathway for stratosphere–troposphere exchange are illustrated in Figure 3.9. In the second pathway, transport occurs from the lower stratosphere to the troposphere in mid-latitudes and from the troposphere to the stratosphere along surfaces of constant potential temperature that cross into the tropopause. Exchange of air tends to occur in association with events known as tropopause folds. In this phenomenon, the tropopause on the poleward side of a jet stream (see Chapter 1) is distorted during development of large weather systems. Large intrusions of stratospheric air occur, which become trapped and eventually mix with the troposphere. Air from the troposphere can also be trapped in the stratosphere during tropopause folds. In a third proposed pathway, air is transported convectively from the troposphere to the stratosphere.

3.1.10   Stratospheric Circulation

Due to strong inversion conditions the stratosphere is very stable, with little vertical air motion. Because of differences in stratospheric temperature between the equator and poles, as well as those caused by diabatic heating (associated with O_3 absorption of ultraviolet (UV) light), zonal (east ↔ west) and meridianal (north ↔ south) flows characterize the stratosphere. Thermal gradients result in strong zonal winds, which reach peak speeds near solstices and reverse after equinoxes.

Despite diabatic heating, circulation in the stratosphere is wave driven. Atmospheric waves transport air poleward in the winter hemisphere. It subsides at the

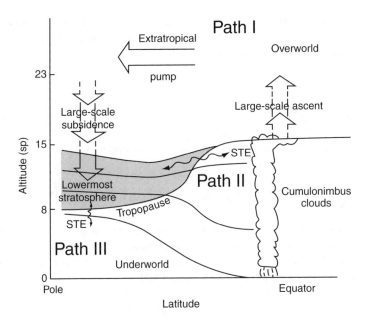

Figure 3.9 Stratosphere–troposphere exchange pathways and processes. (From Holton, J.R. et al., *Rev. Geophys.*, 33, 403, 1995. With permission.)

poles, where it warms by compression. A circulation cell develops throughout the middle and upper stratosphere. There is a slow rising of air in the summer hemisphere and tropics and more rapid sinking over a smaller area in the winter hemisphere.

This stratospheric circulation significantly affects the movement of chemical species. It results in the movement of O_3 (most O_3 is produced in the tropical stratosphere) and gases transported from the troposphere poleward; movement is strong in the winter hemisphere and less so in the summer hemisphere.

Such stratospheric airflows transport trace gases such as CFCs to the Antarctic and O_3-depleted air northward from the Antarctic. They also have significant effects on substances that contribute to the formation of polar stratospheric clouds (see Chapter 4).

3.2 ATMOSPHERIC REMOVAL AND DEPOSITION PROCESSES

3.2.1 Atmospheric Lifetimes

Reference was made to average lifetimes or residence times of various atmospheric pollutants in Chapter 2. All gas- and particulate-phase pollutants have a life history in the atmosphere before they are ultimately removed. By averaging the life histories of all molecules of a substance, or all particles of a particular type or size, one can determine their average residence or lifetime. Residence time can also be described in the context of a pollutant's half-life, i.e., the time required to reduce its concentration by 50% of its initial value.

Atmospheric lifetimes can be calculated using mass balance equations. Under steady-state conditions (emission rate = removal rate), the residence time or lifetime of a substance can be calculated from the following equation:

$$\tau = Q/P \tag{3.1}$$

where τ is the residence time, Q is the total mass of substance in the atmosphere (g), and P is the emission or removal rate (g/year).

If the total mass (Q) of a given atmospheric substance is 2×10^{12} g and $P = 100 \times 10^{12}$ g/year,

$$\tau = 2 \times 10^{12} \text{ g}/(100 \times 10^{12} \text{ g/year}) = {\sim}1 \text{ week}$$

Lifetimes are also calculated on the basis of the substance's reactivity with sink chemicals such as hydroxyl radical (OH·) and nitrate (NO_3^-). In such cases lifetimes are calculated by dividing the product of the OH· concentration and the rate constant (k) into 1. For the reaction between CH_4 and OH·,

$$CH_4 + OH· \rightarrow CH_3· + H_2O \tag{3.2}$$

an atmospheric lifetime of 5 years is calculated:

$$\tau = 1/k[OH·] \tag{3.3}$$

$$= 1/[6.3 \times 10^{-15} \text{ cm}^3 \text{ molecule}^{-1} \text{ sec}^{-1} \times (1 \times 10^6 \text{ OH· molecules cm}^{-3})]$$

$$= 1.5^9 \times 10^8 \text{ sec} = {\sim}5 \text{ year}$$

These atmospheric lifetime calculations are based on chemical kinetics. They assume that there are no competitors for OH· or, if there are, that OH· is in a steady-state concentration. If there are competing loss processes (such as photolysis or reaction with O_3), lifetimes may be shorter.

Though the atmospheric lifetimes of gas-phase substances are determined by chemical reactions with substances such as OH·, their by-products, unreacted molecules, and particulate-phase species will ultimately be removed by deposition processes.

3.2.2 Deposition Processes

Dry deposition is characterized by direct transfer of gas- and particulate-phase substances to vegetation, water, and other Earth surfaces. This transfer may take place by impaction, diffusion to surfaces, and, in the case of plants, physiological uptake of atmospheric contaminants. Particulate-phase substances may also be removed by sedimentation.

Dry deposition is characterized by a deposition velocity (Vg). It is a proportionality constant that relates the flux (F) of a chemical species or particle to a surface and its concentration (C) at some reference height:

$$Vg = -F/C \tag{3.4}$$

The deposition velocity is a positive number despite the fact that the flux is negative; it is given as centimeters per second (cm/sec). Dry deposition velocities are given for coarse and fine particles and for gas-phase substances in Table 3.2. Highest deposition velocities have been observed for coarse particles and gas-phase nitric acid (HNO_3). Because of its high solubility, HNO_3 is readily absorbed into dew and other aqueous surfaces and rapidly taken up by plants.

Wet deposition includes all processes by which airborne gases and particles are transferred to the Earth's surface in aqueous form (rain, snow, fog, clouds, dew). These processes include (1) absorption of gas-phase substances in cloud droplets, raindrops, etc.; (2) in-cloud processes wherein particles serve as nuclei for the condensation of H_2O vapor to form cloud or fog droplets; and (3) collision of rain droplets and particles both within and below clouds. In the last case, collisions with, and subsequent incorporation of, gas-phase substances can also occur. This process is called washout. The removal of particles in the rain-making process is called rainout.

3.3 METEOROLOGICAL APPLICATIONS: AIR POLLUTION CONTROL

As seen in previous discussions in this chapter, a variety of meteorological factors affect the dispersion of pollutants from individual sources as well as urban areas. Thus meteorology has significant applications in air pollution control programs. These include episode prediction, planning, and community responses. These also include determination of whether, under worst-case atmospheric conditions, proposed sources will be in compliance with air quality standards and visibility protection requirements of prevention of significant deterioration provisions of clean air legislation. Compliance is determined from air quality modeling.

Table 3.2 Dry Deposition Velocities (cm/sec) for Particles and Selected Gases

Pollutant	Surface	Range (cm/sec)
Particles > 2 μm	Exterior	0.5–2
Particles ≤ 2 μm	Exterior	<0.5
SO_2	Exterior and interior leaf surfaces	0.2–1 dry foliage, open stomates; >1 for wet surface
HNO_3	Primarily exterior leaf surfaces	1–5 or greater
NO_2	Primarily leaf interiors; also exterior leaf surfaces	0.1–0.5 when stomates open
O_3	Primarily leaf interiors; also exterior leaf surfaces	0.1–0.8

3.3.1   Episode Planning

In the late 1950s and early 1960s, the U.S. Weather Service (now the National Weather Service) observed that elevated pollutant levels in urban areas were associated with slow-moving migratory high-pressure systems. As a consequence, it began a program of issuing air pollution potential forecasts when such systems covered an area of at least 90,000 km^2 (~35,000 mi^2) and were expected to persist at least 36 h.

Despite the fact that regulatory efforts to control air pollution have significantly reduced health threats associated with ambient air pollution, such stagnant high-pressure systems occur periodically, as do the semipermanent high-pressure systems over southern California. As such, they can cause significant increases in ground-level concentrations that may adversely affect the health of individuals at special risk (e.g., asthmatics, those ill with respiratory or cardiovascular disease, the elderly). Therefore, there is a continuing need to forecast air pollution episodes. When the National Weather Service forecasts a developing episode, regulatory authorities at the local, state, and national levels begin to implement episode control plans, which may include requiring phased reductions in emissions of one or more primary pollutants and issuing community health warnings (see Chapter 8).

3.3.2   Air Quality Modeling

New stationary sources regulated under National Ambient Air Quality Standard provisions of clean air legislation are required to demonstrate that they will be in compliance with these standards even under atmospheric conditions that are unusually favorable for elevated ground-level concentrations. Such compliance (in new source reviews) can only be demonstrated by using dispersion models to evaluate the source's impact. Primary inputs to dispersion models are emission and source information, meteorological data, and receptor information. Meteorological data and information required for these models include stability class, wind speed, ambient temperature, and MH. Point source dispersion, as well as other models, is described in detail in Chapter 7.

READINGS

Bourne, N.E., Atmospheric dispersion, in *Handbook of Air Pollution Technology*, Calvert, S. and Englund, H., Eds., John Wiley & Sons, New York, 1984.

Brasseur, G.P., Orlando, J.J., and Tyndall, G.S., Eds., *Atmospheric Chemistry and Global Change*, Oxford University Press, Oxford, 1999.

Briggs, G.A., Plume rise, *AEC Critical Review Series*, Washington, D.C., 1969.

Dobbins, R.A., *Atmospheric Motion and Air Pollution*, John Wiley & Sons, New York, 1979.

Finlayson-Pitts, B.G. and Pitts, J.N., Jr., *Chemistry of the Upper and Lower Atmosphere: Theory, Experiments, and Applications*, Academic Press, Orlando, FL, 2000.

Holton, J.R., *Introduction to Dynamic Meteorology*, 3rd ed., Academic Press, Orlando, FL, 1992.

Lutgens, F.K. and Tarbuck, E.G., *The Atmosphere*, 7th ed., Prentice Hall, Saddlebrook, NJ, 1998.

Scorer, R.S., *Meteorology of Air Pollution: Implications for the Environment and Its Future,* Ellis Horwood, New York, 1990.

Turner, D.B., *Workbook of Atmospheric Dispersion Estimates: An Introduction to Dispersion Modeling*, Lewis Publishers/CRC Press, Boca Raton, FL, 1994.

QUESTIONS

1. What is the planetary boundary layer? What role does it play in tropospheric air motion?
2. What is turbulence and how is it formed?
3. How does turbulence affect the dispersion of pollutants from a source?
4. What is the effect of wind direction on pollutant concentrations downwind of a source?
5. The velocity of wind moving past a constant emission source changes from 1 to 4 m/sec. What is the relative quantitative effect of this change of wind velocity on the pollutant concentration?
6. How is atmospheric stability related to lapse rate conditions?
7. Describe dispersion characteristics of the atmosphere under the following lapse rate conditions: ~2, 0, and 1°C/100 m.
8. Describe radiational inversion formation.
9. Indicate differences in the forms of radiational inversions in river valleys and over cities on flat, open terrain.
10. How do sea, lake, and land breezes affect air quality over a city?
11. Why can't polluted air in most cases penetrate an inversion layer? Indicate the physical principles involved.
12. How are subsidence inversions produced?
13. Characterize subsidence inversions relative to their vertical temperature profile, geographical scale, and persistence.
14. What meteorological factors affect plume rise?
15. Under what lapse rate conditions are looping, coning, and fanning plumes produced?
16. When do maximum ground-level concentrations of pollutants occur in mountain valleys? Why?
17. What is an urban plume? How is it formed?
18. What air quality problems are associated with long-range transport?
19. Generally, how long does it take for a long-lived pollutant to be uniformly mixed vertically in the troposphere? Horizontally throughout the troposphere?
20. By what mechanisms are pollutants transported into the stratosphere?
21. Characterize air circulation in the stratosphere.
22. What factors affect the lifetime of pollutants in the atmosphere?
23. What is dry deposition? What factors contribute to increased deposition rates?
24. Describe processes that result in rainout and washout of pollutants from the atmosphere.
25. What meteorological conditions produce pollution episodes?
26. How is meteorological information used in the implementation of episode plans?

CHAPTER 4

Atmospheric Effects

Although the atmosphere affects the fate of pollutants via various sink processes, it experiences significant short- or long-term pollution-induced changes. These may be local, regional, or global in scale, depending on such factors as sources, long-range transport, movement into the stratosphere, and accumulation of long-lived pollutants. Atmospheric effects may include changes in (1) visibility, (2) urban climate, (3) quantity and frequency of rainfall and associated meteorological phenomena, (4) atmospheric deposition, (5) stratospheric ozone (O_3) depletion, and (6) global climate change. The significance of these effects may be slight (visibility impairment) to potentially very serious (stratospheric O_3 depletion and changes in climate).

4.1 VISIBILITY

The single most recognizable effect of air pollution on the environment is visibility reduction that results from the scattering and absorption of light by suspended particles and, to a lesser extent, absorption of light by atmospheric gases.

Changes in visibility may result from natural processes and phenomena as well as anthropogenic activities.

We humans see objects and distinguish their shape and color characteristics because we have evolved the ability to receive reflected energy in the visible (400 to 700 µm) portion of the electromagnetic spectrum. Our ability to "see" is determined by factors associated with our individual vision system, our unique psychology, and a variety of physical factors in the atmosphere. These include (1) illumination of the observed object or area by the sun as mediated by clouds, ground reflection, and atmosphere; (2) reflection, absorption, and scattering of incoming light by objects and the sky; (3) scattering and absorption of light from objects and the source of illumination by the atmosphere and contaminants; (4) physiological response of the eye and brain to the resulting light energy received; and (5) subjective judgment of images received by the brain.

The human eye perceives the brightest object in its sight path as white, and the brain determines the color of other objects by comparison. The eyes' and brain's ability to perceive contrasts (the color difference between lightest and darkest objects) changes in response to illumination and the setting. Without contrast, or where contrast is limited, an object or scene may not be visible. The threshold contrast is that atmospheric condition at which an object or scene begins to be perceptible against its surroundings. Contrast determines the maximum distance at which an object or the various aspects of a scene can be discerned. Visibility is usually described in the context of light intensity; it can also include color. The human visual system detects color differences in response to the spectral distribution of incoming light.

4.1.1   Natural Visibility Reduction

Our ability to see objects clearly in an "unpolluted" atmosphere is limited by blue-sky scattering, curvature of the Earth's surface, and suspended natural aerosols. At sea level, a particle-free atmosphere scatters light and limits visual range to 330 km (205 mi). The scattering of light by air molecules is called Rayleigh scattering and is responsible for the blue color of the sky on "clear" days. Because Rayleigh scattering decreases with decreasing atmospheric density, visual range increases with altitude. At ~4000 m (12,000 ft), the potential visual range is ~500 km (310 mi).

When dark objects such as distant mountains are viewed through a relatively particle-free atmosphere, they appear blue since blue light is scattered preferentially into the line of sight. Clouds on the horizon may appear yellow–pink as the atmosphere scatters more blue light from bright targets out of the line of sight, resulting in longer wavelengths being seen.

In addition to Rayleigh scattering, which in conjunction with the Earth's curvature serves to place a theoretical limit on how far one can see, other natural factors, e.g., fog, rain, snow, windblown dust, and natural hazes, limit visibility. Natural hazes are formed from aerosols produced from volcanic emissions of sulfur dioxide (SO_2) and from chemical reactions involving hydrocarbons (HCs) and various sulfur (S) species produced bio- and geogenically. Reactions involving terpenes produce particles with diameters of <0.1 μm. Such particles preferentially scatter blue light and are the likely cause of bluish hazes over heavily forested areas such as the Blue Ridge and Great Smoky Mountains.

4.1.2   Visibility Concerns Associated with Human Activities

Pollution-induced changes in visibility that have received both scientific and regulatory interest include intense smog conditions such as those in Los Angeles and Houston and, to a lesser extent, other cities; seasonal regional hazes that affect large areas in the midwestern, eastern, and southeastern U.S.; plume blight and hazes produced by large point sources in the pristine air quality regions of the American Southwest and Rocky Mountains; springtime haze over the Arctic Circle; and, more recently, the Asian brown cloud.

4.1.2.1 Smog

Smog problems in Los Angeles and other cities are long-standing. Despite three decades of pollution control efforts, smog problems persist and, in cases such as Houston, TX, have become worse with time.

4.1.2.2 Regional Haze

The problem of regional haze in the midwestern, eastern, and southern U.S. grew progressively worse from the mid-1950s to the 1990s (visibility has not worsened in the past decade). This was particularly evident during the summer months. Regional haziness in the U.S. during the summer is associated with slow-moving high-pressure systems. These often track up the Ohio Valley, where major point sources emit large quantities of SO_2 that in combination with warm temperatures, abundant sunlight, high relative humidity (RH), and reduced vertical mixing, produce conditions favorable for regional haze production. Such haze is visible from orbiting Earth satellites. The decline in visibility observed during summer months in the Northeast, Midwest, and Southeast is strongly correlated with increased use of coal to produce electrical power to serve air conditioning needs in residential and non-residential buildings.

4.1.2.3 Pristine Air

Visibility impairment has historically been a problem in U.S. urban centers; it is increasingly a problem on a broad regional scale in many areas east of the Mississippi River. In the regulatory sense, however, it has been pristine areas of the western U.S. that have received the most attention. The scenic beauty associated with our national parks, monuments, and wilderness areas is due in great measure to the high air quality, high visibility (visual ranges of 100 to 164 km (62 to 102 mi) are not uncommon), and contrast characteristics of these regions. Major concerns include visibility-reducing and aesthetics-degrading power plant plumes and regional hazes produced by photochemical conversion of SO_2. The former is commonly described as plume blight. In pristine areas, power plant plumes may maintain their distinct character for tens of kilometers downwind. The plume, and haze it will eventually produce, causes blighted vistas, destroying the unique character of some of America's most beautiful landforms and environments.

The problems of visibility degradation of pristine air and protection of areas of high air quality have been addressed in clean air legislation and subsequent regulatory actions by the U.S. Environmental Protection Agency (USEPA) (see Chapter 8).

4.1.2.4 Arctic Haze

The phenomenon of Arctic haze was first reported by weather reconnaissance crews flying over the Arctic in the 1950s. Significant springtime visibility reduction was observed over areas such as Barrow, AK, where the environment was considered to be relatively pristine. The Arctic haze phenomenon has intensified over the past half century. It now covers a linear expanse of 800 to 1300 km (496 to 806 mi). It occurs at altitudes below 9 km (5.8 mi), with a maximum intensity at ~4 to 5 km (2.5 to 3.1 mi).

The primary cause of Arctic haze is industrial emissions of SO_2, elemental carbon (EC), HCs, metals, and other gases and particles from northern European and Eurasian sources. These pollutants are transported northward into an environment where removal processes are less efficient. A steep inversion layer (up to 30 to 40°C (86 to 104°F) differential between cooler surface and warmer air layers several hundred meters above) further contributes to the phenomenon. Sulfates, at concentrations of 2 μg/m³, are on the order of 10 to 20 times greater than would be expected from sources in the affected region.

4.1.2.5 South Asian Haze

While the Arctic haze phenomenon has been known and studied for decades, "discovery" of the South Asian haze or Asian brown cloud was first reported in 1999. The haze, a brownish layer 3 km thick, covers most of the tropical Indian Ocean and South, Southeast, and East Asia (Figure 4.1). This is the most densely populated region in the world. Its discovery and recent understanding of the phenomenon were outcomes of the Indian Ocean Experiment (INDOEX) studies conducted by an international team of scientists.

Haze particles and precursors are transported far beyond their source region, particularly during the dry season (December to April). South Asian haze includes elevated concentrations of sulfates, nitrates, organic compounds, carbon (C), fly ash, and other pollutants. Major sources include biomass burning and industrial emissions.

Figure 4.1   Satellite image of haze over South Asia. (From National Aeronautics and Space Administration, http://www.gsfc.nasa.gov/topstory/2003/0110chinahaze.html.)

Direct impacts include a significant reduction (10 to 20%) in solar radiation reaching the surface, a 50 to 100% increase in solar heating of the lower atmosphere, rainfall suppression, reduced agricultural productivity, and adverse health effects.

4.1.3 Factors Affecting Visibility Reduction

Visibility reduction associated with anthropogenic activities is due primarily to light scattering by particles and, to a lesser extent, absorption of light by atmospheric gases such as nitrogen dioxide (NO_2). Nitrogen dioxide absorbs short-wavelength blue light, with longer wavelengths reaching the eye. Its color, therefore, appears yellow to reddish brown. Light can also be absorbed by particles. Such absorption is important when soot (C) or large amounts of windblown dust are present. In general, most atmospheric particles are relatively inefficient in absorbing visible light.

Fine fraction solid or liquid particle aerosols are responsible for most atmospheric light scattering. Particles with diameters similar to the wavelengths of visible light (0.4 to 0.7 μm) are particularly effective in light scattering (Figure 4.2). The scattering of all visible wavelengths equally is called Mie scattering. This wavelength-dependent interaction of light with the atmosphere can be described mathematically by radiative transfer equations:

$$-dI = BeId_x \qquad (4.1)$$

where $-dI$ is the decrease in intensity (extinction), Be is the extinction coefficient, I is the initial beam intensity, and d_x is the length of light path.

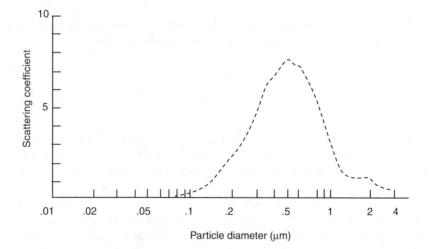

Figure 4.2 Relationship between light scattering and particle size. (From USEPA, EPA/600/8-82-029, EPA, Washington, D.C., 1982.)

Equation 4.1 can be used to describe the effect of the atmosphere on a light beam as it passes from a source to an observer. The value of Be, the extinction coefficient, is determined by the scattering and absorption of light by particles and gases and varies as a function of wavelength and contaminant concentrations.

The intensity of a beam in the direction of an observer decreases with distance as light is scattered or absorbed. Over a short distance, this decrease is proportional to the path length and intensity of a beam at that point. Similarly, when an observer is looking at a distant object such as a mountain, light from the mountain is diminished by absorption and scattering caused by the intervening atmosphere. In addition, light is also scattered into the sight path from the surrounding atmosphere. It is this "sky light" that forms the optical phenomenon we know as haze. These relationships can be seen in Figure 4.3.

With increasing distance, extinction and added sky light cause both dark and bright objects to be "washed out" and approach the brightness of the horizon. Consequently, the contrast of an object relative to the horizon and other objects decreases.

The apparent contrast and visual range of a large, dark target object can be calculated from the following series of exponential equations:

$$C = C_o e^{-BeX} \tag{4.2}$$

where C is the apparent contrast at observer distance, C_o is the initial contrast at the object, e is the natural log base (2.718), Be is the extinction coefficient, and X is the observer-to-object distance.

For a black object, the initial contrast is –1, and Equation 4.2 becomes:

$$C = (-1)e^{-BeX} \tag{4.3}$$

By assuming a standard threshold of perception of 0.02, we can calculate the visual range (V_r), which is equal to the observer-to-object distance.

$$0.02 = -e^{-BeVr} \tag{4.4}$$

$$V_r = 3.92/Be \tag{4.5}$$

Equation 4.2 is called the Koschmieder equation. It is used to determine the visual range from measurements of the extinction coefficient. The visual range is an inverse function of the extinction coefficient, which is directly related to atmospheric aerosol concentrations.

The extinction coefficient (Be) is the sum of both absorption and scattering components expressed as inverse lengths of the atmosphere (megameters). Extinction coefficients range from 10 Mm^{-1} in clean air to 1000 Mm^{-1} in very polluted environments. These correspond to visual ranges of 400 km to 4 km (see Section 4.1.5).

Light scattering by aerosol particles includes a number of physical phenomena: reflection, which causes backscatter; and diffraction and refraction, which cause

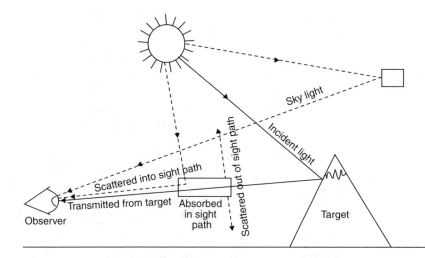

Figure 4.3   Relationships between light scattering and visibility of a target object. (From USEPA, EPA/600/P-99/002aB, EPA, Washington, D.C., 2001.)

scattering in the forward direction (Figure 4.4). In diffractive scattering, a light ray is bent by the edge effect of an aerosol particle into its shadow. In refraction, two effects are produced. For particles with a refractive index of n > 1, the speed of the light wave is reduced, causing a reduction in the wavelength inside the particle. This results in a phase shift, which produces positive and negative interferences. Refraction also bends light waves, causing them to pass through and out of a particle at a different angle from that at which they entered. This refractive bending is responsible for separation of the color components of white light by a prism. Refractive bending of light by aerosol particles can also have chromatic effects. Coarse particles (>2 μm) scatter light in the forward direction primarily by diffraction and refraction. For particles in the accumulation mode (0.1 to 1.0 μm), physical interactions with light are complex and enhanced. Though coarse particles are efficient in scattering light, their contribution to visibility reduction is limited due to their relatively small numbers.

Light scattering may be affected by the chemical composition of aerosol particles. As a rule, particles of differing chemical composition will have different light scattering efficiencies. In addition, hygroscopic substances sorb water (H_2O), which allows them to grow into the critical diameters that scatter light most effectively. Sulfate aerosols are very hygroscopic and therefore absorb light efficiently.

4.1.4   Meteorological Effects

Meteorological factors affect visibility in several ways. Sunlight affects visibility by promoting secondary aerosol formation. Atmospheric photochemistry produces the major visibility-reducing aerosols, i.e., sulfates, nitrates, and oxyhydrocarbons.

Wind speed and atmospheric stability affect visibility since they determine atmospheric dispersion and therefore concentration of aerosol particles. In general, as wind speed increases, visibility improves. This wind-induced atmospheric mixing

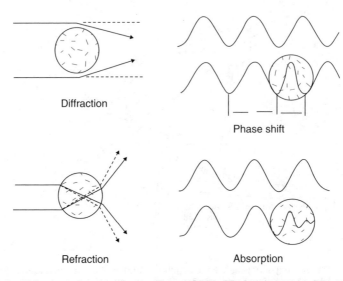

Diffraction

Phase shift

Refraction

Absorption

Figure 4.4 Light-scattering phenomena. (From USEPA, EPA/600/8-82-029, EPA, Washington, D.C., 1982.)

results in lower aerosol concentrations. During periods of atmospheric stagnation (associated with slow-moving high-pressure systems), vertical mixing is suppressed, aerosol concentrations increase, and visibility is significantly reduced. The resultant haze may cover hundreds of thousands of square kilometers. It is not uncommon to have a major portion of the Midwest, Southeast, or East Coast covered by a "blanket" of haze.

Visibility may be significantly affected by RH. Hygroscopic particles such as sulfates (described above) take up moisture and increase in size under humid conditions. Marked visibility reduction has been reported when RH exceeds 70%, though the importance of this effect extends over a range of RHs. In general, regions and periods of low RH are associated with good visibility.

4.1.5 Visibility Measurements

Qualitative visibility measurements (reported as prevailing visibility) have been routinely made (since 1939) by human observers at commercial and military airfields in conjunction with, or in addition to, the National Weather Service. Observers viewed the horizon noting whether dark objects (daytime) or lights (nighttime) could be seen at known distances. These were reported as the greatest visibility attained around at least one half of a horizon circle. Such readings are dependent on the individual observer and the availability of "targets" at varying distances. In most cases, such visibility readings are poorly related to air quality.

In the past two decades, quantitative techniques (using instrumental methods) have been developed that more accurately measure visibility reduction and its relationship to atmospheric pollutants. These methods have been incorporated into visibility monitoring networks.

The largest instrumental visibility monitoring network has been installed at airports in the U.S. to provide real-time data for runway visibilities. The network includes automated surface observing systems (ASOS) and automated weather observing systems (AWOS). These use a forward-scattering visibility meter that measures air clarity. Clarity readings are converted to sensor equivalent visibility values (SEVs), which are a measure of what would be perceived by the human eye. The instrument samples 1.9 cm (0.75 in.) of the atmosphere, providing average 1-min values. By using an algorithm, relatively accurate visibility readings for within 3 to 5 km (2 to 3 mi) of the measurement location can be determined.

Visibility measurements are also routinely made by the National Park Service, U.S. Forest Service, Bureau of Land Management, and the Fish and Wildlife Service in and around Class I visibility protection areas subject to regulation under prevention of significant deterioration (PSD) provisions of clean air legislation (see Chapter 8). Visibility measurements in Class I regions (see Chapter 8) have been made using instrumental methods for more than a decade. The USEPA has developed a collaborative program with these federal agencies in a large monitoring network that includes visibility and air quality measurements. This network is called the Interagency Monitoring of Protected Visual Environments (IMPROVE).

The most frequently used approach to characterize the relationship between visibility and air quality is measurement of the extinction coefficient using a transmissometer. Visual range is then calculated from the Koschmieder equation (Equation 4.2) by assuming that the atmosphere and daytime illumination is uniform over the sight path and the threshold is 2%. Recall that visual range is inversely related to the extinction coefficient, which is directly related to atmospheric aerosol concentrations.

The extinction coefficient, B_e, can be used to calculate a deciview index. The deciview index expresses uniform changes in haziness in increments from pristine conditions to very impaired visibility. The relationship between extinction coefficients, deciviews, and visual range can be seen in Figure 4.5. A 10 to 20% change in the extinction coefficient represents a 1- to 2-deciview change; a doubling of the extinction coefficient increases the deciview value by 7 units and decreases the visual range by 50%.

4.1.6 Visibility Patterns and Trends

Visibility patterns have been evaluated at approximately 280 airports over the past 40 years using visual range data. These data indicate that summertime visibility

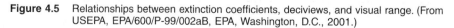

Figure 4.5 Relationships between extinction coefficients, deciviews, and visual range. (From USEPA, EPA/600/P-99/002aB, EPA, Washington, D.C., 2001.)

impairment has been particularly significant in areas east of the Mississippi River in the lower Midwest, in middle Atlantic and southeastern states, and on the southern coast of California, which, for the most part, includes the Los Angeles Basin. Summertime haze in the eastern U.S. increased significantly from 1970 to 1980 and decreased slightly between 1980 and 1990. In 1970, the highest summertime haze conditions were reported for Indiana, Ohio, Pennsylvania, New Jersey, southeast New York, Delaware, Maryland, Kentucky, West Virginia, Virginia, and North Carolina. By 1980, high haze conditions expanded through most of the southeastern states, including Florida and westward into Missouri. Haziness in the Southeast has increased 80% since the late 1950s.

In addition, visibility trends have been evaluated from instrumental monitoring of light extinction at national parks and wilderness areas over the past 20 years. These monitoring efforts were expanded under the IMPROVE program, which began in 1987. Monitoring data include visibility and pollutant measurements.

IMPROVE data show that currently, visibility impairment is significantly worse in the rural East than in the rural West. The average visual range in eastern Class I regions is ~24 to 40 km (15 to 25 mi) compared to an estimated natural visibility of 145 km (90 mi). The average annual visibility is 56 to 145 km (35 to 90 mi) in most Class I areas (which are mostly in the western states) compared to an estimated natural visibility of 226 km (140 mi).

No significant declines in visibility and air quality were observed at IMPROVE sites during the period 1988–1996. Seventy percent of these sites showed improvements for annual best and median visibility days. Little or no visibility improvement was observed in the number of worst visibility days at sites such as the Badlands, Great Smoky Mountains, Mesa Verde, Shenandoah, and Yosemite National Parks.

Air quality data from the IMPROVE program indicate that visibility reduction is associated with particulate-phase sulfates, nitrates, elemental C, and crustal material. The relative contribution of each of these particle components can be seen for eastern and western rural sites in Table 4.1. Note the dominance of sulfates in causing visibility impairment in the eastern U.S. and their significant contribution even in western states (65% on the worst days). Both organic C and nitrates are seen to be significant causes of visibility reduction at western sites. Organic C contributes up to 40 to 45% of light extinction in the Pacific Northwest, with nitrates being the primary cause of light extinction (~40%) in southern California.

Table 4.1 Pollutant Contributions (%) to Visibility Impairment at IMPROVE Sites in the Eastern and Western U.S.

Pollutant	Eastern U.S.	Western U.S.
Sulfates	60–80	25–65
Organic carbon	10–30	15–35
Nitrates	5–15	5–45
Elemental carbon	5–15	5–15
Crustal material	5–15	5–20

Source: From USEPA, EPA-454/F-00-002, EPA, Washington, D.C., 2000.

4.2 TURBIDITY

Visibility is a measure of horizontal light scattering. Turbidity, on the other hand, is a measure of vertical extinction. Vertical extinction is dependent on the same atmospheric light-scattering characteristics of aerosols associated with visibility reduction. The principal effect of atmospheric turbidity is diminution of sunlight received at the ground. Though most light scattering is in a forward direction, intensity is reduced.

Turbidity, or optical thickness of the atmosphere, is directly related to aerosol concentrations. Aerosols in both the tropospheric and stratospheric layers of the atmosphere contribute to turbidity. In general, tropospheric extinction can be directly attributed to aerosols from human activity. Stratospheric turbidity is largely attributable to aerosols produced from volcanoes and, increasingly, aerosols associated with human activities.

The turbid air over major cities is often described as a dust dome. In studies in the Cincinnati, OH, metropolitan area, a complex stratification of the dust dome has been observed, with alternating polluted and cleaner air layers that correlate with different lapse rate conditions (Figure 3.4). High turbidity occurs between 11 A.M. and 12 P.M. on both relatively clean and polluted days. On relatively clean days, the lowest 1000 m (~3000 ft) contributed 30% of light extinction; on polluted days, 65%. Other turbidity-related phenomena in the troposphere include (1) a commonly observed turbid air mass centered over the eastern part of the U.S.; (2) maximum turbidity during the summer and minimum turbidity in winter; (3) losses of direct radiation of up to 50% in the rural midwestern U.S. during summer, resulting in up to a 20% reduction of total radiation received at the ground; and (4) an increasing upward trend in turbidity over Washington, D.C., and Davos, Switzerland, over the past four decades.

In the stratosphere, an aerosol layer described as the Jange layer forms from volcanic eruptions and is maintained by particle formation and transport in the upper tropical troposphere. This stratospheric aerosol layer varies in height from 9 to 28 km (5.5 to 17 mi), with maximum concentrations at ~18 km (11 mi). This aerosol consists of highly concentrated liquid sulfuric acid (H_2SO_4) droplets, typically 0.1 to 0.3 μm in diameter. Most of the mass of this aerosol layer is formed from SO_2 and reduced S gases such as carbon disulfide (CS_2) and carbonyl sulfide (COS). They are oxidized in the stratosphere to H_2SO_4 that condenses on preexisting nuclei.

The stratospheric aerosol layer can have a variety of atmospheric effects. In backscattering sunlight, it reduces the amount of solar radiation reaching the ground, thus cooling the atmosphere. Cooling of portions of the troposphere by such turbidity has been observed after major volcanic eruptions, particularly those that have high S gas emissions. In addition, aerosol particles provide sites for heterogeneous chemical reactions that repartition nitrogen (N) species and decrease O_3 concentrations. In polar regions, reactions on these sulfate particles can result in catalytic destruction of O_3 by chlorine (Cl) atoms.

4.3 THERMAL AIR POLLUTION

Historically, only gases and aerosol particles were considered to be atmospheric pollutants. However, because of the effect of surfaces and waste heat on urban heat balance and the resultant alteration of the atmospheric environment, heat or thermal energy is certainly a form of air pollution.

In making temperature measurements on a clear night (with light winds) in an urban area and also in the surrounding countryside, it is possible to detect and characterize a phenomenon known as the urban "heat island." In this heat island, areas of elevated temperatures (Figure 4.6) can be identified, with the entire urban area being warmer than the surrounding countryside. Urban–rural differences of 1 to 2°C (1.8 to 3.6°F) are common, and differences may be as great as 5°C (9°F). The principal contributors to heat island formation include (1) waste heat dumped into the environment from energy-utilizing processes, (2) solar energy absorption and heat storage properties of urban surfaces, and (3) decreased urban ventilation resulting from the increased surface roughness of buildings. Urban surfaces such as buildings, sidewalks, and streets more readily absorb and store heat from sunlight during daylight hours than do vegetated areas external to the city. In addition, urban surfaces cool more gradually at night.

In cities such as New York, anthropogenic heat may be 2.5 times greater than that received from the sun during winter; during summer it represents a significant portion of the total city energy budget (on the order of 17%). On an annual basis, anthropogenic heat in many cities may exceed 10% of that received from the sun.

Figure 4.6 Urban heat island associated with Tokyo, Japan. (From Saitoh, T.S., Shimade, T., and Hoshi, H., *Atmos. Environ.*, 30, 3431, 1996. With permission.)

Urban heat islands may have significant effects on weather and local climate within urban centers and, in some cases, downwind. These may include (1) longer frost-free seasons, (2) less frequent fog, (3) decreased snow accumulation, (4) lower RH, (5) decreased likelihood of surface-based nocturnal inversions, (6) increased rainfall downwind, and (7) a distinctive nocturnal metropolitan air circulation pattern.

This nocturnal circulation pattern is much like a sea breeze on an island. As warm air rises from the urban center it flows outward toward rural areas. Cooler air outside the urban center sinks and flows inward toward the city. This thermally induced airflow creates a turbid mass over the city that tends to be highest in the middle of the heat island. The resultant "dust dome" forms best under conditions of light wind and compact city form.

Because of thermal turbulence and vertical mixing, surface-based inversions generally do not occur over urban heat islands. However, elevated nocturnal inversions of longer duration may frequently occur. These elevated inversions may result, in part, from the dust dome absorption of thermal energy. They may also result from the advection of stable rural air over warmer urban surfaces. As this air moves over the city, it is heated from below, resulting in an inversion layer above the lower, relatively warm air.

Increased rainfall downwind of cities (discussed in a subsequent section) may be due in part to the heat island effect. Enhanced vertical convective mixing tends to make precipitation more probable.

One of the more notable potential effects of the heat island phenomenon is the virtual disappearance of the fogs that had been a common part of the environment of London, England.

4.4 EFFECTS ON PRECIPITATION AND PRECIPITATION PROCESSES

Pollutants emitted in urban areas may increase precipitation and affect other related meteorological phenomena. In some cases, the opposite effect has been observed; i.e., pollution sources have decreased downwind precipitation.

Atmospheric particles can serve as sites for condensation of H_2O vapor. Condensation around nucleating particles is an important factor in cloud development and precipitation. Only a small fraction of the total aerosol load of the atmosphere serves as weather-active or cloud condensation nuclei (CCN).

The concentration of CCN determines the initial size and number of H_2O droplets, which determine the frequency and amount of precipitation. An inverse relationship exists between CCN concentrations and initial droplet size; i.e., as CCN concentration increases, droplet size decreases. Water droplets increase in size by coalescence or agglomeration. Since larger droplets coalesce more effectively, they have a higher probability of falling as rain. The more numerous, and smaller, droplets produce less rainfall. This phenomenon may explain the decreased rainfall reported downwind of sources such as sugarcane fires in Australia and forest fires in the northwestern U.S. These sources apparently contribute large numbers of CCN, which decrease droplet size.

Where pollution sources contribute significant numbers of giant condensation nuclei, small CCN may be less important in affecting precipitation. The increase in rainfall observed downwind of some cities may be due to the contribution of giant nuclei (>1 μm) that promote coalescence and increase precipitation probability.

Pollutants may also affect precipitation in cold cloud processes by influencing the concentration of freezing nuclei. Since atmospheric levels of freezing nuclei are frequently low, the addition of freezing nuclei in the form of pollution particles may enhance precipitation under cold cloud conditions. An application of this principle is cloud seeding with silver iodide crystals.

In previous paragraphs it was noted that increased precipitation has been observed downwind of some U.S. cities. Studies downwind of St. Louis, MO, indicate that summer rainfall in the 1970s increased 6 to 15% as a consequence of urban effects. Precipitation increases appear to be associated with elevated particulate pollutant levels.

4.5 ATMOSPHERIC DEPOSITION

Pollutants affect not only the frequency and quantity of precipitation, but also its chemical composition. Historically, the primary focus of environmental and regulatory concern related to changes in precipitation chemistry has been acidification. The effects of acidic deposition have been extensively studied. As a result of its known adverse environmental effects, a significant regulatory effort to reduce emissions of acid precursors has been implemented under provisions of the Clean Air Act Amendments of 1990 (see Chapter 8). Though some improvements in precipitation acidity are already evident in North America, acidic deposition remains a significant environmental problem in northern Europe and is an increasing problem in Asia.

Environmental concerns associated with the effects of acidic deposition have, to some degree, overshadowed other major atmospheric deposition phenomena. These include the deposition of nitrate N, mercury (Hg), and long-lived organic compounds such as pesticides, polychlorinated biphenyls (PCBs), and dioxins. These emerging atmospheric deposition concerns are treated here, with potential effects on ecosystems discussed in Chapter 6.

4.5.1 Acidic Deposition

4.5.1.1 Characteristics of the Phenomenon

Changes in the hydrogen ion (H^+) concentration or pH of precipitation associated with atmospheric pollution have been reported for over a century. However, significant scientific attention to the phenomenon only began in Scandinavian countries in the mid-1960s and in North America in the early to late 1970s.

When pollutant-caused acidification of precipitation was first described, it was called acid rain. As additional studies were conducted, it became apparent that snow was also being affected. Thus, it became more appropriate to describe the phenomenon as acid precipitation. Since acidic materials are deposited on land and water surfaces in the absence of precipitation (dry deposition), the concept was further broadened. It also includes acid fog (with a pH as low as 1.8), acid clouds, acid dew, and probably acid frost as well. To take all of these into account, the term *acidic deposition* is used in the scientific community to describe the phenomenon. The term *acid rain*, however, is commonly used by political leaders and the general public. Thus, in public discourse, the terms *acid rain* and *acidic deposition* are synonymous.

When H_2O condenses after evaporation or distillation, it has a pH of 7 and is neither acidic nor alkaline. As it comes into equilibrium with carbon dioxide (CO_2) in the atmosphere, it rapidly changes to a pH of 5.65. Cloud H_2O, rainwater, and snow, in the absence of pollution and in equilibrium with CO_2, would naturally be acidic (pH = 5.65).

The term *pH* refers to the negative log of the hydrogen ion (H^+) concentration. A solution with a pH of 6 has a H^+ concentration of 0.000001 M, and one with a pH of 5 has a H^+ concentration of 0.00001 M. Values of pH are geometric; i.e., a pH of 5 is 10 times more acidic than a pH of 6, and a pH of 4 is 100 times more acidic than a pH of 6.

A pH of 5.65 has been used as a reference to assess the significance of pollution-related precipitation acidification. Studies in remote areas of the world where anthropogenic effects on precipitation are likely to be minimal indicate that the average background pH value is closer to 5.0 than 5.65.

Average precipitation deposition values (pH) for the U.S. in 1999 are illustrated in Figure 4.7. Areas with high acidic deposition are clustered around the Ohio Valley, where high-S coal is used to generate electricity. In this area, pH values of rainfall average approximately 4.2 to 4.4, with individual values as low as 3.0. Using a reference value of pH = 5.0, precipitation in the Ohio Valley in 1994 was approximately 6 to 7 times more acidic than it would normally be.

In the northeastern U.S., increased precipitation acidity appears to be due to sulfuric (60%) and nitric (35%) acids. The apparent major sources of precursors (S and N oxides) for these strong acids include fossil fuel-fired power plants, industrial boilers, metal smelters, and automobiles. In the western states where sulfur oxide (SO_x) sources are few, precipitation acidity is predominantly due to nitric acid (HNO_3), with motor vehicles the primary source of acid precursors.

The major source region for acid precursors in the eastern U.S. is the middle Ohio Valley. Approximately 60 to 70% of SO_x emissions have been associated with large electric power plants that burn high-S coal. These plants, with their tall stacks (~200 to 400 m, 660 to 1260 ft), disperse emissions at relatively high altitudes (>300 m, ~1000 ft), providing longer residence times for the conversion of acid precursors to strong acids. Consequently, emission of SO_x from the Midwest and northern Appalachian states has affected acidic deposition over large geographical areas downwind, including trans-boundary transport into Canada. In the 1980s and early 1990s, approximately 50% of the acid-forming pollution burden falling on eastern Canada originated in the U.S. Canadian sources, to a lesser extent, contribute to acidic deposition in the U.S.

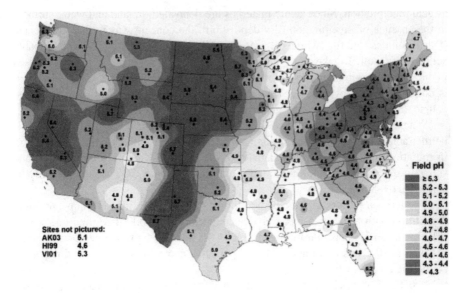

Figure 4.7   Geographical distribution of precipitation H⁺ (expressed as pH) concentration over the continental U.S. in 1999. (From National Atmospheric Deposition Program/National Trends Network, http://nadp.sws.uicu.edu/.)

Long-range transport results in acidic deposition on land and water surfaces at considerable distances downwind of sources of acid precursors. However, acidic deposition at a particular receptor site is likely to be more strongly affected by local sources. The relative contribution to acidic deposition of distant, compared to local, sources has been a matter of intense debate. Nevertheless, it is apparent that long-range transport of acid precursors has a profound effect on acidic deposition hundreds of kilometers downwind of precursor sources.

Although HNO_3 is responsible for only about 35% of the H⁺ concentration in acidic precipitation and dry deposition in the eastern U.S., the importance of nitrogen oxides (NO_x), and their contribution to acidic deposition, may be far more significant than this number indicates. In the northeastern U.S., S emissions have not increased much in the past quarter century, although significant increases have occurred in the Southeast. What has changed substantially since 1960 has been emissions of NO_x from automobiles and electrical power generation. It is the absorption of sunlight by NO_2 that initiates photochemical reactions that promote atmospheric production of H_2SO_4 and HNO_3. The problem of acidic deposition is not simply a matter of emissions of acid precursors, but also atmospheric processes that result in their conversion to strong acids.

4.5.1.2 Environmental Impacts

Acidic deposition poses a threat to ecosystems in which, because of local or regional geology (crystalline or metamorphic rock), soils and surface waters cannot adequately neutralize acidified rain, snow, and dry deposited materials. Such soils and waters contain little or no calcium or magnesium carbonates and therefore have low acid-neutralizing capacity (ANC). Acid-sensitive ecosystems in North America are illustrated in Figure 4.8. These include the mountainous regions of New York and New England; the Appalachian Mountain chain, including portions of Pennsylvania, West Virginia, Virginia, Kentucky, and Tennessee; upper Michigan, Wisconsin, and Minnesota; the Pre-Cambrian Shield of Canada, including Ontario, Quebec, and New Brunswick; and various mountainous regions of the North American West.

4.5.1.2.1 Chemical Changes in Aquatic Systems

Acidic deposition has been linked to the chronic acidification of thousands of lakes and episodic acidification of thousands of streams in the Scandinavian peninsula and southeastern Canada. In the U.S., surface H_2O acidification by acidic deposition has been more limited. The National Surface Water Survey (NSWS), conducted in acid-sensitive regions in the 1980s, revealed that the chemical composition of approximately 75% (880) of acidic lakes and 47% (2200) of acidic streams was dominated by sulfate from atmospheric deposition, indicating that acidic deposition was the primary cause of acidification. Most surface waters acidified by acidic deposition were found in six regions: New England, the southwest Adirondacks, small forested watersheds in the mid-Atlantic mountains, the mid-Atlantic coastal plain, the north-central portion of Florida, and the upper peninsula of Michigan and northeastern Wisconsin. Virtually no acidic surface waters were observed in the western U.S. or the southeastern mountains.

Though a lake or stream may have low ANC and therefore be acid sensitive, acidification, whether it is chronic or episodic, will depend on source factors as well

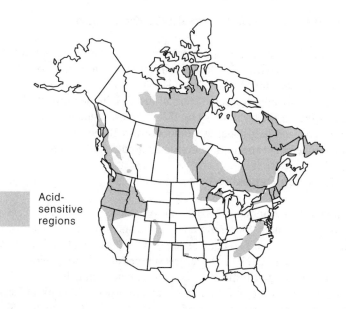

Acid-sensitive regions

Figure 4.8   Acid-sensitive ecosystems in North America. (From USEPA, EPA/600/8-79-028, EPA, Washington, D.C., 1979.)

as processes occurring within a watershed. As rain or snowmelt waters move through a watershed, their chemistry is altered by processes that neutralize acids. The likelihood of neutralization increases as the distance to receiving waters increases.

Surface H_2O chemistry changes associated with acidic deposition may include (1) increases in sulfate concentration, (2) little change in nitrate levels (since watersheds effectively retain nitrate), (3) increases in base cations such as calcium (Ca^{+2}) and magnesium (Mg^{+2}) associated with acid neutralization processes, (4) decreases in pH and ANC, (5) increases in aluminum Al^{+3}, and (6) decreases in dissolved organic carbon (DOC).

Though chronic acidification of lakes has received the most attention, it is important to note that episodic acidification is a common phenomenon in regions that receive high levels of acidic deposition. In lakes and streams where chronic acidification has lowered ANC levels, small decreases in ANC caused by pulses of nitrate, sulfate, or organic anions can lower pH to critical levels. Pulses of nitrate commonly contribute to acidic episodes in portions of New England, the mid-Atlantic mountains, and the Adirondacks. Pulses of sulfate associated with acidic deposition have been shown to be significant contributors to episodic acidification in some streams in the mid-Atlantic mountains.

4.5.1.2.2 Changes in Soil Chemistry

Acidic deposition can affect soil chemistry, particularly in regions with low ANC. Changes include leaching of base cations such as Ca^{+2}, Mg^{+2}, and potassium (K^+). These changes result in reduced base cation exchange capacity of the soil. Acidity also results in the leaching of toxic metals such as Al, lead (Pb), Hg, etc.

4.5.1.3 Trends in Sulfate and Acidic Deposition

From 1990 to 1995, SO_2 emissions from fossil fuel-fired utilities decreased by approximately 50% (from 7.9 to 3.9 million metric tons, 8.7 to 4.3 million tons). As a consequence, the deposition and concentration of sulfates in precipitation decreased dramatically over large areas of the U.S. in the years 1995 to 1998. These decreases were on the order of 10 to 25%. Largest reductions were observed along the Ohio River and its tributaries. Reductions in H^+ concentrations were similar to those of sulfate concentrations in both magnitude and location.

4.5.2 Nitrogen Deposition

Although NO_x, ammonia (NH_3), and organic N compounds occur naturally in the atmosphere, their concentrations significantly increase as a consequence of human activities. Fossil fuel combustion and combustion of fuels by mobile sources are the major sources of NO_x. The largest sources of anthropogenic-related emissions of NH_3 are agricultural fertilizers and domesticated animals.

Nitrogen oxides are removed from the atmosphere by sink processes wherein NO_x is converted to nitrate compounds such as HNO_3 and ammonium nitrate (NH_3NO_3). Ammonia neutralizes strong acids such as HNO_3 and H_2SO_4.

Atmospheric deposition of N compounds is significant. Highest nitrate deposition rates (>14 kg/ha, 77 lb/acre) have been reported in Illinois, Indiana, Ohio, Pennsylvania, Maryland, Delaware, and portions of New York, Ohio, Missouri, and Florida (Figure 4.9). Elevated nitrate deposition is common in most states east of the Rocky Mountains. Maximum NH_3 deposition is relatively low (<4 kg/ha, 22 lb/acre). Highest NH_3 occurs over farming states.

Anthropogenic emissions of NO_x are approximately three times those that occur naturally. These emissions are expected to increase by about 60% in the next 50 years. Presumably, most NO_x is returned to the Earth's surface as nitrates.

Outside agricultural areas, atmospheric deposition is the major source of N in surface soils and H_2O. It is a significant source of nitrate in many watersheds and estuaries. Representative atmospheric N inputs to the Albemarle-Pimlico Sound in North Carolina, Chesapeake Bay, Delaware Bay, Massachusetts bays, Tampa Bay, FL, and the Potomac River are 38, 21, 20, 5 to 27, 28, and 5%, respectively.

Because N is a major plant nutrient, atmospherically deposited N has the potential to cause significant ecological changes in forest and freshwater ecosystems and brackish H_2O estuaries. Potential ecosystem effects are discussed in Chapter 6.

4.5.3 Mercury Deposition

Atmospheric deposition of Hg is increasingly becoming a major environmental and public health concern because of its acute and chronic toxicity and movement through food chains.

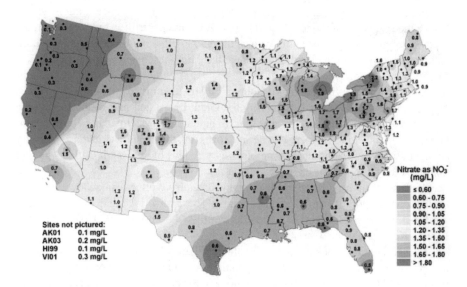

Figure 4.9 Geographical distribution of nitrate (NO_3^-) nitrogen over the continental U.S. in 1999. (From National Atmospheric Deposition Program/National Trends Network, http://nadp.sws.uicu.edu/.)

Mercury is found in small quantities in many rocks and minerals. Though natural sources are responsible for some atmospheric contamination, anthropogenic sources account for 75% of all emissions to the atmosphere. These sources include utility and industrial combustion of coal (44%); municipal, medical, and hazardous waste incineration (33%); manufacturing processes (10%); and a variety of miscellaneous uses (13%).

Mercury's unique chemistry affects its movement between the Earth's surface and the atmosphere. Elemental mercury (Hg^o) is environmentally mobile, traveling significant distances. It is not readily absorbed by tissues and is relatively nontoxic. Divalent mercury (Hg^{+2}) is typically found bound to particles. The atmospheric lifetime of particle-bound Hg varies as a function of particle size. The smaller the particle of which it is a constituent, the longer it remains airborne and the farther it travels.

Mercury emissions from most sources are likely to be in the form of Hg^o, Hg^{+2}, and particle-bound Hg. Ionic mercury dissolves readily in water. As such, it is the most common form found in surface waters. Microorganisms transform Hg^{+2} to methylmercury (CH_3Hg), a much more toxic form. Methylmercury tends to remain dissolved in water. On conversion back to Hg^o, it is readily emitted to the atmosphere. By being transformed from one form to another over and over again (Equation 4.6), Hg can be transported over great distances.

$$Hg^o \leftrightarrow Hg^{+2} \leftrightarrow CH_3Hg \qquad (4.6)$$

Mercury moves in the environment until it is deposited at either high altitudes or high latitudes, where its mobility is markedly decreased.

Weekly wet deposition measurements of total and CH_3Hg were being made at 50 sites in the U.S. during 2000 as part of the Mercury Deposition Network, a component of the National Atmospheric Deposition Program. Concentrations varied from a median value of 9.6 to 120 ng/l (9.6 to 120 parts per trillion by weight/volume (pptw/v)), with average deposition values of ~250 to 3000 ng/m². Highest deposition values were reported for Florida, Louisiana, the Carolinas, and regions around Lakes Michigan and Superior.

Atmospheric Hg is the major source of Hg loading to aquatic ecosystems. Approximately 85% of the Hg entering the Great Lakes and half of the load to the Chesapeake Bay is atmospheric in origin. In aquatic ecosystems, Hg is bioaccumulated; i.e., concentration increases as it moves up through food chains. Mercury concentrations in fish and fish-eating birds may be tens of thousands of times greater than in waters supporting their prey species.

Because of high Hg levels in fish, many states have issued statewide advisories on fish consumption from lakes and streams. These have included 11 midwestern, northeastern, and mid-Atlantic states, as well as the entire Gulf Coast from the Florida Keys to Corpus Christi, TX.

Though Hg emissions per unit of coal burned are very small, such emissions are a concern because of the enormous quantities of coal that are consumed and the tendency of Hg to both accumulate and move in the environment. As a consequence, the USEPA in 2001 began initial steps to develop and promulgate Hg emission reduction requirements from coal-fired power plants under hazardous pollutant provisions of clean air legislation.

4.5.4 Deposition of Pesticides, PCBs, Dioxins, and Furans

A variety of semivolatile organic compounds are found in atmospheric deposition samples at relatively low concentrations. The more important of these are organochlorine pesticides, PCBs, and families of related compounds known as dioxins and furans.

These compounds share important characteristics. They are highly persistent, which means they have long environmental lifetimes (half-lives on the order of 10 to 20 years). Because they are semivolatile, they go through many cycles of volatilization and subsequent condensation. Like Hg, this process facilitates the movement of these substances around the planet.

Organochlorine pesticides, PCBs, dioxins, and furans can move up through food chains. As such, highest concentrations are found in species at the top of food chains, most notably fish and fish-eating birds. A number of states have fish advisories for PCBs.

These organochlorine substances are neurotoxins, potential human carcinogens, and endocrine disruptors. Dioxins are reported to be very potent teratogens; i.e., they can cause congenital birth defects. The most toxic dioxin compound (2,3,7,8-tetrachlorodibenzo-p-dioxin (TCDD)) has an LD_{50} (dose required to kill 50% of animals under test) of 2 ppbw/w.

Though most organochlorine pesticides either have been banned or are subject to use restrictions in the U.S., they continue to be significant contaminants due to their long lifetimes and movement from countries that continue their use. Though PCBs have not been used in the U.S. since the late 1970s, they continue to be a major environmental contamination problem. Dioxins and furans differ from organochlorine pesticides and PCBs in that their production is unintentional. They are produced as by-products in the (1) manufacture of certain pesticides, (2) bleaching of wood pulp in the making of paper, and (3) combustion incineration of wastes that contain chlorinated plastics and paper. As such, medical and municipal incinerators have been major sources of dioxin and furan emissions. Though emission rates are low, such substances accumulate over time.

4.6 STRATOSPHERIC OZONE DEPLETION

4.6.1 Ozone Depletion Concerns

The specter of anthropogenically-induced stratospheric O_3 depletion was first raised by atmospheric scientists in the late 1960s and early to mid-1970s. Initial concern was expressed about the building, and subsequent extensive use, of high-altitude supersonic transport (SST) planes in commercial aviation. Such aircraft would fly in the lower stratosphere (in order to reduce friction), where they would release substantial quantities of nitric oxide (NO) that can catalytically destroy O_3. Scientific and public concern associated with potential O_3 depletion resulted in cancellation of U.S. government support for development of an SST commercial aviation capability in the U.S.

About this time, atmospheric scientists proposed that nuclear weapons testing in the atmosphere could cause stratospheric O_3 depletion. High temperatures associated with nuclear detonations can produce large quantities of O_3-destroying NO that will flow into the stratosphere.

After cancellation of the U.S. SST program, stratospheric O_3 concerns temporarily abated. In 1974, the O_3 depletion specter arose anew when two scientists, Sherwood Rowland and Mario Molina, proposed that chlorofluorocarbons (CFCs) could be broken down in the stratosphere, releasing Cl atoms that catalytically destroy O_3 molecules.

The Rowland–Molina theory of O_3 depletion received considerable scientific and public policy attention and was in a major way responsible for the ban on CFC use as aerosol propellants by the U.S., Canada, and Sweden in 1978. Other developed countries that used CFCs industrially and commercially viewed the Rowland–Molina hypothesis as an unverified theory and thus continued CFC use without limit.

In the late 1970s, Rowland proposed that the O_3 layer may be threatened by the increasing release of nitrous oxide (N_2O) to the atmosphere from denitrification processes of ammonium- and nitrate-based fertilizers. In this new theory, N_2O would be transported to the stratosphere where it would be photolytically destroyed to produce NO, an O_3-destroying chemical associated with previous concerns such as SSTs and atmospheric nuclear testing.

In the period 1985–1988, the potential for anthropogenically caused stratospheric O_3 depletion went from the realm of an unverified theory to the reality that significant O_3 depletion was occurring over the South Pole in the austral (southern hemispheric) spring. The phenomenon called the ozone hole and other reported downward trends in column O_3 on a global scale led many atmospheric scientists to conclude that emissions of CFCs and other halogenated compounds into the troposphere, and their transport into the stratosphere and subsequent photolytic destruction with production of Cl and bromine (Br) atoms, were an environmental problem of enormous gravity. This consensus among scientists and policy makers resulted in global regulatory initiatives to phase out and ban the use of long-lived halocarbons containing Cl and Br.

4.6.2   Ozone Layer Dynamics

The O_3 layer represents an atmospheric environment in which O_3 is continually produced and destroyed. The natural formation and destruction of O_3 in the stratosphere is initiated by high-intensity, short-wave solar ultraviolet (UV) radiation (<242 nm), as shown in the following equations:

$$O_2 + hv \rightarrow O + O \qquad (4.7)$$

$$O + O_2 + M \rightarrow O_3 + M \qquad (4.8)$$

$$O + O_3 \rightarrow 2\,O_2 \qquad (4.9)$$

$$O_3 + hv \rightarrow O_2 + O(^1D) \qquad (4.10)$$

where hv is a photon of UV light and M is an energy-absorbing molecule. This series of reactions is called the Chapman cycle. Ozone formation and destruction in the stratosphere include reactions involving molecular oxygen (O_2), atomic oxygen (O), and O_3. Ozone is initially formed by the UV photolysis of O_2 into two O atoms that react with O_2 to produce O_3 (Equation 4.7 and Equation 4.8). Reactions between O_3 and O result in the breakdown of O_3 and the regeneration of O_2. Ozone absorption of UV light also results in destruction of O_3 and regeneration of O_2 (Equation 4.10). Heat produced by the absorption of UV radiation is responsible for the inverted temperature profile that characterizes the stratosphere (Figure 1.7). Chapman cycle reactions (above) produce more O_3 than is present in the stratosphere. As a consequence, stratospheric O_3 levels must be maintained at observed levels by loss mechanisms. The O_3 loss mechanisms responsible for maintaining stable stratospheric O_3 levels are reactions with NO_x (described below).

Considerable latitudinal and seasonal differences in stratospheric O_3 levels occur (Figure 1.2). Though more O_3 is produced over the tropics than at the poles, total column O_3 levels (because of transport mechanisms) have historically been higher in polar regions than over the tropics. Variability in hemispheric O_3 levels results from what is described as the quasibiennial oscillation, a cycle of varying wind direction in the equatorial stratosphere that affects northward or southward movement of O_3 in different phases of the cycle.

4.6.3 Role of Nitrogen Oxides

Nitrogen oxides (primarily NO) associated with human activities were perceived to be a threat to the O_3 layer because of emissions from SSTs and from nuclear weapons testing. They are primarily responsible for maintaining stratospheric O_3 at equilibrium levels. Nitric oxide can react with O_3 to produce NO_2 and O_2. Nitrogen dioxide can subsequently react with $O(^3P)$ to regenerate NO. Thus, NO catalytically destroys O_3 because NO is repeatedly regenerated:

$$NO + O_3 \rightarrow NO_2 + O_2 \qquad (4.11)$$

$$NO_2 + O(^3P) \rightarrow NO + O_2 \qquad (4.12)$$

$$NO_2 + hv \rightarrow NO + O(^3P) \qquad (4.13)$$

Nitrogen oxides such as NO_2 and NO are emitted into the lower atmosphere in large quantities. Because of a variety of chemical reactions and sink processes, these emissions pose little threat to the O_3 layer. Ozone depletion associated with NO_x would require significant direct emissions to the stratosphere from extensive high-altitude aircraft traffic, nuclear explosions, or destruction of stable N compounds such as N_2O. Nitrous oxide is transported into the stratosphere, where it is oxidized by singlet oxygen, $O(^1D)$, yielding two molecules of NO:

$$O(^1D) + N_2O \rightarrow 2\ NO \qquad (4.14)$$

It can also be formed by reactions initiated by absorption of cosmic rays and solar protons. Nitrous oxide is the major source of NO_x in the stratosphere.

A major sink mechanism for NO is reaction with hydroperoxy radical ($HO_2\cdot$) to produce hydroxyl radical ($OH\cdot$) and NO_2 (which is further oxidized to HNO_3):

$$HO_2\cdot + NO \rightarrow OH\cdot + NO_2 \tag{4.15}$$

$$NO_2 + OH\cdot + M \rightarrow HNO_3 + M \tag{4.16}$$

Nitric acid serves as a means by which NO_x is removed from the stratosphere and also serves as a reservoir for subsequent NO_x regeneration.

The reaction of NO_x with chlorine monoxide (ClO) scavenges Cl in the stratosphere. The resultant product can serve as a reservoir for Cl atoms (see following section).

4.6.4   Role of Chlorofluorocarbons and Other Halogenated Hydrocarbons

Chlorofluorocarbons were first identified as a potential threat to the O_3 layer in the early 1970s when it was observed that free Cl catalytically destroys O_3 and that CFCs (which have no tropospheric sink) can be photolytically destroyed to yield free Cl and fluorine (F) by UV light in the spectral region of <230 nm.

The wavelength distribution of solar radiation and UV light shifts to shorter wavelengths with altitude. As a consequence, CFC destruction occurs primarily in the upper stratosphere where high irradiance of CFCs by UV wavelengths of <230 nm occurs. Because these UV wavelengths do not pass into the lower stratosphere, CFCs must be transported to the upper stratosphere by molecular diffusion. Since Cl production from CFC destruction occurs above the altitude where most O_3 and O_2 is present, Cl atoms must diffuse downward into the middle of the O_3 layer where they participate in catalytic chain reactions that destroy O_3:

$$Cl + O_3 \rightarrow ClO + O_2 \tag{4.17}$$

$$ClO + O(^1D) \rightarrow Cl + O_2 \tag{4.18}$$

When O_3 reacts with Cl, it is converted to O_2 and ClO. Chlorine monoxide reacts with $O(^1D)$ to regenerate Cl. Continual regeneration of Cl by this and other reaction pathways can, in theory, result in destruction of 100,000 O_3 molecules before Cl is removed from the stratosphere as hydrochloric acid (HCl) or chlorine nitrate ($ClONO_2$).

The reaction cycle in Equation 4.17 and Equation 4.18 is important primarily in the middle and upper stratospheres where $O(^1D)$ levels are high. At 15 km (9.3 mi) it accounts for only 5% of halogen-associated loss; at 21 km (13 mi) it accounts for approximately 25%.

Most O_3 depletion in the lower stratosphere associated with halogens occurs as a result of the following series of chemical reactions:

$$Cl + O_3 \rightarrow ClO + O_2 \tag{4.19}$$

$$ClO + HO_2\cdot \rightarrow HOCl + O_2 \tag{4.20}$$

$$HOCl + h\nu \rightarrow Cl + OH\cdot \tag{4.21}$$

$$\frac{OH\cdot + O_3 \rightarrow HO_2\cdot + O_2}{\text{Net: } 2\,O_3 \rightarrow 3\,O_2} \tag{4.22}$$

In this cycle, ClO reacts with $HO_2\cdot$, with O_3 destruction also resulting from reactions with OH·. It accounts for about 30% of the halogen-associated loss in the lower stratosphere.

A number of other halogens have the potential to destroy stratospheric O_3. Major O_3-destroying chemicals include carbon tetrachloride (CCl_4), methyl chloroform (CH_3CCl_3), methyl chloride (CH_3Cl), methyl bromide (CH_3Br), and halon fire extinguishing agents such as $CBrClF_2$ and $CBrF_3$.

Brominated compounds are photolytically destroyed in the stratosphere yielding Br atoms that react catalytically with O_3 in a manner similar to that in Equation 4.19 and Equation 4.20:

$$Br + O_3 \rightarrow BrO + O_2 \tag{4.23}$$

$$\frac{BrO + O(^1D) \rightarrow Br + O_2}{\text{Net: } O_3 + O \rightarrow 2\,O_2} \tag{4.24}$$

This Br cycle results in an ~20 to 30% halogen-related O_3 loss in the lower stratosphere.

The ClO and BrO cycles are interconnected by the following reactions:

$$ClO + BrO \rightarrow Br + ClOO \tag{4.25}$$

$$ClO + BrO \rightarrow BrCl + O_2 \tag{4.26}$$

The ClO–BrO cycles appear to be responsible for ~20 to 25% of O_3 loss due to halogen chemistry at altitudes of 16 to 20 km (9.9 to 12.4 mi) in the middle latitudes and are significant in the lower polar stratosphere.

Sink processes for Cl and Br involve reactions that yield HCl, bromic acid (HBr), $ClONO_2$, and bromine nitrate ($BrONO_2$). Hydrochloric acid forms as a result of the reaction of Cl with methane (CH_4) transported from the troposphere:

$$Cl + CH_4 \rightarrow HCl + CH_3\cdot \tag{4.27}$$

Bromic acid is formed on reaction with $HO_2\cdot$:

$$Br + HO_2\cdot \rightarrow HBr + O_2 \tag{4.28}$$

ClO forms $CLONO_2$ by reaction with NO_2:

$$ClO + NO_2 + M \rightarrow ClONO_2 + M \tag{4.29}$$

BrO forms $BrONO_2$ in the same fashion:

$$BrO + NO_2 + M \rightarrow BrONO_2 + M \tag{4.30}$$

Both HCl and $ClONO_2$ are temporary reservoirs for Cl, since it can be easily regenerated via the following reactions:

$$HCl + OH\cdot \rightarrow Cl + H_2O \tag{4.31}$$

$$ClONO_2 + hv \rightarrow Cl + NO_3 \tag{4.32}$$

Bromine atoms are similarly regenerated from HBr and $BrONO_2$.

On an atom-to-atom basis, Br is more efficient than Cl in destroying stratospheric O_3. Unlike Cl, it reacts relatively slowly with organic compounds such as CH_4. Consequently, Br sink processes are slower. In addition, photolysis of brominated compounds to produce Br is much faster than the corresponding photolysis of chlorinated compounds.

4.6.5   The Antarctic Ozone Hole

In 1985, British scientists conducting ground-based measurements of total column O_3 above Halley Bay (76°S latitude) observed unusually large declines in O_3 levels associated with the beginning of the southern hemispheric spring. Since such declines had not been previously reported (measurements were also being made by satellites), observed O_3 declines were thought to be anomalous. After careful reevaluation of the accuracy and precision of their instruments, British scientists concluded observed O_3 declines were real. Examination of raw NASA satellite data (which excluded extreme results in data summaries) revealed that the phenomenon began to occur in the mid-1970s. Satellite data also indicated that the area affected by the Antarctic O_3 hole extended over several million square kilometers (~1.5 million mi^2). Since its discovery, the geographical area affected by the Antarctic O_3 hole has increased from a few million square kilometers in 1979 to >25 million km^2 (15 million mi^2) in 2000, with lowest column values of <90 dobson units and O_3 depletions of nearly 100% (~0 dobson units) at altitudes of 15 to 20 km (~9-12 mi). Its persistence has increased from weeks to months.

The Antarctic O_3 hole begins to form as the sun comes up over the Antarctic in the beginning of the austral spring. Changes in O_3 levels reported as partial pressure

Figure 4.10 Changes in altitudinal O_3 concentrations over the South Pole in winter–early spring in 1997. (From USEPA, http://www.epa.gov/ozone/science/hole/size.html.)

from August to October can be seen in Figure 4.10. Recent observations indicate that the O_3 hole has persisted into early December. Sulfate aerosol from the Mt. Pinatubo volcanic eruption in the Philippines in 1991 is believed to have contributed to its expansion and to historically low values in 1992 and 1993.

The Antarctic O_3 hole surprised atmospheric scientists. Their O_3 depletion models were based on gas-phase chemistry and focused on the stratosphere over and near source regions where temperatures were, in theory, more favorable for O_3-destroying chemistry. Several theories were initially proposed to explain its formation. Some scientists suggested that the unique polar circulation patterns (i.e., the polar vortex) were responsible, whereas others hypothesized that it was being produced by chemical reactions involving anthropogenic pollutants.

Numerous studies since 1986 have demonstrated the presence of elevated concentrations of Cl and ClO in the region of the O_3 hole, providing strong evidence that it is caused by chemical reactions involving Cl in conjunction with unique meteorological conditions in the Antarctic. Ozone destruction occurs as a consequence of complex heterogeneous chemical reactions involving polar stratospheric clouds (PSCs), HCl, $ClONO_2$, molecular chlorine (Cl_2), and hypochlorous acid (HOCl). Both HCl and $ClONO_2$ are produced in sink processes that remove Cl from O_3-destroying processes.

Polar stratospheric clouds form in the very cold darkness of the long Antarctic night. The PSC particles, which consist of frozen solutions of HNO_3 and H_2O (Type 1 PSC) or frozen water (Type 2 PSC), absorb sink substances such as $ClONO_2$, HCl, HOCl, and nitrogen pentoxide (N_2O_5), which are then activated to release Cl_2 and HOCl. These heterogeneous-phase processes and subsequent chemical reactions are illustrated in Figure 4.11 and the following equations:

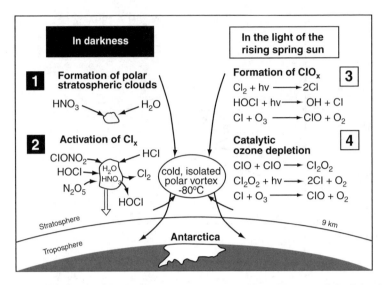

Figure 4.11 Ozone depletion processes over Antarctica. (From Brasseur, G.P., Orlando, J.J., and Tyndall, G.S., *Atmospheric Chemistry and Global Change*, Oxford University Press, Oxford, 1999. With permission.)

$$ClONO_2 + HCl_{(s)} \rightarrow HNO_{3(s)} + Cl_2 \qquad (4.33)$$

$$ClONO_2 + H_2O_{(s)} \rightarrow HNO_3 + HOCl \qquad (4.34)$$

$$HOCl + HCl_{(s)} \rightarrow Cl_2 + H_2O_{(s)} \qquad (4.35)$$

In addition, NO_x (in the form of N_2O_5) is removed from the system by production of HNO_3.

$$N_2O_5 + H_2O_{(s)} \rightarrow 2\ HNO_{3(s)} \qquad (4.36)$$

$$N_2O_5 + HCl_{(s)} \rightarrow ClNO_2 + HNO_{3(s)} \qquad (4.37)$$

The subscript (s) indicates that the substance is in the solid phase.

When the sun comes up over the Antarctic in the beginning of the austral spring, O_3 destruction occurs in two different pathways. In the first of these, Cl_2 and $HOCl$ absorb sunlight that results in production of Cl atoms.

$$Cl_2 + hv \rightarrow 2\ Cl \qquad (4.38)$$

$$HOCl + hv \rightarrow OH + Cl \qquad (4.39)$$

$$Cl + O_3 \rightarrow ClO + O_2 \qquad (4.40)$$

Much greater O_3 destruction or loss results from reactions involving ClO dimers (ClOClO):

$$ClO + ClO + M \rightarrow ClOClO + M \tag{4.41}$$

$$ClOClO + hv \rightarrow Cl + ClOO \tag{4.42}$$

$$\frac{ClOO + M \rightarrow Cl + O_2 + M}{Net: 2\,(Cl + O_3 \rightarrow ClO + O_2)} \tag{4.43}$$

As indicated, PSCs play a critical role in O_3 depletion chemistry in the Antarctic. Type 1 PSCs form at temperatures 2 to 5 K above the ice frost point. These Type 1 particles are believed to contain large quantities of HNO_3 and H_2O. There is evidence that they may be ternary solutions of HNO_3, H_2O, and H_2SO_4. They range in size from 0.1 to ~5 μm in diameter. Type 2 PSCs are believed to consist mostly of H_2O. They are formed at temperatures corresponding to the frost point of H_2O (–188 K) or below. Particle sizes are large, ranging from 5 to 50 μm.

In addition to PSC particles, stratospheric sulfate aerosol (SSA) produced by volcanos such as El Chinchon (1982) and Mt. Pinatubo (1991) plays a significant role in stratospheric O_3 depletion chemistry. Stratospheric sulfate aerosol particles provide surfaces for heterogeneous-phase chemistry and serve as nuclei for PSC formation.

4.6.6   Ozone Depletion in the Arctic

Though O_3 depletion occurs in the Arctic as a result of chemical reactions that are similar to those in the Antarctic, an O_3 hole does not form. This is due to the fact that Arctic stratospheric temperatures are warmer by 10 K and the polar vortex is less stable and less isolated from mixing with external air masses than in the Antarctic. In addition, O_3 concentrations usually increase in December and early spring as a result of long-range transport from the tropics. Any decreases associated with O_3 destruction would be superimposed on normal increases and may be masked.

Nevertheless, significant reductions in stratospheric O_3 were observed in the Arctic in six of the last nine years of the 20th century during late winter and early spring (January to March). Reductions on the order of 20 to 25% were considerably smaller than those observed each spring in the Antarctic. These Arctic O_3 losses have been associated with unusually cold Arctic winters. Significant Arctic O_3 depletion occurs primarily in the lower stratosphere.

4.6.7   Trends

Ozone depletion occurs at high and mid-latitudes. It has not been observed in the latitudinal zone 30°N to 30°S. Though the Antarctic O_3 hole continues unabated, the extent of depletion has remained relatively unchanged since the early 1990s.

The rate of O_3 depletion over the mid-latitudes has slowed from an estimated linear downward trend of 4.0, 1.8, and 3.8% per decade for northern mid-latitudes in winter–spring, northern mid-latitudes in summer–fall, and year-round in southern latitudes, respectively, over the period 1979–1991. Since the recovery from the Mt. Pinatubo eruption, total column O_3 levels have been nearly constant in mid-latitudes, as reported in the late 1990s.

Total levels of O_3-depleting compounds in the troposphere have been declining since 1994. Though total Cl was declining, total Br was still increasing as of 1998. The peak total Cl abundance was 3.7 ppbv between 1992 and 1994. The combined abundance of stratospheric Cl and Br peaked at the beginning of the millennium.

Ozone depletion has resulted in cooling of the lower stratosphere since 1980. The trend was 0.6°C (~1°F) per decade during the period 1979–1994. Model simulations indicate that this cooling trend was strongly correlated with O_3 depletion at these altitudes.

4.6.8   Changes in Surface UV Light Radiation

UV light in the UV-A and UV-B spectral regions received at the surface of the Earth varies geographically. As would be expected, higher irradiance of the ground occurs in the tropics, with progressive decreases toward the poles. The amount of UV light received at a particular location depends on the angle of incidence of the sun, column O_3 levels, cloud cover, and atmospheric pollution.

Increases in UV radiation associated with the Antarctic O_3 hole have been well documented. In the late southern hemisphere spring, UV-B levels over the Antarctic exceed those that occur in San Diego, CA, where the sun's angle of incidence is considerably less.

In regions of the planet where smaller decreases in stratospheric O_3 depletion have occurred, it has been more difficult to measure increases in surface UV-B irradiance. Surprisingly, between 1974 and 1985, a decrease in UV-B irradiance was observed in eight geographical areas in the U.S. This anomaly has only recently been determined to have been due to instrument error.

A subsequent study conducted near Toronto, Canada, demonstrated that UV-B irradiance increased significantly in the period 1989–1993. The intensity of UV at 300 nm increased by 35% per year in winter and 7% per year in summer.

Results of the Canadian study have been confirmed and firmly established by numerous ground-based measurements. In other mid-latitude studies (~ 40°N), increases in UV-B of 1.5% per year at 300 nm and 0.8% per year at 305 nm have been reported for the period 1989–1997.

Ground-based measurements of UV irradiation increases were consistent with estimates from satellite data and radiative forcing models that include cloud, surface reflectivity, and aerosol effects. Increases in UV-B irradiance per decade from 1979 to 1992 were estimated to be 3.7 ± 3% at 60°N and 3 ± 2.8% at 40°N latitude, and 9 ± 6% at 60°S and 3.6 ± 2% at 40°S latitude.

The relationship between decreased column O_3 levels and increases in UV-B irradiance for clear-sky measurements can be seen from the model line (Figure 4.12) fitted to data from Canada, Germany, Greece, Hawaii, New Zealand, and the South

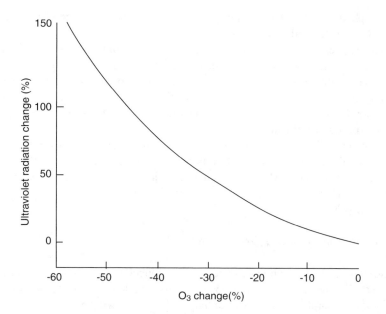

Figure 4.12 Relationship between change in stratospheric O_3 concentrations and change in surface-received UV light. (From World Meteorological Organization, *Scientific Assessment of Ozone Depletion*, WHO Global Ozone Research Monitoring Project, Report 44, WMO, Geneva, Switzerland, 1998.)

Pole. The portion of the model line in the left third of the graph was based on data collected at the South Pole.

4.6.9 Ozone Depletion Effects on Human Health

The O_3 layer absorbs 99% of UV wavelengths less than 320 nm. Ozone absorption in the UV-B region (290 to 320 nm) is, however, less efficient than at shorter wavelengths, and relatively significant fluxes of both UV-B and UV-A (320 to 400 nm) reach the Earth's surface. UV-B, on absorption by organisms, can cause significant adverse biological effects. As a consequence, it is often described as damaging UV radiation (DUV).

Significant short- and long-term environmental exposures to DUV may cause sunburn, cataracts, or skin cancer. There are three basic forms of skin cancer: basal cell carcinoma, squamous cell carcinoma, and melanoma. Carcinoma-type skin cancers have clearly been demonstrated to be associated with exposure to sunlight, with the greatest prevalence observed among light-complected individuals living in so-called sunbelt latitudes in North America. Skin carcinomas, fortunately, are the least life threatening of all forms of cancer, with a mortality rate of <5%. Because tumors must be surgically removed, skin cancers are often disfiguring.

Melanoma tumors develop from the darkly pigmented melanin structures or moles that many Caucasians develop on their skin, particularly during midlife. Melanoma is a much more deadly form of skin cancer because the tumors metastasize rapidly, spreading malignant cells to other parts of the body through the bloodstream and lymphatic system. The mortality rate is ~30%. The incidence of melanoma in the U.S. has increased eightfold since 1960.

No dose–response relationship between melanoma and exposure to sunlight and UV radiation has been demonstrated. Rather, melanoma appears to be associated with severe sunburn experienced early in life. Melanoma has also been linked to building environments with fluorescent lighting. Paradoxically, melanoma tumors apparently associated with UV radiation exposures from fluorescent lighting occur on nonexposed parts of the body.

Increases in UV-B radiation fluxes at the Earth's surface as a result of stratospheric O_3 depletion by even a few percent are expected to cause a significant increase in all forms of skin cancer among Caucasians and other light-pigmented individuals. This would mean tens of thousands of new cases of skin cancer in the U.S. each year.

4.6.10 Recovery of the Ozone Layer

Stratospheric O_3 depletion caused by anthropogenic emissions of O_3-destroying chemicals is expected to gradually abate over the next half century as they are slowly removed by natural processes. This anticipated recovery will be due in great measure to the reduction in emissions of Cl- and Br-containing gases that has resulted from a series of international agreements that began with the Montreal Protocol in 1987. In the early austral spring of 2002, the Antarctic O_3 hole was smaller (15 vs. 25 million km^2) than any other year in over a decade. Whether this was a beginning of a trend or due to normal meteorological variations (the polar vortex was also much smaller) was not known at the time of this writing.

The stratospheric O_3 layer is unlikely to return to what it was prior to 1980. It will be affected by increasing concentrations of CH_4, N_2O, sulfate particles, and H_2O vapor. It may also be affected by changes in the Earth's climate.

4.7 GLOBAL WARMING

4.7.1 Climate and Global Climate

The atmosphere, driven by the energy of the sun, is in continual motion and change. This change or flux viewed in the context of hours, days, or weeks is called weather. Weather includes precipitation events; fluctuations in temperature, humidity, and wind speed; the degree of cloud cover; and incoming solar radiation. These phenomena, viewed within the context of geographical areas and extended periods of time (years, decades, centuries, millennia), describe climate. Climate is characterized by average temperatures, temperature extremes, precipitation, the quantity and periodicity of solar radiation, and other atmospheric phenomena. Though climate is often described as being average weather, it is, in fact, a much broader concept.

Weather, because it is based on short-term events, can be quite variable. Climate is less variable because it is based on averaged values. Climate is subject to change, but the time frame of such change may be relatively long. Evidence of small to large climatic changes may be found in historical accounts, various aspects of the geological record, and analytical results of measurements made on deep-sea sediments, polar ice, and tree rings.

Climate is usually viewed in the context of different geographical regions. It can be viewed in the global context as well. Global climate is described by the average global surface temperature. Primary determinants are the (1) absolute amount of energy received from the sun, (2) amount of solar energy reflected back to space (albedo), and (3) absorption of incoming solar radiation and outgoing thermal radiation by atmospheric constituents.

Global surface temperatures, and thereby climate, have changed significantly over time. Studies of oxygen isotope ratios in carbonate ocean sediments indicate that the Earth's temperature decreased by 15°C (59°F) in the past 65 million years. This decrease has been attributed to changes in positions of continents and oceans (continental drift), formation of high mountains, intensification of volcanism, and a decrease in atmospheric CO_2 levels. Changes in global climate may also occur by a variety of other physical mechanisms, e.g., (1) quasi-periodic ocean temperature changes, (2) increased sea salt formation associated with decreased Arctic Ocean salinity, (3) variation in solar radiation, (4) changes in surface albedo, and (5) changes in the Earth's orbital geometry.

The effects of orbital variation and variations in solar irradiance are discussed below to illustrate effects of natural phenomena on the Earth's climate system.

4.7.1.1 Orbital Changes and Climatic Variation

The Earth's orbit relative to the sun varies in three distinctively different ways. First, the Earth's axis is tilted 23.5° from the perpendicular to the plane described by the Earth's orbit around the sun. In summer, the northern hemisphere tilts toward the sun; in winter, it tilts away from the sun.

The obliquity or tilting of the Earth's axis varies from 22.1 to 24.5°, with a complete cycle every 41,000 years. The obliquity of the Earth's axis is responsible for seasonal climatic changes over the Earth's surface. As it changes from 22.1 to 24.5°, contrasts between seasons are likely to become greater.

Second, the Earth's axis changes direction over time. This precession or wobbling of the Earth describes a circle among the stars, with the center at due north. At present, the axis points toward Polaris, the North Star. In about 12,000 years, it will point at Vega, which will become the North Star. When the axis is tilted toward Vega, the orbital positions at which summer and winter solstices occur will be reversed. The northern hemisphere will experience summer near perihelion (when the Earth is closest to the sun) and winter near aphelion (when the Earth is farthest from the sun). Winters can be expected to be colder and summers warmer than at present. The precession cycle is about 20,000 years.

Lastly, the nature of the Earth's orbit changes from a somewhat circular to a more oblong ellipse over a period of 100,000 years. This orbital change is called eccentricity. Because the orbit is not exactly symmetrical, the Earth at present is farthest away from the sun (aphelion) on July 4 and closest to the sun on January 3 (perihelion). This difference in distance is about 3%. As such, the Earth receives

6% more solar energy in January than in July. When the orbit is more elliptical, the amount of solar energy received at perihelion would be on the order of 20 to 30% more than at aphelion. Orbital variations, obliquity (a), precession (b), and eccentricity (c), are illustrated in Figure 4.13. These variations do not increase or decrease the amount of solar energy received by the Earth; they do determine how it is distributed over Earth's surface and change the degree of contrast between the seasons.

Orbital variations have been linked to cycles of glaciation experienced by the Earth over the past 2.5 million years. Serbian scientist Milutin Milankovitch proposed such a linkage in the early 20th century. Therefore, the 20,000-, 40,000-, and 100,000-year periodicities associated with precession, obliquity, and eccentricity are called Milankovitch cycles. Evidence of periods of significant global surface cooling, with periodicities of 20,000, 40,000, and 100,000 years, have been reported from O isotope studies of marine carbonate sediments. As a consequence, a number of scientists have concluded that these orbital changes were the fundamental cause of glaciations in the past 2.5 million years and that they are "pacemakers of the ice ages."

Glacial periods lasting thousands of years were characterized by the development and movement of massive ice sheets (several kilometers thick) over parts of the northern hemisphere. They were followed by climatic warming and glacial retreat. The last large glacier over North America receded approximately 12,000 years ago. The Earth is presently in an interglacial warming period.

4.7.1.2 Variations in Solar Radiation

Average global surface temperatures are determined in part by the amount of solar radiation reaching the Earth's atmosphere and surface. Astronomers have theorized that at least for long time periods, solar output, and hence the quantity of solar radiation received by the Earth, is essentially constant. The solar constant (1370 $W/m^2/sec$) is the total solar radiation received on a flat surface oriented perpendicular to the sun external to the Earth's atmosphere. Its value is 1.94 $cal/cm^2/sec$. Satellite measurements have recently shown that solar radiation does vary and that the solar constant is not constant when viewed in the short term. Short-term changes in solar output are directly related to sunspot cycles. The peak value of solar irradiances occurs near the maximum of solar activity. The peak-to-peak variation in the sun's total irradiance during the sunspot cycle is ~0.08%.

Sunspots are dark, relatively cool regions on the sun's surface with limited heat flow. Interestingly, the sun's irradiance is greater rather than less during peak periods of sunspot activity. Apparently, there is a concurrently active source of solar emission that more than compensates for the decreased output from sunspot regions. Increased solar emissions are associated with faculae and plages, i.e., bright, active regions where emissions of solar radiation are enhanced relative to those for the surrounding quieter regions of the sun. Plages make up an order of magnitude or more of the sun's surface and are longer-lived than sunspots.

Changes in the sun's luminosity associated with sunspots and other sun surface activity have been proposed as a cause of significant climatic change. The Little Ice

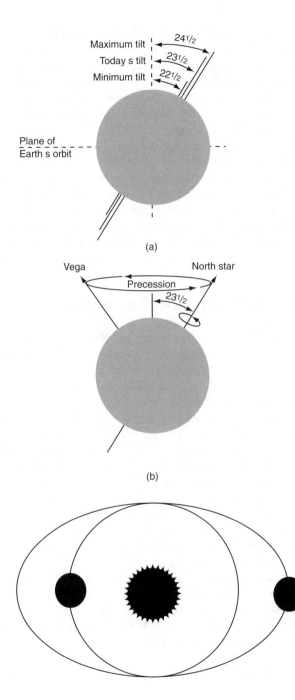

Figure 4.13 Variations in the Earth's orbit.

Age, a period of global cooling observed worldwide during the 14th to 19th centuries, was associated with a significant decline in sunspot activity. The Little Ice Age was accompanied by (1) the advance of European mountain glaciers, (2) lower average winter temperatures in central Britain, (3) abandonment of cereal cultivation in Iceland and vineyards in England, (4) abandonment of settlements in Greenland, and (5) winter freezing of the North Sea and canals in Venice. It is probable that global surface temperatures were lower by 1°C (1.8°F) or more during that period, and that precipitation patterns were also changed.

Changes in the sun's luminosity associated with sunspot cycles have from time to time captured the interest of climatologists in their attempts to explain climatic change. Ice core and ocean temperature analyses show significant correlations between temperature changes and the 11- and 22-year sunspot cycles. These correlations suggest that changes in solar luminosity may, in part, have contributed to the global warming observed in the last century. However, the potential relationship between changes in sunspot activity and climate changes has been subject to considerable controversy.

4.7.2   The Greenhouse Effect and Global Warming

As indicated in Chapter 1, solar energy not absorbed by the atmosphere or reflected by clouds passes through the atmosphere; most of it is then absorbed by the Earth's surface. On absorption, it is converted into longer-wavelength infrared radiation. Though the atmosphere is relatively transparent to visible light, it is considerably less so for infrared energy.

The lower atmosphere is warmed by sorption of this infrared energy and reradiation from atmospheric gases and clouds. The atmosphere, acting much like the panes of a greenhouse, allows short-wave visible light to pass through while retarding the flow of longer-wave infrared energy back to space. This is called the greenhouse effect (Figure 4.14) and is responsible for the relatively moderate temperatures that occur near the Earth's surface (~15°C, 59°F).

The greenhouse effect is primarily due to the sorption and reradiation of thermal energy by CO_2 and H_2O vapor. Other trace gases also have a greenhouse role. These include CH_4, N_2O, tropospheric O_3, and man-made substances such as CFCs, perfluorocarbons, and sulfur hexafluoride.

The ability of the atmosphere to retard thermal flows back to space is described as emissivity. Atmospheric emissivity varies across the infrared spectrum. As can be seen in Figure 1.5, most infrared absorption is due to H_2O vapor and CO_2. These gases, as well as O_3, absorb relatively little infrared energy in the spectral region of 7 to 13 μm, the so-called atmospheric window.

4.7.3   Radiative Forcing

Though the atmosphere retains thermal energy as a result of the greenhouse effect, incoming radiation is balanced by outgoing terrestrial radiation. Factors that alter electromagnetic radiation received from the sun and lost to space, or alter its redistribution within the atmosphere and among the atmosphere, land, and oceans,

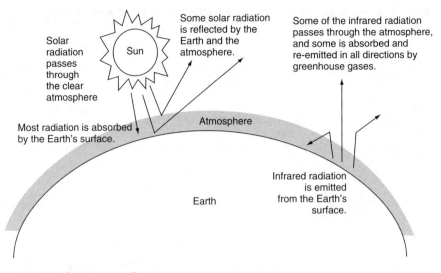

Figure 4.14 Greenhouse effect.

have the potential to affect global and regional climates. As defined by the Intergovernmental Panel on Climate Change (IPCC), any change in the net radiative energy available to the global Earth atmosphere is termed *radiative forcing*. Positive forcings warm the Earth; negative forcings cool it.

An increase in greenhouse gases reduces the Earth's ability to radiate heat back to space. The atmosphere absorbs more outgoing thermal energy and reemits it at higher altitudes and lower temperatures. Consequently, this positive radiative forcing warms the lower atmosphere and surface. Thus, the greenhouse effect is enhanced. The degree to which radiative forcing occurs depends on the magnitude of increases in greenhouse gases, and the concentrations and radiative properties of individual gases.

Some aerosols, either naturally or anthropogenically produced, reflect sunlight. As such, they can cause negative forcing and climatic cooling. Black carbon aerosol, on the other hand, absorbs solar radiation and warms the climate system. Changes in aerosol concentrations can affect the amount of clouds and their reflectivity. In most cases, tropospheric aerosols contribute to negative radiative forcing. However, aerosols have much shorter atmospheric lifetimes than greenhouse gases.

Climate forcing is any change imposed on the Earth's energy balance. It can be measured and expressed in watts per square meter. Climate responses are the changes that result from changes in climate forcing.

4.7.4 Greenhouse Gases

Greenhouse gas concentrations affected by human activities (for 2000 and the preindustrial period prior to 1800) are summarized in Table 4.2. Also presented are annual percentage increases (based on year 2000), radiative effectiveness (ability to absorb thermal energy) of greenhouse gases relative to CO_2, and estimated contributions to global warming (radiative forcing).

Table 4.2 Greenhouse Gases, Radiative Effectiveness, and Radiative Forcing

	CO_2	CH_4	Halocarbons (as CFCs)	N_2O	O_3
Concentration					
Preindustrial	~280 ppmv	~0.665 ppmv		~270 ppbv	13 ppbv
2000	368 ppmv	1.76 ppmv	~800 pptv	316 ppbv	20 ppbv
Annual increase					
Percent/year, 2000	0.4	~0.5	~0	0.25	?
Concentration/year	1.5 ppmv	7 ppbv	—	0.8 ppbv	—
Radiative effectiveness	1	21	13,000–16,000	206	2000
Relative contribution to global forcing (%)	56	18	13	6	8

Source: Adapted from Elsom, D.M., *Atmospheric Pollution: A Global Problem*, 2nd ed., Blackwell Publishers, Oxford, 1992. With permission.

Carbon dioxide concentrations have increased by more than 31% since 1750. The rate of increase has averaged 1.5 ppmv over the past two decades (0.4%/year). Much of this increase is due to combustion of fossil fuels (70 to 90%). Land use changes (where high-C-density ecosystems such as forests are replaced by lower-C-density agricultural or grazing land) have also had significant effects on CO_2 emissions to the atmosphere (10 to 30%). Current CO_2 levels are the highest they have been in almost a half million years.

Most CO_2 removed from the atmosphere is absorbed by the upper 70 to 100 m (230 to 330 ft) of the Earth's oceans. Forests appear to be an important sink as well. As a consequence, only ~40 to 50% of CO_2 emitted to the atmosphere remains there. The atmospheric lifetime for CO_2 is estimated to be 50 to 100 years.

Though CH_4 and N_2O are strong infrared absorbers, their role as greenhouse gases compared to CO_2 and H_2O vapor has historically been relatively limited. However, their importance has increased as their atmospheric concentrations have increased. Methane and N_2O absorb strongly in the atmospheric window with relative radiative effectiveness of 21 and 206, respectively; i.e., on a molecule-to-molecule basis, they absorb and retain heat 21 and 206 times, respectively, more effectively than CO_2.

Methane concentrations in the atmosphere have increased 150% from preindustrial times. The average annual increase was on the order of 10 to 15 ppbv/year from 1980 to 1992, but has varied from 0 ppbv in 1993 and 2000 to 15 ppbv in 1998. Major emission sources of CH_4 appear to include anaerobic bacterial activity in swamps and rice paddies, digestive systems of ruminant animals (particularly livestock) and termites, and landfill sites. Other sources include coal mines and leaking natural gas pipelines.

Nitrous oxide is produced by biological processes in soil. It is also emitted to the atmosphere as a result of fossil fuel and biomass burning, soil disturbance, and from animal and human wastes. Use of nitrogen fertilizers may also be a significant source of atmospheric N_2O. Atmospheric N_2O levels have increased by about 46 ppbv, or 16%, since 1750, with a current annual average increase of 0.8 ppbv, or 0.25%. Present-day N_2O concentrations are the highest they have been in 1000 years.

Chlorofluorocarbons are man-made and therefore new to the atmosphere. Though most environmental concern associated with increasing atmospheric CFC levels has focused on stratospheric O_3 depletion, CFCs are strong thermal absorbers (not surprising, given their widespread use as refrigerants). Though concentrations of the two most widely used CFCs (CFC-11 and CFC-12) are in the pptv range, the potential contribution of CFCs to global warming is significant. Chlorofluorocarbons absorb strongly in the atmospheric window. On a molecule-to-molecule basis, they absorb infrared energy 12,000 to 16,000 times more effectively than CO_2. With the phaseout of CFC production and use, CFC levels in the troposphere have been declining.

As seen in Table 4.2, tropospheric O_3 has significant potential to absorb thermal radiation, with a radiative effectiveness of 2000. Tropospheric O_3 levels have increased by ~36% since 1750.

The radiative forcing associated with increases in well-mixed greenhouse gases since 1750 is estimated to be 2.43 W/m^2. Of this, 1.46 W/m^2 has been apportioned to CO_2, 0.48 W/m^2 to CH_4, 0.34 W/m^2 to halocarbons, and 0.15 W/m^2 to N_2O. A positive radiative forcing of 0.35 W/m^2 has been estimated for the tropospheric increase in O_3, and –0.15 W/m^2 for stratospheric O_3 depletion. Including O_3, the increase in radiative forcing would be 2.63 W/m^2. This would correspond to percentage contributions to global climatic warming indicated in Table 4.2. Though CO_2 has the lowest radiative effectiveness, its relatively high emissions and increasing atmospheric levels are apparently responsible for more than 50% of the radiative forcing that has taken place over the last two and a half centuries. As CFC emissions continue to be restricted and CFCs destroyed by stratospheric processes, their contribution to radiative forcing is expected to diminish significantly.

Radiative forcing for aerosols is estimated to range from –0.3 to –1.0 for sulfate and 0.2 to 0.8 for elemental C.

4.7.5 Evidence of Climate Change

The IPCC, in its 2001 assessment report, unequivocally concluded that the global climate had undergone significant changes in the 20th century. These changes included an increase in global surface temperature (adjusted for heat island effects) of 0.45 to 0.6°C (0.8 to 1°F). Most of this increase has occurred in two distinct periods: 1910–1945 and since 1976 (Figure 4.15). The rate of warming for both periods was 0.15°C (0.27°F)/decade. Warming over land surfaces between 1950 and 1993 has been twice that over the oceans.

The year 2000 was the 22nd consecutive year that the average global surface temperature was above the climate mean temperature for the period 1960–1990. The seven warmest years in the past 140 years occurred in the1990s and 2000. The combined land surface–air and sea surface temperature in 2000 was 0.29°C (0.52°F) above the 1961–1990 climatic average.

Other temperature changes have been reported. Between 1950 and 1993, average nighttime minimum air temperatures over land surfaces increased by ~0.2°C (0.38°F)/decade. This is twice the rate of increase in daytime maximum air temperatures (0.1°C (0.18°F)/decade). These changes have increased the freeze-free season in many mid- and high-latitude regions.

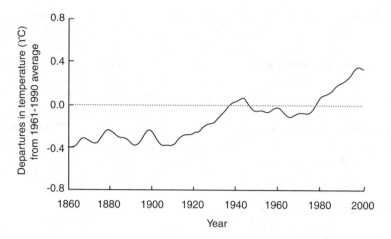

Figure 4.15 Changes in global surface temperatures since 1860 from the 1961 to 1990 average. (From IPCC, Summary for Policy Makers Working Group, http://www.
ipcc.ch/spm22-01.pdf.)

The largest increases in global surface temperatures have occurred over continents in the mid- and high latitudes of the northern hemisphere. These increases
have been greater than at any time in the past 1000 years.

Satellite temperature data (collected since 1979) for the troposphere up to an
altitude of 7.7 km (4.8 mi) show little evidence of the warming trend observed at
the ground. Though balloon measurements show a tropospheric warming trend since
1958 (Figure 4.16), balloon data in the period 1979–2000 resemble satellite data
(with balloon data somewhat higher). This anomaly is cited by opponents of proposed global warming control programs and contrarian scientists who contend that
human-induced global warming is not occurring. Causes of these apparent discrep-
ancies in atmospheric and surface temperature trends are not known. Suggestions
have included short-term temperature variations due to stratospheric O_3 depletion,
El Nino, and the short time period of measurement involved.

Changes in atmospheric temperature are supported by new analyses that the heat
content of oceans has significantly increased since the late 1950s. More than half
of this increase has occurred in the upper 300 m (~990 ft). The equivalent rate of
temperature increase in this layer is about 0.05°C (0.09°F)/decade. The oceans
represent the largest reservoir of heat in the world's climate system.

Changes in other climate indicators have been reported. These include an increase
in annual land precipitation in the northern hemisphere in the mid- and high latitudes.
Atmospheric H_2O vapor has reportedly increased several percent per decade over
many regions of the northern hemisphere over the past 25 years.

Increases in cloud cover on the order of 2% for mid- to high-latitude continental
regions over the northern hemisphere have also been reported for the 20th century.
This change has been positively correlated with decreases in the range of diurnal
temperatures.

Increases in land surface temperatures have been positively correlated with
decreasing snow cover and land ice. There is increasing evidence that the retreat of

Figure 4.16 Surface and tropospheric temperature (satellite data) trends for the period 1979–2000. (From USEPA, Global Warming: Climate, http://Yosemite.epa.gov/OAR/globalwarming.nsf/content/ClimateTrendsTemperature.html.)

alpine and continental glaciers is occurring in response to 20th-century global warming. Additionally, ground-based measurements have shown an approximate 2-week reduction in the annual duration in lake and river ice in the mid- to high latitudes in the northern hemisphere.

Decreases in sea ice have been reported in the Arctic. The extent of sea ice in the Arctic spring and summer has decreased on the order of 10 to 15% since 1950. This decrease is consistent with an increase in springtime and, to a lesser extent, summer temperatures. Recent data indicate an approximate 40% decline in Arctic sea ice thickness in late summer to early autumn between 1958 and 1976 and again in the mid-1990s, and a much smaller decline in winter.

Sea level changes have also occurred. These have been on the order of 10 to 20 cm (8 to 10 in.) above levels in the early 20th century. The annual rate of increase has been 1 to 2 mm (0.004 to 0.008 in.)/year.

The strongest evidence to date to support the occurrence of global warming is reported changes in plants and animals in terms of (1) shifts in species' ranges, and (2) timing of such events as migration, egg laying, and flowering. Recent analyses of hundreds of studies (and more than 1700 species) have revealed consistent temperature-related shifts in species distribution poleward and up in elevation (6.1 km/decade toward the poles) and advancement of spring events by 2.3 days/decade. Such biological changes are further described in Chapter 6.

4.7.6 Linking Climate Change to Anthropogenic Influences

Changes in global climate are becoming well documented. The question before us is "Why is it changing?" Is it changing because of natural oscillations in the

Earth's climatic system, increased solar radiation associated with sunspot activity, increases in greenhouse gases, or some combination of these?

Significant greenhouse gas emissions have occurred and will continue to occur. Since the principle of greenhouse warming of the Earth's atmosphere is well established, it is logical to hypothesize that climate changes that have taken place in the past century are at least in part due to radiative forcing associated with greenhouse gas emissions. Such a hypothesis can only be "proven" by (1) evaluating past changes in the Earth's climate in relationship to geological indicators of past atmospheric temperatures and greenhouse gas concentrations determined from air trapped in Arctic and Antarctic ice, and (2) simulating effects of greenhouse gas increases on global and regional climate by using sophisticated models.

Climate models are based on physical laws that are represented by mathematical equations. Models are designed to predict changes in both global and regional climate in response to perturbations such as increases in one or more greenhouse gases, increased solar radiation, increased albedo, etc.

Climate models have been undergoing development, use, and continued refinement for almost a decade and a half. To simulate climate, major components of the Earth's climate system are represented in submodels — e.g., atmosphere, land surfaces, oceans, cryosphere (frozen ground), and biosphere — as well as the processes that go on within them. Models in which the atmosphere and ocean components have been coupled are known as atmosphere–ocean general circulation models (AOGCMs). Comprehensive AOGCMs are very complex and require supercomputers to run.

Coupled models provide credible simulations of climate down to subcontinental scales and over temporal scales from seasons to decades. They cannot simulate all aspects of climate and do not fully account for surface–air temperature differences observed since 1979. Clouds and humidity are sources of significant uncertainty. Nevertheless, AOGCMs have had a relatively good track record in reproducing warming trends in the 20th century when driven by greenhouse gases and sulfate aerosols.

Various lines of evidence indicate that temperature changes in the past century are not due to natural variability. These include (1) reconstructions of temperatures in the past 1000 years, (2) new model estimates of internal variability and responses to natural forcing, (3) qualitative consistencies between observed climate changes and model responses to anthropogenic forcing, (4) evidence of a human influence using a wide range of detection techniques, and (5) the fact that most attribution (cause-and-effect relationships with defined levels of statistical confidence) studies have found that the estimated rate and magnitude of global warming due to increasing greenhouse gases alone are comparable to or greater than observed global warming over the past 50 years.

4.7.7 Projections of the Earth's Future Climate

Using estimates of greenhouse gas emissions and climate models, the IPCC has projected a suite of future climate changes to the year 2100. These scenarios lead to very different estimates of CO_2 concentrations in 2100 (from 540 to 770 ppmv). Estimated increases in concentrations of other greenhouse gases vary considerably. In most scenarios, radiative forcing due to CO_2, CH_4, N_2O, and tropospheric O_3 continues to increase. However, the fraction of total forcing due to CO_2 is projected to increase from approximately one half to three quarters.

Climate sensitivity (equilibrium response of global surface temperature to an equivalent doubling of the CO_2 concentration) is predicted to be in the range of 1.5 to 4.5°C (2.7 to 8.1°F). This range of potential global surface temperatures reflects uncertainties, particularly those related to clouds and other processes.

Global climate models predict significant changes in other broad-scale climate variables. These include:

1. Greater relative warming at high latitudes due to faster warming of land surfaces compared to oceans
2. Decrease in daily temperature differences, with higher nighttime low temperatures
3. General decrease in daily surface temperature variability in winter and increases in summer in the northern hemisphere
4. Global increase in H_2O vapor, evaporation, and precipitation, with precipitation increases over mid- to high latitudes and Antarctica and changes in the variability of Asian monsoon precipitation
5. Decrease in snow and ice cover over the northern hemisphere, recession of glaciers and ice caps, decreased Greenland ice sheet mass, and increase in the mass of Antarctic ice sheets due to increased precipitation
6. Rise in sea level due to thermal expansion and melting of glaciers and polar ice caps, with projected increases of 0.09 to 0.88 m (0.30 to 2.8 ft)
7. Increase in extreme weather and climate events, e.g., higher maximum temperatures, more hot days, more intense precipitation, increased risk of drought, and increases in tropical cyclones and peak and mean precipitation

4.7.8   Global Warming Uncertainties

The recently confirmed role of sulfate aerosol associated with regional haze over developed areas of the world in producing a cooling effect points out how difficult it is to predict what the effects of anthropogenic activities may be on something as complex as the Earth's climatic system. As a consequence, a fair amount of uncertainty exists in predicting climatic changes on global and regional scales. There are, nevertheless, a number of remaining uncertainties. These include (1) anomolous tropospheric vertical temperature profiles in balloon and satellite observations, (2) estimates of internal climate variability from models and observations, (3) reconstruction of solar and volcanic forcing, (4) effects of aerosols and cloud cover, and (5) differences in response of different models to the same radiative forcing potential.

One major uncertainty is cloud behavior. Global warming is expected to increase cloud formation as well as change cloud distribution and characteristics. Differences in the nature of cloud cover can affect the predicted increase in global warming. By increasing cloud cover, particularly low-lying clouds over the oceans, more sunlight would be reflected back to space. An increase of the Earth's albedo of 0.5% would be expected to halve the greenhouse warming associated with a doubling of CO_2. If cirrus clouds were to increase in abundance, they would absorb heat, thus warming the atmosphere.

Other uncertainties are associated with the various feedback mechanisms among land, oceans, atmosphere, and the Earth's climatic system. Global warming would, for example, increase evaporation, resulting in increased H_2O vapor concentrations. This is expected to enhance the greenhouse effect. Global warming is expected to melt sea ice and snow cover, allowing underlying surfaces to absorb more solar radiation and thus decreasing the Earth's albedo and increasing evaporation. By increasing the temperature of the world's oceans, the solubility of CO_2 would be decreased. As upper layers of the world's oceans become saturated with CO_2, less CO_2 would be removed from the atmosphere. In addition, as global temperatures rise, there is a potential for large amounts of CH_4 to be released from frozen tundra in the northern hemisphere.

A variety of both positive and negative feedback mechanisms exist that can either enhance or mitigate global warming. Most proposed feedback mechanisms would enhance warming.

4.7.9 Environmental Impacts

A variety of systems may be sensitive to climate change. These include H_2O resources, agriculture and forestry, coastal zones and marine systems, human settlements, and biological communities.

Changes in rainfall, stream flow, and groundwater recharge are predicted for many regions as the planet warms. Consequently, human populations living in H_2O-stressed countries would increase from 1.7 to 5 billion by 2025 (depending on the rate of population growth). Flood frequency and magnitude would also increase in many regions as a result of the greater frequency of heavy rainfall events.

Crop yield reductions are predicted to occur as a result of increased drought conditions and expansion of the world's arid and semiarid zones. These reductions in crop yields could be counterbalanced in part by CO_2 stimulation of plant growth.

Coastal zones are predicted to be adversely affected as a result of sea level rises as well as the increase in storm frequency and severity. As a consequence, many coastal areas are expected to experience increased flooding, accelerated erosion, loss of wetlands, and saltwater intrusion into freshwater sources.

Human settlements may be adversely affected, particularly in low-lying coastal areas. Of particular concern is flooding from a rise in sea level, coastal storm surges, and runoff from more intense storms. Such changes pose risks to homes, businesses, infrastructure, and humans themselves.

Significant changes in biological communities are already occurring, and the number of species affected is expected to increase.

4.7.10 Stratospheric Gases and Atmospheric Temperature Changes

While the atmosphere has been warming near the Earth's surface, a significant cooling trend (~0.6°C (1.08°F)/decade) has been occurring in the lower stratosphere (~16-21 km (~9.6-12.6 mi.)) since 1980. At middle latitudes it has been even larger (~0.75°C (1.35°F)/decade), with an even more substantial cooling trend (~3°C (5.4°F)/decade) in the polar regions during late winter and spring. This polar trend has been observed in the Antarctic since the early 1980s and in the Arctic since the early 1990s. Model simulations indicate that O_3 depletion is the primary contributing factor. Greenhouse gases (e.g., CO_2) may also be contributing to stratospheric cooling (~25% of that observed).

Methane, on photolytic decomposition, is a major source of stratospheric H_2O. Increased stratospheric H_2O is expected to significantly enhance global warming. A doubling of stratospheric H_2O vapor has the same global warming potential as a 70-ppmv rise in CO_2 in the troposphere.

Since coupling between climatic phenomena in the stratosphere and troposphere is known to occur, observed stratospheric cooling and potential future stratospheric warming are likely to have some effect on global surface temperatures. The nature and magnitude of these effects are not known.

READINGS

Adams, D.D. and Page, W.P., Eds., *Acidic Deposition: Environmental, Economic and Policy Issues,* Plenum Publishers, New York, 1985.

Barrie, L.A., Arctic air pollution: an overview of current knowledge, *Atmos. Environ.,* 20, 643, 1986.

Beaty, C.B., The causes of glaciation, *Am. Sci.,* 66, 452, 1978.

Brasseur, G.P., Orlando, J.J., and Tyndall, G.S., *Atmospheric Chemistry and Global Change,* Oxford University Press, Oxford, 1999.

Bridgman, H.A., *Global Air Pollution: Problems in the 1990s,* John Wiley & Sons, New York, 1994.

Brown, C.A., Hamill, P., and Wilson, J.C., Particle formation in the upper tropical troposphere: a source of nuclei for the stratosphere, *Science,* 270, 1650, 1995.

Charlson, R.J., Schwartz, S.E., and Hales, S.E., Climate forcing by anthropogenic aerosols, *Science,* 255, 423, 1992.

Deming, D., Climate warming in North America: analysis of borehole temperatures, *Science,* 268, 1576, 1995.

Douglas, J., Mercury and the global environment, *EPRI J.,* Jan./Feb., 14, 1994.

Elsom, D.M., *Atmospheric Pollution: A Global Problem,* 2nd ed., Blackwell Publishers, Oxford, 1992.

Finlayson-Pitts, B.J. and Pitts, J.N., Jr., *Chemistry of the Upper and Lower Atmosphere: Theory, Experiments and Applications,* Academic Press, San Diego, CA, 2000.

Gates, D.M., *Climate Change and Its Biological Consequences,* Sinuer Associates, Sunderland, MA, 1993.

Graedel, T.E. and Crutzen, P.J., *Atmospheric Change: An Earth System Perspective,* Freeman, New York, 1993.

Hidy, G.M., Source-receptor relationships for acid deposition: pure and simple? A critical review, *JAPCA*, 34, 518, 1984.

Hidy, G.M., Hansen, D.A., Henry, R.C., Ganesan, K., and Collins, J., Trends in historical acid precursor emissions and their airborne and precipitation products, *JAPCA*, 34, 333, 1984.

Houghton, J.T., Ding, Y., Griggs, D.J., Noguer, M., van der Linden, P.J., and Xiaosu, D., Eds., *Climate Change 2001: The Scientific Basis: Third Assessment Report of the Intergovernmental Panel on Climate Change*, Cambridge University Press, Cambridge, U.K., 2001.

Husar, R.B. and Patterson, D.E., *Haze Climate of the United States*, National Technical Information Service PB 87-141057/AS, Springfield, VA, 1987.

Jennings, S.G., Ed., *Aerosol Effects on Climate*, University of Arizona Press, Tucson, AZ, 1993.

Kahn, J.R., The surface temperature of the sun and changes in the solar constant, *Science*, 242, 908, 1998.

Kerr, R.A., Global pollution: is the Arctic haze actually industrial smog? *Science*, 205, 290, 1979.

Kiehl, J.T. and Briegleb, B.B., The relative roles of sulfate aerosols and greenhouse gases in climate forcings, *Science*, 260, 311, 1993.

Kraljic, M.A., *The Greenhouse Effect*, H.W. Wilson Co., New York, 1992.

Legge, A.H. and Srupa, S.V., Eds., *Acidic Deposition: Sulphur and Nitrogen Oxides*, Lewis Publishers, Chelsea, MI, 1990.

Lucier, A.A. and Haines, S.G., Eds., *Mechanisms of Forest Response to Acidic Deposition*, Springer-Verlag, New York, 1990.

McCarthy, J.J., Canziani, O.F., Leary, N.A., Dokken, D.J., and White, K.S., Eds., *Climate Change 2001: Impacts, Adaptation and Vulnerability: Third Assessment Report of the Intergovernmental Panel on Climate Change*, Cambridge University Press, Cambridge, U.K., 2001.

McElroy, M.B. and Salawitch, R.J., Changing composition of the global stratosphere, *Science*, 243, 763, 1989.

Metz, B., Davidson, O., Swort, R., and Pan, J., Eds., *Climate Change 2001: Mitigation: Third Assessment Report of the Intergovernmental Panel on Climate Change*, Cambridge University Press, Cambridge, U.K., 2001.

National Acid Precipitation Assessment Program, *Interim Assessment: The Causes and Effects of Acidic Deposition*, U.S. Government Printing Office, Washington, D.C., 1987.

National Research Council, *Acidic Deposition: Long-Term Trends*, National Academy Press, Washington, D.C., 1988.

National Research Council, Ozone depletion, greenhouse gases and climate change, in *Proceedings of the Joint Symposium Board of Atmospheric Science & Climate & Committee on Global Change*, National Academy Press, Washington, D.C., 1989.

National Research Council, *Rethinking the Ozone Problem in Urban and Regional Air Pollution*, National Academy Press, Washington, D.C., 1991.

National Research Council, *Protecting Visibility in National Parks and Wilderness Areas*, National Academy Press, Washington, D.C., 1993.

Peters, R.L. and Lovejoy, T.E., *Global Warming and Biological Diversity*, Yale University Press, New Haven, CT, 1992.

Rodriguez, J.M., Probing stratospheric ozone, *Science*, 261, 1128, 1993.

Rowland, F.S., Chlorofluorocarbons and the depletion of stratospheric ozone, *Am. Sci.*, 77, 36, 1989.

Saitoh, T.S., Shimade, T., and Hoshi, H., Modeling and simulations of the Tokyo urban heat island, *Atmos. Environ.*, 30, 3431, 1996.

Schneider, S.H., The greenhouse effect: science and policy, *Science,* 243, 771, 1989.

Schneider, S.H., Detecting climate change signals: are there any "fingerprints"? *Science,* 263, 341, 1994.

Schwartz, S.E., Acidic deposition: unraveling a regional phenomenon, *Science,* 243, 753, 1989.

Seinfeld, J.H. and Pandes, S.N., *Atmospheric Chemistry: From Air Pollution to Climate Change,* John Wiley & Sons, New York, 1998.

Stolarski, R., Bojkov, R., Bishop, L., Zerefos, C., Staehlin, J., and Zawodny, J., Measured trends in stratospheric ozone, *Science,* 256, 342, 1992.

Thompson, D.J., The seasons, global temperature, and precession, *Science,* 268, 59, 1995.

Tolbert, M.A., Sulfate aerosols and polar stratospheric cloud formation, *Science,* 264, 527, 1994.

Turekian, K.K., *Global Environmental Change: Past, Present and Future,* Prentice Hall, Saddlebrook, NJ, 1996.

USEPA, *Protecting Visibility: An EPA Report to Congress,* EPA/450/5-79-008, EPA, Washington, D.C., 1979.

USEPA, *Research Summary: Acid Rain,* EPA/600/8-79-028, EPA, Washington, D.C., 1979.

USEPA, *Air Quality Criteria for Particulate Matter and Sulfur Oxides,* EPA/600/8-82-029a-c, EPA, Washington, D.C., 1982.

USEPA, *Air Quality Criteria for Particulate Matter,* Vol. 1, EPA/600/AP-95/001a, EPA, Washington, D.C., 1995.

USEPA, *Latest Findings on National Air Quality: 1999 Status and Trends,* EPA/454/F-00-002, EPA, Washington, D.C., 2000.

USEPA, *Air Quality Criteria for Particulate Matter,* EPA/600/P-99/002bB, EPA, Washington, D.C., 2001.

USEPA, Third United States Climate Action Report, 2002, http://Yosemite.epa.gov/global-warming.nsf/contract/ResourceCenterPublicationsUSClimateActionReport.html.

Van de Kamp, *Health Effects of Global Warming,* Dept. of Health and Human Services, Bethesda, MD, 1992.

Van Loon, G.W. and Duffey, S.J., *Environmental Chemistry: A Global Perspective,* Oxford University Press, Oxford, 2000.

Warnek, P., *Chemistry of the Natural Atmosphere,* Academic Press, New York, 1988.

World Meteorological Organization, *Scientific Assessment of Ozone Depletion*, WMO Global Ozone Research and Monitoring Project, Report 44, WMO, Geneva, Switzerland, 1998.

Wuebbles, D.J., Grant, K.E., Cornell, P.S., and Penner, J.E., The role of atmospheric chemistry in climate change, *JAPCA,* 39, 22, 1989.

QUESTIONS

1. Significant changes in visibility have occurred in the eastern part of the U.S. in the past half century. Describe their cause.
2. Even if western states had particle aerosol levels similar to those in the east, visibility reduction would be less. Why?
3. Describe methods by which visibility reduction can be quantified.
4. How do atmospheric particles reduce our ability to see objects clearly?
5. What is atmospheric turbidity? How is turbidity related to visibility?

6. What is Arctic haze and how is it formed?
7. What is a heat island? What is its environmental significance?
8. How do particle size and number affect the probability of precipitation?
9. Why is the term *acid rain* not appropriate to describe acid phenomena in the atmosphere?
10. Describe the relationship between pH and H^+ concentrations.
11. Why are some ecosystems acid sensitive while others are not?
12. Emissions of mercury to the atmosphere are relatively small. Why are these emissions increasingly becoming an environmental concern?
13. Why is atmospheric deposition of pesticides, PCBs, and dioxin an environmental concern?
14. Why is the atmospheric deposition of nitrogen an environmental concern?
15. What are the effects of stratospheric O_3 on the atmosphere?
16. Chlorine atoms in the stratosphere are present at an average concentration of approximately 4 ppbv. Relatively speaking, this is a very low concentration. Why is this chlorine such an environmental concern?
17. What proof is there that chlorine and bromine atoms are causing stratospheric O_3 depletion?
18. Why does the Antarctic O_3 hole occur in the southern hemispheric spring?
19. What evidence is there that stratospheric O_3 depletion is occurring outside the Antarctic?
20. What are polar stratospheric clouds? What role do they have in O_3 depletion?
21. Describe two direct effects of stratospheric O_3 depletion.
22. How do variations in the Earth's orbit cause changes in climate?
23. What is the greenhouse effect? Describe the role of major greenhouse gases.
24. What is radiative forcing?
25. Describe the radiative effectiveness of CFCs, CH_4, N_2O, and tropospheric O_3 relative to CO_2.
26. Describe evidence that global warming associated with greenhouse gas emissions is already occurring.
27. What pollutant has the greatest impact on radiative forcing? Why?
28. What are some of the potential impacts of global warming on coastal and agricultural areas?
29. Though global surface temperatures have increased, the lower stratosphere has cooled. Why?
30. Why are global warming effects greater in the Arctic than at middle latitudes?

CHAPTER 5

Health Effects

5.1 AIR POLLUTION EPISODES

Air pollution episodes are characterized by significant short-term increases in atmospheric pollutant concentrations above normal daily levels. Episodes vary from those that pose relatively limited public health concern to those more extreme. Extreme air pollution episodes were reported for the Meuse Valley, Belgium, in 1930; Donora, PA, and the Monongehela River Valley in 1948; and London in 1952. In each of these cases, a persistent (3 to 6 days) thermal inversion combined with significant industrial and, in the case of London, domestic pollutant emissions resulted in high ground-level concentrations that caused acute illness and, in some cases, death in exposed populations.

Sixty deaths were reported in the Meuse Valley disaster, 20 in Donora, and ~4000 in London. In each case, death claimed individuals with existing respiratory or cardiovascular disease. Pneumonia was the primary cause of death in the 1952 London disaster. Based on retrospective studies, "excess mortality" was associated with a number of London "fogs," going back to as early as 1873. In each of these extreme episodes, many individuals reported becoming ill. Illness characterized by cough, shortness of breath, chest pain, and eye and nose irritation was reported by people of all ages. In the Donora area, over half of the 14,000 residents reported experiencing acute illness symptoms.

Other somewhat less extreme episodes have been reported for London in 1956, 1957, 1959, 1962, 1975, and 1991; Dublin in 1982; the Ruhr Valley of Germany in 1962, 1979, 1982, and 1985; the Netherlands in 1985, 1987, and 1991; and New York City in 1953, 1962, and 1966.

These episodes are significant in that they provided solid scientific documentation that exposure to elevated ambient pollutant levels can cause acute illness and even death. The more disastrous episodes contributed to significant efforts to control ambient air pollution. This was particularly the case in London, where clean air legislation was enacted as early as 1956.

Despite significant control efforts, air pollution episodes continue to occur in developed countries. Though pollutant levels are now considerably lower than in the disastrous episodes of the past, application of increasingly more sophisticated epidemiological techniques to what could be described as historically minor episodes shows significant associations between elevated pollutant levels, acute illness such as asthmatic attacks, and even increased death rates.

Air pollution episodes are not limited to developed countries. Increased industrialization, limited pollution control efforts, and population growth pose significant exposure risks on a day-to-day basis, as well as during episode periods, in many cities in China, India, Mexico, and other countries.

5.2 POLLUTANT EXPOSURES AND CAUSE–EFFECT RELATIONSHIPS

Ambient air pollution exposures that result in severe community-wide illness and death are, for the most part, relatively rare in the U.S. and other developed countries. Pollutant exposures in our cities, towns, and even countrysides vary considerably, both in types of pollutants present and prevailing concentrations. Exposures reflect the nature and extent of industrial activities, dispersion conditions of the surrounding environment, motor vehicle emissions, atmospheric chemistry, and, in some mountain valleys, wood space heating.

Pollutant exposures to the "normal" range of concentrations may result in a variety of health effects. These include acute but transient symptoms such as eye, nose, and throat irritation and, in some sensitive individuals, asthmatic attacks. Acute asthmatic responses to ambient pollutants are a major regulatory concern. The primary focus of most regulatory activity in North America and other developed countries has been to protect exposed or potentially exposed populations from the chronic effects of relatively low-level exposures. These may include respiratory and cardiovascular disease, neurotoxic effects, and cancer.

A good understanding of the observed or potential health effects of pollutants at levels typical of urban–suburban communities is requisite to all health-based regulatory programs. Ideally, such regulatory activities are supported by scientific data that adequately establish a cause-and-effect relationship between pollutant exposures and illness symptoms among those exposed so that an adequate margin of safety can be incorporated into air quality standards. Since humans experience a variety of exposures and other stresses, establishing definitive cause–effect relationships between adverse health effects and one or more atmospheric pollutants is a difficult undertaking.

A strong relationship between individual or combined pollutant exposures and health effects is indicated when there is a convergence of evidence from epidemiological, toxicological, and occupational exposure studies. Information from such studies is of particular importance in evaluating potential health effects of chronic exposures to ambient pollutants.

5.2.1 Epidemiological Studies

Epidemiological studies are conducted to determine potential relationships between a variety of environmental factors and human disease. They are characterized by application of statistical analysis to data collected on the health status of populations of individuals, pollutant exposures, and potential confounding factors. Such studies often provide evidence of possible causal relationships between pollutant exposures and observed or reported health effects. In general, epidemiological studies become more important as the risk attributable to atmospheric pollutants becomes smaller and the duration of exposure required to produce effects becomes longer. They have been particularly useful in identifying acute effects of elevated short-term pollutant exposures. These effects may include pulmonary function changes, asthmatic attacks, and increased mortality.

Epidemiological studies vary in design. They may be cross-sectional (assess the relationship between pollutant exposures and health effects over a cross section of a population), longitudinal (assess exposures and health effects over a period of time), or case-control (exposed and non- or less-exposed populations are compared). They may be prospective (data to be collected) or retrospective (evaluation of existing data). Study designs differ in their power to identify potential causal relationships and statistical confidence in results obtained. Over the past decade, increasingly sophisticated statistical techniques have been applied to the study of short-term pollutant exposures. These have included time-series analyses of daily health data collected over a period of days or weeks. Such studies are longitudinal, and as such, each participant serves as his or her own control. Consequently, the effect of confounding variables (described below) is significantly reduced when data are statistically analyzed. Such studies are characterized by a strong study design, good exposure and health effects data, and large study populations.

5.2.1.1 Confounding Factors

Epidemiological assessment of potential relationships between pollutant exposures and observed parameters of human health has, in many cases, been confounded by a variety of coexisting factors. These have included individual sensitivity, age, existing disease, gender, race, socioeconomic status, tobacco smoking, lifestyle, occupation, and meteorological conditions. Assessment of potential effects may also be confounded by interaction between (1) two or more pollutants; (2) pollutants and meteorological variables such as temperature and relative humidity; (3) pollutants and other exposures (occupational, smoking, and indoor air pollution); and (4) pollutants and infectious disease. Such interactions may, in part, explain differences observed between epidemiological and toxicological studies.

5.2.1.2 Interaction Effects

5.2.1.2.1 Interactions between Pollutants

Since community air is commonly contaminated by complex mixtures of gases and particulate matter (PM), it is likely that some of these may interact to modify the physiological effects of others. This may occur in several different ways. For example, one pollutant may affect the site of deposition of another. This is apparently the case for gases such as sulfur dioxide (SO_2). Because of its solubility in watery fluids, SO_2 is normally removed in the upper respiratory system. Absorption and adsorption of SO_2 on particles provides a transport mechanism for SO_2 into the pulmonary system, where SO_2 toxicity occurs. Consequently, particles can potentiate the effect of SO_2. Harmful aerosols such as sulfuric acid (H_2SO_4) can be produced as a result of interaction of gaseous pollutants with the moist environment of the lungs.

Pollutant interactions may result in health and physiological effects that are additive, antagonistic, or synergistic. Exposure to pollutants with similar toxic effects may result in an additive response. The total toxic response is the sum of individual responses to each pollutant. Antagonism occurs when one pollutant interferes with toxic effects of another, thus mitigating effects of the more toxic pollutant. Reactions of ammonia (NH_3) and SO_2 in human breath are an example of such an effect. Synergism describes health responses that are significantly greater than the sum of the individual effects of toxic pollutants. This is particularly true for pollutants that inhibit pulmonary defense mechanisms (e.g., tobacco smoke). Inhibition of respiratory clearance increases the risk of disease (e.g., cancer) from pollutants such as asbestos and radon (Rn). In many cases, synergistic effects are multiplicative; i.e., risk is increased by the product of each individual exposure risk.

Epidemiological and toxicological information on additive, antagonistic, and synergistic effects is limited. Because of the large variety of pollutants present in metropolitan atmospheres, it is probable that each of these interaction effects occurs. In day-to-day pollutant exposures, they cannot, for the most part, be evaluated using scientific tools currently available. It is standard regulatory practice to set exposure limits on individual pollutants as if they were acting alone. Such exposure limits are subject to intense legal, political, and scientific challenges from those whose activities would be subject to regulation. Use of exposure limits based on multiple pollutant exposures would pose an enormous regulatory challenge.

5.2.1.2.2 Interactions between Pollutants and Meteorological Factors

Most major air pollution episodes were associated with meteorological extremes that confounded assessment of the causal role of ambient air pollution in the genesis of mortality and morbidity. During the many London "fogs" and disasters in the Meuse Valley and Donora, weather was cold and damp. The independent effect of temperature extremes on mortality has been well established. Indeed, studies in the U.K. have indicated that mortality was more closely related to cold and humidity than to fog frequency. Studies in Los Angeles have noted a marked increase in mortality during periods of elevated photochemical oxidant levels and high temperature. In such cases, investigators commonly concluded that observed excess mortality was due to high temperatures rather than pollutant exposures. Heat waves, in general, are major stressors of the elderly and infirm and usually result in excess mortality. With the application of more powerful research designs and statistical techniques, epidemiologists are increasingly able to separate the independent effects of temperature extremes and pollutant exposures in evaluations of excess mortality.

5.2.1.2.3 Interaction between Ambient Air Pollutants and Other Pollutant Exposures

In addition to community air pollution exposures, many individuals are exposed to elevated pollutant levels from cigarette smoking and in their occupational and domestic environments. Such exposures may mask health risks associated with ambient pollutants. They may also increase health risks associated with ambient pollution.

Cigarette smokers may be at higher risk for adverse effects associated with exposures to ambient pollution. This increased risk may be due to one or more factors. First, cigarette smoking may initiate respiratory and cardiovascular disease. Individuals with such disease, no matter what the cause, are at increased risk of illness or death when exposed to elevated pollutant levels. Smoking may increase the effect of community air pollution exposures (Figure 5.1) by impairing respiratory clearance mechanisms.

Many individuals are exposed to relatively high levels of gaseous and particulate-phase pollutants in their place of employment. These exposures are usually much greater than those associated with breathing polluted community air. Because they are occupationally exposed, some individuals may be at higher risk from ambient pollutant exposures. This increased risk may be due to occupationally-induced respiratory or cardiovascular disease or impaired defense mechanisms.

Assessment of health effects of community air pollutants is confounded, to some degree, by the fact that individuals spend more time indoors than in the ambient environment. On average, Americans only spend 2 h/day outdoors. Therefore, their ambient pollution exposure may be quite different than a police officer's or construction worker's. An expanded discussion of indoor air pollution and its health implications can be found in Chapter 11.

5.2.1.3 Exposure Assessment

Most early epidemiological studies of air pollution and related health effects were limited by inadequate information on pollutants present and data on population exposures. As a result, many of these studies used the place of residence as an index of exposure. The place of residence often introduces bias into statistical analyses since it may be associated with unique ethnic or cultural traits, living standards, occupations, and exposure to infectious agents. Exposure assessment in many recent longitudinal studies is carried out using personal exposure monitors, time activity diaries, and exposure models based on fixed-site monitors and dispersion modeling of various emission sources. Exposure assessment, like other aspects of epidemiological studies, has become more sophisticated and better serves the needs of pollutant effects studies.

5.2.1.4 Population Susceptibility

Epidemiological assessment of air pollution-related health effects has historically been complicated by variations in population susceptibility. Such assessments are

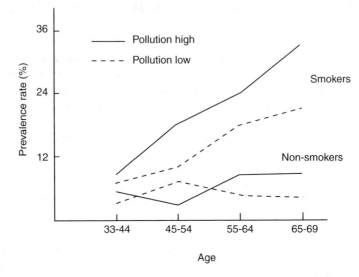

Figure 5.1   Prevalence of chronic bronchitis associated with tobacco smoking and exposure to ambient air pollution. (From Lambert, P.M. and Reid, D.D., *Lancet*, 5, 853, 1970. With permission.)

now simplified by identifying populations at risk, e.g., the aged, the very young, those with existing respiratory or cardiovascular disease, those occupationally exposed, and cigarette smokers. These may be more sensitive to normally encountered pollutant levels than the average healthy individual. Epidemiological studies of sensitive populations are more powerful in identifying air pollution–health effects associations when they indeed exist. Present air quality standards (Chapter 8) promulgated in the U.S. are intended to protect those at special risk as well as the general population.

Despite confounding factors and the historical limitations described, epidemiological studies have been able to provide valuable insight into defining the probability that air pollution exposures contribute to the incidence or prevalence of specific community health problems. A strong association or high probability of disease causation can be inferred when (1) there are a number of different populations in which a similar association is observed, including different population groups, locations, climate, and times of year; (2) incidence or severity of the health effect increases with increasing exposure and, conversely, decreases with decreasing exposure; and (3) a plausible biological mechanism can be hypothesized for the observed association.

The value of epidemiological studies is increased when they are used in conjunction with controlled biological (toxicological) studies on animals and humans. They are important in identifying possible associations that can be tested under controlled laboratory conditions. In addition, they can be used to evaluate potential human health risks identified in laboratory exposures.

5.2.2 Toxicological Studies

Toxicological, or controlled biological, studies are conducted on animals and, less commonly, humans to determine the functional, structural, and biochemical effects of toxic substances. Historically, toxicological studies have been conducted to determine the toxicity of a substance administered in varying doses, with dose being a function of the concentration of the substance and duration of exposure. Significant differences in toxic response occur when the same dose of toxic material is administered over different exposure periods. Acute exposures result when an organism is subjected to a high dose rate, i.e., a high concentration of a substance, over a relatively short period of time. Response to the exposure is sudden, severe, and usually lasts for a brief period. If the dose rate is sufficiently high, death may result. Indeed, the term *toxic* implies that a substance has the potential to cause death. Lower doses, i.e., lower concentrations over longer periods, generally do not directly cause mortality. As dose rate decreases, the response is less severe, may take longer to develop, and may be more persistent. In chronic exposures, adverse effects may take years, if not decades, to develop.

In traditional toxicology, animals are exposed to varied doses of toxic substances to determine their ability to cause physiological and pathological changes and to characterize dose–response relationships. In addition to evaluating such physiological and pathological changes, the scope of toxicology in recent years has been expanded to include studies of carcinogenesis (cancer induction), teratogenesis (induction of birth defects), mutagenesis (production of mutations), gametotoxicity (damage to sex cells), and endocrine disruption.

One of the major advantages of toxicological studies is that the investigator can control many variables that may confound epidemiological studies. Controllable variables in animal studies include age, sex, environmental conditions such as temperature and relative humidity, genetic constitution, nutrition, and pollutant dose. By controlling dose as a function of exposure duration, dose–response relationships can be developed. Human exposures are also controllable, but to a more limited degree. In addition to those variables described for animals, factors peculiar to humans such as smoking, occupation, and health status are to some degree under the experimenter's control.

5.2.2.1 Human Studies

Ideally, toxicological studies of humans should provide the strongest scientific evidence available for establishing a cause–effect relationship between air pollutant exposures and adverse health effects. In such studies it is possible to expose humans under controlled conditions and monitor them for signs indicating the onset of disease. Most human studies are, however, limited to short-term exposures. Because of long exposure periods required to induce chronic responses and the irreversibility of some of these, most human studies are not conducted at realistic concentrations of pollutants that occur in ambient atmospheres. In addition, ethical considerations, which are imperative when designing human exposure studies, limit the kind of information that can be acquired.

In theory, a true cause–effect relationship can only be developed by human experimentation. Ethical and other limitations make it virtually impossible to conduct such studies.

5.2.2.2 Animal Studies

Animal exposures can provide valuable information on the effects of pollutants when no other acceptable means of obtaining such information are available. Since animals can be sacrificed for pathological and biochemical analyses, they are ideally suited for toxicological studies. Additionally, chronic exposures can easily be conducted.

Use of animals as models for humans does have limitations; animals are not humans. Though they may have the same general organ systems, they may be structurally and physiologically different. In addition, species differ in their sensitivity to pollutants. Dogs, for example, are 10 times more resistant to NO_2 than rodents. Also, most animals are not as long-lived as humans. Consequently, results obtained from animal studies can only be extrapolated to humans with some degree of uncertainty. Uncertainty is reduced when similar results can be demonstrated in more than one mammalian species, especially primates, and the organ systems under study are closely analogous to those of humans.

To provide observable toxicological responses, exposures used in controlled studies are usually much higher than those occurring in ambient environments. In general, controlled exposures are limited to one toxic pollutant.

5.2.3 Occupational Exposure Studies

Although ethical and legal constraints limit use of human volunteers for experimental evaluation of the effects of toxic substances, such constraints have not limited the voluntary or involuntary exposure of industrial workers. For many air pollutants of interest, exposures in the work environment occur at much higher levels than in ambient air. Because of potential health risks and legal concerns, some industries maintain detailed exposure and health data on employees. As a consequence, occupational exposures may be relatively well defined and often provide an important source of information on illness development and expression, job-related disease, and dose–response relationships. In the absence of human toxicological studies, occupational exposures often provide the only available human dose–response data.

Studies of occupationally-exposed workers have limitations. Although one or more pollutants may be common to both occupational and ambient environments, the overall pollutant mix varies considerably. In addition, occupational exposures are limited to workday and workweek cycles. Community exposures are not only more heterogeneous and variable; they are further complicated by differences in indoor and ambient air quality and the amount of time individuals spend in such environments.

A major limitation of occupational studies is that the population of workers does not reflect the general population. In many industries, workers are nominally healthy males between the ages of 18 and 65. Studies of such workers may provide important information on disease syndromes induced by specific pollutants or pollutant combinations. They cannot, however, provide adequate information on those who are not industrially exposed and who may be at special risk. In such instances, information on health effects on healthy individuals may not be as important as responses of individuals who are more sensitive. Such individuals may be excluded from the work force or, because of existing health problems, be excluded from specific occupations.

5.3 IMPACT OF POLLUTANTS ON THE HUMAN BODY

Pollutant effects are normally manifested in specific target organs. These may be direct; i.e., pollutants come in intimate contact with the organ affected. Such is the case for eye and respiratory irritation. Effects may be indirect. For example, pollutants may enter the bloodstream from the lungs or gastrointestinal system by respiratory clearance. Effects may then be distant from the immediate organ of contact. Indeed, a target organ may not have immediate intimate contact with air contaminants. The principal target organs or organ systems are eyes and the respiratory and cardiovascular systems (Figure 5.2).

5.3.1   Eye Irritation

Eye irritation is one of the more prevalent manifestations of pollutant effects on the human body. It is most often associated with exposure to aldehydes and photochemical oxidants. The threshold for eye irritation by oxidants is approximately 0.10 to 0.15 ppmv (reported as ozone (O_3)). Eye irritation increases with increasing oxidant concentration, although O_3 and nitrogen dioxide (NO_2), the two principal oxidants, do not cause eye irritation. Apparently, oxidant levels are indicators of the eye irritation potential of photochemical pollutants such as peroxyacyl nitrate (PAN), acrolein, formaldehyde (HCHO), and other photochemically-produced compounds. There is some question as to whether eye irritation should be categorized as a significant health effect since no other physiological changes are detectable and eye irritation resolves quickly after exposure ceases.

5.3.2   Effects on the Cardiovascular System

Pollutants such as carbon monoxide (CO) and lead (Pb) are absorbed into the bloodstream and may have both direct and indirect effects on the cardiovascular system. These effects will be discussed in subsequent sections. Cardiovascular disease may also result from the indirect effects of other air pollution-incited disease. For example, some individuals die of *cor pulmonale*, heart failure resulting from the stress of severe chronic respiratory disease. Recent epidemiological studies have

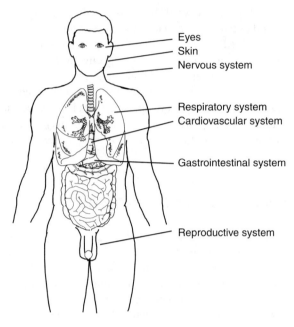

Figure 5.2 Target organs for air pollutant exposures.

shown that premature cardiovascular system-related mortality is strongly associated with exposures to small (≤2.5 μm) particles.

5.3.3   Effects on the Respiratory System

The respiratory system is responsible for gas exchange and, therefore, receives direct exposure to airborne contaminants. Effects on the respiratory system are determined in great measure by its structural and functional anatomy as well as respiratory defense mechanisms.

The principal function of the respiratory system is to supply the body with oxygen (O_2) for metabolism. In addition, it removes waste carbon dioxide (CO_2) from the bloodstream and tissues. These functions are facilitated by the three major units of the respiratory tract: — the nasopharyngeal, tracheobronchial, and pulmonary regions (Figure 5.3).

5.3.3.1   Units of the Respiratory System

5.3.3.1.1   The Nasopharyngeal Region

The nasopharyngeal region, or upper airway, consists of nasal passages, nasopharynx, oropharynx, and glottis. It begins at the terminus of the bony turbinates and extends to the region of the vocal cords. The nasal area is subdivided into two cavities by a septum. These cavities, just beyond the nostrils, are ringed with coarse hairs. Projecting from the wall of the cavities are bony turbinates that force inspired

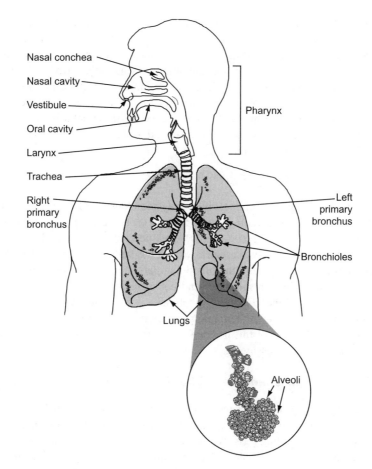

Figure 5.3 General anatomy of the human respiratory system.

air to flow along a winding course where, on contact with the large surface area of the nasal passages, it is warmed and moistened. Nasal surfaces are covered by a mucus layer and hairlike projections called cilia that serve to remove the larger particles in the thoracic range (≤ 10 μm).

The oropharynx extends from the soft palate to the glottis, located at the junction of the trachea and esophagus, where eating and breathing functions are separated. The glottis is the space between the vocal cords that are found in the larynx, or voice box.

5.3.3.1.2 The Tracheobronchial Region

The tracheobronchial region of the respiratory system is a series of tubes or ducts that transport inspired air to lung tissue, where O_2 and CO_2 are exchanged. This region consists of a large tube, the trachea, which subdivides to form smaller-diameter tubes, bronchi. Bronchi in turn branch to form smaller and smaller airways. From the trachea to the site of gas exchange there may be as many as 23 to 32 generations of branching, with each generation being smaller in diameter and length. Bronchi comprise the first 12 to 22 branching generations. Their walls have cartilaginous plates and a muscular layer that constricts when foreign substances enter bronchi. Air flows from bronchi into the smaller bronchioles (which do not have cartilaginous plates and are usually less than 1 mm in diameter). Interior surfaces of the trachea, bronchi, and bronchioles are lined by ciliated cells interspersed with mucus-secreting cells. The mucus layer produced by these cells is moved toward the oropharynx by the rhythmically beating cilia. Terminal bronchioles are the smallest bronchioles to have cilia.

5.3.3.1.3 The Pulmonary Region

The pulmonary region of the respiratory system consists of respiratory bronchioles, alveolar ducts, and alveoli (Figure 5.3). Alveoli are membranous air sacs at the terminus of the alveolar ducts. A fully developed lung is estimated to have ~300 million alveoli. The large alveolar surface area facilitates efficient exchange of O_2 and CO_2 between the lungs and the blood capillaries that cover the alveoli.

Movement of air into (inspiration) and out of (expiration) the lungs is accomplished by muscles of the chest and diaphragm, which alternately compress and expand the lungs. During inspiration, air pressure in the alveoli decreases, becoming slightly negative. Because of this negative pressure, air flows into the lungs. During expiration, alveolar air pressure rises, causing air to flow out of the lungs.

5.3.3.2 Defense Mechanisms of the Respiratory System

The respiratory system is protected from airborne contaminants by a variety of defense mechanisms. In the nasal region, the stiff nasal hairs or impaction on the mucus layer of the winding passages of the turbinates may remove large particles. Cilia sweep the mucus layer and entrapped particles toward the back of the throat where they are swallowed or expectorated. Contaminants may also be removed from upper airways by the sneeze reflex.

Other defense mechanisms serve to remove or prevent contaminants (mainly particles) from entering the tracheobronchial and pulmonary regions. When, for example, particles enter the upper portion of the bronchial tree, muscle layers constrict the bronchi, thus narrowing bronchial diameter and reducing the amount of particles entering the lungs. The cough reflex, similar to the sneeze reflex, assists this process by eliminating particles trapped in mucus in the tracheobronchial region. Cilia that line the tracheobronchial region clean airways by propelling mucus and particles upward where they are removed by coughing and expectoration or swallowing. In the latter way, some particulate-phase pollutants may enter the gastrointestinal system.

Despite these defense mechanisms, many fine particles and some relatively insoluble gases enter pulmonary tissue. Within the alveoli and bronchioles of the pulmonary region, specialized cells called phagocytes and macrophages ingest deposited matter (phagocytes and macrophages, derived from white blood cells, consume bacteria, viruses, and other particles and, as a result, serve as a pulmonary defense mechanism). These phagocytes and the matter they contain are normally transported out of the lungs in mucus by the cilia. However, they may also move through the alveolar membrane and enter either the lymphatic or circulatory system. Additionally, soluble components deposited in the alveoli may be absorbed by adjacent tissue and enter the lymphatic or circulatory system without being phagocytized. Although these mechanisms affect clearance from the respiratory system, they may expose other body systems to toxic materials. They may also result in tissue damage when phagocytes are destroyed and spill their contents when attempting to ingest toxic particles.

5.3.3.3 Air Pollution and Respiratory Disease

Community air pollution has been implicated as a causal or aggravating agent in diseases of the respiratory system such as chronic bronchitis, pulmonary emphysema, lung cancer, bronchial asthma, and infections.

5.3.3.3.1 Chronic Bronchitis

Bronchitis is a respiratory disease characterized by inflammation of the membrane lining bronchial airways. Bronchitis may be caused by pathogenic infections or respiratory irritants, e.g., those that occur in cigarette smoke, industrial exposures, and ambient and indoor air pollution.

When bronchial inflammation persists for 3 months or longer, it is classified as chronic bronchitis. Extended irritation of the bronchial membrane is the primary cause of chronic bronchitis. It is characterized by a persistent cough and excessive mucus or sputum production, often accompanied by destruction of cilia and thickening of bronchial epithelium. Chronic bronchitis significantly affects respiratory function, as airway resistance is increased by occlusion of the bronchi, swelling of the inflamed membrane, and concomitant excess mucus production. Consequently, persons afflicted with the disease have difficulty breathing.

Severe cases of chronic bronchitis are often followed by development of pulmonary emphysema. Combined chronic bronchitis and pulmonary emphysema are usually diagnosed and described as chronic obstructive lung disease (COLD) or chronic obstructive pulmonary disease (COPD).

Some epidemiological studies suggest that ambient pollutants may initiate or aggravate chronic bronchitis. The problem of identifying ambient air pollutants as causal agents in the etiology of chronic bronchitis is confounded by the significant causal role of cigarette smoking in initiating this disease. It is further obscured by the effects of occupational exposures. However, available epidemiological and toxicological evidence suggest that long-term community exposures to a combination of ambient pollutants such as PM and SO_2 may contribute to initiation of chronic bronchitis.

5.3.3.3.2 Pulmonary Emphysema

Whereas chronic bronchitis is a disease that affects the upper portion of the respiratory airway system, pulmonary emphysema is a disease of lung tissue that normally facilitates gas exchange with blood.

Emphysema is often a disease of older individuals. It is characterized by destruction or degeneration of walls of the alveolar sacs. As a consequence, the total surface area of pulmonary tissue is reduced, diminishing aeration of blood. Individuals afflicted with emphysema usually develop pulmonary hypertension. In addition to elevation of arterial pressure, pulmonary resistance is increased. Pulmonary emphysema is consequently accompanied by shortness of breath and other breathing difficulties.

Emphysema patients usually have problems exhaling, since air remains trapped in the lungs and overinflates damaged lung tissue. This overinflation contributes to the "barrel chest" characteristic of most emphysema patients. Exhalation difficulties are due to the compression and collapse of some of the smaller airways and the overall decrease in lung elasticity common to emphysema victims.

Little epidemiological evidence exists to implicate ambient air pollutants as a contributing factor to the initiation and development of pulmonary emphysema. Animal studies suggest, however, that chronic exposures to NO_2 can initiate preemphysematous lung changes.

5.3.3.3.3 Lung Cancer

Cancer is one of the leading causes of death in the U.S., accounting for ~370,000, or one of five, deaths each year. Of these, ~150,000 are due to lung cancer, which is the leading cause of cancer deaths. Common to all cancers is the characteristic unrestrained cell growth that produces malignant tumors, which have a higher rate of growth than normal tissue that surrounds them. Most cancers metastasize, i.e., spread to other parts of the body. The tendency for lung cancer cells to metastasize is particularly great; as a consequence, the prognosis for lung cancer patients is poor, with an approximate 90% mortality.

Cancer is usually a disease of the elderly and individuals in late middle age. The pattern is due in part to the fact that cancers are commonly latent diseases; i.e., there is a delay of several decades or more between the initiation of exposures to a carcinogenic agent and development of the disease syndrome. For lung cancer, the latency period may be as long as 30 years or more, with peak incidence at age 50. The long interval between initial exposure to the carcinogenic (i.e., cancer-producing) agent and disease onset makes it especially difficult to conclusively identify a specific causal agent. During this latency period an individual may undergo changes in occupation, socioeconomic status, place of residency, nutrition, and personal habits such as smoking. These changes tend to confound the epidemiological evaluation of urban air pollution in the etiology of lung cancer. The latency period not only confounds identification of causal agents in diseases such as lung cancer, but also tends to give individuals a false sense of security relative to the safety of a variety of chemical or physical exposures.

The most common form of lung cancer is bronchogenic; i.e., it originates in the bronchial lining and invades tissues of the bronchial tree. The malignancy may spread to the rest of the lung and eventually to other parts of the body. Available epidemiological evidence indicates that ~90% or more of the ~150,000 lung cancer deaths in the U.S. each year are caused by chronic exposure to tobacco smoke. Lung cancer has also been linked to passive tobacco smoke exposures and occupational exposures to asbestos, arsenic, and radioactive gases (Rn) and dusts.

A causal role for ambient air pollution in development of some lung cancers is suggested from several types of epidemiological investigations, including (1) comparisons of urban vs. rural populations and migrants from high- to low-air pollution countries, and (2) statistical regression analyses of the relationships between lung cancer deaths and various indices of air pollution.

Comparisons of rural and urban nonsmoking populations indicate that urban lung cancer death rates are approximately twice those found in rural areas. These studies suggest that there is an apparent urban factor that contributes to the overall incidence of lung cancer. However, they do not identify characteristics of the urban environment that are primarily responsible.

Studies of individuals migrating from countries that had high ambient air pollution levels to those where air pollution levels were much lower indicate that lung cancer rates in such migrants tend to be intermediate between those of the country of origin and their new homeland. Even when differences in smoking habits are considered, lung cancer rates of migrants from historically high-air pollution countries such as the U.K. were greater than those from countries such as New Zealand. These data suggest that ambient air pollution exposures may produce effects that persist long after exposures have ceased.

The strongest evidence for a link between ambient air pollution and lung cancer comes from an ongoing prospective mortality study of 1.2 million adults started in 1982. Based on an epidemiological risk factor analysis of 500,000 of these adults, lung cancer mortality was shown to be significantly linked with exposures to fine particles ($PM_{2.5}$), with an 8% increase in lung cancer mortality for each 10 $\mu g/m^3$ increase in $PM_{2.5}$.

5.3.3.3.4 Bronchial Asthma

Asthma is an acute respiratory ailment characterized by constriction of muscles and swelling of the lining of respiratory airways, excessive mucus production, and increased resistance to airflow. Such reactions are episodic, being described as asthmatic attacks. They may occur suddenly and without warning. Overt symptoms include severe shortness of breath, wheeze, chest tightness, and cough. Attacks may occur for a few minutes to several hours.

Asthma is a relatively common respiratory ailment with a prevalence rate of ~4.3% in the U.S., or ~10 to 12 million individuals. Asthma prevalence in children is relatively high, with a rate of 8.3% in midwestern children and 11.8% in southern children ages 3 to 11 years. Highest asthma rates are reported for African-American children (13.4%). The overall prevalence rate of asthma has increased 50% since 1980, while the annual death rate has doubled to 5000 per year.

Asthma has been strongly linked with exposure and sensitization to inhalant allergens. There is limited evidence that exposure to air pollutants may potentiate allergenic responses that cause asthma.

Asthmatic attacks may be caused by exposure to inhalant allergens and nonspecific irritant gas- and particulate-phase substances. In controlled exposures, SO_2 has been observed to cause asthmatic attacks, and a number of epidemiological studies have linked admissions to hospital emergency rooms for treatment of asthmatic symptoms with high-O_3 pollution days. Increased asthmatic symptoms are reported for children in "asthma camps" on such days.

Recent studies indicate that O_3 exposures not only initiate asthmatic attacks, but also contribute to asthma development.

5.3.3.3.5   Respiratory System Infections

A number of animal studies over the past three decades have shown that exposure to pollutants such as O_3, NO_2, SO_2, and PM increases the risk of respiratory system infections. Such infections include pneumonia and bronchitis and are primarily caused by bacteria. The effect that pollutants have on viral infections such as influenza is less clear.

Primary factors involved in increased infection risks may include modulation of pulmonary immune cells (e.g., macrophages), lung surfactant, and physical factors such as ciliary beating and particle clearance.

Because of differences in exposure conditions and sensitivity, it is difficult to extrapolate infection risks from animals to humans. Nevertheless, results from human exposure studies and retrospective epidemiological studies of air pollution episodes strongly suggest that young and aged members of an exposed population are at greatest risk of infection, particularly of the upper respiratory tract.

5.4 HEALTH EFFECTS OF REGULATED AIR POLLUTANTS

Though polluted atmospheres may contain hundreds of different substances, only a relatively small number have been identified as having the potential to cause adverse health effects under community-wide exposure conditions encountered in ambient environments. Most notable are those for which National Ambient Air Quality Standards (NAAQSs) have been promulgated. These are CO, sulfur oxides (SO_x), O_3, PM with aerodynamic diameters of ≤ 10 and ≤ 2.5 μm (PM_{10} and $PM_{2.5}$, respectively), NO_2, and Pb. Though nonmethane hydrocarbons (NMHCs) are regulated under NAAQSs, their regulation is not based on any human health risks. Rather, they are regulated in order to achieve the O_3 standard. A large number of other substances, commonly referred to as air toxics, are regulated under National Emissions Standards for Hazardous Air Pollutants (NESHAP) and Air Toxics provisions of clean air legislation. Most pollutants regulated under NESHAP and Air Toxics provisions are associated with sources that result in localized, rather than community-wide, exposures. Because of their ubiquitous presence, population exposure potential, and potential to cause adverse health effects, pollutants regulated under NAAQS provisions are discussed here in detail.

5.4.1   Carbon Monoxide

Potentially harmful exposures to CO occur from a variety of sources and environments. At very high concentrations (>1000 ppmv), CO exposures may be lethal, with death resulting from asphyxia. At lower concentrations (several hundred ppmv), CO may cause neurological-type symptoms, including headache, fatigue, nausea, and, in some cases, vomiting. Asphyxiation and sublethal symptoms are usually caused by exposures associated with poorly vented combustion appliances, idling motor vehicles in closed environments, excessive CO production and inadequate ventilation associated with occupational industrial activities, and smoke inhalation from structural fires. Carbon monoxide poisonings, though not uncommon, represent exposures that are considerably higher than those that occur in ambient air of our cities and towns, the kind of exposure conditions addressed by the NAAQS for CO. Such community exposures rarely exceed the 35-ppmv, 1-h standard.

The principal mechanism of CO toxicity is tissue hypoxia associated with preferential bonding of CO with hemoglobin (Hb) to form carboxyhemoglobin (COHb). Carbon monoxide has an Hb affinity 240 times that of O_2. In addition, CO binds more tightly with Hb than does O_2. As a consequence, the blood's capacity to carry O_2 is decreased. Carbon monoxide also binds to several intracellular proteins, which may result in extravascular effects.

The quantity of COHb formed depends on a variety of factors, e.g., the concentration and duration of CO exposure, exercise, temperature, health status, and metabolism of the individual exposed. At low exposure concentrations, the quantity (percent) of Hb tied up by CO is a linear function of exposure concentration and duration (Figure 5.4). Note the effect of exercise on lung ventilation rate and COHb formation in blood.

Formation of COHb is reversible. Depending on initial COHb levels, the elimination half-time is approximately 2 to 6.5 h. Because elimination is slow, COHb levels in blood may increase on continuous exposure. Approximately 10 to 50% of the body store of CO is external to the vascular system.

The primary health concerns associated with low-level ambient (<50 ppmv) exposures to CO are effects on the cardiovascular system (particularly in sensitive individuals) and neurobehavior. There are over 500,000 deaths in the U.S. each year from heart attacks, and 5 to 7 million people are estimated to have a history of heart attack, angina (heart-related chest pain), or both. Scientific support for the current CO NAAQS comes from studies of patients with stable angina associated with cardiovascular disease. The lowest observed physiological effect level in patients with exercise-induced ischemia (tissue oxygen deficiency) is somewhere between 3 and 4% COHb. Baseline values in nonexposed nonsmokers are ~0.5 to 0.7%. Because nonsmokers may be exposed from a variety of sources, average COHb levels are ~1%; COHb levels in smokers vary from 3 to 8%, with an average of ~4%.

The literature on effects of CO on the cardiovascular system is extensive. Many studies have been confounded by the effects of cigarette smoking. Cigarette smoking significantly increases the risk of atherosclerosis and myocardial infarction (heart attack). It is not certain whether chronically elevated COHb levels in smokers are major causal factors for these effects. It has been suggested that chronic exposure to CO may cause atherosclerosis, while acute exposures may trigger myocardial arrythmias and infarction. Evidence that CO exposure from sources other than smoking can produce chronic effects is limited.

Figure 5.4   COHb levels in the blood as a function of exposure concentration, duration and ventilation rate. (From USEPA, EPA/600/8-90/045f, EPA, Washington, D.C., 1991.)

A number of clinical studies of volunteers have been conducted over the past several decades. Of note are results of the Health Effects Institute's (HEI) multicenter study. Carbon monoxide exposures sufficient to elevate COHb to 2% produced significant angina-related effects in exercising patients with coronary artery disease. A dose–response relationship was observed, with no obvious threshold. The significance of the HEI study is that exposure at the 8-h CO NAAQS of 9 ppmv is sufficient to produce a COHb level of 2.1%. If one assumes that 10% of the U.S. population may reach a COHb concentration of 2.1% (independent of tobacco smoking), and couples this with the widespread prevalence of coronary disease, it is apparent that a significant number of people may be at risk even when CO concentrations do not exceed the 8-h NAAQS. Effects of environmental CO exposures on smokers are likely to be additive to those associated with smoking. Smokers, who are, as a group, at greater risk of coronary disease, would therefore be subject to an increased risk of developing coronary symptoms.

A number of epidemiological studies in the 1970s and better-designed studies conducted in the 1990s have shown significant associations between CO levels and daily mortality, frequency of admission to hospital emergency departments for cardiorespiratory complaints, and hospital admissions for congestive heart failure. Other potential health effects such as chronic flu-like illness symptoms have been associated with short-term repeated exposure to CO. Complaints of headache, irritability, and malaise have been described for children and adults. Symptoms have been shown at measured COHb concentrations of only 5%. It has also been suggested that 3 to 5% of individuals who seek medical attention for headache or dizziness in urban hospital emergency rooms may be suffering from effects of CO exposure. Such effects have been described as "chronic occult CO poisoning syndrome." The nature of exposures responsible for this syndrome is unknown. It is probable that they may be associated with malfunction or improper operation of gas furnaces, stoves, or ovens, augmented by ambient exposures.

The brain is the body organ most sensitive to reduced tissue O_2 levels associated with CO exposures. Though it can, under normal circumstances, initially compensate for CO-induced hypoxia by increased blood flow and tissue O_2 extraction, some neurological or neurobehavioral effects can be anticipated. Behaviors that require sustained attention or performance appear to be very sensitive to COHb-induced effects. Human studies of hand–eye coordination, vigilance, and continuous performance have reported consistent relationships with COHb as low as 5%. Effects at lower COHb values appear to be small.

Neurological and cardiovascular effects clearly demonstrated to be associated with different COHb blood concentrations are summarized in Table 5.1. Levels of COHb required to produce these changes are likely to occur only in worst-case atmospheric conditions in the U.S., i.e., in downtown areas with heavy traffic and very poor atmospheric mixing conditions. They are also likely to occur in under-regulated environments of large cities in developing countries. High-population, high-altitude Mexico City would be particularly at risk.

The NAAQS for CO, like other NAAQSs, is designed to protect the most sensitive populations, in this case, individuals with existing cardiovascular disease. Based on both theoretical and experimental research, other populations may be at risk. These would include the fetus, infants, the elderly, individuals with preexisting disease that decreases the availability of O_2, and individuals using certain medications and recreational drugs. Exposure to other pollutants and residence at high altitude may also increase CO exposure health risks.

5.4.2 Sulfur Oxides

The sulfur oxides include both gas- and particulate-phase chemical species. Of these, only SO_2 is present in sufficient ambient concentrations to be a public health concern. Particulate-phase SO_x includes strongly to weakly acidic substances such as H_2SO_4 and its neutralization products, e.g., acid ammonium sulfate (NH_4HSO_4) and ammonium sulfate (($NH_4)_2SO_4$); most acid sulfate studies have focused on H_2SO_4.

5.4.2.1 Sulfur Dioxide

Much of the available information on acute and chronic effects of SO_2 is based on laboratory animal studies conducted at exposure concentrations considerably in excess of those found in polluted atmospheres. Because of its solubility in watery fluids, SO_2 is efficiently removed in the upper respiratory tract. It forms sulfurous acid (H_2SO_3), which readily dissociates into bisulfite and sulfite ions that move into systemic circulation. Less than 1% of inspired SO_2 reaches alveoli. Exposure of the lower airways and alveoli increases during exercise, mouth breathing, and under low ambient SO_2 levels. The major physiological effects of SO_2 are changes in mechanical function of the upper airways, including an increase in nasal flow resistance and a decrease in nasal mucus flow rate. Animal studies indicate that SO_2 exposures can adversely affect pulmonary defense mechanisms such as mucociliary transport,

Table 5.1   Neurological and Cardiovascular Responses of Humans at Various COHb Saturation Levels

Blood COHb (%)	Effect
0–1	None known
2.0	Quicker onset of angina pain in exercising patients; pain duration lengthened
2.5	Impairment of time interval discrimination
3.0	Change in relative brightness thresholds
4.5	Increased reaction time to visual stimuli
10	Performance changes in driving simulation
10-20	Headache, fatigue, dizziness, coordination loss

Source: From NACPA, USDHEW, Publication AP-62, Washington, D.C., 1970.

alveolar clearance of particles, and macrophage function. There is some evidence that exposures may cause chronic bronchitis.

When the primary air quality standard for SO_2 (0.03-ppmv (80 $\mu g/m^3$) annual mean, 0.14 ppmv (365 $\mu g/m^3$) maximum 24-h concentration) was promulgated in 1971, no health effects had been reported for short-term exposures (<1 h) at levels observed in the ambient environment (≤1 ppmv, 2.67 mg/m^3). Since 1980, numerous challenge studies conducted on asthmatic individuals have shown that when asthmatics and others with hyperactive airways are briefly exposed to SO_2 at concentrations of 0.25 to 0.50 ppmv (0.66 to 1.3 mg/m^3), they exhibit acute responses characterized by bronchoconstriction (airway narrowing) with increased airway resistance and decreased forced expiratory flow rate, and clinical symptoms of shortness of breath and wheezing. Bronchoconstriction occurs within 5 to 10 min of exposure and is brief in duration, with lung function returning to normal for most subjects within an hour of exposure. Such responses occur even at lower SO_2 levels during moderate exercise. This results from increases in breathing rate and oral breathing and thus exposure. Oral breathing increases SO_2 penetration to the lower respiratory tract and results in airway drying.

Mild and moderately asthmatic children and adults are at greatest risk for short-term SO_2-induced respiratory effects. Individuals with more severe asthma are at lower risk because their low exercise tolerance deters them from engaging in exercise sufficiently intense to cause effects.

A substantial percentage (20 to 25%) of mild and moderately asthmatic individuals exposed for 5 to 10 min to concentrations of 0.6 to 1.0 ppmv (1.60 to 2.67 mg/m^3) SO_2 during moderate exercise or activity would be expected to have changes in respiratory function and symptom induction that clearly would exceed those associated with daily variation in lung function or in response to other stimuli. At SO_2 exposures of less than 0.5 ppmv, the severity of effects may be sufficient to cause disruption of ongoing activities, increase use of bronchodilator medication, and possibly increase the need for medical attention. Less than 20% of mild and moderately asthmatic individuals exposed to 0.2 to 0.5 ppmv SO_2 during moderate exercise are likely to experience lung function changes greater than they experience daily. Such responses are less likely to be perceptible and of immediate public health concern.

Available epidemiological data do not show any evidence of excess asthma mortality associated with SO_2 exposures in urban areas. Viewed from a national perspective, it appears that the probability of asthmatic individuals being exposed to peak levels of SO_2 greater than 0.25 ppmv for 5 to 10 min is low. However, some monitoring data indicate that peak SO_2 concentrations of >0.25 ppmv do occasionally occur, suggesting that mild and moderately asthmatic individuals residing in the vicinity of major sources may be at increased risk.

5.4.2.2 Acid Sulfate

Sulfuric acid is a strong irritant. However, consistent effects on pulmonary function or respiratory symptoms have not been shown in acute exposure studies of healthy adults. There is some evidence that asthmatic individuals are more sensitive to the effects of acid sulfate on lung mechanical function. They may experience moderate bronchoconstriction following exposure to H_2SO_4 at concentrations of <1 mg/m.^{3} These effects are not as consistent or dramatic as those reported for SO_2 exposures. Children with allergic asthma appear to be more sensitive to H_2SO_4 than older asthmatics.

Sulfuric acid, like SO_2, can affect pulmonary defense mechanisms. Acute exposures as low as 0.1 mg/m^3 have been shown to alter mucociliary transport in normal humans. Clearance of deposited particles appears to be stimulated at low concentrations and retarded at higher concentrations. Sulfate exposure can also adversely affect macrophage functioning. There is also some evidence that acid sulfates can enhance sensitization to potentially allergenic substances and modulate the activity of cells involved in allergic responses.

Biological responses to acid sulfates are likely due to deposition of hydrogen ions (H^+) on surface tissues of respiratory airways. The relative potency of acid sulfates has been shown in toxicological studies to be directly related to its degree of acidity (increased potency with increased H^+ concentration). Since H^+ may be responsible for acid sulfate toxicity, the biological response is likely to be affected by how it is inhaled. Acid sulfates are more likely to be neutralized by ammonia (NH_3) as a result of mouth, as compared to nasal, breathing.

There is insufficient evidence to definitively establish a causal relationship between acid aerosol exposure and adverse health effects. This may reflect, in part, the fact that few epidemiological studies have included measurements of H^+ and that acid aerosols vary colinearly with PM_{10} and $PM_{2.5}$ concentrations. Long-term and population-based annual assessments of mortality (comparing different communities) indicate only a weak effect of acid aerosol. Recent studies with direct measurements of acid aerosols and other major pollutants have shown consistent relationships between acid aerosol levels and hospital admissions for respiratory disease. Acid aerosol has also been related to a significantly higher prevalence of bronchitis in children and reduced lung growth (determined from lung function measurements) as a result of chronic exposures.

5.4.3   Particulate Matter

Before atmospheric particles can cause adverse health effects, they must enter and be deposited in the human respiratory system. Respiratory penetration and deposition depend on the aerodynamic size of particles, respiratory defense mechanisms, and breathing patterns. Particles can be described as inhalable (IPM), thoracic (TPM), or respirable (RPM), based on their penetration and potential for deposition. Size ranges and fractional penetration of IPM, TPM, and RPM are illustrated in Figure 5.5. Inhalable PM can enter the upper airways of the head; TPM, the airways and gas exchange regions of the lung. Respirable particles are a subset of TPM with a high probability of being deposited in the pulmonary region.

Particles more than 10 μm generally do not pass through the nasal hairs and defense mechanisms of the upper respiratory system. Save for allergens, they are not a public health concern. Thoracic and respirable particles are significant public health concerns since they enter respiratory airways and are deposited in lung tissue.

The aerodynamic diameter of particles is the primary determinant of respiratory deposition. Deposition occurs by inertial impaction, sedimentation, and diffusion. Impaction is particularly effective in the particle size range of 5 to 10 μm; sedimentation, 3 to 5 μm; and diffusion, <0.1 μm.

Inertial impaction is the process by which larger particles traveling in an airstream collide with a surface when the airstream changes direction; sedimentation occurs when large particles fall to a surface under the force of gravity. In diffusion, very small particles come into contact with surfaces as a result of their random motion. These deposition mechanisms are affected by breathing patterns. Deep breathing increases the number of particles that come into contact with respiratory epithelia and, as a consequence, increases deposition by impaction, sedimentation, and diffusion. Breathing depth and rate influence the (1) size of particles deposited, (2) regional location of deposition, and (3) deposition mechanisms.

Regional deposition patterns are particularly important since sites of particle deposition determine mechanisms and rates of clearance, potential for retention and particle dissolution, and the location and nature of tissue injury that may occur.

Deposition is described in the context of regions of the respiratory system (nasopharyngeal, tracheobronchial, and pulmonary) and, based on particle size, the fraction deposited. Fractional deposition efficiency for three TPM particle sizes can be seen in Figure 5.6. Particle deposition is greatest in the pulmonary region for the smallest particle size (2.5 μm), with high nasopharyngeal deposition for the largest particle size (7.5 μm).

Because of changes in flow patterns in the tracheobronchial zone, many particles tend to be deposited at or near airway bifurcations (branchings). As nerve endings are concentrated at these sites, deposited particles mechanically stimulate reflex coughing and bronchoconstriction. The sensitivity of nerve endings to chemical stimuli causes an increase in breathing rate and reduced pulmonary compliance (ability of the lungs to yield to increases in pressure without disruption).

Deposition of particles is also influenced by mass concentrations, molecular composition, pH, and solubility. Deposition varies among nonsmokers, smokers, and individuals with lung disease. Tracheobronchial deposition is slightly higher in

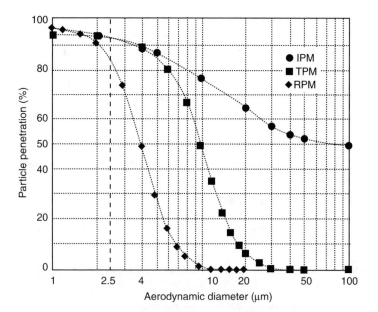

Figure 5.5 Particle penetration (percent) for inhalable, thoracic, and respirable particles through an ideal inlet. (From USEPA, Third External Review Draft of Air Quality Criteria for Particulate Matter (April 2002), Vol. 1, http://www.epa.gov/ncea/pdfs/partmatt/Vol_1_AQCD_PM_3rd_Review_pdf.)

smokers and greatly increased in individuals with lung disease. It is also increased in individuals with varying levels of airway obstruction, such as smokers, asthmatics, and individuals with COPD.

Particle retention depends on clearance rates, which vary among different regions of the respiratory tract. In the ciliated airways of the nose and upper tracheobronchial zone, clearance in healthy individuals is achieved in less than 1 day. In general, clearance time increases as the site of deposition becomes deeper. Clearance from the alveoli may take weeks to months. Slow particle clearance is considered to be detrimental since harmful substances are in contact with sensitive tissue for longer periods.

Until the late 1970s and early 1980s, it was widely held that exposures to high concentrations of PM prevalent in the U.S. and Europe before 1970 were a significant health hazard and that evidence for adverse effects at the much lower concentrations present in the atmosphere after 1970 was relatively weak.

By the mid-1980s and 1990s, investigators began to report significant associations between ambient PM levels and various adverse health effects. These reports included (1) the Harvard Six Cities Study, which showed that long-term PM exposure was associated with increased risk of respiratory illness in children and cardiopulmonary mortality in adults; (2) Utah Valley studies, which showed that PM levels were significantly associated with various illness indices, including respiratory hospitalizations, lung function changes, respiratory symptoms, school absenteeism, and premature mortality; and (3) a series of studies that showed a relationship between daily changes in ambient PM levels and mortality.

Figure 5.6   Deposition of particles of varying size in nasopharyngeal, tracheobronchial, and pulmonary regions of the lung. (From Foster, W.M. in, *Air Pollution and Health*, Holgate, S.T., Samet, J.M., Koren, H.S., and Maynard, R.L., Eds., Academic Press, San Diego, 1991, p. 304. With permission.)

Most recent epidemiological investigations have focused on short-term, acute exposures. They show consistent effects across a range of related health outcomes. A summary of acute effect estimates as a percentage change with a 10 μg/m³ increase in PM₁₀ is presented in Figure 5.7. These indicate that a 10 μg/m³ increase in average 24-h PM₁₀ concentration is associated with an increase of ~0.8% in daily mortality, a somewhat larger association with cardiovascular mortality, and a much larger association with respiratory mortality. Increased hospitalizations and related health care visits are significant for various respiratory diseases and, to a lesser extent, cardiovascular disease. Increased symptom prevalence is associated with PM₁₀ levels; these include lower respiratory system symptoms, asthma, and cough.

Chronic studies indicate that long-term exposure to PM, particularly in the small particle size range (PM₂.₅), is associated with various cardiac and pulmonary health effects (Figure 5.8). Total mortality appears to increase 2 to 4% for every 5 μg/m³ increase in PM₂.₅, with somewhat larger associations for cardiopulmonary mortality. Chronic exposure studies have also reported significant increases in respiratory disease (bronchitis) and small declines in lung function.

Though epidemiological evidence linking both acute and chronic PM exposures to significant health effects is strong, toxicological mechanisms for these effects are less clear. Considerable research has focused on the biological mechanisms that may be responsible for the cardiopulmonary and respiratory effects reported in epidemiological studies.

Associations of mortality and morbidity (illness) with ambient PM levels suggest that a variety of tissues may be affected by particles in the PM₁₀ range in ways that cause lung disease and systemic effects. Several biologically plausible mechanisms have been proposed to explain these results. They consistently include some form of inflammatory response. Since production of clots in coronary vessels (coronary

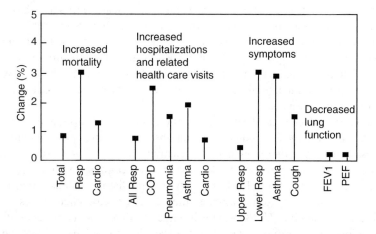

Figure 5.7   Estimated percent change in health endpoints associated with a 10 µg/m³ increase in PM₁₀ levels in acute exposure epidemiological studies. (From Pope, C.A., in *Air Pollution and Health*, Holgate, S.T., Samet, J.M., Koren, H.S., and Maynard, R.L., Eds., Academic Press, San Diego, 1999, p. 688. With permission.)

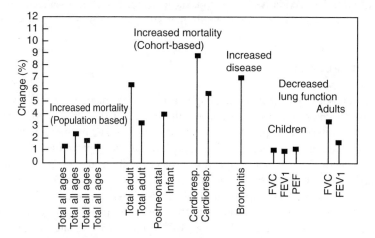

Figure 5.8   Estimated percent change in health endpoints associated with a 5 µg/m³ increase in PM₂.₅ levels in chronic exposure studies. (From Pope, C.A., in *Air Pollution and Health*, Holgate, S.T., Samet, J.M., Koren, H.S., and Maynard, R.L., Eds., Academic Press, San Diego, 1999, p. 694. With permission.)

thrombosis) is the cause of myocardial infarction (heart attack), the apparent causal relationship between coronary thrombosis and particle deposition in the airways is somewhat problematic. It has been proposed that inflammatory responses in the lungs produce procoagulant factors locally or as a result of mediators that, when released, act on the liver to increase levels of procoagulant factors. There is limited epidemiological evidence to support this, as increased blood viscosity has been observed at elevated pollutant levels.

Ultrafine particle (UFP) ($\leq$0.01 μm in diameter) research provides a biologically plausible explanation for the toxicity of ambient particles. Such particles induce significant toxic effects in lung tissue even when they are formed from materials that are not toxic as larger respirable particles. Factors that affect or are associated with UFP toxicity include surface area and chemistry, particle number, oxidative stress, and particle interstitialization (i.e., movement of particles from lung spaces into surrounding tissues).

The large surface area of UFPs appears to facilitate adsorption and absorption of substances from the environment or from the fluid of lung epithelia onto particle surfaces, which then increases particle reactivity. Iron (Fe), one such substance, produces reactive oxygen species. Macrophages attempting to phagocytize UFPs appear to be stimulated to release inflammatory mediators such as tumor necrosis factor, and on sustained stimulation, epithelial cells release chemokines that contribute to inflammation. UFPs present in large numbers commonly undergo interstitialization and therefore cannot be removed by normal pulmonary clearance mechanisms.

Particles may cause lung tissue injury as a result of oxidative stress. Production of free radicals (including hydroxyl radical, OH·) may occur in response to exposures to small particles. Oxidative stress appears to result from the localized release of high concentrations of transition metals that stimulate release of reactive oxygen species.

5.4.4   Hydrocarbons

Hydrocarbons (HCs) in the atmosphere pose two major exposure and health concerns. The more important of these is the precursor role that reactive (and less reactive) HCs have on tropospheric O_3 production. The HC or, more appropriately, the nonmethane hydrocarbon (NMHC) standard has been promulgated to achieve the NAAQS for O_3; it is not designed to protect public health from exposure risks that may be associated with specific NMHCs.

As a general rule, exposures to individual NMHC species are relatively low and health risks on a community-wide basis are considered to be small to negligible.

NMHCs such as the aldehydes (e.g., HCHO) and acrolein are potent irritants. Exposure to these substances, individually or collectively, may cause eye, nose, throat, and sinus irritation. Such symptoms are transitory, resulting in no apparent long-term adverse health effects.

Polycyclic aromatic hydrocarbons (PAHs) are of some health concern. The PAHs include group of substances characterized by multiple benzene rings. They are commonly produced as a result of incomplete combustion of coal and wood and, to a lesser extent, other fuels. The PAHs are carcinogenic, with benzo-α-pyrene being one of the most potent carcinogens and common PAHs found in aerosol samples.

As a result of fuel use changes, PAH levels in urban areas in the U.S. have declined to about one third of the levels present in the mid-1950s. Similar declines have been reported for northern Europe. Though PAHs have been suspected of contributing to increased lung cancer rates in urban areas, epidemiological studies have not definitively shown such a relationship.

Many NMHCs are regulated as hazardous or toxic air pollutants. One of these is benzene, one of the seven pollutants initially regulated under the hazardous pollutant provision of the 1970 Clean Air Act (CAA) Amendments. They also include most of the 182 other substances regulated under the Air Toxics provision of the 1990 CAA Amendments.

Pollutants regulated under Hazardous/Toxics provisions include substances that vary in their toxicological properties and potential risks to humans and the environment. Many are suspected human carcinogens, mutagens, teratogens, etc.; some are irritants, and some neurotoxic.

5.4.5   Nitrogen Oxides

Nitric oxide is a relatively nonirritating gas believed to pose little or no health risk at ambient exposure levels. Its importance lies in the fact that it is rapidly oxidized to NO_2, which has a much higher toxicity.

Unlike SO_2, which is rapidly absorbed in fluids of the upper tracheobronchial region, NO_2 is less soluble and thus passes into the pulmonary region where tissue damage may occur. In individuals occupationally exposed to high NO_2 levels, adverse effects such as pulmonary edema only manifest themselves hours after exposure has ended. Elevated NO_2 exposures have been known to occur in the manufacture of nitric acid (HNO_3), in farm silos, from electric arc welding, and from use of explosives in mining.

In toxicological studies of animals, abnormal pathological and physiological changes have only been observed at concentrations in excess of those found in ambient environments. At the lowest NO_2 exposure levels at which adverse effects have been detected (0.5 ppmv, 1000 $\mu g/m^3$), pathological changes have included destruction of cilia, alveolar tissue disruption, and obstruction of respiratory bronchioles. Exposures of rats and rabbits at higher levels have caused severe tissue damage resembling emphysema.

Results of animal toxicological studies support the hypothesis that NO_2 may be a causal or aggravating agent in respiratory infections. There is evidence that NO_2 can damage respiratory defense mechanisms, allowing bacteria to multiply and invade lung tissues. In mice, hamsters, and monkeys infected with pneumonia-causing bacteria, exposure to elevated NO_2 levels resulted in reduced ability to clear bacteria from the lungs, reduced life span, and increased mortality.

Epidemiological studies have been conducted to determine whether NO_2 exposures at ambient levels are associated with respiratory symptoms and disease or other adverse health effects. Most have focused on the potential effects of either outdoor or indoor NO_2 exposures on children's health. A few have attempted to evaluate the potential adverse effects of combined indoor and outdoor exposures.

Results from a variety of outdoor exposure studies on children's health have been reported. The earliest of these (known as the Chattanooga School Children Study) investigated potential health effects associated with emissions of NO_2 from a trinitrotoluene (TNT)-manufacturing facility. Lung function (ventilation rate) of second-grade children in the high-NO_2 area was significantly lower than that of children in the control area. Respiratory illness rates of families in the exposed area were also higher. Later studies indicated that respiratory symptom duration increased with increasing annual mean NO_2 concentrations, more frequent hospitalizations with increasing 2-year means, and more treatments in hospitals for lower respiratory illness with higher lifelong NO_2 exposures in female children. Similar results have also been reported for Swedish school children in 10 communities where chronic cough and respiratory infections (bronchitis or pneumonia) were associated with increased annual mean NO_2 exposures.

Studies of adults have yielded mixed results. Positive associations between NO_2 and respiratory illness have been reported in some investigations but not others. In the European study on air pollution and health (APHEA), significant positive associations were observed between daily death rates and NO_2 concentrations. An increase of 25 ppbv (50 $\mu g/m^3$) 1-h maximum concentration was associated with a 1.3% increase in the daily number of deaths. Other less definitive studies indicate that increased death rates are associated with increased NO_2 exposures.

Significant public health concern was expressed relative to potential health effects of NO_2 exposures on children's health following publication of a 1970s' epidemiological study in the U.K. that observed increased respiratory symptoms among children living in homes using gas cooking stoves (compared to those using electric stoves).

In 1992, the USEPA conducted a review and meta-analysis of 11 epidemiological studies that evaluated potential effects of indoor NO_2 exposures on respiratory illness among children under age 12. Relatively good agreement was observed among these studies, with an approximate 20% increase in respiratory illness associated with the presence of gas cooking stoves in households. This increase in respiratory symptoms was described as being equivalent to a 15-ppbv (30 $\mu g/m^3$) increase in NO_2.

In a recent residential Australian study, the presence of a gas stove and exposure to NO_2 were observed to be significant risk factors for asthma and respiratory symptoms. The presence of a gas cooking stove in a household increased the likelihood of asthma threefold and respiratory illness twofold.

Several studies indicate that NO_2 exposures associated with kerosene heater usage increase the risk of respiratory illness in children under age 7. Nitrogen dioxide exposures of >16 ppbv (32 $\mu g/m^3$) were reported to more than double the risk of lower respiratory symptoms and illness (fever, chest pain, productive cough, wheeze, chest cold, bronchitis, pneumonia, or asthma) compared to children at lower levels of exposure in nonkerosene-stove-using households. Increased risk of upper respiratory symptoms (sore throat, nasal congestion, dry cough, croup, or head cold) was also reported.

The NAAQS for NO_2 in the U.S. is 50 ppbv (100 $\mu g/m^3$) averaged over 1 year. The World Health Organization (WHO), however, recommends a much lower annual average guideline (21 ppbv, 42 $\mu g/m^3$), as well as a 1-h guideline of 100 ppbv (200 $\mu g/m^3$). These recommendations indicate that the U.S. NAAQS for NO_2 may not be sufficiently protective of public health and that a short-term standard (1 h) is needed to protect individuals (particularly children) from peak exposures.

5.4.6 Ozone

Ozone is one of the most ubiquitous and toxic pollutants found in ambient atmospheres. Depending on the year, upwards of 20% or more of the U.S. population lives in areas where the 1-h NAAQS (regulatory requirement prior to 2004) of 120 ppbv (235 $\mu g/m^3$) was exceeded.

Pathological and physiological effects of O_3 exposures (at the high end (or higher) of those reported in urban areas) have been evaluated in animal (usually rodent) exposure studies. Ciliated cells in respiratory airways as well as squamous epithelial cells in alveoli are injured or killed. This is followed by cell proliferation and tissue repair, with a loss of damaged epithelial cells and formation of exudates (fluid) and inflammatory cells. Longer-term exposures are characterized by continued hyperplasia (excess epithelial cell proliferation), low-grade inflammation with exudate, and synthesis of collagen (scar-forming substance).

Ozone may injure tissue membranes by oxidizing amino acids, sulfhydryl (SH) groups on enzymes and other proteins, and polyunsaturated fatty acids (lipids). This lipid peroxidation and resultant production of free radicals increases membrane permeability, with a concomitant leakage of essential electrolytes and enzymes. It also results in swelling and disintegration of cell organelles such as lysosomes and mitochondria, and inhibition of metabolic pathways. Significant increases in activity of various antioxidant enzymes and enzyme systems have been reported at continuous exposure concentrations in the range of 0.4 to 0.8 ppmv (783 to 1567 $\mu g/m^3$).

At the physiological level, O_3 exposures result in significant lung function changes. These include increased respiratory rates and pulmonary resistance, decreased tidal volume, and changes in respiratory mechanics. Changes appear to be transient, returning to normal after exposure ceases or after several days of adaptation to continuous exposures. Such changes are dose dependent and increase with exercise-associated lung ventilation.

Ozone exposures have a variety of effects that impair the body's ability to defend itself against infection. These include (1) interference with mucociliary transport as a result of injury to the ciliated epithelium lining the upper respiratory system, (2) impaired killing of bacteria, and (3) impaired macrophage function. On the other hand, O_3 can inactivate certain viruses, including those that cause influenza.

There is limited information available on O_3's potential to cause cancer. Studies on microorganisms, plant root tips, and tissue culture indicate that O_3 is mutagenic and therefore genotoxic. Mutagenic and genotoxic substances, as a rule, have a high potential for causing cancer. There is, however, no conclusive evidence that O_3 causes cancer in laboratory animals or humans.

There appears to be an adaptation to, or attenuation of, the effects of O_3 with intermittent, repeated exposures to health-affecting O_3 levels. In humans, the effect of O_3 is greatest on the second day of exposure, with normal function returning by the fifth day. In California, studies of human pulmonary function changes associated with relatively high exposures (~0.50 ppmv, 979 $\mu g/m^3$) showed a more severe response when these levels occurred in a high-pollution season rather than during a low-O_3 season.

Recent epidemiological studies have shown significant associations between daily O_3 levels and a range of adverse health effects. These include decreases in pulmonary function, aggravation of preexisting disease such as asthma, increases in daily hospital admissions and hospital admission for respiratory ailments, and premature mortality.

Most recent O_3 epidemiological investigations are longitudinal and follow a single population over time. Such studies, with time-series statistical analyses, have been able to discriminate between ambient temperature and O_3 exposure effects. This is especially true for mortality.

Significant associations between ambient O_3 levels and pulmonary function changes and other respiratory responses have been reported from studies of children attending summer camps. These have included significant decreases in FEV_1 (forced expiratory volume in the first second after a maximal inhalation), asthma exacerbation, chest symptoms, and decreased lung function.

Studies conducted in the U.S., Canada, and Europe designed to evaluate potential relationships between hospital admissions for respiratory disease and atmospheric pollutants reported significant associations between respiratory (e.g., pneumonia, COPD) and asthma admissions and daily 1-h maximum O_3 levels.

Emergency room visits for treatment of respiratory symptoms (which occur an order of magnitude more frequently than hospital admissions) significantly increase on high-O_3 days.

Acute morbidity effects experienced by the general public include minor restricted activity days (due to illness), acute respiratory symptoms days, and impaired athletic performance. Though less severe than health effects described above, they are likely to occur more frequently and affect a larger percentage of the population. Reduced health impacts from these and more serious health effects have been calculated for New York City as health impacts avoided if the O_3 NAAQS 8-h average of 80 ppb (160 $\mu g/m^3$) were achieved (Figure 5.9).

Several recent epidemiological studies indicate that chronic exposure to ambient O_3 is associated with an increased prevalence of asthma in humans as well as declines in mid-expiratory lung function. Thus, O_3 not only exacerbates asthma, but also appears to cause it.

As is apparent, recent epidemiological studies have shown that exposure to O_3 at current ambient levels has significant adverse health effects on human respiratory health and includes a small risk of premature mortality. There appears to be no threshold dose for these effects.

5.4.7 Lead

Lead is a heavy metal that occurs naturally in soil, water, and air. It has been used for thousands of years because of its properties and relative ease of smelting from mineral ores. As a consequence, humans have been, and continue to be, exposed to Pb levels significantly above background. Notable historical exposures have included ingestion of (1) liquids stored in Pb containers (particularly during Roman times), (2) food from Pb-soldered cans, and (3) Pb paint and dusts by children; and inhalation of (1) airborne Pb dusts from motor vehicles, and (2) primary and

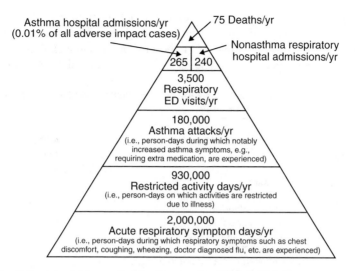

Asthma hospital admissions/yr
(0.01% of all adverse impact cases)

75 Deaths/yr

Nonasthma respiratory
hospital admissions/yr

265 | 240

3,500
Respiratory
ED visits/yr

180,000
Asthma attacks/yr
(i.e., person-days during which notably
increased asthma symptoms, e.g.,
requiring extra medication, are experienced)

930,000
Restricted activity days/yr
(i.e., person-days on which activities are restricted
due to illness)

2,000,000
Acute respiratory symptom days/yr
(i.e., person-days during which respiratory symptoms such as chest
discomfort, coughing, wheezing, doctor diagnosed flu, etc. are experienced)

Figure 5.9 Estimates of adverse health effects avoided in New York City as a consequence of compliance with the 80-ppbv, 8-h O_3 NAAQS. (From Thurston, G.D. and Ito, K., in *Air Pollution and Health*, Holgate, S.T., Samet, J.M., Koren, H.S., and Maynard, R.L., Eds, Academic Press, San Diego, 1999, p. 504. With permission.)

secondary Pb as well as other metal smelter emissions. Significant exposures have also occurred (and continue to occur) among Pb workers. Much of our understanding of its adverse effects on humans is based on worker exposures as well as exposures of children to ingested Pb paint dust.

Unlike other trace elements, Pb is not required by the body to carry out any metabolic or physiological functions. Because of its widespread use, total human body burdens today are ~300 to 500 times greater than those of our preindustrial ancestors.

On exposure, Pb is absorbed and transported by the bloodstream to all parts of the body. Though commonly found in blood and soft tissues, because of its similarities to calcium (Ca), it tends to be deposited in bone; bone thus serves as a storage depot.

The amount of Pb extracted from food, water, ingested dust, and airborne particles depends on its form and an individual's nutritional status (poorly nourished individuals absorb more Pb), metabolic activity, and prior exposure history. On inhalation exposure, ~20 to 40% of inhaled Pb particles are deposited in lung tissue, where 50% or more is absorbed and enters the bloodstream. Some inhaled particles are removed by pulmonary clearance mechanisms and, on swallowing, enter the bloodstream through the gastrointestinal tract. Lead absorption in the gastrointestinal system averages ~40% in children and 10% in adults.

Acute or chronic Pb poisoning may occur as a result of significant exposures. Principal target organs are blood, the brain and nervous system, and the renal and reproductive systems. Acute exposures (blood levels of >60 µg/dl) may produce colic, shock, severe anemia, nervousness, kidney damage, irreversible brain damage, and death. Chronic poisoning may result in a variety of symptoms depending on the exposure level. Acute and chronic symptom responses based on blood lead levels (BLLs) are summarized in Table 5.2. Exposures on the more extreme end of BLLs may cause severe brain damage and damage to kidneys and blood-forming tissue.

Table 5.2 Blood Lead Levels and Associated Health and Physiological Effects in Children and Adults

Blood Lead Levels (µg/dl)	Children	Adults
<10		Early signs of hypertension ALA-D inhibition
10–15	Crosses placenta Neurodevelopmental effects ALA-D inhibition Impairment of IQ Increased erythrocyte protoporphyrin Reduced gestational age and birth weight	
15–20		Increased erythrocyte protoporphyrin
20–30	Altered CNS electrophysical response Interference with vitamin D metabolism	
30–40	Reduced Hb synthesis Peripheral nerve dysfunction	Systolic hypertension
40–50		Reduced Hb synthesis
60	Peripheral neuropathies	Reproductive effects in females
70	Anemia	
80		Anemia Encephalopathy symptoms
80–100	Encephalopathy symptoms Chronic nephropathy Colic and other gastrointestinal symptoms	Chronic nephropathy

Note: CNS = central nervous system.

   ALA-D = aminolevulinic acid dehydratase

Source: From CDC, *Preventing Lead Poisoning in Children*, DDHS, Washington, D.C., October 1991.

At the lower end, health effects may include neurodevelopmental changes in children, increased blood pressure and related cardiovascular effects in adults, and possibly cancer.

Since Pb accumulates and is only slowly removed from the body, repeated exposures (over months to years) commonly result in elevated BLLs. Since it is stored in bone, BLLs reflect relatively recent exposures (past 1 to 3 months) and mobilization from bone and other depots. Though not indicative of the total body burden, blood Pb is relatively closely correlated with exposure levels.

Hematological changes, i.e., effects on blood chemistry and associated physiological changes, are the earliest manifestations of chronic exposure. Lead interferes with the synthesis of heme (the O_2-carrying component of Hb in red blood cells), which may result in anemia, a symptom commonly associated with chronic Pb poisoning. Inhibition of enzymes involved in heme synthesis has been observed at BLLs as low as 10 μg/dl. Lead also inhibits heme-dependent liver enzymes, increasing an individual's vulnerability to the harmful effects of other toxic substances. It can also interfere with vitamin D production and cause neurological effects. Inhibition of vitamin D production has been observed at BLLs of 30 μg/dl in adults and as low as 12 μg/dl in children.

Lead exposures may cause adverse sexual and reproductive effects. Sperm abnormalities, reduced fertility, and altered testicular function have been observed in male industrial workers at BLLs of 40 to 50 μg/dl. Since it readily crosses the placenta, Pb may also pose significant risks to the fetus in exposed pregnant females. Lead is mobilized from bone during pregnancy; BLLs in pregnant females are reportedly higher than population averages. Associations between maternal BLLs and preterm delivery and low birth weight have been reported.

The nervous system is adversely affected by Pb. At high blood levels (>80 μg/dl), it causes encephalopathy (brain damage). Children with BLLs in this range may experience permanent neurological damage such as severe mental retardation and recurrent convulsions. There is evidence to suggest that Pb may impair peripheral nerve conduction in children at BLLs as low as 20 to 30 μg/dl. Brain wave changes have been observed at levels as low as 15 μg/dl, with no apparent threshold.

Prospective epidemiological studies have shown an association between general measures of intelligence (intelligence quotient (IQ)) in children and pre- and postnatal blood levels as low as 10 to 15 μg/dl. These studies suggest that up to at least age 7, exposure to relatively low Pb levels (10 to 40 μg/dl) may result in neurodevelopmental effects: decreased intelligence, short-term memory loss, reading and spelling underachievement, impairment of visual motor function, poor perception integration, disruptive classroom behavior, and impaired reaction time. As a consequence, the U.S. Centers for Disease Control (CDC) has issued a guideline value of 10 μg/dl as the lowest BLL of public health concern in children.

An association between Pb exposure (at low BLLs) and elevated blood pressure in adults has been reported. A doubling of blood Pb appears to increase systolic pressure by 1 mmHg and diastolic pressure by 0.6 mmHg. In other studies of systolic blood pressure in men, a doubling of blood Pb from 5 to 10 μg/dl was associated with an increase of 1.25 mmHg systolic pressure.

Lead has been shown to be genotoxic; i.e., it can cause gene mutations. It can also cause cell transformation and interfere with DNA synthesis in mammalian tissue culture. Lead compounds can induce kidney tumors in rodents. Though not conclusive, studies are suggestive of a causal relationship between Pb exposures and cancer. Based on available evidence, the USEPA has designated Pb a Group B2 human carcinogen; i.e., there is sufficient evidence of carcinogenicity from animal studies, but epidemiologic studies are not conclusive.

Young children (<7 years old) are subject to the greatest health risk from Pb exposures because they have (1) a less developed blood–brain barrier and therefore greater neurologic sensitivity, (2) a faster resting inhalation rate, (3) a tendency to breathe through their mouth during play, (4) a greater intake dose rate due to their smaller weight, and (5) hand-to-mouth behaviors that result in ingestion of Pb from soil and dust. They also absorb more Pb through the intestines, particularly under the age of 2. Children from low-income neighborhoods are frequently at very high risk because of diets low in calcium (Ca) and iron (Fe), elements that suppress Pb absorption.

Lead inhalation exposures have declined considerably since the mid-1970s. In survey studies conducted by the CDC from 1976 to 1980 and 1989 to 1991, BLLs decreased by approximately 78% in all age groups. Decreases in the period 1976–1980 are illustrated in Figure 5.10. During that time, 88% of children ages 1 to 5 were estimated to have had BLLs of ≥10 µg/dl; in the period 1988–1991, less than 10% of children were determined to have BLLs of ≥10 µg/dl. Decreases in BLLs have been observed across subgroups stratified by race/ethnicity, gender, urban status, and income levels. Declines in BLLs similar to those in the U.S. have been reported in Sweden, Belgium, and the U.K.

Observed declines in BLLs coincided in time with the phasedown of Pb in gasoline, reduction in use of Pb-soldered cans for food and soft drinks (beginning in 1980), and limits on Pb in paint (1978). The CDC has concluded that the major contributor to the observed decline in BLLs has been reduced Pb use in gasoline.

A strong and consistent association exists between ambient concentrations of Pb and BLLs in children and adults. The blood Pb:air Pb exposure relationship has been estimated to be 1.8 µg/dl per µg/m^3 for adults and 4.2 µg/dl per µg/m^3 for children. Assuming no threshold exists for neurodevelopmental effects in children and hypertension in adults, even airborne lead exposures of 1 µg/m^3 may be expected to have some adverse effects on children and adults. Fortunately, ambient concentrations in urban areas unaffected by emissions from stationary sources are now relatively low in the U.S. In California, a state with a significant motor vehicle population, average ambient Pb levels of 0.06 µg/m^3 are at their lowest values in a half century.

Though both ambient concentrations and BLLs have decreased significantly in the U.S. over the past two decades, Pb exposure from ambient air pollution is likely to be a significant public health threat in both developed and developing countries where it is still widely used as a gasoline additive and major stationary sources are unregulated.

5.4.8 Hazardous Air Pollutants

Under Section 112 of the 1970 CAA Amendments, the USEPA was authorized to promulgate national emission standards for substances designated as hazardous air pollutants. Such substances were given special regulatory status since they were deemed to pose health risks of a more serious nature than pollutants regulated under NAAQSs. These substances were thought to cause unique exposure risks because of their innate toxicity and no apparent threshold for adverse effects. Additionally, exposures from such substances were usually limited to the near vicinity of a source and were, in general, not considered to be community-wide concerns.

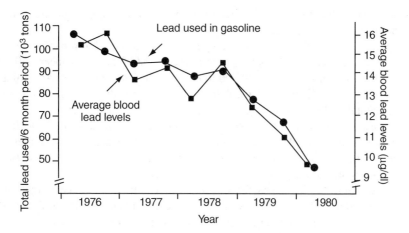

Figure 5.10 Declines in gasoline lead use and BLLs in the U.S., 1976–1980. (From CDC, *Preventing Lead Poisoning in Children*, DDHS, Washington, D.C., October 1991.)

By the time the CAA Amendments were passed in 1990, only seven hazardous pollutants had been designated and emission standards promulgated. These were mercury (Hg), beryllium (Be), asbestos, vinyl chloride, benzene, inorganic arsenic (As), and radionuclides. Of these, only Hg and Be are uniquely toxic in the classical sense. The other five are presumed human carcinogens. Asbestos exposures, for example, have been definitively linked to increased risk of lung cancer and mesothelioma (a cancer of the chest cavity) in asbestos workers and their families; vinyl chloride with angiosarcoma (a form of liver cancer) and other cancers among rubber and plastics workers; benzene with leukemia among a variety of workers exposed to it; as with lung cancer among smelter workers; and radionuclides with a number of different cancers.

Congress, in the 1990 CAA Amendments, listed another 182 substances and groups of substances for regulation as hazardous and toxic pollutants using technology-based rather than health standards. Major health concerns associated with these newly listed hazardous and toxic air pollutants are acute and chronic toxicity, neurotoxicity, reproductive toxicity, carcinogenicity, mutagenicity, and teratogenicity.

Though the toxic effects of many of these substances are well known from animal studies and workplace human exposures, little is known about ambient concentrations and health risks associated with ambient exposures.

5.5 PERSONAL AIR POLLUTION

It is paradoxical that as a nation the U.S. has expended more than a trillion dollars over the past 30+ years in an effort to control ambient air pollution and thereby safeguard public health, while nearly 40 million Americans voluntarily expose themselves to a form of air pollution whose health consequences are orders of magnitude greater than those from ambient exposures. This is more of a paradox when one considers that scientific evidence for a causal role of cigarette smoking in a number of health problems is overwhelming.

Approximately 500,000 deaths each year can be attributed to tobacco smoking. Of these, 130,000 can be attributed to lung cancer, 25,000 to other cancers, 300,000 to cardiovascular disease, and ~20,000 to chronic pulmonary disease. According to reports issued by the Surgeon General, tobacco smoking is the single most important environmental factor contributing to premature mortality in the U.S.

The excess mortality of smokers (compared to nonsmokers) is 70%. It is a function of quantity smoked and smoking duration. Life expectancy decreases with increased tobacco consumption and years of smoking; it is decreased by 8 to 9 years for a male smoking two packs of cigarettes per day.

In addition to the increased risk of premature death, smokers have higher morbidity rates. Before clinical disease is manifested, smokers usually suffer the discomfort of excessive coughing and phlegm production, and hoarseness.

One of the major effects of tobacco smoking is induction of chronic respiratory diseases such as chronic bronchitis and emphysema. It is probable that tobacco smoking has caused a high percentage of diagnosed cases of such diseases as COPD because smokers have a higher incidence of these diseases than nonsmokers.

Tobacco smoking is a major cause of human cancer. Strong epidemiological evidence exists to implicate smoking as the causal factor in lung cancer and cancers of the larynx, oral cavity, esophagus, urinary bladder, kidney, and pancreas. Overall, lung cancer risk for smokers is about 10 to 12 times greater than that for nonsmokers.

In the public mind, lung cancer is the most important health consequence of smoking. It is not generally realized that smoking-related mortality from cardiovascular disease is approximately twice that for all forms of smoking-related cancer. Smoking is a major independent risk factor for heart attacks in adults. It also appears to contribute to more severe and extensive atherosclerosis of the aorta and coronary arteries.

The voluntary risk accepted by smokers becomes less than voluntary to the fetus of a pregnant female. Significant scientific evidence exists that maternal smoking retards fetal growth and increases the risk of spontaneous abortion and fetal and neonatal death in otherwise normal infants. There is also increasing evidence that children of smoking mothers may have measurable deficiencies in physical growth and intellectual and emotional development.

Of the thousands of identified chemical constituents in tobacco smoke, CO, nicotine, and tar are likely to be the major contributors to health hazards associated with smoking. Carbon monoxide levels are especially high, with levels in mainstream smoke of 1.5 to 5.5% (15,000 to 55,000 ppmv). An increased risk of cardiovascular patients to myocardial infarction and sudden death appears to be associated with tobacco smoke CO exposures. However, because tobacco smoke contains numerous harmful substances, it is difficult to specifically assess the harmful effects of CO and its role in cardiovascular disease. Indeed, effects of tobacco smoking on the incidence of clinical cardiovascular disease may not be due to CO alone; other factors such as nicotine, cyanide, and trace elements may be important. Nicotine is considered to be a major contributor to coronary atherosclerotic disease. Tar and nicotine make up the particulate phase of cigarette smoke. Tar consists of a variety of PAHs considered to be the likely causes of smoking-related cancers.

5.6 RISK ASSESSMENT AND MANAGEMENT

Risk assessment is a tool used by the USEPA and several other federal agencies to quantify risk associated with exposure to substances in the environment. It is a systematic process whereby hazardous properties of a substance, human exposure to it, and dose–response relationships are evaluated to characterize risks of human environmental exposures; it serves as a basis for risk management decisions (Figure 5.11). The term *risk assessment* implies that environmental exposures pose a probability of an unwanted event taking place, e.g., disease or even death. The assessment process has the following components: hazard identification, dose–response assessment, and exposure assessment. These three lead to a final risk characterization.

Hazard identification involves delineation of whether exposure to a substance can cause a particular health effect, as well as an increased incidence of that effect (e.g., cancer, birth defects, liver toxicity), and characterization of the nature and strength of a causal relationship.

Dose–response assessments attempt to characterize the relationship between an exposure dose and the severity and incidence of observed health effects. Factors that influence dose–response relationships are considered, including intensity and pattern of exposure, age, and lifestyle variables that may affect susceptibility. They also include extrapolation from responses at high doses to those at lower ones, and from animals to humans.

Exposure assessments involve determinations of the level, frequency, and duration of exposures of human populations in question. They may include direct measurements of personal exposures, measurements from fixed-area monitoring sites, or modeling from source emissions. Modeling is used when monitoring data are not available.

Risk characterization combines the results of hazard identification and dose–response and exposure assessments to estimate the probability of a specific adverse health effect to an exposed individual or human population.

Historically, risk management has been seen as a process distinct from risk assessment. In risk management, results of risk assessments are integrated with political, social, economic, and technological factors to determine the need and methods for reducing risk. In the early history of risk assessment, the National Research Council (NRC) stressed the importance of maintaining a distinction between risk assessment and risk management to prevent biasing risk assessments based on the political feasibility of regulating substances in question.

Risk management involves decision making. This includes decisions on whether to regulate the substance and, if so, what specific regulatory options should be applied. At this stage, the process includes risk communication, the process by which results of risk assessments and any proposed risk management options are conveyed to those who are likely to be participants in the public policy-making process and may be directly or indirectly affected by regulatory decisions. Risk communication

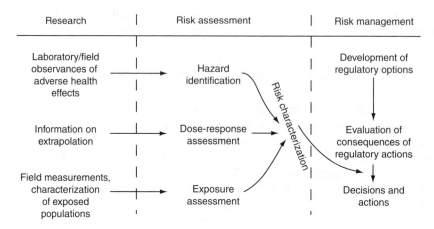

Figure 5.11 Relationships between different components of risk assessment and risk management. (Adapted from Samet, J., in *Air Pollution and Health*, Holgate, S.T., Samet, J.M., Koren, H.S., and Maynard, R.L., Eds., Academic Press, San Diego, 1999, p. 841. With permission.)

is a particularly difficult process due to the inherent limitations and uncertainties associated with scientific data, uncertainties associated with risk estimates, technical complexity, and strong views held by those to be regulated (and their supporters) and those who believe more stringent regulation is required. Risk communication, even when handled well, will in many, if not most, cases subject risk assessment and subsequent risk management processes to considerable controversy.

The USEPA in 1986 developed and issued risk assessment guidelines consistent with earlier National Research Council (NRC) recommendations. These, for the most part, focus on assessing risks of carcinogenicity, mutagenicity, developmental toxicity, and effects of chemical mixtures. They include default options that are policy judgments of how to accommodate uncertainties associated with risk characterization. These options are applied when scientific evidence is not sufficiently convincing as to which of several competing toxicological models and theories are correct. Historically, the USEPA has conducted risk assessments that incorporate conservative default options that are more likely to overstate than understate human health risks. This is particularly true with suspected human carcinogens. In such cases, the USEPA assumes a no-threshold relationship between dose and response, much like that used for exposure to radioactivity. As some cancer-causing substances may have thresholds, use of the most conservative toxicological model in the face of uncertainty, not surprisingly, is strongly criticized by the regulated community.

As a result of the controversies that risk assessments and subsequent regulatory decision making give rise to, as well as obvious limitations of the risk assessment process, an NRC panel in 1996 redefined risk characterization as "a synthesis and summary of information about a potentially hazardous situation that addresses the needs and interests of decision makers and interested and affected parties. Risk characterization is a prelude to decision making and depends on an iterative, analytical-deliberative process."

Risk assessment, until relatively recently, received limited application in ambient air pollution regulatory activities. Congress, in revising the hazardous pollutant provisions of the 1990 CAA Amendments, required the USEPA to conduct risk assessments secondarily to technology-based regulation of 182 designated hazardous air pollutants. After applying technology-based emission limits, the USEPA must set residual risk standards that protect public health (with an adequate margin of safety) if technology-based standards are determined not to do so. A residual risk standard is required if the technology-based standard leaves a lifetime risk greater than one in a million for the most exposed person.

As indicated, many stakeholders are involved in risk assessment processes that may result in regulation of substances for which a quantifiable and potentially significant public health risk has been identified. These span the spectrum of concerns about risk, from the regulated community, who may believe that risks are negligible and control requirements onerous, to environmental activists who believe that public health is being unduly compromised. All stakeholders publicly agree that policy decisions should be based on "good science." However, good science, like beauty, is in the eyes of the beholder. Risk assessments by their very nature contain uncertainties. Because of such uncertainties, risk characterization and resulting policy decisions are heavily criticized.

The risk assessment process, as it has evolved over the past two decades, attempts to quantify risk, i.e., describe the probability of disease in a population associated with specific pollutants and exposed populations. Those who oppose environmental regulation specific to a given situation or more generally often propose that the risk of a specific pollutant be compared to other risks in society (e.g., death from operating a motor vehicle). In such comparative risk evaluations, relatively low risks are associated with exposures to environmental pollutants.

The implication of comparative risk evaluations is that environmental regulation of many substances is not justified. Regulators, however, view risk in a broader context. Risk involves activities that are, in many cases, voluntary. Driving a motor vehicle and smoking tobacco are voluntary risks. Breathing air is essential for life and is not a voluntary action. Breathing polluted ambient air is therefore an involuntary risk. It is an involuntary risk to healthy adults and children, asthmatics, individuals with respiratory or cardiovascular disease, the old and infirm, and infant children.

READINGS

Ackermann-Liebuch, U. and Rapp, R., Epidemiological effects of oxides of nitrogen, especially NO_2, in *Air Pollution and Health,* Holgate, S.T., Samet, J.M., Koren, H.S., and Maynard, R.L., Eds., Academic Press, San Diego, 1999, p. 559.

Boushey, H.A., Ozone and asthma, in *Susceptibility to Inhaled Pollutants,* ASTM STP 1024, Utill, M.Z. and Frank, R., Eds., American Society for Testing Materials, Philadelphia, 1989, p. 214.

Brunekreef, B., Dockery, D.W., and Krzyzanowski, M., Epidemiological studies on short-term effects of low levels of major ambient air pollution components, *Environ. Health Perspect.*, 103 (Suppl.), 3, 1995.

California Air Resources Board/California Environmental Protection Agency, *Proposed Identification of Inorganic Lead as a Toxic Air Contaminant,* Part B Health Assessment, California Air Resources Board/California Environmental Protection Agency, Sacramento, 1996.

Carnow, B.W. and Meir, P., Air pollution and pulmonary cancer, *Arch. Environ. Health,* 27, 207, 1973.

Centers for Disease Control, *Preventing Lead Poisoning in Children,* DHHS, Washington, D.C., October 1991.

Cohen, A.J. and Nikula, K., The health effects of diesel exhaust: laboratory and epidemiologic studies, in *Air Pollution and Health,* Holgate, S.T., Samet, J.M., Koren, H.S., and Maynard, R.L., Academic Press, San Diego, 1999, p. 707.

Frampton, M.W. and Roberts, N.J., Respiratory infection and oxidants, in *Susceptibility to Inhaled Pollutants,* ASTM STP 1024, Utill, M.Z. and Frank, R., Eds., American Society for Testing Materials, Philadelphia, 1989, p. 182.

Goldstein, B., Toxic substances in the atmospheric environment: a critical review, *JAPCA,* 33, 454, 1983.

Horstmann, D.H. and Folinsbee, L.J., Sulfur dioxide-induced bronchoconstriction in asthmatics exposed for short durations under controlled conditions: a selected review, in *Susceptibility to Inhaled Pollutants,* ASTM STP 1024, Utill, M.Z. and Frank, R., Eds., American Society for Testing Materials, Philadelphia, 1989, p. 195.

Koenig, Z.Q., Covert, D.S., and Pierson, W.E., Acid aerosols and asthma: a review, in *Susceptibility to Inhaled Pollutants,* ASTM STP 1024, Utill, M.Z. and Frank, R., Eds., American Society for Testing Materials, Philadelphia, 1989, p. 207.

Maynard, R.L. and Waller, R., Carbon monoxide, in *Air Pollution and Health,* Holgate, S.T., Samet, J.M., Koren, H.S., and Maynard, R.L., Eds., Academic Press, San Diego, 1999, p. 749.

McClellan, R.O. and Jackson, T.E., Carcinogenic responses to air pollutants, in *Air Pollution and Health,* Holgate, S.T., Samet, J.M., Koren, H.S., and Maynard, R.L., Eds., Academic Press, San Diego, 1999, p. 381.

McNee, W. and Donaldson, K., Particulate air pollution: injurious and protective mechanisms in the lungs, in *Air Pollution and Health,* Holgate, S.T., Samet, J.M., Koren, H.S., and Maynard, R.L., Academic Press, San Diego, 1999, p. 431.

National Research Council, Committee on Risk Assessment of Hazardous Air Pollutants, *Science and Judgment in Risk Assessment*, National Academy Press, Washington, D.C., 1994.

National Research Council, Committee on Risk Characterization, *Understanding Risk: Informing Decisions in a Democratic Society,* Stern, P.C. and Fineberg, H.V., Eds., National Academy Press, Washington, D.C., 1996.

Paige, R.C. and Plopper, C.G., Acute and chronic effects of ozone in animal models, in *Air Pollution and Health,* Holgate, S.T., Samet, J.M., Koren, H.S., and Maynard, R.L., Eds., Academic Press, San Diego, 1999, p. 531.

Pope, C.A. and Dockery, D.W., Epidemiology of particle effects, in *Air Pollution and Health,* Holgate, S.T., Samet, J.M., Koren, H.S., and Maynard, R.L., Eds., Academic Press, San Diego, 1999, p. 673.

Samet, J., Risk assessment and air pollution, in *Air Pollution and Health,* Holgate, S.T., Samet, J.M., Koren, H.S., and Maynard, R.L., Eds., Academic Press, San Diego, 1999, p. 881.

Samet, J.M. and Cohen, A.J., Air pollution and lung cancer, in *Air Pollution and Health,* Holgate, S.T., Samet, J.M., Koren, H.S., and Maynard, R.L., Eds., Academic Press, San Diego, 1999, p. 841.

Samet, J.M. and Jaakkola, J.J.K., The epidemiological approach to investigating outdoor air pollution, in *Air Pollution and Health,* Holgate, S.T., Samet, J.M., Koren, H.S., and Maynard, R.L., Eds., Academic Press, San Diego, 1999, p. 431.

Schlesinger, R.B., Toxicology of sulfur oxides, in *Air Pollution and Health,* Holgate, S.T., Samet, J.M., Koren, H.S., and Maynard, R.L., Eds., Academic Press, San Diego, 1999, p. 585.

Speizer, F.E., Acid sulfate aerosols and human health, in *Air Pollution and Health,* Holgate, S.T., Samet, J.M., Koren, H.S., and Maynard, R.L., Eds., Academic Press, San Diego, 1999, p. 603.

Thurston, G.D. and Ito, K., Epidemiological studies of ozone exposure effects, in *Air Pollution and Health,* Holgate, S.T., Samet, J.M., Koren, H.S., and Maynard, R.L., Eds., Academic Press, San Diego, 1999, p. 485.

USEPA, *Air Quality Criteria for Particulate Matter and Sulfur Oxides*, EPA/600/8-82-029a-c, EPA, Washington, D.C., 1982.

USEPA, *Air Quality Criteria for Lead,* EPA/600/8-83-028a-f, EPA, Washington, D.C., 1986.

USEPA, *Air Quality Criteria for Lead,* supplement to the 1986 addendum, EPA/600/8-90/049f, EPA, Washington, D.C., 1990.

USEPA, *Air Quality Criteria for Carbon Monoxide,* EPA/600/8-90/045f, EPA, Washington, D.C., 1991.

USEPA, *Review of the National Ambient Air Quality Standards for Sulfur Oxides: Assessment of Scientific and Technical Information*, supplement to the 1986 OAQPs Staff Paper addendum, EPA/452/R-94-013, EPA, Washington, D.C., 1994.

USEPA, *Air Quality Criteria for Particulate Matter,* Vol. III, EPA/600/AP-95-001c, EPA, Washington, D.C., 1995.

Wedge, A., Lead, in *Air Pollution and Health,* Holgate, S.T., Samet, J.M., Koren, H.S., and Maynard, R.L., Eds., Academic Press, San Diego, 1999, p. 797.

QUESTIONS

1. What is the public health significance of air pollution episodes?
2. Describe differences between acute and chronic health effects.
3. How do environmental and public health authorities establish a cause-and-effect relationship between observed symptoms and disease in a community and exposure to atmospheric pollutants?
4. Describe additive, synergistic, and antagonistic interactions associated with human exposures to pollutants.
5. The human respiratory system has defense mechanisms that protect it from airborne particles. Describe each of these mechanisms and how they function.
6. Asthma is a very common and severe respiratory disease. How is asthma affected by exposure to ambient pollutants?
7. What, if any, is the relationship between exposure to atmospheric pollution and infectious respiratory disease?
8. Describe evidence to support a causal relationship between exposure to atmospheric pollutants and the development of lung cancer in humans.
9. Describe adverse health effects associated with exposure to CO.
10. Describe health effects associated with exposures to SO_2 and acid sulfates.

11. Describe relationships reported for particle size and human health effects.
12. Distinguish between inhalable, respirable, and thoracic particulate matter.
13. What is chronic bronchitis? What causes and exacerbates it?
14. Exposure to elevated levels of fine particles has been observed to be statistically associated with premature mortality. What biological mechanisms are likely responsible?
15. What are the health concerns associated with regulation of emissions of non-methane hydrocarbons under air quality standard provisions of clean air legislation?
16. What are the primary health concerns associated with exposures to hazardous or toxic air pollutants?
17. Describe health concerns associated with exposures to ambient O_3 levels.
18. Describe how lead exposures have changed over the past three decades in North America and the health benefits of such changes.
19. Asbestos is regulated as a hazardous air pollutant. Why?
20. Compare health risks associated with exposures to ambient air pollution and tobacco smoking.
21. What are the differences between risk assessment and risk management?
22. Describe how exposure and disease risks are evaluated and quantified.
23. Describe the public health and regulatory significance of voluntary and involuntary risk.
24. What is a comparative risk assessment?

CHAPTER 6

Welfare Effects

Under the Clean Air Act (CAA) Amendments of 1970, National Ambient Air Quality Standards (NAAQSs) were to be promulgated to protect public health and welfare. The former were to be primary standards, and the latter secondary. Public health protection has had the highest priority. Public welfare is less well defined. Welfare effects are air pollution effects not related to human health. Though it has always been clear what some welfare effects are (e.g., damage to crops, plants, and materials; malodors), others have evolved over time. In the latter case, these include visibility concerns, atmospheric deposition, possible ecosystem changes associated with stratospheric ozone (O_3) depletion, and global warming.

Welfare effects within the regulatory context are limited to pollutants for which air quality standards have been promulgated, the so-called criteria pollutants: carbon monoxide (CO), particulate matter (PM), sulfur dioxide (SO_2), nitrogen oxides (NO_x), O_3, nonmethane hydrocarbons (NMHCs), and lead (Pb). The concept of welfare effects as it is discussed in this chapter is used more broadly. It includes nonhealth effects associated with other pollutants as well.

Most of the issues discussed in Chapter 4 (atmospheric effects) are welfare effects. Ecological changes and concerns associated with most of those issues, save visibility, are discussed in this chapter.

6.1 EFFECTS ON AGRICULTURAL CROPS, ORNAMENTAL PLANTS, AND TREES

Plants have long served as sentinels of the biological injury that air pollutants are capable of producing as a result of acute and chronic exposures. Phytotoxic responses to pollutants such as SO_2, hydrogen chloride (HCl), and hydrogen fluoride (HF) have been reported in Europe since the middle of the 19th century. Particularly severe injury to, and destruction of, vegetation was associated with emissions of SO_2 and heavy metals from primary nonferrous metal smelters. In the U.S., this included the Copper Hill Smelter in Ducktown, TN, a copper (Cu) smelter in

Kellogg, ID, a zinc (Zn) smelter in eastern Pennsylvania, and other smaller smelters that have been long closed and would have been forgotten save for the despoiling of the surrounding environment that persists though men and their machines are long gone.

Smelter emissions have been a problem in Canada. In the 1920s, emissions from a smelter in Trail, British Columbia, were carried down the Columbia River Valley into the U.S., causing significant damage to agricultural crops. This was one of the earliest recorded instances of a trans-boundary air pollution problem. The once-forested and subsequently devastated landscape downwind of the giant nickel (Ni) smelter near Subdury, Ontario, is to this day a reminder of the significant impact that atmospheric pollutants can have on the environment. (On a historical note, emissions from the Sudbury smelter have been significantly reduced over the past two decades.)

Injury to agricultural plants and forests has been reported near uncontrolled nonferrous metal smelters in many parts of the world. Significant plant injury also occurs in developing countries around industrial sources, including coal- and oil-fired power plants, phosphate fertilizer mills, and glass plants.

Until the 1940s and 1950s, damage to agricultural crops, ornamental plants, and forests was, for the most part, a problem associated with point sources. As a result of intensive scientific investigation, the widespread injury to agricultural crops observed (as early as the mid-1940s) in the Los Angeles Basin was determined to be due to phytotoxic air pollutants such as O_3 and peroxyacyl nitrate (PAN), produced in the atmosphere as a result of photochemical reactions. Plant biochemist Dr. Arie Hagen-Smit, often referred to as the father of air pollution, initiated studies to unravel the chemical complexities of smog formation.

Ozone injury on sensitive vegetation has since been observed throughout the U.S. and many parts of the world. Ozone levels sufficient to cause injury on very sensitive vegetation are reported in most areas east of the Mississippi River. Because of its ubiquitous distribution and high phytotoxicity, O_3 is the most important phytotoxic air contaminant.

Control efforts and changes in operating practices have resulted in a significant reduction in the localized plant damage that had been associated with many point sources. Paradoxically, one of these changes, the use of tall stacks for more effective dilution of emissions from coal-fired power plants, has, as expected, resulted in decreased injury to vegetation in the vicinity of these sources, but inadvertently contributed to the problem of long-range transport and atmospheric deposition of strong acids and other pollutants.

Pollutants that have a history of causing significant plant injury under ambient exposure conditions include O_3, SO_2, PAN, fluoride (F), and ethylene (C_2H_4). Although it has been suspected of being a major plant-injuring pollution problem for some time, acidic deposition has only recently been determined to be, at least in part, causally related to plant injury under ambient conditions (e.g., the decline of red spruce in high-elevation forests).

Other pollutants are also known to cause injury to plants. They include nitrogen dioxide (NO_2), chlorine (Cl_2), hydrochloric acid (HCl), ammonia (NH_3), and PM. They are classified as minor pollutants since their emissions from continuous point sources at levels sufficient to cause injury have, for the most part, been relatively limited. Plant injury associated with exposures to these pollutants is commonly associated with accidental releases in industrial operations or transportation. On the other hand, injury induced by particulate dusts is usually associated with continuous emissions from point sources.

6.1.1   Plant Injury

Injury to plants can be manifested as visible or subtle effects. The former are identifiable changes in leaf structure, which may include chlorophyll destruction (chlorosis), tissue death (necrosis), and pigment formation. Visible symptoms may result from acute or chronic exposures. Acute injury occurs from brief exposures (several hours) to elevated levels of a phytotoxic pollutant. Tissue necrosis is the dominant symptom pattern associated with acute exposures, with necrotic foliar patterns usually specific for a given pollutant.

Chronic injury results from intermittent or long-term exposures to relatively low pollutant concentrations, with chlorophyll destruction and chlorosis the major symptoms. Chlorosis, however, is a nonspecific symptom associated with a number of pollutants, natural senescence (aging), and injury caused by other factors such as nutritional deficiencies.

Some pollutants produce symptom patterns other than, or in addition to, the classic patterns of necrosis and chlorosis associated with acute and chronic injury. These are described as growth abnormalities. They may be subtle, such as growth reduction, or visible, such as accelerated senescence of flowers, bolting (breaking) of flower buds, abscission (breaking off) of plant parts, and a curvature of the leaf petiole called epinasty.

The severity of injury, i.e., the amount of leaf surface affected, depends on pollutant concentration and duration of exposure. The greater the exposure, the more severe the injury. Entire leaves or, in extreme cases, the whole plant may be killed. The extent of injury also depends on sensitivity of the species, variety, or strain to the pollutant. Considerable variability in the response of plants to different phytotoxic pollutants, as well as responses to a given pollutant, occurs. As a general rule, actively growing plants and the most actively growing tissue are most sensitive to pollutant exposures and more likely to experience injury.

6.1.1.1   Sulfur Dioxide

Plants exposed to elevated SO_2 levels develop characteristic acute symptoms within 48 h of exposure. Tissue in the internal center of leaves of broad-leaved species is killed, resulting in a pattern of ivory-to-white, red, or dead brown tissue between the veins. This interveinal necrosis extends outward from the veins to the leaf margin and is observable on both surfaces. In narrow-leaved species, e.g., cereal grains, lilies, irises, and gladiolas, injury appears as irregular bifacial necrotic streaks between longer veins, and in some species, injury occurs at the tip of the leaf. Injury on coniferous plants is a reddish brown to brown tip necrosis with a banded appearance from repeated exposure. Chlorosis may occur on older needles under moderate exposures; such chlorotic needles are often prematurely shed. Interveinal necrosis on SO_2-injured pinto bean leaves is illustrated in Figure 6.1.

Figure 6.1 Interveinal necrosis on pinto bean associated with exposure to SO_2.

6.1.1.2 Ozone

Unlike SO_2 that initially injures tissues in the internal center of leaves, O_3 preferentially injures cells close to the upper surface (in broad-leaved species). The most common O_3-induced symptom is "flecks" visible on the upper surface (Figure 6.2). These flecks are produced when groups of cells below the epidermis (surface cell layer) are injured or killed. They may be white, tan, or yellow in color. The upper surface may look chlorotic or "bronzed" if flecking is extensive. A variation of this symptom is stipple, wherein injured cells become unevenly pigmented, producing a red–purple to black–brown appearance (Figure 6.3). This injury is confined to interveinal areas. Injury in narrow-leaved species appears as small necrotic or chlorotic flecks between veins.

Ozone injury on coniferous species such as pine is common. It is the causal factor of emergence tipburn of eastern white pine, a disease observed on this species

Figure 6.2 Ozone injury on tobacco.

Figure 6.3   Ozone injury on grape.

since the early 20th century. The disease syndrome is characterized by death of the tips of young elongating needles in spring and early summer. Killed needle tips become reddish brown. Ozone not only causes tip necrosis, but also other symptoms, ranging from chlorotic flecks to chlorotic mottling (small chlorotic patches). On O_3-affected trees, older needles prematurely age, discolor, and are shed. Consequently, affected trees may possess only the current year's needles.

Ozone, along with SO_2, appears to be responsible for chlorotic dwarf disease of eastern white pine, observed in the northeastern U.S. for most of the 20th century. In young, sensitive trees, current-year needles develop chlorotic flecks and mottling. Older needles become prematurely chlorotic and are shed before the new set of needles is fully developed. These trees are severely stunted and usually die before they reach 15 years of age.

In the San Bernardino Mountains of southern California, large numbers of ponderosa pines have been affected by what is called ozone needle mottle. The earliest symptoms are small chlorotic tissue patches that develop from the tip to the base in older needles. Death of the needle tip may also occur. As a consequence, O_3-injured trees may possess only 1-year-old needles during summer months rather than the 3- to 5-year normal complement.

6.1.1.3 Peroxyacyl Nitrate

Peroxyacyl nitrate is the causal agent of what was described as smog injury on vegetable crops in the Los Angeles Basin from the 1940s to the early 1960s, where it was a major problem. Injury due to PAN has been observed around other western cities, but to a lesser degree; it has also been found in New Jersey and southeastern Canada.

PAN destroys tissue near the lower surface of leaves, resulting in what has been described as glazing on the underside of the leaf. A distinct pattern of banding is observed after repeated intermittent exposures. Injury is not commonly reported on narrow-leaved or coniferous species.

6.1.1.4 Fluorides

Fluoride injury may result from the uptake of gaseous hydrogen fluoride (HF) through leaves or from soluble particulate fluorides absorbed through leaves or roots. Whatever the route of exposure, fluorides are transported through veins to leaf margins and leaf tips, where they accumulate. Fluoride injury results from exposures to elevated tissue concentrations that occur as a result of the accumulation of F over a period of weeks to months. Severity of injury is directly related to the level of F accumulated in marginal leaf tip tissue.

In broad-leaved species, F injury appears as marginal (Figure 6.4) or tip necrosis. Necrotic tissue is often separated from healthy tissue by a narrow, sharply defined reddish band. It may have a characteristic wavy color pattern due to successive F-induced tissue death. Necrotic tissue may break away from uninjured tissue near the reddish band, giving the leaf a ragged appearance.

Fluoride injury on narrow-leaved species is characterized by tip necrosis that extends in irregular streaks down the leaf. Necrotic tissue may vary in color from ivory to various shades of brown. A band of dark brown tissue sharply demarks injured from healthy tissues. Symptoms consist of chlorotic mottle at the margins and leaf tips. Small irregular chlorotic patches form between the veins and merge to form continuous bands as injury becomes extensive. Necrosis occurs when tissue injury is severe.

Tip necrosis is the characteristic symptom of F injury in conifers. Necrosis begins at the tip of current-year needles and progresses downward. The necrotic tip is usually reddish brown. This tip necrosis usually results in premature shedding of affected needles.

Figure 6.4 Fluoride injury on Oregon grape.

6.1.1.5 Ethylene

Plant injury caused by C_2H_4 was first reported when C_2H_4 was used in illuminating gas in greenhouses at the turn of the 20th century. Since it is emitted in automobile exhaust, C_2H_4 is a ubiquitous pollutant in urban and suburban areas. Its concentrations in ambient air, though relatively low, may be sufficient to cause significant effects on plants.

Ethylene is a natural maturation hormone in plants, controlling ripening and aging of fruits. Effects of ambient levels of C_2H_4 on plants growing in the urban environment are potentially great, as C_2H_4 has significant biological activity at very low concentrations (e.g., ppbv).

One of the best-documented effects of C_2H_4 exposure is dry sepal of orchids, associated with use of C_2H_4 illuminating gas in greenhouses. Dry sepal has also been reported in urban areas, presumably from elevated levels of C_2H_4 associated with automobile emissions.

Other plants in the flowering stage may be sensitive to C_2H_4 exposures. Common C_2H_4-induced responses include premature bud break, inhibition of flowering, and accelerated flower aging.

Because of its hormonal properties, C_2H_4 has the potential to cause widespread adverse effects on plants. However, field identification of injury resulting from such exposure may be nearly impossible, as injury mimics natural senescence and changes induced by a variety of environmental factors. An interesting example of this problem is the widely observed premature pigmentation of tree and shrub leaves along many major highways.

6.1.1.6 Acidic Deposition

Acidic substances deposited on plants or the plant soil environment have the potential to cause direct injury or measurable changes in plant growth. Laboratory and greenhouse studies with simulated acidic rain events have shown that a wide variety of plants can be injured by exposure to acidic rainfall with a pH of ~3.0. Reported symptoms include (1) small (<1 mm) necrotic leaf lesions or marginal tip necrosis; (2) pitting, curling, wrinkling, shortening, and death of leaves; (3) premature abscission; (4) erosion of the waxy upper leaf surface; (5) chlorosis; (6) excessive adventitious (side) budding; (7) premature senescence; and (8) galls, hypertrophy (tissue swelling), and hyperplasia (excessive cell production) in the specific species exposed.

The most commonly reported symptom of acidic deposition injury on plants in laboratory studies is small necrotic lesions. These appear to be a result of deposition of acidified water on the leaf surface and its subsequent evaporation. Repeated exposure to simulated acid rain events results in an increased number of lesions on affected leaves. Foliar injury has been observed at pH levels associated with recorded severe ambient deposition events. However, to this author's knowledge, there have been no confirmed reports of atmospheric deposition causing similar symptoms on plants under field conditions. Other effects not manifested as visible injury include reduced yield and leaching of metabolites and inorganic nutrients from plant surfaces. Leaching of such substances appears to be a common response to acid deposition. Leachates include amino acids, free sugars, organic acids, vitamins, a variety of essential mineral nutrients — nitrogen (N), calcium (Ca), phosphorous (P), magnesium (Mg) — alkaloids, growth regulators, and pectic substances.

A variety of potential impacts of acidic deposition on forest element cycles and tree nutrition have been proposed. These include:

1. Accelerated leaching of base cations from foliage and soils that may result in nutrient deficiency, root damage, and reduced drought tolerance
2. Increased mobilization of aluminum (Al) and other metals, also resulting in nutrient deficiency, root damage, and reduced drought tolerance
3. Inhibited microbiological processing of soil, possibly leading to reduced decomposition of organic matter, damage to mycorrhizae, nutrient deficiency, and altered pathogen–host relationships
4. Increased bioavailability of N, leading to accelerated organic matter decomposition and increased susceptibility to natural stresses such as freezing temperatures, drought, etc.

The importance of these mechanisms in altering tree and forest growth is not known.

6.1.1.7 Interaction Effects

Pollutants do not occur in the atmosphere singly. Therefore, it is possible, and even probable, that simultaneous, sequential, or intermittent exposures to several phytotoxic pollutants may result in plant injury. Available evidence indicates that simultaneous exposures to gaseous mixtures can produce synergistic, additive, or antagonistic effects. The type of interaction response is related to plant species exposed, pollutant combinations and concentrations, duration of exposure, and amount of injury caused by pollutants when exposed singly.

Plant exposures to mixtures of $SO_2 + O_3$ and $SO_2 + NO_2$ have reportedly decreased injury thresholds. Injury produced by exposure to gaseous mixtures may be more severe than the sum of the injury caused by exposure to individual pollutants (a synergistic response). Antagonistic responses are generally observed when injury caused by pollutants applied singly is severe; in these cases, the effect of mixtures is to reduce the severity of injury. Visible symptoms produced by gaseous mixtures are usually characteristic of a single pollutant.

6.1.2 Economic Losses

Numerous cases of significant pollutant-caused plant injury were reported in the 20th century. Excluding the photochemical oxidant problem in southern California and some areas of the northeast, most reports of air pollution-induced plant injury have been associated with point sources. With the exception of primary metal smelters, economic losses associated with point sources have often been insignificant. Many offending point sources in the U.S. have since been controlled. Nevertheless, because of photochemical oxidants such as O_3 and PAN, significant air pollution injury to vegetation is still widespread; associated economic losses, however, are often undetermined.

Scientists generally agree that O_3 causes 90% or more of the air pollution injury to crops in the U.S. This recognition has led to the establishment of a National Crop Loss Assessment Network (NCLAN). From summaries of O_3 monitoring data; determination of the O_3 sensitivity of major crop plants such as corn, soy beans, wheat, cotton, grain, sorghum, and barley; and economic data, NCLAN-participating scientists have modeled the economic impact of O_3 on U.S. agriculture. They have estimated that a 25% reduction in ambient tropospheric O_3 would result in a $1.71 billion annual increase in agricultural production; a 40% reduction would result in a $2.52 billion annual increase.

Quantification of economic loss is a difficult task. It requires surveys of suspected air pollution injury, confirmation of the causal factor, and estimation of the dollar costs. Loss estimates may be confounded by a number of factors. For example, presence of visible injury on plants may not be translatable into economic loss. In many cases, plants with severe foliar injury may quickly recover, replacing injured leaves with younger healthy ones with no apparent lasting effects. In other instances, even a slight amount of injury can decrease the marketability of a crop. To deal with this reality, plant scientists use the concept of plant damage rather than plant injury when it relates to potential economic losses. It is implicit in the damage concept that economic losses have resulted. Plant injury does not have this connotation.

It is reasonable to expect that when a phytotoxic pollution problem continues for a period of time, economic losses will decline, as no grower can sustain such losses indefinitely. He or she may discontinue agricultural production altogether or utilize pollution-resistant varieties or other pollutant-tolerant crops.

6.1.3 Forest Declines

In many real-world cases, particularly forest ecosystems, the relationship between observed injury and exposures to atmospheric pollutants has not been clearly established. This has been particularly true of many "forest declines."

The term *decline* is used to describe the process by which large numbers of trees die. In a decline, tree death occurs progressively; i.e., trees are weakened, become less vigorous, and eventually die.

Declines may occur as a result of a variety of natural or anthropogenic stress factors. Natural phenomena initiating forest declines are drought, insects, and freezing temperatures. Weakened trees may succumb to other factors, including root rot, insects, and disease. In many cases, no single factor can explain the observed death of trees.

A number of forest declines have occurred in North America and central Europe in the past five decades. In some cases, forest declines have been definitively linked to atmospheric pollutants. In others, one or more atmospheric pollutants are suspected, at least in part, of contributing to the death of forest species.

6.1.3.1 California Pines

Declines in Ponderosa and Jeffery pines were first noted in the San Bernadino National Forest east of Los Angeles in the 1950s. Older needles on affected trees lost chlorophyll and died prematurely. Injured pines had reduced radial growth and tolerance to the western pine beetle and other stresses. Similar declines were later seen in the Laguna Mountains east of San Diego and the Sierra Nevada and San Gabriel Mountains. Both Ponderosa and Jeffrey pine are sensitive to elevated O_3 levels; decline of these California pines has been definitively linked to O_3.

6.1.3.2 Eastern White Pine

Sensitive white pine genotypes have shown evidence of decline throughout their range in eastern North America. This decline has been a selective one, with only the most sensitive trees showing evidence of foliar injury, as well as reduced height and diameter growth. These effects are most pronounced in high-O_3 regions, and a causal connection between elevated O_3 levels and decline of eastern white pine appears evident.

6.1.3.3 Red Spruce and Fraser Fir

Declines in red spruce forests have been reported in the eastern mountains of the U.S. from New England to North Carolina. Radial growth has decreased sharply, with reductions in the Northeast observed as early as the 1960s. The most dramatic diebacks occurred at elevations above 800 m (2400 ft), with declines of 50% or more occurring in the past three decades. Symptoms include dieback of branch tips from the top and inward from newly grown shoots and yellowing of the upper surfaces of older needles.

As a general rule, the percentage of severely damaged and dying trees increases with elevation. On three New England mountains, approximately 60% of red spruce were reported to be dead or dying at 1000 m (~3300 ft) compared to 20% at 700 m (~2300 ft). At lower elevations, foliar symptoms, reduced tree growth, and tree death have been reported but are less prevalent than at higher elevations.

In the mountains of North Carolina, forest declines include both red spruce and Fraser fir. Growth of red spruce and, to a lesser extent, Fraser fir began to decrease in the 1960s. By 1987, most red spruce and Fraser firs on Mt. Mitchell's (located in the Great Smoky Mountains) western-facing slope had died. These coniferous species have also experienced high mortality on mountaintops in Tennessee and Virginia.

In the high-elevation mountains where declines are occurring, pollutant levels (particularly O_3) are relatively high compared to those at lower elevations. In addition, deposition rates of hydrogen ion (H^+) are often 10 times greater than at lower elevations. Where forest damage is greatest, trees are exposed to high-acidity, high-O_3 fogs and clouds for upwards of 100 days/year.

6.1.3.4 Pines in the Southeast

Surveys conducted by the U.S. Forest Service in the southeastern states of Virginia, North and South Carolina, Georgia, and Florida have shown that average

Figure 6.5 Decline of coniferous species in Harz Mountains, Germany, 1981. (Courtesy of Krupa, S.R., Dept. of Plant Pathology, University of Minnesota.)

radial growth rates for loblolly and slash pines less than 63 cm (16 in.) in diameter have decreased by 30 to 50% over the past 40 years.

6.1.3.5 Hardwoods in the Northeast

Crown dieback and increased mortality rates in sugar maple were reported in southeastern Canada in the late 1970s. Similar dieback symptoms have been reported for sugar maples in Pennsylvania, New York, Massachusetts, and Vermont. Yellow birch, American beech, and white birch have also shown decline symptoms.

6.1.3.6 Declines in Central Europe

Multispecies forest declines were first reported in central Europe on low mountains in the 1970s. Silver fir in the Black Forest of Germany lost needles from the inside of branches outward and from the bottom of the tree upward toward the crown and died in large numbers. Initially needles yellowed, followed by defoliation. Similar symptoms were subsequently observed on pines, beeches, and oaks. Discoloration and premature leaf fall have now been reported in all major forest species in western Germany. Decline effects on a coniferous forest can be seen in Figure 6.5.

Though damage to trees was first reported at 800 m (~2600 ft) and above, it also occurs at lower elevations. Older trees on west-facing slopes that face prevailing winds are most affected.

A survey conducted in 1983 in western Germany indicated that approximately 34% of the country's trees were affected; by 1986, 54%. Damage to firs was extensive, with 83% showing some symptoms and more than 60% having moderate to severe damage. There was also major injury to spruce, beech, and oak.

Similar injury to forest trees was reported throughout central Europe at all elevations and on all soil types. In surveys conducted between 1983 and 1986, symptoms of decline were seen in 14% of Swiss forests, 22% of Austrian forests, 22% of coniferous and 4% of deciduous trees in France, and 24% of forest areas in Holland.

Research studies indicate that air pollutants, through mechanisms that vary by site, are a significant causal factor in forest declines. A link to air pollution is suggested by the large number of species affected, rapid onset of symptoms, large geographical areas affected, and wide range of associated climates and soil conditions involved. The scientific consensus on these declines is that they are caused by a combination of direct foliar damage and nutrient imbalance, both due to pollutant exposures. Ozone, along with acid fogs, appears to be the principal cause of tree damage and decline of European forests.

6.2 EFFECTS ON DOMESTICATED ANIMALS

Retrospective studies of air pollution disasters, such as that which occurred in Donora, PA, in 1948, indicate that in addition to human deaths and illnesses, domesticated animals such as dogs and cats were affected. From epidemiological studies, it is evident that humans chronically exposed to high community air pollution levels have a higher incidence of respiratory disease. Although similar studies have not been done on domesticated animals, dogs and cats, which share the same basic environment as humans, are likely to be affected by such exposures.

6.2.1 Fluoride

Though it is no longer a problem in North America, F once caused more air pollution injury to domesticated animals that any other pollutant. Cases of F toxicity (fluorosis in cattle, sheep, horses, and pigs) resulted primarily from ingestion of forage contaminated by gas, particulate, and soil sources. Fluorosis in livestock once was prevalent in Florida, Tennessee, Utah, Washington, and Oregon.

Fluorosis can be acute or chronic. Acute fluorosis is rare since livestock do not voluntarily consume heavily contaminated forage. Chronic fluorosis is still observed in livestock that ingest F-contaminated forage over time.

Because F interferes with normal Ca metabolism, chronic toxicity is characterized by dental and skeletal changes. There may be incomplete formation of enamel, dentine, or teeth themselves; teeth soften so that excessive dental wear occurs, interfering with proper mastication of food. These dental changes take place during tooth development and are not reversible.

Skeletal changes are usually manifested as diffuse thickening of long bones and calcification of ligaments, resulting in stiffness and lameness. Lame animals eat less because they have difficulty standing for any length of time. Affected animals become emaciated as a result of reduced food intake and difficult mastication. In dairy cattle this results in decreased milk production.

Though F poisoning is no longer a problem in North America and other developed countries, it may be occurring in developing countries where there is little or no environmental regulation.

6.2.2   Lead

Lead emissions from uncontrolled sources were a cause of livestock poisoning in North America in the recent past. Such Pb poisoning was also common in the early part of the 20th century when lead arsenate insecticide was used as an orchard spray.

Air pollution-related poisoning of livestock in the U.S. was reported in the vicinity of primary and secondary Pb smelting operations up until the late 1960s. Poisoning was, for the most part, due to ingestion of Pb dust-contaminated forage. Lead poisoning may be acute or chronic depending on exposure level and duration. In cattle, early stages of acute Pb poisoning are characterized by excessive thirst, salivation, loss of appetite, constipation, delirium, and reduced milk production. Affected animals die when exposures are very high. Chronic poisoning results in a spectrum of symptoms, including diarrhea, colic, nervous disorders, swollen joints, lethargy, incoordination, bellowing stupor, rough coat, and emaciation, as well as a variety of metabolic changes similar to those observed in chronically exposed humans.

6.2.3   Other Toxic Exposures

In most cases when air pollution injury to livestock is confirmed, such injury has been associated with stationary sources processing minerals. These minerals contained biologically active elements that, when ingested in excessive amounts, produce toxic affects. Other elements in mineral dusts such as arsenic (As), selenium (Se), and molybdenum (Mb) have reportedly poisoned livestock and other animals.

6.3 ECOLOGICAL EFFECTS

Phytotoxic pollutants can cause injury to individual plants of a species, widespread injury to individual species (e.g., Ponderosa pine, red spruce, etc.), and, by inference, whole ecosystems. In an ecosystem, the existence of all species is to some extent affected by the presence of all others. If a species such as Ponderosa pine is eliminated from forest stands in the San Bernadino Mountains, significant ecosystem changes will take place. Species that depend on Ponderosa pine for food and shelter must adapt to the new ecological order as Ponderosa pine is replaced by more O_3-tolerant species. Significant ecological changes would also occur as a result of other forest declines described previously.

When a population of a species is subject to an environmental stress, it responds in typical Darwinian fashion. The most sensitive, or least tolerant, individuals die and are eliminated. If the stress continues, the species survives, but individuals tolerant to the stress dominate the population. As such, the impact on ecosystems may be limited. Such responses can be expected for many species subject to low-level exposure to pollutants such as O_3, atmospheric deposition pollutants such as sulfuric and nitric acids (H_2SO_4 and HNO_3), increased exposure to ultraviolet (UV) light associated with stratospheric O_3 depletion, and climate changes associated with greenhouse gases.

Individual species, as well as ecosystems of which they are members, are relatively resistant to environmental stressors. If the stress is limited, they may genetically adapt without significant ecological changes. If the environmental stress is sufficient to eliminate one or more species or provide a competitive advantage to other species, significant ecological changes may take place. Major ecosystem changes are characteristic of aquatic systems undergoing significant acidification as a result of atmospheric deposition. They affect all levels of food chains, including plants and animals.

With the exception of acidic deposition, effects of atmospheric pollutants on ecosystems have been little studied. The potential exists for subtle to major ecosystem changes as a result of:

1. Direct exposures to phytotoxic pollutants such as O_3
2. Responses to components of atmospheric deposition (in addition to strong acids) such as nitrate N, elemental mercury (Hg), and methyl mercury (CH_3Hg); pesticides; endocrine disruptors such as polychlorinated biphenyls (PCBs) and dioxins; and other pollutants that accumulate in watersheds and estuaries
3. Increased UV radiation over portions of the southern and northern hemispheres
4. Global warming

6.3.1 Phytotoxic Pollutants

With the exception of major forest declines linked to O_3 exposures, few studies have been conducted to evaluate potential changes in plant, animal, and microbial populations that may be occurring as a result of pollutant exposures on a regional level. This has been due, in part, to limited support for such studies and the inherent complexity of associating changes in species composition with exposure to atmospheric pollutants or other potential stress factors. For example, a progressive decline in flowering dogwood has been observed at Mammoth Cave National Park. It is not known whether such changes are natural or due to other factors such as phytotoxic air pollutants, acidic deposition, or a variety of other abiotic environmental stresses.

A change is taking place, but the cause is unknown. Other unstudied changes in forest ecosystems are likely to be taking place as well.

6.3.2 Atmospheric Deposition

Atmospheric deposition poses significant ecological concerns. Major attention is given here to acidic deposition because of its well-recognized and documented environmental effects.

6.3.2.1 Acidic Deposition

Acidic deposition poses a threat to ecosystems that, because of local or regional geology (crystalline or metamorphic rock), have soils and surface waters that cannot adequately neutralize acidified rain, snow, and dry deposited materials. Such soils and waters contain little or no Ca or Mg carbonates and therefore have low acid-neutralizing capacity (ANC). Acid-sensitive ecosystems in North America are illustrated in Figure 4.8. These include mountainous regions of New York and New England; the Appalachian Mountain chain (including portions of Pennsylvania, West Virginia, Virginia, Kentucky, and Tennessee); upper Michigan, Wisconsin, and Minnesota; the Pre-Cambrian Shield of Canada, including Ontario, Quebec, and New Brunswick; and various mountainous regions of the North American West.

6.3.2.1.1 Aquatic Ecosystems

Chemical constituents that cause significant biological changes in aquatic ecosystems include H^+, aluminum (Al^{+3}), and calcium (Ca^{+2}) ions. Increased H^+ (decreased pH) and Al^{+3} cause a variety of adverse biological effects that may be exacerbated by low Ca^{+2} concentrations. As pH decreases below 5.5, Al^{+3} concentrations increase to what are often toxic levels.

Changes in water chemistry associated with chronic or episodic acidification affect aquatic organisms at all ecological levels. Because of their large size and recreational importance, significant scientific attention has been given to the effects of acidification on fish populations. Declines in fish populations in acidic and low ANC lakes in the Adirondacks have been directly linked to atmospheric deposition-induced acidification. Over 200 lakes are now known to be fishless. Declines in fish populations in lakes and streams in New England and mid-Atlantic mountains have also been reported. In Canada, thousands of lakes in Ontario, Quebec, and New Brunswick are fishless, apparently due to increased H^+ concentrations associated with acidic deposition.

Extinction of fish populations in freshwater lakes and decline of fish populations (and fish kills) in streams appear to be a result of episodic as well as chronic acidification. Sudden shifts in pH following spring snowmelt have reportedly caused fish kills. Long-term increases in H^+ and Al^{+3} concentrations result in recruitment failure; i.e., few young fish reach maturity. Recruitment failure occurs because of the high mortality of eggs, larvae, and young fish that are particularly vulnerable to elevated H^+ and Al^{+3} concentrations. As a consequence, acidified waters show decreased population densities, with a shift of size and age toward larger, older fish. As acidification continues, a lake becomes fishless. Since acidification is a progressive phenomenon, the most sensitive fish species are affected first. Based on toxicity models developed from bioassay data, about 20% of lakes in the Adirondacks, 6% in New England, and 3% in the upper Midwest have H^+, Al^{+3}, and Ca^{+2} index values unsuitable for survival of sensitive fish species such as the common minnow and rainbow trout. Almost 10% of lakes in the Adirondacks have H^+, Al^{+3}, and Ca^{+2} index values that make them unsuitable for the more acid-tolerant brook trout. In the mid-Atlantic coastal plain, an estimated 60% of upstream sites have H^+ and Al^{+3} levels during spring that would kill 50% of anadromous (ocean species that spawn in freshwater rivers and streams) fish species such as the larval stage of the blueback herring.

As with other stressed environments, species diversity reduction is one of the initial significant biological changes that occur in acidified waters. Though total biomass may initially remain unchanged, fewer species are present. Shifts in species composition occur at pH levels in the range of 6.0 to 6.5. Unicellular phytoplankton species decrease and certain filamentous species increase in abundance (and in severely acidified lakes cover all surfaces with a thick mat). Acidification decreases the diversity and abundance of benthic (bottom-dwelling) invertebrates, which, because of their role as microdecomposers, are essential to nutrient cycling in aquatic systems. Reduced invertebrate populations result in fewer prey species for higher aquatic organisms.

Decreased decomposition by benthic organisms and bacteria results in accumulation of organic debris, with less mineralization and release of organically bound nutrients. Strong acidity causes nutrient ions to bind to particles, decreasing their uptake by phytoplankton.

Acidification effects are profound, whether associated with acidic deposition or natural processes. As lakes become increasingly acidified, significant species changes and decreases in biomass production occur.

6.3.2.1.2 Terrestrial Ecosystems

Acidic deposition has the potential to directly and indirectly affect terrestrial plants and other organisms in natural ecosystems. Major scientific attention has been given to evaluating potential causal connections between acidic deposition and "declines" of Norway spruce and beech in Germany and other countries of central Europe, high-elevation spruce and fir forests in the northern and southern Appalachian Mountains of the eastern U.S., low-elevation spruce and fir forests in the northeastern U.S., southern pines in the southeastern U.S., and sugar maple in the northeastern U.S. and southeastern Canada.

As indicated in Section 6.1.3, acidic deposition has been implicated as a contributing factor to forest declines of high-altitude red spruce in the Appalachian Mountains and apparent widespread growth reductions in southern Appalachian tree species. Acidic deposition has a significant potential to affect high-elevation eastern forest species because of acid cloud water (pH 2.8 to 3.8) exposures that occur upwards of 100 days/year. Such exposures result in alterations in plant nutrition, cold hardening of buds, and a variety of other physiological processes. Acidic deposition may cause reduced growth as a result of (1) foliar nutrient leaching; (2) reduced soil nutrient availability; (3) interference with normal root function associated with increased mobilization of Al^{+3} and inhibition of Ca^{+2}, magnesium (Mg^{+2}), and phosphorous (P^{+2}) uptake; (4) reduced carbon (C) levels due to reduced photosynthetic activity and increased respiratory C consumption; and (5) increased sensitivity to cold damage. Acidic deposition may also affect tree growth as a result of soil mobilization of Al^{+3} that causes direct toxic effects to roots. Acidification of forest soils results in interference with development of mycorrhizae, the symbiotic growth of roots and fungi essential for normal growth of many forest species.

Direct effects of acidic deposition on forest species, as well as indirect effects resulting from changes in soil chemistry and microbiology, have the potential to cause significant changes in forest ecosystems. The magnitude of such effects over the long term (next half century) may be large.

6.3.2.2 Nitrogen

The atmosphere is a significant source of N input into terrestrial ecosystems, watersheds, and estuaries. Nitrogen is an important plant nutrient, as it is required for synthesis of amino acids, proteins, nucleic acids, and chlorophyll. To a large extent, it governs utilization of potassium (K^+), P^{+2}, and other nutrients.

6.3.2.2.1 Terrestrial Ecosystems

Until the industrial revolution, availability of soil N was the limiting factor in the growth of trees and possibly other plants in forest ecosystems. Modern-day atmospheric inputs of N to forest environments is on the order of 5 to 20 times greater than that in the preindustrial era.

Nitrogen associated with atmospheric deposition serves as a plant nutrient in N-poor soils, particularly in forests. Such N inputs to forest ecosystems may result in increased growth of some species and altered species composition and diversity because of changes in competitive relationships.

Soil N increases can change the vegetative structure of an ecosystem. When N becomes readily available in environments previously characterized by low N availability, plants adapted to such an environment will be replaced by those capable of using now-increased N levels. Such effects have been reported for heathlands in the Netherlands, where nitrophilous (nitrogen-loving) grass species have replaced slower-growing species. Herbaceous forest plant composition in the Netherlands has shifted in the past several decades to species commonly found on N-rich soils.

In other European countries, N deposition has altered the composition of non-woody plants in mixed oak forests. Twenty of 30 species closely associated with high N deposition have increased in abundance.

Excessive N deposition may adversely affect the growth of forest trees. Decline of some tree species in German forests has been attributed to Mg^{+2} deficiencies caused by excessive N levels. Chronic N addition studies indicate that excessive N levels may result in decreased forest tree productivity and increased mortality.

Studies in Vermont suggest that coniferous forest stands undergoing decline are being replaced by fast-growing deciduous forests that cycle N more rapidly. Plant succession and biodiversity are being significantly affected by chronic inputs of N into historically low-N-input environments.

A major evolving environmental concern is "N saturation" that results when atmospheric inputs to what are normally background soil N levels exceed the capacity of plants and soil microorganisms to utilize and retain it. Excess N "runs off," resulting in increased stream, river, and lake nitrate levels. Nitrogen saturation has been reported in spruce–fir ecosystems in the Appalachians, eastern hardwood watersheds in West Virginia, and mixed conifer and chaparral forests in the Los Angeles Basin.

A variety of ecosystem changes resulting from N saturation have been suggested. These include (1) permanent increase in foliage N and reduced P^{+2} and lignin in wood, (2) reduced productivity due to physiological disruptions, (3) decreased root growth, (4) increased nitrification and nitrate (NO_3^-) leaching, and (5) reduced soil fertility resulting from increased leaching.

Increased soil N can significantly alter nutrient cycling in terrestrial ecosystems. Processes affected include (1) plant nutrient uptake and allocation, (2) litter production, (3) release of N as NH_3 (ammonification) and conversion of NH_3 to NO_3^- (nitrification), and (4) leaching of NO_3^- and release of trace gases such as nitrous oxide (N_2O). In addition, under N saturation, soil microbial communities change from predominantly fungal (mycorrhizal) to those dominated by bacteria.

6.3.2.2.2 Aquatic Ecosystems

Nitrogen associated with atmospheric deposition changes the fertility, or nutrient condition, of lakes, rivers, streams, and estuaries (shallow water areas where rivers enter the sea). Increased plant nutrients in aquatic ecosystems significantly increase growth of algae and aquatic weeds. This increase in plant productivity results in increased biomass production at all levels of the food chain, including fish, and increases turbidity of lakes and streams. Plant nutrients (e.g., NO_3^-), when present at elevated levels, may cause population explosions of algae called algal blooms. As these massive algal populations die, their decomposition results in episodic oxygen (O_2) depletion, with significant adverse ecological consequences. Plant nutrients contribute to the process of eutrophication, wherein freshwater lakes mature and fill in more rapidly as a result of increased biomass production and organic sediment deposition.

The atmosphere is but one source of plant nutrients that significantly affects the ecology of lakes, rivers, streams, and estuaries. Other sources such as agricultural runoff and wastewater treatment dominate the N budget of such systems. Nevertheless, in some stream, river, and lake waters, the atmosphere may be the source of 40% of nitrate N. Atmospheric deposition of NO_3^- to estuaries such as those of the Chesapeake Bay may be more significant in terms of ecological effects than in freshwater aquatic ecosystems. In the latter, the limiting factor for plant growth is P^{+2}, not N. This is not the case in estuaries.

6.3.2.3 Mercury and Other Trace Metals

Mercury deposition on land and water surfaces, and its accumulation in the muds of freshwater lakes, is an increasing environmental and public health concern. Though in a relative sense environmental Hg levels are low, their significance is greatly increased because Hg is biomagnified in food chains either as ionic Hg^{+2} or organic CH_3Hg. Methylmercury, produced by microorganisms as they transform Hg from its elemental form, tends to remain dissolved in water. As such, it moves upward through food chains and accumulates in fish tissues. This process is called biomagnification. It can result in CH_3Hg or Hg^{+2} levels in fish that are orders of magnitude higher than those found in water.

Because fish are at the top of most aquatic food chains, Hg levels in fish may be high. They are so high in some cases that the U.S. government issues fish consumption advisories for many bodies of water, particularly in the Great Lakes Basin and Florida Everglades.

The primary environmental concern associated with Hg pollution has been the potential exposure of the fish-consuming public to unsafe Hg levels. Mercury is a very toxic substance in its elemental, ionic, and organic forms.

Emissions to the atmosphere from major Hg sources have been regulated under hazardous air pollutant standards since 1973, and significant limits have been placed on industrial water sources. Nevertheless, because of emissions from unregulated sources such as coal-fired power plants, Hg pollution of the environment is a major concern.

Though regulatory efforts have focused on public health, Hg poses an exposure/toxicity risk to a number of wildlife species that consume large quantities of fish. These include bald eagles, ospreys, loons, kingfishers, herons, and mink. It also poses a poisoning risk to sharks in near-shore waters. There is only limited evidence available on the chronic effects of Hg exposure on wildlife species. Significant neurological and reproductive system effects in populations of fish-eating birds and mammals may be occurring.

Atmospheric deposition contains a number of other trace metals that can accumulate in the environment. These include Zn, copper (Cu), Pb, Ni, tin (Sn), chromium (Cr), cadmium (Cd), silver (Ag), Mb, and vanadium (V). Heavy metals tend to accumulate in forest soil humus or soil layers immediately below where root activity occurs. As a consequence, they may be toxic to roots and soil microorganisms and interfere with nutrient cycles. Heavy metal deposition has affected root growth of plants in heavily contaminated soils around smelters. Effects of low-level heavy metal contamination associated with atmospheric deposition have received only limited study. Some invertebrates (e.g., earthworms that consume leaf litter) are known to accumulate Cd, Pb, Zn, and Ni.

6.3.2.4 Organochlorine Compounds

6.3.2.4.1 Pesticides

The era of widespread use of persistent organochlorine pesticides in the U.S. began coming to a close with a 1972 ban on the use of DDT; cancellation of the registrations of other persistent pesticides or restrictions on their use followed in the next decade.

Because of their low volatilities ($\sim 10^{-4}$ mmHg), organochlorine pesticides have never been major atmospheric pollutants of concern. Nevertheless, as semivolatile and very persistent compounds, they have rather remarkable environmental mobility. For example, they have been found in fat tissue of Antarctic penguins 6450 km (4000 miles) from the nearest site of application.

Pesticides move through the atmosphere by a continual process of volatilization and deposition on particles and other surfaces. As such, they are commonly found in atmospheric deposition samples, albeit at very low levels.

Organochlorine pesticides, like Hg, are bioaccumulated. Consequently, they are biomagnified in food chains, with highest levels in fish and predator birds. The effects of pesticides on predator birds in the period 1950–1975 were dramatic, resulting in a precipitous decline in such species in North America. With the ban of DDT and other organochlorine pesticides, and restrictions in use of other similar organochlorine compounds, pesticide levels in fish and birds began to decline, with the latter slowly increasing from endangered species status (e.g., bald eagles, ospreys, peregrine falcons).

Though organochlorine pesticides are, for the most part, no longer used in the U.S., Canada, and Europe, they are still used in many parts of the world. Because of their persistence (half-life = ~12 to 15 years) and continued atmospheric transport, organochlorine pesticides are commonly measured in North American precipitation samples.

The environmental significance of this low-level contamination of the environment and movement through food chains is unknown. These substances have been reported to be endocrine disruptors in that they mimic estrogen, a female sex hormone. Major concerns have been expressed relative to their potential effect on male reproductive organs and processes, such as those observed in alligators associated with a spill of a large quantity of DDT.

The organochlorine pesticides, because of their persistence, mobility in the environment, and passage through food chains, have very likely contaminated every living thing on the planet and will likely continue to do so for some time.

6.3.2.4.2   *Polychlorinated Biphenyls, Dioxins, and Furans*

Though their use in the U.S. was discontinued in the 1970s, PCBs remain widespread contaminants of the Earth's environment. Like organochlorine pesticides, they are semivolatile and persistent. They are also biomagnified in food chains, with concentrations in aquatic organisms orders of magnitude greater than in the surrounding water. They affect the liver and nervous and reproductive systems and are animal carcinogens. Effects of PCBs on wildlife are largely unknown. Exposure through aquatic food chains is believed to be responsible for genetic or congenital abnormalities observed among fish-eating birds (particularly cormorants) in the Great Lakes Basin. Like other organochlorine compounds, PCBs appear to be endocrine disruptors.

Dioxins and furans are chlorinated organic compounds widely present in the environment at very low concentrations. They are characterized by their very high mammalian toxicity (LD_{50} [dose required to kill 50% of animals under test] = 2 ppbw/w), persistence, movement, and accumulation in food chains. 2,3,7,8-Tetra-chlorodibenzo-p-dioxin (TCDD) has been shown to cause endocrine disruption, weaken immune systems, and cause adverse reproductive changes in wildlife species.

6.3.3   Stratospheric Ozone Depletion

The primary environmental concern associated with stratospheric O_3 depletion is increased UV-B radiation over Antarctica and potential increases over the northern hemisphere. UV light is a plant growth regulator; it controls cell elongation in terrestrial plants. As a consequence, plants tend to grow "shorter" in full sunlight. As a result of increases in UV-B exposure, plants may not grow as tall, and therefore, regions subject to increased UV-B levels would not be as biologically productive. This could be a concern for annual crop plants and forest tree species.

Plants of species located at different latitudes are adapted to UV-B levels characteristic of their region. As such, it is likely that tolerant forms would, in a matter of a decade or so, begin to replace less tolerant forms in response to increased UV-B levels. Species would survive by adapting to the changed UV light conditions. Less tolerant or less genetically variable species may, however, be under severe environmental stress and be replaced by more tolerant species.

There is a significant potential for adverse biological changes in phytoplankton and other parts of the food chain in the upper zone of ocean water surrounding Antarctica. Exposures to UV-B can decrease photosynthesis and productivity in several marine species of algae and adversely affect various forms of larvae and other organisms. Algae or phytoplankton are the base of the Antarctic food chain that includes the very abundant krill (a small, shrimp-like animal) and higher-order species such as fish, birds, and sea mammals (e.g., seals and whales). As such, any major UV-B-related effects on phytoplankton could have major consequences for species in these food chains.

A worldwide decline in many amphibian species has been observed in the past several decades. The cause of this decline is not known. A few studies have shown that a number of amphibian species are very sensitive to near-surface UV-B levels. On this limited basis, a potential causal relationship between increased UV levels associated with stratospheric O_3 depletion and amphibian mortality has been proposed.

6.3.4   Global Warming

Atmospheric temperature is one of the most important environmental factors that affect the distribution of plant and animal communities around the planet. Temperature effects on individual plant or animal species and ecosystems are both direct and indirect (e.g., effects on water (H_2O) availability). Even slight changes in average atmospheric temperature can result in changes in the presence of a particular species in a given environment, its abundance, and its distribution. Limited evidence already exists that the near-surface planetary warming that has occurred in the past two decades has caused ecological changes. Projections of future global atmospheric warming, with associated effects in the polar regions (thinning of ice cover), coastal areas (sea level rises), and desert zones (increased desertification), as well as changes in precipitation patterns and migration of climatic zones, indicate that global warming will result in ecological changes, the likes of which the Earth has not experienced in thousands of years. These changes will affect, in varying degrees, most species that exist on our planet. Effects of global warming are expected to be particularly significant in polar regions and coastal areas.

6.3.4.1 Polar Regions

Based on climate models, polar regions will experience the greatest temperature increases (due to decreased albedo (see Chapter 1) associated with melting of ice and snow). Significant evidence exists that changes in the Arctic and Antarctic environments have occurred. These include (1) satellite data that indicate snow cover over the polar region has declined by 10% since the 1960s, (2) an average 2°C (3.6°F) warming of Alaska since the 1950s, (3) the decline of lake ice duration in middle and high latitudes, (4) declines in the extent and thickness of Arctic sea ice, and (5) the thinning of the west Antarctic ice sheet.

Sea ice is a major factor in regulating heat and moisture exchange and ocean salinity. It provides habitat for major polar species such as polar bears, seals, and penguins. As it thins, coastal areas become subject to erosion, altering marine mammal habitat.

During the past two decades, the health of polar bears in Arctic regions has been declining. Canadian scientists have suggested that the breakup of sea ice 3 weeks sooner than normal reduces the hunting season for seals (bears' primary food source) on sea ice. A 50% decline in Adelie penguins on the Antarctic peninsula has been observed over the past 25 years, presumably because of regional changes in sea ice and snowfall related to global warming.

6.3.4.2 Coastal Zones

Climatic models predict significant sea level rise associated with thermal expansion of seawater and melting of polar ice caps. Sea level rises would inundate coastal wetlands, erode shorelines, and increase the salinity of estuaries, bays, and rivers. As coastal wetlands are inundated, previously dry areas would be flooded. Overall, the size of coastal wetlands would decrease significantly.

Coastal wetlands, which include estuaries, are among the most biologically productive environments in the world. Their loss would result in enormous changes in biological diversity, and abundance and distribution of individual species.

6.3.4.3 Plant and Animal Communities

Plant and animal communities are characterized by the major plant types found in them. These include forests, grasslands, deserts, savannahs, and tundra. The nature of these plant communities, called biomes, depends on the combination of environmental factors that exist in a given region or location. Global warming can be expected to result in significant changes in climatic zones; therefore, individual biomes will expand or contract, depending on environmental changes that occur.

Before the advent of modern man, forests covered vast areas of the Earth's surface. Though a large percentage of forests has been cleared for agricultural purposes, many forested areas still exist. A projected 2°C (3.6°F) warming of the Earth's atmosphere is expected to shift the ideal range for many North America forest species northward by ~300 km (180 mi). As global warming would occur progressively, forest species would slowly begin to colonize more northerly locations and be replaced in their current southern range by species that are more tolerant of hotter, and in some cases drier, conditions. If the climate were to become wetter, forests would expand into what are now rangelands.

Increased carbon dioxide (CO_2) levels are expected to have a positive effect on forest growth as well as other plant species since CO_2 is required for photosynthesis; it also enables plants to use H_2O more efficiently. As is the case with N, some species would respond more favorably to increased CO_2 levels and thus have a competitive advantage. Consequently, changes in species composition and abundance are likely to occur.

Each biome, and the ecosystems that comprise it, is characterized by plant and animal species that have evolved over time. Some are endemic to a single region, while others migrate across climatic and biological zones, e.g., migratory birds.

Climate changes associated with global warming can have both direct and indirect effects on plants and animals. As wetlands are lost, some species are likely to experience a significant loss of habitat, and thus a significant reduction in population size. As temperatures warm, many species can be expected to expand their range poleward and up in elevation. Other expected changes include (1) increases or decreases in population density; (2) timing of events such as migration, breeding and nesting, and flowering; (3) changes in body size and behavior; and (4) shifts in genetic frequency.

As indicated in Chapter 4, significant changes in species range and phenology (timing) in the direction expected if global warming were occurring have been documented for hundreds of plant and animal species. The average shift in spring phenology for temperate zone species has averaged 5.1 days earlier per decade. An example of this is the migration and nesting of birds in Michigan (2 to 3 weeks earlier for several species compared to 1980, with 20 species nesting up to 9 days earlier than in 1971). Notable expansion of species ranges poleward in North America includes the opossum, mockingbird, and armadillo.

6.4 EFFECTS ON MATERIALS

Gaseous and particulate-phase pollutants are known to have adverse effects on materials. Of particular importance are effects on metals, carbonate building stones, paints, textiles, fabric dyes, rubber, leather, and paper. In western Europe, a repository for many monuments of history and fine works of art, air pollutant-induced damage has been incalculable. Because these cultural treasures are irreplaceable, their preservation from destructive effects of airborne contaminants poses a significant challenge to their present guardians. In a sense, destruction of western European antiquities symbolizes the seriousness of the materials damage problem.

Materials can be affected by both physical and chemical mechanisms. Physical damage may result from the abrasive effect of wind-driven particles impacting surfaces and the soiling effect of dust deposition. Chemical reactions may occur when pollutants and materials come into direct contact. Absorbed gases may act directly on the material, or they may first be converted to new substances responsible for observed effects. The action of chemicals usually results in irreversible changes. Consequently, chemical damage is a more serious problem than the physical changes caused by particulate dusts.

6.4.1 Metal Corrosion

Corrosion of metals in industrialized areas represents one of the most ubiquitous effects of atmospheric pollutants. Since corrosion is natural, it is difficult to assess the relative contribution of pollutants to this process. As the ferrous metals, iron (Fe) and steel, account for about 90% of all metal usage, their pollution-induced corrosion is of particular importance. When ferrous metals corrode, they take on a characteristic rusty appearance. Corrosion is rapid at first and then slows as a partially protective film develops. Acceleration of corrosion in industrial communities has been reported to be due, at least in part, to SO_2. Corrosive effects of SO_2 on Fe, steel, and steel alloys increase with increased SO_2 concentration and deposition rates. Deposition is facilitated by surface moisture, which promotes corrosion by electrochemical reactions. The corrosive effect of SO_2 appears to be reduced in atmospheres with elevated oxidant levels.

Nonferrous metals can also experience significant pollution-induced corrosion. For example, Zn, widely used to protect steel from atmospheric corrosion, will corrode when acidic gases destroy the basic carbonate coating that normally forms on it. The reaction of sulfur oxides (SO_x) with Cu results in the familiar green patina of copper sulfate ($CuSO_4$) that forms on the surface of Cu materials. This coating of $CuSO_4$ makes the metal resistant to further corrosion. The green discoloration of SO_x-induced Cu corrosion is most commonly observed on bronze statues in city parks and squares. Copper may also be discolored by exposures to hydrogen sulfide (H_2S), which blackens it.

Because nonferrous metals are used to form electrical connections in electronic equipment, corrosion of these connections can result in serious operational and maintenance problems for equipment users. The stress corrosion and breakage of Ni–brass spring relays that used to occur in the central offices of telephone companies in numerous major cities is such an example. Breakage occurred when Ni–brass connections were under moderate stress and positive electrical potential. Stress corrosion and breakage of the Ni–brass relay springs were caused by the high NO_3^- concentrations in dust that accumulated on these metal components. The problem was resolved by using Cu–Ni spring relays not susceptible to similar stress corrosion.

By forming thin insulating films over contacts, pollutants can damage low-power electrical contacts used in computers, communications, and other electronic equipment. Such films may result in open circuits, causing equipment to malfunction. They are produced by chemical reactions between the contact metals and contaminants such as SO_2 and H_2S. Equipment malfunction may also result from contamination of electrical contacts by PM. Particles may physically prevent closing or may result in chemical corrosion of contact metals.

6.4.2 Building Materials

Unlike metal corrosion, whose true cause is difficult to discern, pollutant effects on building surfaces are very evident. Soiling and staining of buildings in urban areas that have or have had high atmospheric aerosol levels is a particularly striking example. The "dirtiness" associated with many cities is due to the dark coating of particulate dusts deposited by the smoke of industries and coal-fired space heating over many years.

In addition to soiling, building materials such as marble, limestone, and carbonate cement are chemically eroded by SO_2 and acid aerosols. The reaction of SO_2 with carbonate building stones results in the formation of hydrated Ca salts, $CaSO_3 \cdot 2$ H_2O and $CaSO_4 \cdot 2$ H_2O (gypsum), both of which are H_2O soluble. The soluble salts produced may precipitate from solution and form encrustations; as they are washed away, the stone surface becomes more susceptible to pollutant effects. Stone damage is further accelerated by sulfur (S)-oxidizing bacteria and fungi.

Deterioration rates are affected by pollutant concentration, moisture content and permeability of stone, and deposition rates. Dry deposition of SO_2 between rain events has been reported as the major causal factor in carbonate building stone deterioration. Pollutant-induced erosion on a carbonate stone street monument can be seen in Figure 6.6.

Chemical erosion of priceless, irreplaceable historical monuments and works of art in western and southern Europe can be described as no less than catastrophic. Although primarily a European problem, similar destruction is occurring on monuments in the U.S. In the late 19th century, the obelisk, Cleopatra's Needle, was moved from Egypt to a park in New York City. The monument, which had withstood the ravages of desert heat and sands for thousands of years without significant deterioration, has had hieroglyphics obliterated on two of its faces in the past half century. The other two faces have eroded less, as they were away from the direction of prevailing winds.

A significant concern is the Acropolis, located downwind of Athens, Greece. Although chemical erosion of marble structures in the Acropolis has been occurring for more than 100 years, rapid population growth and use of high-S heating oils since World War II have greatly accelerated their destruction.

In addition to the Acropolis, the Colosseum in Rome and Taj Mahal in India are in various stages of dissolution. In many European cities, marble statues have had to be moved indoors. Cities such as Venice have enacted legislation to preserve and restore hundreds of damaged piazzi, churches, and historical buildings. The grand cathedrals of Europe (e.g., Notre Dame, Chartes, and Cologne) are constantly

Figure 6.6   Erosion of carbonate column in Maltese square. (Courtesy of DHD Multimedia Gallery, http://aj.hd.org.)

undergoing restoration of acid-ravaged limestone statuary and other features on their external façades.

6.4.3   Paints

Paints consist of two functional components: vehicles and pigments. The vehicle consists of a binder and additives that hold the pigment to a substrate surface. Principal materials used as vehicles include hydrocarbon solvents, glycols, and resins, which may be natural, modified natural, or synthetic. Pigments, classified as white or colored, provide aesthetic appeal, hiding power, and durability. White Pb, titanium dioxide (TiO_2), and zinc oxide (ZnO) have been widely used white pigments. Colored pigments are a variety of mineral, metal, and organic compounds. The function of paints is to provide an aesthetically appealing surface coating that protects underlying material from deterioration.

The appearance and durability of paints are affected by natural weathering processes. Paint weathering is accelerated by environmental factors such as moisture, temperature, sunlight (particularly UV light), and fungi. Paint appearance and durability may also be affected by air pollutants such as PM, H_2S, SO_x, NH_3, and O_3.

Air pollutants affect the durability of painted surfaces by promoting chalking, discoloration, loss of gloss, erosion, blistering, and peeling. They may also decrease paint adhesion and strength and increase drying time. Sulfur dioxide can increase the drying time of some paints by reacting with drying oils and competing with auto-oxidative curing reactions responsible for cross-linking the binder. Erosion of oil-based house paint is reportedly enhanced by exposure to SO_2 and high relative humidity. Apparently, SO_2 reacts with extender pigments such as calcium carbonate ($CaCO_3$) and ZnO. Particles can also damage painted surfaces by serving as carriers for corrosive pollutants or by staining or pitting painted surfaces.

One of the more obvious effects of pollutants on paints is the dirty appearance that results from accumulation of aerosol particles. These may cause chemical deterioration of freshly applied paints that have not completely cured. Particles may also serve as wicks that allow chemically reactive substances to migrate into underlying surfaces, resulting in corrosion (if the underlying material is metallic) and subsequent paint peeling.

Exterior paints pigmented with white Pb may be discolored by reaction of Pb with low atmospheric levels of H_2S. Blackening of the paint surface by formation of lead sulfide (PbS) can occur at H_2S exposures as low as 0.05 ppmv for several hours. The intensity of discoloration is related not only to the concentration and duration of H_2S exposure, but also to Pb content of the paint and the presence of moisture.

There have been numerous reports over the last several decades of damage to automobile paints and coatings. Damage occurs on horizontal surfaces of freshly painted vehicles and appears as irregularly shaped and permanently etched areas. It is easily observed on dark-colored vehicles and appears after evaporation of moisture droplets. The consensus in the automobile industry is that this paint damage problem is due to environmental "fallout." The most likely cause is believed to be acid rain, as chemical analyses of exposed test panels have shown elevated levels of sulfate.

6.4.4 Textiles and Textile Dyes

Textiles consist of a basic fiber component and additives such as dyes, water repellents, and finishes. Each of these is subject to pollutant damage.

Exposures to atmospheric pollutants may result in significant deterioration and weakening of textile fibers. Fabrics such as cotton, hemp, linen, and rayon are particularly sensitive to acid-forming gases and aerosols. Acids chemically break their cellulose chains, producing water-soluble products that have low tensile strength. Synthetic fibers such as nylon may also incur significant acid damage as well as undergo oxidation by NO_2, which reduces the affinity of nylon fibers for certain dyes.

A number of pollutants can react with fabric dyes by oxidation or reduction reactions, causing them to fade. Fading of textile dyes associated with NO_x exposure, particularly NO_2, has had a long history. For example, NO_x emissions from open electric-arc lamps and incandescent gas mantles caused fading of stored woolen goods in Germany prior to World War I. Nitrogen oxides emissions from gas heaters were also believed to be responsible for a serious fading problem on acetate rayon fabrics during the 1930s. During the 1950s, NO_x emissions from home gas dryers caused fading of colored cotton fabrics. Nitrogen dioxide-induced fading has also been reported for fabrics exposed to urban environments.

Reports of NO_2-induced fading of textile materials from ambient exposures were soon followed by observations that ambient O_3 levels were also a major fading cause. In the early 1960s, O_3 fading was reported for polyester–cotton permanent press slacks stored in warehouses or on retail shelves. Some manufacturers suffered significant economic losses. Fading of nylon carpets was a problem along the Gulf Coast from Florida to Texas. Blue dyes were particularly sensitive to a combination of ambient O_3 exposure and high humidity. This Gulf Coast fading was overcome by using O_3-resistant dyes and modifying nylon fibers to decrease exposure to O_3. Pollutant-induced fading of fabric dyes has posed significant challenges for textile manufacturers. Consequently, the industry had to develop resistant dyes and chemical inhibitors to mitigate such problems.

6.4.5 Paper and Leather

Papers that came into use after 1750 are very sensitive to SO_2 because of its conversion to H_2SO_4 by metallic impurities in paper. Sulfuric acid causes paper to become brittle, decreasing its service life. This embrittlement is a major concern to libraries and museums in cities with elevated SO_2 concentrations, as it makes preservation of historical books and documents more difficult.

Historical book preservation involves more than the paper component. Air pollution damage to leather-bound books is also a major concern. Damage to leather is initially seen as cracking. Cracking results in increased leather exposure, thus accelerating deterioration. Leather eventually loses its resiliency and turns to a reddish brown dust.

6.4.6 Rubber

Ozone can induce cracking in rubber and plastic compounds, most notably those under stress. The depth and nature of cracking depends on O_3 concentration, rubber or plastic formulation, and degree of stress. Unsaturated natural and synthetic rubbers, e.g., butadiene-styrene and butadiene-acrylonitrile, are especially vulnerable to O_3 cracking. Ozone reacts with double bonds in such products. Saturated compounds such as thiokol, butyl, and silicon polymers are relatively resistant to O_3 cracking. Rubber compounds are often formulated by the addition of antioxidants to reduce cracking. This is particularly important in rubber tires.

Rubber cracking was one of the first effects of smog observed in the Los Angeles area. During the 1950s it was used to monitor O_3 concentrations. Under standardized conditions, O_3 concentrations could be directly related to the depth of O_3-induced cracks.

6.4.7 Glass

With the exception of F, glass is resistant to most air pollutants. Fluoride reacts with silicon (Si) compounds to produce a familiar etched appearance. Etching of glass has been observed in the vicinity of major F sources such as phosphate fertilizer mills, Al smelters, enamel frit plants, and even steel mills. Continuous exposure to F may render glass opaque.

6.4.8 Economic Losses

Economic losses due to air pollution-induced damage to materials are difficult to assess; it is very difficult to distinguish between natural deterioration and that caused or accelerated by air pollutants. Only limited efforts have been made to quantify costs associated with air pollution damage to materials. It is probable, however, that such costs are in the tens of billions of dollars.

6.5 ODOR POLLUTION

The welfare effects described in this chapter generally do not impact people directly. Odor, on the other hand, does. Though exposure to malodorous substances may cause symptoms in some individuals, the problem of odor, or odor air pollution, is usually viewed as being one of annoyance.

The term *odor*, as used in everyday speech, commonly carries with it an unpleasant connotation. The term defined by scientists is neutral; it connotes only that some volatile or semivolatile substance is sensed by the human olfactory system.

The ability to smell chemical substances is common among members of the animal kingdom. In many species, the olfactory sense plays an important role in locating food, attracting individuals of the opposite sex for breeding, and warning of approaching danger. In humans, the olfactory sense plays a lesser and apparently less obvious role.

The olfactory function in humans consists of two different organs found in the nose. The olfactory epithelium (found in the highest part of the nose) consists of millions of receptor cells that connect directly to the olfactory bulbs of the brain. The free endings of trigeminal nerves distributed throughout the nasal cavity serve as a secondary organ of smell. Trigeminal nerves respond to odoriferous substances that cause irritation, tingling, or burning. The chemical senses that correspond to these two olfactory organs are not easily separated, with many odoriferous substances or odorants stimulating both systems. They stimulate different parts of the brain and therefore have different effects. The major function of the trigeminal nerve system is to stimulate reflex actions such as sneezing or the interruption of breathing when the body is exposed to potentially harmful substances.

6.5.1 Odor Measurement

Odor perception is a psychophysiological response to the inhalation of odoriferous chemicals. Odors cannot be chemically measured. However, some sensory attributes of odors, e.g., detectability, intensity, character (quality), and hedonic tone (pleasantness, unpleasantness), can be measured by exposing individuals in controlled environments.

The limit of detection (the odor threshold) may be characterized in two ways. It may be the concentration of a substance that is detectably different from background. Alternatively, it may be defined as the first concentration at which an observer can positively identify an odor. The former is the detection threshold; the latter, a recognition threshold.

Though the olfactory sense in humans is not as acute as it is in many animals, it nevertheless has the ability to detect many substances at very low concentrations. Odor thresholds for a variety of chemicals are summarized in Table 6.1. Note that humans can detect H_2S at ~0.5 ppbv.

Odors differ in their character or quality. This odor parameter allows us to distinguish odors of different substances by prior odor associations. The characters of a variety of selected chemicals are summarized in Table 6.1. For example, dimethyl amine is described as fishy, phenol as medicinal, paracresol as tar-like, etc.

The olfactory sense is highly subjective, with some odors being pleasant to some individuals and unpleasant to others. The difficulty of defining odor problems lies in part with the lack of unanimity as to what is unpleasant.

Table 6.1 Odor Thresholds and Characteristics of Selected Chemical Compounds

Chemical	Odor Threshold (ppmv)	Odor Character
Acetaldehyde	0.21	Green, sweet
Acetone	100.0	Chemical sweet, pungent
Dimethyl amine	0.047	Fishy
Ammonia	46.8	Pungent
Benzene	4.68	Solvent
Butyric acid	0.001	Sour
Dimethyl sulfide	0.001	Vegetable sulfide
Ethanol	10.0	Sweet
Ethyl mercaptan	0.001	Earthy, sulfide-like
Formaldehyde	1.0	Hay, straw-like, pungent
Hydrogen sulfide	0.00047	Eggy sulfide
Methanol	100.0	Sweet
Methylethyl ketone	10.0	Sweet
Paracresol	0.001	Tar-like, pungent
Perchloroethylene	4.68	Chlorinated solvent
Phenol	0.047	Medicinal
Sulfur dioxide	0.47	Sulfury
Toluene	2.14	Mothballs, rubbery

Unpleasantness may also be related to odor intensity. Some otherwise unpleasant odors may elicit a pleasant sensation in an individual when they are in relatively low to moderate concentrations. The olfactory response to an odorant decreases nonlinearly as the concentration decreases. For a substance such as amyl butyrate, the perceived odor intensity decreases by 50% for a tenfold reduction in concentration. This logarithmic relationship is common for most odorants.

The perceived intensity of an odor rapidly decreases after initial exposure. Within a few minutes, it may not be perceived at all. This phenomenon is called olfactory fatigue. Sensitivity to the odor is restored within minutes after exposure ceases.

A person may become habituated to unpleasant odors ("get used to it" or become "more tolerant"). Habituation operates over much longer time periods than olfactory fatigue.

Unpleasant odors may affect our sense of well-being. Responses to a variety of malodors (bad odors) can include nausea, vomiting, headache, coughing, sneezing, shallow breathing, disturbed sleep, appetite disturbance, sensory irritation, annoyance, and depression. Effects may be physiological, psychological, or both.

6.5.2 Odor Problems

It is likely that malodors from nearby sources are responsible for more complaints to regulatory agencies than any other form of air pollution. Particularly notable sources of malodors (and, in many instances, citizen complaints) are rendering plants; soap-making facilities; petrochemical plants; refineries; pulp and kraft paper mills; fish-processing plants; diesel exhaust; sewage treatment plants; and agricultural operations, including feedlots, poultry houses, and hog confinements. Malodors associated with such sources include a variety of amines, sulfur gases (such as H_2S, methyl and ethyl mercaptan, and carbon disulfide (CS_2)), phenol, NH_3, aldehydes, fatty acids, etc.

In addition to the annoyance caused by malodors, the presence of a continuous source may result in decreases of property values as prospective homebuyers avoid such areas. Nevertheless, as urban–suburban populations expand, many new homes are built in once-rural areas where malodorous agricultural operations exist. Not surprisingly, new residents often find barnyard odors objectionable. In many cases, the nature of the "barnyard" is changing. Many, if not most, agricultural livestock operations today involve large numbers of animals. These "factory farms" produce enormous quantities of agricultural wastes and, of course, stench. In agricultural states, such farms may be exempt from air pollution regulations and even zoning requirements.

READINGS

Adams, R.M., Hamilton, S.A., and McCord, B.A., An assessment of economic effects of O_3 on U.S. agriculture, *JAPCA*, 35, 938, 1985.

Alscher, R.G. and Wellburn, A.R., Eds., *Plant Responses to the Gaseous Environment: Molecular, Metabolic, and Physiological Aspects*, Third International Symposium on Air Pollutants and Metabolism, Blacksburg, VA, Chapman & Hall, London, 1992.

Ambosht, R.S., Ed., *Modern Trends in Ecology and Environment*, Backhuys Publishers, Leiden, Netherlands, 1998.

Butlin, R.N., Effects of air pollutants on buildings and materials, *Proc. R. Soc. Edinburgh*, 97B, 255, 1991.

Graedel, T.E. and McGill, R., Degradation of materials in the atmosphere, *Environ. Sci. Technol.*, 20, 1093, 1986.

Hellman, T.M. and Small, F.H., Characterization of the odor properties of 101 petrochemicals using sensory methods, *JAPCA*, 24, 979, 1974.

Irving, P.M., Ed., *Acidic Deposition: State of Science and Technology*, Vols. I–IV, U.S. National Acid Precipitation Assessment Program, Washington, D.C., 1991.

Jacobson, J.S. and Hill, A.C., Eds., *Recognition of Air Pollution Injury to Vegetation: A Pictorial Atlas*, Air Pollution Control Association, Pittsburgh, PA, 1970.

Johnson, D.W. and Taylor, G.E., Role of air pollution in forest decline in eastern North America, *J. Water Air Soil Pollut.*, 48, 21, 1989.

Krupa, S.R., *Air Pollution, People, and Plants*, APS Press, St. Paul, MN, 1997.

Legge, A.H. and Krupa, S.V., Eds., *Air Pollutants and Effects on Terrestrial Ecosystems*, John Wiley & Sons, New York, 1986.

Lepp, N.W., Effect of heavy metal pollution on plants, in *Metals in the Environment*, Vol. 2, Lepp, N.W., Ed., Applied Science Publishers, London, 1981.

Lucier, A.A. and Haines, S.G., Eds., *Mechanisms of Forest Response to Acidic Deposition*, Springer-Verlag, New York, 1990.

MacKenzie, J.J. and El-Ashry, M.T., Eds., *Ill Winds: Airborne Pollution's Toll on Trees and Crops*, Yale University Press, New Haven, CT, 1989.

Manion, P.D. and Lachance, D., *Forest Decline Concepts*, APS Press, St. Paul, MN, 1992.

National Research Council, Committee on Biological Effects of Atmospheric Pollutants, *Fluorides*, National Academy of Sciences, Washington, D.C., 1971.

Olszyk, D.M., Bytnerowicz, A., and Takemoto, B.K., Photochemical oxidant pollution and vegetation: effects of gases, fog and particles, *Environ. Pollut.*, 61, 11, 1989.

Prokop, W.H., Odors, in *Air Pollution Engineering Manual*, 2nd ed., Davis, W.T., Ed., Air and Waste Management Association, Pittsburgh, 2000, p. 136.

Schutze, E.D., Air pollution and forest decline in a spruce (*Picea abies*) forest, *Science*, 244, 776, 1989.

Skiffington, R.A. and Wilson, E.J., Excess nitrogen: issues for consideration, *Environ. Pollut.*, 54, 159, 1988.

Smith, W.H., Ed., *Air Pollution and Forests: Interactions between Air Contaminants and Forest Ecosystems*, 2nd ed., Springer-Verlag, New York, 1990.

Tasdemiroglu, E., Costs of materials damage due to air pollution in the United States, *Renewable Energy*, 1, 639, 1991.

Taylor, G.E., Johnson, D.W., and Andersen, C.P., Air pollution and forest ecosystems: a regional to global perspective, *Ecol. Appl.*, 4, 662, 1994.

USEPA, *Air Quality Criteria for Ozone and Related Photochemical Oxidants*, EPA/600/AP-93/004AF-CF, EPA, Washington, D.C., 1996.

USEPA, *Nitrogen Oxides: Impacts on Public Health and the Environment*, http://ttncaaal/t/ria.html, 1997.

USEPA, *Air Quality Criteria for Particulate Matter*, EPA/600/P-99/002C, EPA, Washington, D.C., 1999.

Winner, W.E., Mooney, H.A., and Goldstein, R.A., Eds., *Sulfur Dioxide and Vegetation: Physiology, Ecology and Policy Issues*, Stanford University Press, Palo Alto, CA, 1985.

World Health Organization, *Nitrogen Oxides*, 2nd ed., Environmental Health Criteria 188, WHO, Geneva, Switzerland, 1997.

QUESTIONS

1. Describe major injury types caused by phytotoxic pollutants.
2. Indicate pollutants that cause the following symptoms in plants: (a) interveinal necrosis, (b) stipple and chlorotic mottle, (c) undersurface glazing.
3. What are the distinguishing characteristics of forest declines?
4. What are the likely causes of the following forest declines: (a) Ponderosa pine in California, (b) red spruce in Vermont, (c) forest tree species in Germany?
5. How does acidic deposition affect forest trees?
6. In many developing countries lead is smelted in facilities without any emission controls. What welfare concern(s) does this pose?
7. Describe adverse effects of lake acidification as pH progressively decreases from 6.5 to 5.0.
8. Acidic deposition has adverse effects on aquatic organisms as a result of the mobilization of metals. Describe how this occurs.
9. What are the potential adverse effects of nitrogen deposition on forest ecosystems?
10. Why is nitrate deposition in estuaries an important environmental concern?
11. What is biomagnification and what is its significance relative to mercury and PCB contamination of the environment?
12. What is the environmental significance of dioxin emissions from incinerators?
13. What are some of the potential ecological effects associated with stratospheric O_3 depletion?
14. Describe potential or existing effects of global warming on Arctic ecosystems and wildlife.
15. What effect would global warming have on the distribution of forest trees in North America?
16. Describe pollutant effects on metals and building stones.
17. Describe pollutant effects on paper, leather, and rubber.
18. Describe malodors as an air quality problem.
19. Considering the various welfare effects discussed in this chapter, what pollutant or pollutants have the potential to cause the most environmental harm?
20. In 2001, USEPA began the process of developing limits on mercury emissions from coal-fired power plants. Why?

CHAPTER 7

Air Quality and Emissions Assessment

The quality of air in our communities, countryside, and even remote locations changes from hour to hour, day to day, and over longer timescales. Concentrations of pollutants depend on the magnitude of emissions from individual sources, source emission density, topography, and state of the atmosphere.

Air quality can be defined qualitatively. It is poor when pollutants (1) cause a reduction in visibility, (2) soil building surfaces and damage materials, (3) damage crops and other plants, or (4) cause adverse health effects. It is deemed good when the sky appears clean and no adverse environmental effects are evident.

Qualitative assessments of air quality, though indicative of the nature of the atmosphere relative to pollution concerns, cannot be used to support regulatory programs designed to protect the environment. Air quality must therefore be characterized quantitatively. As such, it is defined in the context of concentrations of specific target pollutants and observed environmental and public health impacts.

Protection of human health and the environment from adverse effects of pollutants is the primary goal of all air pollution control programs. Protection and enhancement of air quality require that good data on ambient concentrations of major pollutants and emissions from individual sources and groups of sources be available to regulatory authorities. Systematic efforts to monitor ambient pollutant concentrations both temporally and spatially, as well as characterize and quantify emissions, are vital to the success of all air quality protection and pollution control programs. These efforts include ambient air quality monitoring, source emissions assessment, and modeling.

7.1 AIR QUALITY MONITORING

Pollutant levels in North America are assessed by means of systematically conducted, long-term ambient air quality monitoring efforts. In the U.S., monitoring provides data to:

1. Determine compliance with National Ambient Air Quality Standards (NAAQSs) for seven pollutant categories (criteria pollutants) in air quality control regions (AQCRs) or portions thereof
2. Determine long-term trends
3. Determine human exposures
4. Support the Air Quality Index program
5. Support emissions reduction programs
6. Determine effectiveness of emission control programs
7. Support environmental assessments such as visibility impairment and degradation of watersheds
8. Support research efforts designed to determine potential associations between pollutant levels and adverse health and environmental effects

Air quality monitoring efforts are important in identifying new pollutant-related environmental problems (e.g., mercury (Hg) deposition) and tracking changes in tropospheric and stratospheric concentrations of ozone (O_3)-destroying chemicals (as well as O_3 itself) and atmospheric concentrations of greenhouse gases such as carbon dioxide (CO_2), methane (CH_4), nitrous oxide (N_2O), and chlorofluorocarbons (CFCs).

7.1.1 Monitoring Considerations

Air quality monitoring involves numerous measurements of individual pollutants over time at a number of locations in organized, systematic programs. Concentration measurements are made from samples of small volumes of ambient air.

7.1.1.1 Sampling

Because of the vastness and dynamic nature of the atmosphere, it is not possible to determine the infinite number of concentration values typical of individual pollutants. As in other instances where a population of values is potentially large, pollutant concentrations must be inferred from a relatively limited number of samples. Consequently, they must be determined from fixed-site monitors whose locations reflect objectives of air quality monitoring programs.

Concentrations are determined by collecting pollutants in or on a sampling medium or in automated continuous systems, where they are drawn through a sensing device. In manual methods, sampling and analysis are discrete events, with sample analysis occurring days to weeks after collection. In automated systems, sampling and analysis are simultaneous or near-simultaneous events and concentrations are measured in real time.

In manual methods, a sufficient amount of pollutant must be collected to meet the lower limit of detection (LOD) requirements of the analytical procedure used. This quantity will depend on the atmospheric concentration and sample size (the volume of air that passes through or comes in contact with the sampling medium). Sample size is a direct function of sampling rate and duration; samples are collected intermittently.

Sampling rates for individual pollutants using manual methods reflect collection efficiencies and instrument limitations. Collection efficiency is the ratio of pollutant collected to the actual quantity present in the air volume sampled. It depends on the sampling technique employed and chemical and physical properties of individual pollutants. For gases, collection efficiency decreases as flow rate increases above the optimum value. Optimum collection efficiencies are often achieved at relatively low sampling rates for gas-phase contaminants (<1 l/min). Because of low flow rates or low collected pollutant mass, sampling duration may be extended to collect a sufficient quantity of pollutant for analysis. For some gas-phase substances, refrigeration or use of preservatives is necessary to minimize pollutant loss before samples are analyzed.

In continuous real-time sampling, the analytical technique is usually sensitive enough that the sample size or volume needed to accurately detect and quantify specific pollutants is relatively small (typically milliliters). As a consequence, flow rates used in such instruments are also relatively low (milliliters per minute), and optimal rates of flow are used to achieve the sensitivity and accuracy for which instruments are designed.

7.1.1.2 Averaging Times

Collection and analytical limitations associated with manual, intermittent methodologies often require extended sampling durations; thus, concentration values are integrated or averaged over the sampling period. Sampling durations required to collect samples, and intended use of the data, determine the averaging time. For example, 24-h averages are appropriate for long-term trends.

Real-time air quality monitoring instruments used in the past three-plus decades provide a continuous record of fluctuating concentrations. Though such data are deemed more useful than integrated data obtained from intermittent samples, real-time data are so voluminous that they are, for the most part, uninterpretable. As a consequence, data from real-time monitoring instruments are integrated to provided hourly average concentrations or concentrations reflective of the needs of NAAQSs. For pollutants such as O_3, where peak levels occur for a limited time period, 1- and 8-h averaging times are employed. Differences in average concentrations associated with real-time data are illustrated in Figure 7.1.

Averaging times are specified for regulated air pollutants. For example, the averaging times for particulate matter with aerodynamic diameters of ≤2.5 μm ($PM_{2.5}$) are 24 h and 1 year, but for those with diameters of ≤10 μm (PM_{10}), it is 1 year; for carbon monoxide (CO) and O_3, 1 and 8 h; for nonmethane hydrocarbons (NMHCs), 3 h; for sulfur dioxide (SO_2), 3 and 24 h and 1 year; for nitrogen dioxide (NO_2), 1 year; and for lead (Pb), 3 months. Concentrations averaged over the entire year may be based on a continuous record of measurements or, in the case of PM_{10} and Pb, a more limited number of samples.

7.1.1.3 Sampling Techniques

The principal objective of sampling is to collect a pollutant or pollutants for subsequent analysis or provide a sensing environment for real-time measurements. Both require a system whereby gases or particles are drawn to the surface of a

Figure 7.1 Pollutant concentrations associated with different averaging times applied to real-time data.

collecting medium or into a sensing environment. These functions are accomplished by sampling trains that include a vacuum pump, flow regulator, and collecting device or sensing unit. Sampling trains for gases may also utilize filters to prevent particles from entering the collection unit, scrubbers for interfering gases, etc.

Based on the type of information desired, as well as collection and analytical limitations, sampling may be conducted by passive, intermittent, or continuous procedures. Such procedures provide air quality data representing a range of averaging times, from the instantaneous values of continuous systems to the 7-day average employed for some passive samplers.

Because of low cost and ease of use, passive sampling techniques were widely used in the early years of pollution control. These early techniques included the capture of particles on sticky paper, collection of settlable particles in dustfall jars or buckets, and measurement of sulfation rates on lead dioxide-impregnated gauze-covered cylinders (Pb candles). These techniques were relatively crude and abandoned with the advent of dynamic sampling methods.

New developments in passive sampling technology in the past three decades have made passive sampling a widely accepted and used methodology for personal exposure monitoring in occupational environments, measurement of indoor air contaminants such as formaldehyde (HCHO) and radon (Rn), and air quality and health research studies. Such sampling devices include small badge-type devices, tubes, canisters, etc. (Figure 7.2). Because of low cost, passive sampling techniques would be appropriate for use in ambient monitoring activities of developing nations.

In grab sampling, air samples are collected in a relatively short time (seconds or minutes) using a variety of devices. Grab sampling is typically used in problem environments (e.g., chemical spills) as a means of identifying pollutants and their concentrations present at the time samples were collected. Grab sampling using evacuated canisters for collection and identification of volatile organic compounds (VOCs) is an integral part of Photochemical Assessment Monitoring Stations (PAMS) network activities.

Figure 7.2   Passive ozone sampler. (Courtesy of Ogawa & Company, USA Inc.)

Intermittent sampling using dynamic sampling methods (i.e., manual methods) came into widespread use in air quality monitoring networks in the U.S. in the 1950s through early 1970s, both for PM and gases such as SO_2 and nitrogen oxides (NO_x). Intermittent sampling has the advantage of relatively low cost and provides reasonably useable data. It has the disadvantage of relatively long averaging times (typically 24 h), historically viewed by regulators as undesirable for gas-phase pollutants. Intermittent sampling continues to be the sampling method of choice for PM_{10}, $PM_{2.5}$, Pb, and atmospheric deposition networks. Because of its relatively low cost, intermittent sampling is often used to supplement continuous monitoring systems. Intermittent sample collections at multiple locations may provide more useful data than a continuous record at a few locations. It is particularly useful where resources for pollutant monitoring are limited.

Continuous monitoring devices provide instantaneous or near-instantaneous sampling results. Because data are acquired in real time, continuous monitors are the systems of choice in air quality monitoring programs. Their major disadvantages are high capital equipment and system maintenance costs and a need to reduce data to a manageable form.

7.1.1.3.1   Gas-Phase Pollutants

Manual collection of gas-phase substances may be conducted by employing the principles of absorption, adsorption, condensation, or capture. Of these, absorption has been the most widely used. It involves drawing contaminated air through a liquid reagent. The contaminant is collected when it dissolves in the absorbing reagent contained in a bubbler or impinger.

Adsorption is a process wherein gas-phase substances are collected on a solid surface. In adsorption, gases and vapors are physically attracted to a solid that has a large surface area, such as activated carbon (C), silica gel, molecular sieve, and others.

Collected gases may be desorbed from the collecting medium and subsequently analyzed. Adsorption sampling is common for a variety of organic gases. Because adsorbents are not very specific, a mixture of gases is collected, making analysis more involved. Gases may be collected by condensation or freezing. In such sampling, air is drawn through a collection vessel maintained at subambient temperatures. The low temperature causes vapors to condense or freeze and be retained in the collecting vessel. The sample must be removed and subsequently analyzed. Such cryogenic, or freeze-out, sampling has been used to condense and isolate a variety of organic gases.

In grab sampling, air is collected or captured for subsequent laboratory analysis using gas-tight syringes, gas sampling bags, or evacuated tubes, bottles, or canisters. Sample volumes vary from less than 1 ml using a gas syringe to 100 l in a gas sampling bag.

7.1.1.3.2 Particulate-Phase Pollutants

Particulate-phase pollutants may be collected by gravitational settling, filtration, inertial impaction, electrostatic precipitation, or thermostatic deposition. In some continuous sampling devices, they are drawn through a chamber that contains an optical sensor. For the most part, ambient air quality monitoring for PM is conducted using filtration or inertial impaction.

Filtration has been the primary PM collection technique used in air quality monitoring. To determine concentrations gravimetrically (i.e., by determining the weight of exposed and clean filters), large volume flow rates are used to collect sufficient particle mass. A typical sampling duration is 24 h.

The standard sampling device for monitoring atmospheric PM concentrations in the U.S. until 1988 was the high-volume sampler (commonly referred to as a Hi-Vol). A glass fiber filter was located upstream of a heavy-duty vacuum cleaner-type motor operated at a rate of 1.13 to 1.17 m^3/min (40 to 60 cubic feet/min (CFM)). The sampler was mounted in a protective shelter with the filter parallel to the ground. It efficiently collected suspended particles in the size range of 0.3 to 100 μm. Concentrations were reported as micrograms per cubic meter ($μg/m^3$).

The Hi-Vol collected particles in a wider size range than was appropriate for a health-based air quality standard. As such, its use was superceded by size-selective devices that collect particles that can easily enter the human respiratory system. Hi-Vol use in the U.S. is now limited to sampling airborne particles in conformance with Pb monitoring requirements.

Various devices are used to collect particles in distinct size ranges. These include those in which a size-selective inlet module is mounted atop a Hi-Vol-type device. The module removes particles of >10 μm by impaction on a greased surface. Particles with cutoff diameters of ≤10 μm are then collected on a quartz glass fiber filter. A PM_{10} sampling device is illustrated in Figure 7.3.

Cascade impactors are fractionating devices that collect particles on filter surfaces by impaction using orifice plates with smaller and smaller diameter holes. They are designed to collect particles using up to six fractionating stages. Air is drawn through orifice plates that deflect it from its flow path. Due to inertial forces, particles strike a filter surface where they are collected.

Figure 7.3   PM₁₀ sampler with size-selective inlet. (Courtesy of ThermoAnderson.)

Dichotomous impactors are designed to collect particles in two size ranges: those with aerodynamic diameters less than 10 μm but greater than 2.5 μm, and those with diameters less than 2.5 μm. Both fractions are collected on individual membrane filters. Though dichotomous, or virtual, impactors operate like typical inertial units, large particles are impacted into a void rather than onto an impervious surface.

Various techniques are used to measure particle concentrations continuously in automated systems. These devices collect particles on filter tape, with concentrations determined hourly or integrated over 24 h by beta-particle attenuation or tapered element oscillating monitors (TEOMs). An automated dichotomous sampler can be seen in Figure 7.4.

TEOMs measure particle concentrations on a continuous basis. Particle mass collected on an exchangeable filter cartridge is determined from frequency changes in a tapered element. As particle mass collects on the exchangeable filter, the element's natural oscillation frequency changes. Since a direct relationship exists between the element's change in frequency and particle mass, this methodology provides PM measurements of high accuracy and precision.

Figure 7.4   Automated dichotomous sampler. (Courtesy of Thermo Andersen.)

7.1.2   Reporting Pollutant Concentrations

In the U.S., concentrations of gas-phase substances in air are commonly expressed as mixing ratios and reported in parts per million by volume (ppmv) or whole number fractions thereof, such as parts per hundred million (pphmv), parts per billion (ppbv), or parts per trillion (pptv).

One part per million by volume is equal to a volume of gas mixed in a million parts of air. A microliter of a gas mixed in a liter of air would be equal to 1 ppmv:

$$1 \text{ ppmv} = 1 \text{ µl gas/1 air} \qquad (7.1)$$

Mixing ratios commonly used to express air concentrations must not be confused with those used for weight–volume mixtures (milligrams per liter) for water (H_2O) and weight–weight ratios (micrograms per gram, milligrams per kilogram) for solids. Though all are expressed as parts per million, they do not express equal quantities.

Concentrations of gas-phase substances mixed in air are, in many countries, expressed in metric units as mass per unit volume (micrograms per cubic meter ($\mu g/m^3$), milligrams per cubic meter (mg/m^3)), in conformance with international practice. However, air concentrations in metric-using countries are increasingly being expressed as mixing ratios. This reflects the fact that concentrations across the chemical spectrum can be more easily compared when mixing ratios are used; it also reflects the impact American science has on the world community.

Concentrations can be converted from one form of expression to another by application of the gas laws. Assume that a manual instrument sampled SO_2 at an average flow rate of 0.2 l/min for 24 h. Concentration is determined from the mass of SO_2 (in this case, 40 µg) collected in the sample solution using the following equation:

$$\text{ppmv} = [(m)(24.45)]/MV \qquad (7.2)$$

where m is the mass of collected gas (µg), V is the sample volume (l), M is the molecular weight of the gas (e.g., $SO_2 = 64$), and 24.45 is the microliters of a gas produced from a micromole of a substance at reference temperature and pressure (25°C and 760 mmHg).

Therefore,

$$\text{ppmv} = [(40 \ \mu g)(24.45 \ \mu l)]/[(64)(288 \ l)]$$

$$\text{ppmv} = 0.053$$

The ppmv concentration of a gas in air can be converted into its respective System International (SI) concentration by means of Equation 7.3, derived from Equation 7.2:

$$mg/m^3 = (M/24.45) \ \text{ppmv} \qquad (7.3)$$

For purposes of illustration, let us convert the ambient air quality standard for O_3 (0.08 ppmv, 8-h average) to its corresponding concentration in $\mu g/m^3$:

$$mg/m^3 = (48/24.45)(0.08)$$

$$= 0.157 \ mg/m^3$$

$$= 157 \ \mu g/m^3$$

If the concentration of a gas in milligrams per cubic meter is known, it can be converted to parts per million by volume using Equation 7.4:

$$\text{ppmv} = (24.45/M)(mg/m^3) \qquad (7.4)$$

For purposes of illustration, let us convert the 24-h standard for SO_2 expressed in SI units as 0.365 mg/m³ to ppmv:

$$ppmv = (24.45/64)(0.365)$$

$$= 0.14 \text{ ppmv}$$

Concentrations of gas-phase substances are commonly reported for reference conditions of temperature and pressure, 25°C and 760 mmHg. A measured volume of air may be standardized (V_s) by means of the following equation:

$$V_s = V_m \ [(P_1/P_2)(T_2/T_1)] \qquad (7.5)$$

where V_m is the measured volume, P_1 is the field atmospheric pressure (mmHg), P_2 is the reference pressure (mmHg), T_1 is the field ambient temperature (K), and T_2 is the reference temperature (K).

Assuming that field sampling conditions were 22°C and 740 mmHg and $V_m = 288$ l, the standardized volume for the SO_2 sample above would be

$$V_s = 288 \ [(740/760)(298/295)]$$

$$V_s = 283.1$$

This will change the concentration calculated from Equation 7.2 from 0.053 to 0.054 ppmv, an approximate 2% difference.

Particle concentrations in air quality monitoring programs are always reported as weight per unit volume, i.e., $\mu g/m^3$.

7.1.2.1 Accuracy, Precision, Bias

Measured concentrations of pollutants must reliably reflect actual concentrations. Three concepts are important in determining such reliability: accuracy, precision, and bias.

Accuracy is the relative agreement between a measured value and an accepted reference or true value. Though 100% accuracy is desirable, such accuracy is difficult to achieve since all sampling and analysis techniques are subject to errors associated with the sampling instrument, analytical method, personal errors by those conducting measurements, and extending a measurement system beyond its limits.

Sampling and analytical techniques used to measure concentrations of the same chemical species vary in accuracy. As such, their utility and acceptability are determined by meeting an accuracy performance objective, typically within ±10% of the reference or true value.

Accuracy is determined and reported in the context of absolute or relative error. Absolute error is defined as

$$E = O - A \qquad (7.6)$$

where O is the observed value and A is the actual value.

If in making a series of measurements of a known concentration (1 ppmv) of SO_2 the observed average value were 0.925 ppmv, then

$$E = 0.925 - 1.0$$

$$= -0.075$$

Absolute error is -0.075. The measured concentration in this case is less than the actual concentration. Consequently, the error is negative. Had the measured concentration been 1.075 ppmv, the error would have been positive. The relative error (E_r) determined as a percentage can be calculated from the following equation:

$$E_r = [(E)(100)]/1 \text{ ppmv} \tag{7.7}$$

$$= -7.5\%$$

Accuracy is commonly reported as a percentage of the true value. In this case, the sampling or analytical procedure has an accuracy of 92.5%. If the measured value were 1.075 ppmv, it would have an accuracy of 107.5%.

Error and accuracy values are based on average values determined from a series of measurements. In the case above, the mean of 10 measurements (0.850, 1.000, 0.950, 0.960, 0.870, 0.900, 0.880, 0.920, 0.940, and 0.990 ppmv) from a 1-ppmv reference concentration of SO_2 was 0.925 ppmv. Variation of measured values around the mean value describes the scientific/statistical concept of precision, a measure of the reproducibility of results. (Note: Dictionaries define precision as the exactness of a result, and as such, concepts of accuracy and precision are indistinguishable and are commonly used interchangeably.)

The precision of a sampling/analytical technique is determined from calculations of the mean of repeated measurements and its standard deviation. In a normally distributed population, ~67% of sample values are within ±1 standard deviation (σ) of the mean; 95% are within 2 σ values.

Standard deviation can be calculated from the following equation:

$$\sigma = \left[\sum_{i=1}^{n} \left(X_i = \overline{X}^2 \right) / N \right]^{0.5} \tag{7.8}$$

where $\overline{X}$ is the mean, X_i is the individual measured value, and N is the total number of measurements.

By using a programmable calculator or spreadsheet software, 1 σ is determined to be 0.048. Therefore, the precision of our sampling results is ±0.048 ppmv. Precision, however, is usually reported as a percentage by calculating the coefficient of variation (CV), where X is the sample mean:

$$CV = (\sigma / \overline{X})(100) \tag{7.9}$$

$$= 5.2\%$$

Using ±1 σ as a reference, the precision of our measurements was ±5.2%.

In this case, our sampling/analytical technique had high relative accuracy (92.5%) and good precision (low variability). Sample precision values within ±10% of the measured value are considered to be good.

Differences in the concepts of accuracy and precision can be seen in Figure 7.5. In the first target (a), values are scattered and do not come close to the target's center. In the second (b), all values are close to the middle of the target. In the former, accuracy is relatively low, with poor precision. In the latter, results are both accurate and precise.

The third target (c) illustrates the concept of bias. Results show good to excellent precision but deviate from the true value in a systematic way. Bias occurs as a result of calibration errors (the instrument may be out of calibration), but may also be due to hardware problems associated with instrument electronics or systematic software problems.

Accuracy in the context of air quality monitoring activities includes a combination of random (precision) and systematic (bias) errors. The term *accuracy* is used when these two types of error cannot be separated. As a consequence, accuracy values are stated in the same format as precision. Reported accuracy values include precision determinations. For most ambient air quality monitoring activities, accuracies (which take precision into account) within ±10% of the true value are considered acceptable. For passive sampling techniques, accuracy values within ±25% are considered acceptable.

7.1.2.2 Calibration

All sampling and monitoring instruments must be periodically calibrated to reduce bias-associated errors and assure data reliability. Calibration is a process in which measured values are compared to standard reference values (in the case of pollutant concentrations) or volume airflows are measured using standard airflow measuring techniques and devices. When bias-type errors are identified, instrument adjustments are made to reduce observed errors, i.e., bring the instrument back into calibration.

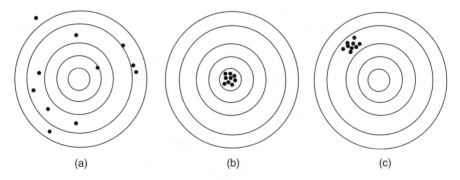

| (a)   (b)   (c) |

Figure 7.5 Graphical illustration of accuracy, precision, and bias.

In some manual sampling methods, instrument calibration is limited to measuring airflow rates. Such calibration may be conducted using flow-measuring instruments based on primary or secondary standards. A primary standard is one that is directly traceable to the National Institute of Standards and Technology (NIST). A primary standard would be, e.g., a bubble meter (gas burette and soap solution used for low flows) or volumetric flask. In the latter case, the volumetric flask may be used to calibrate wet or dry test meters that may be used to calibrate rotometers. Wet and dry test meters and rotometers are secondary standards. Because of their portability and ease of use, field instruments are commonly calibrated with secondary standard devices (such as rotometers). Based on USEPA regulations, airflow rate measurements must be made using instruments traceable to an authoritative volume (e.g., NIST).

Monitoring instruments for gas-phase pollutants such as O_3, NO_2, CO, and SO_2 must be periodically calibrated to determine whether measured values are consistent with reference values. Concentration calibrations are made using gas standards traceable to a NIST reference material or a NIST-certified gas manufacturer's internal standard.

Standards for gases such as CO, SO_2, NO_2, and nitric oxide (NO) are available in gas cylinders or permeation tubes. Gas permeation tubes contain a sealed volume of calibration gas that passes through a membrane at a known rate. Permeation tubes can be inserted into a gas sampling stream at a constant flow rate to obtain standardized concentrations for calibration.

Due to its high reactivity, gas standards are not available for O_3 calibration. Ozone monitors must therefore be calibrated by producing test atmospheres using an O_3 generator and gas dilution system. Some O_3 monitors are also used to calibrate other O_3 monitors.

7.1.3 Federal Monitoring Requirements

Under regulations promulgated by the USEPA, ambient air quality monitoring programs must be developed and operated by states as part of control programs for assessing and determining compliance with NAAQSs. They must meet minimum design and quality assurance requirements.

State and local air monitoring stations (SLAMS) must have sites that monitor (1) highest pollutant concentrations, (2) representative concentrations in areas of high population density, (3) the impact of major emission sources, and (4) regional background concentrations. Recently, the USEPA added two new objectives for air quality monitoring efforts: determination of (1) extent of pollutant transport among populated areas, and (2) welfare-related impacts in more rural and remote areas.

In siting air quality monitoring stations, states are required to consider spatial scales consistent with federal monitoring objectives. Spatial scales are estimates of the sizes of areas around monitoring locations that experience similar pollutant concentrations.

Spatial scale categories are (1) microscale, (2) middle scale, (3) neighborhood scale, (4) urban scale, and (5) regional scale. The microscale ranges from a few meters to 100 m (~300 ft); the middle scale from 100 m to 0.5 km (~0.3 mi); the neighborhood scale from 0.5 to 4.0 km (~2.5 mi); the urban scale from 4 to 50 km (~30 mi); and the regional scale from tens to hundreds of kilometers.

Spatial scales for the highest concentration or source impact are micro-, middle-, neighborhood, and, less frequently, urban scales. Appropriate spatial scales for high population densities and representative concentrations are middle, neighborhood, and urban. Neighborhood or regional scales are appropriate for background levels. Urban and regional scales are also appropriate for determination of pollutant transport and welfare-related impacts in rural remote areas.

States that have AQCRs that exceed air quality standards for O_3 must also establish a Photochemical Assessment Monitoring Stations (PAMS) network in conformance with federal requirements.

All federally mandated air quality monitoring programs must use monitoring techniques based on federal reference methods (FRMs) or approved equivalents, meet minimum quality assurance requirements, and report air monitoring results to the USEPA annually.

7.1.3.1 Air Quality Monitoring Networks

Operation of SLAMS and PAMS networks is supported in part by federal grant monies. They are operated in conjunction with federal monitoring activities that include National Air Monitoring Stations (NAMS). The NAMS program is an urban-area, long-term air monitoring network that provides a systematic, consistent database for air quality comparisons and trends analysis. The NAMS are, in most cases, similar to SLAMS. The SLAMS program, however, has more flexibility in siting and operation to reflect state, local, and tribal needs. In many cases, NAMS are operated by SLAMS programs under the USEPA contract.

The NAMS and SLAMS data, as well as those of other sites (e.g., industrial monitors), are reported annually to the USEPA's Aerometric Information Retrieval System (AIRS), where data are compiled and published.

The PAMS network is designed to monitor O_3 and photochemical air pollutants in areas with persistently high ground-level O_3 concentrations. Sites are located along the east coast from Portsmouth, NH, to Washington, D.C., and in California from San Diego to Sacramento. They are also in Phoenix, AZ; El Paso, Houston, and Dallas–Ft. Worth, TX; Baton Rouge, LA; Atlanta, GA; and the Lake Michigan lakefront from Milwaukee to northwest Indiana. The PAMS sites monitor NO_x, total nonmethane organic compounds (NMOCs), target VOCs, and upper air meteorology during the summer months.

The Interagency Monitoring of Protected Visual Environments (IMPROVE) network was set up in the mid-1990s. It represents a cooperative effort between the USEPA and other federal resource management agencies such as the National Park Service (NPS). Its focus is monitoring $PM_{2.5}$ and visibility, and assessing the relationship between them.

Other federal monitoring networks are associated with intensive research-oriented studies and atmospheric deposition. Included in the former are the North American Research Strategy for Tropospheric Ozone (NARSTO), $PM_{2.5}$ supersites, and PM health centers.

The Clean Air Status and Trends Network (CASTNet) is this country's primary atmospheric data source for rural O_3 levels and dry atmospheric deposition. It includes 79 monitoring stations operated by the USEPA or the NPS that report weekly average concentrations of sulfate (SO_4^{-2}), nitrate (NO_3^-), ammonium (NH_4^+), SO_2, and nitric acid (HNO_3) and hourly O_3 concentrations. The National Oceanic and Atmospheric Administration (NOAA) operates a smaller dry deposition network called the Atmospheric Integrated Monitoring Network (AIRMON), which focuses on research questions related to atmospheric deposition.

Other atmospheric deposition networks include the Integrated Atmospheric Deposition Network (IADN) and the National Atmospheric Deposition Program/National Trends Network (NADP/NTN). The NADP began in 1978 as part of the National Acid Precipitation Assessment Program (NAPAP) in an effort to assess the problem of atmospheric deposition and its effects on aquatic and terrestrial ecosystems, as well as agricultural crops. It has been a cooperative program between federal and state agencies, universities, and utilities and other industries to measure concentrations of hydrogen ion (H^+), NH_4^+, SO_4^{-2}, NO_3^-, chloride (Cl^-), calcium (Ca^{+2}), magnesium (Mg^{+2}), and potassium (K^+) in wet deposition. Its mission has been expanded to include Hg. At the time of this writing, the Mercury Deposition Network (MDN) is located at 40 NADP sites in 16 states and 2 Canadian provinces. Its focus is to measure Hg levels in wet deposition to assess seasonal and annual fluxes into sensitive ecosystems.

The IADN network was established in 1990 as a joint effort between the U.S. and Canada to conduct air and precipitation monitoring in the Great Lakes Basin. It measures target substances in wet and dry deposition and gas-phase organic vapors. Of particular concern are semivolatile organic compounds (SVOCs) such as polychlorinated biphenyls (PCBs), dioxins, polycyclic aromatic hydrocarbons (PAHs), and a variety of organochlorine pesticides. In the year 2000, the network included one master station on each of the Great Lakes plus 14 satellite stations.

The SLAMS program, and less so the NAMS program, has been undergoing a significant reevaluation. Considerable progress has been made in controlling emissions of Pb, SO_2, and CO. In many AQCRs, levels of these pollutants as well as NO_x are well below NAAQSs and have been for a long time. As such, opportunities exist to discontinue monitoring activities for these pollutants at selected sites in order to redirect resources for evolving pollutant concerns (e.g., $PM_{2.5}$ and air toxics).

7.1.3.2 Federal Reference Methods

All air quality monitoring activities conducted at NAMS, SLAMS, and PAMS sites for criteria pollutants (those regulated under NAAQSs) must use methodologies approved by the USEPA as reference methods or their equivalent. All reference and equivalent methods are subject to intensive performance evaluations before approval.

Federal Reference Methods (FRMs) for the seven criteria pollutants are summarized in Table 7.1. Though the FRM has primary regulatory status, it is not uncommon for one or more equivalent methods to be more widely used in day-to-day monitoring activities.

Table 7.1   Federal Reference Methods for Criteria Pollutants

Pollutant	Reference Method
Sulfur dioxide	Spectrophotometry (pararosaniline method)
Nitrogen dioxide	Gas-phase chemiluminescence
Carbon monoxide	Nondispersive infrared photometry
Ozone	Chemiluminescence
Total nonmethane hydrocarbons	Gas chromatography — FID
PM_{10}	Performance-approved product
$PM_{2.5}$	Performance-approved product
Lead	TSP sampling; analysis by wet chemistry, atomic absorption spectrometry

Note: FID = Flame ionization detection

7.1.3.2.1   Sulfur Dioxide

The FRM for SO_2 is the spectrophotometric pararosaniline method. In its manual application, SO_2 is collected by bubbling air through a potassium tetrachloromercurate solution to form a $HgCl_2SO_3^{-2}$ complex. Subsequent reactions with HCHO and colorless pararosaniline hydrochloride produce a red–violet product that can be measured spectrophotometrically and converted to an SO_2 concentration using a standardized concentration curve.

Manual collection and analysis is both cumbersome and limiting. As a consequence, the FRM is more often used in automated continuous monitoring devices. Even as an automated procedure, such monitors require considerable maintenance. As such, many air quality monitoring stations use a USEPA-approved equivalent method.

In the year 2000, a multigas Longpath Monitoring System was approved as an automated equivalent method for SO_2. This device can provide remote sensing data over an open path (total path length up to 1 km) using a Fourier transform infrared (FT-IR) spectrometer. In these systems, a source generates a beam of infrared energy across the air medium to be measured by a retroreflector (Figure 7.6) in a monostatic system or a detector at the other end of the path in a bistatic system. Infrared energy absorbed at specific wavelengths can be directly related to the total concentration of a target gas (e.g., SO_2) along the path length.

Other equivalent SO_2 monitoring methods include conductimetry (collection of SO_2 in a hydrogen peroxide (H_2O_2) solution and measurement of increased conductance of sulfuric acid (H_2SO_4) formed), amperometry or electrochemistry (measurement of an electric current produced on SO_2's reaction with the collecting solution), ultraviolet (UV) fluorescence detection, flame photometry, and gas chromatography.

7.1.3.2.2   Nitrogen Oxides

The FRM for measurement of NO_x is gas-phase chemiluminescence. Chemiluminscence occurs as a result of the emission of light from electronically excited

Figure 7.6 Schematic of FT-IR spectrophotometer use in the field. (Courtesy of USEPA.)

chemical species formed in chemical reactions. The reaction of O_3 with NO produces an electronically excited NO_2 molecule:

$$NO + O_3 \rightarrow NO_2 \cdot + O_2 \tag{7.10}$$

When $NO_2 \cdot$ returns to its ground state, it emits light in the 600- to 3000-nm range. Emitted light is amplified by a photomultiplier tube; the intensity of emitted light is proportional to the NO concentration.

Measurement of NO_2 is based on its conversion to NO and subsequent detection and measurement by chemiluminescence. Such analyses are subject to interference by other nitrogen (N)-containing compounds (e.g., peroxyacyl nitrate (PAN)), resulting in higher measured NO_2 concentrations than may actually be present.

The manual equivalent method (including an automated version) uses a sodium arsenite reagent to collect NO_2, which on analysis forms a colored solution that is measured spectrophotometrically.

The open-path FT-IR long-path monitoring system has also been approved for use as an automated reference method for NO_2.

7.1.3.2.3 Carbon Monoxide

The FRM for CO is nondispersive infrared (NDIR) spectrometry. Measurement is based on the ability of CO to strongly absorb infrared energy at certain wavelengths. NDIR spectrophotometric devices utilize two cylindrical cells (usually 100 cm (~39 in.) long), a sample and a reference cell, in which a source emits CO-absorbing infrared wavelengths. As CO passes through the sample cell, it absorbs infrared energy in direct proportion to its concentration. Differences in infrared energy (or heating) in the two cells cause a slight movement in a diaphragm, which is detected and recorded. Concentrations can be measured with a relative accuracy of ±5% in the optimum range (0 to 150 ppmv).

7.1.3.2.4 Ozone

The FRM for O_3 is chemiluminescence. It is based on measurement of light emissions produced on reaction of O_3 with ethylene (C_2H_4). Because of C_2H_4's flammability, the FRM was amended in 1979 to replace C_2H_4 with Rhodamine B dye embedded in a disk. Because Rhodamine B does not attain a stable baseline rapidly after exposure to O_3, UV analyzers approved as equivalent methods are used in almost all ambient air quality monitoring stations in the U.S.

The measurement principle of UV analysers is based on absorption of UV by O_3 and subsequent use of photometry to measure the reduction of UV energy at a detector at 254 nm; reduction is dependent on the path length of the sample cell, O_3 concentration, and wavelength of UV light.

The FT-IR long-path monitoring system has also been approved as an automated equivalent method for ambient O_3 monitoring.

7.1.3.2.5 Particulate Matter: PM_{10} and $PM_{2.5}$

The NAAQSs for PM include standards for PM_{10} and $PM_{2.5}$. The FRMs for PM_{10} and $PM_{2.5}$ are defined by a combination of design and performance specifications. Consequently, manufacturers design and fabricate samplers to meet reference method requirements. Each approved sampling instrument can be designated as an FRM.

Manual reference samplers have included instruments that (1) collect PM on a filter medium or collecting surface using size-selective inlets on modified types of Hi-Vols, and (2) dichotomous virtual impactors. Samples are collected over a period of 24 h and concentrations determined gravimetrically. Equivalent methods must provide results that are comparable in performance to reference methods.

Several automated commercial PM_{10} samplers have been approved as equivalent methods and will likely be approved as equivalent methods for $PM_{2.5}$. These devices collect particles on filter tape, with concentrations determined hourly or integrated over 24 h by beta-particle attenuation or TEOMs.

7.1.3.2.6 Lead

Both the FRM and equivalent methods for Pb require manual collection of total suspended particulate (TSP) mass by a Hi-Vol particulate collector, operated at a flow rate of 1.12 to 1.70 m^3/min (40 to 60 CFM), on glass fiber filters for 24 h. Filter samples are collected, dried, weighed, and analyzed for Pb. Lead analysis reference methods are based on wet chemistry and spectrophotometry.

Commonly used equivalent analytical techniques include flame atomic absorption spectrometry, graphite furnace atomic absorption spectrometry, inductively coupled plasma emission spectrometry, and energy-dispersive x-ray fluorescence spectrometry.

7.1.3.3 Quality Assurance Programs

All environmental monitoring and measurement programs must have a quality assurance and quality control (QA/QC) program in place that provides for acquisition of both valid and reliable air quality monitoring data. All data submitted to the USEPA under SLAMS must meet minimum QA/QC program requirements.

Quality assurance and quality control are distinct and interrelated quality system functions. Quality assurance is a program that sets policy and oversees management controls that include planning, review of data collection activities, and data use. It also involves setting data quality objectives, assigning responsibilities, conducting reviews, and implementing corrective actions.

Quality control, on the other hand, focuses on the technical aspects of data quality programs. It includes implementation of specific QC procedures such as calibrations, checks, replicate samples, routine self-assessment, and audits.

Under federal rules, QA/QC efforts must be documented so that users can evaluate data quality before any conclusions are drawn from collected data. Under SLAMS programs, federally required quality system-monitoring efforts must meet a well-defined need; take cost into consideration; and implement a QA program whose policies, procedures, standards, and documentation provide data of acceptable quality with minimal data loss.

All monitoring methods used in SLAMS must be assessed periodically to quantitatively determine accuracy, precision, and bias independent of routine instrument calibrations for airflow and concentration determinations. These technical system audits involve on-site inspection of SLAMS programs to assess compliance with requirements that govern collection, analysis, validation, and reporting of air quality data. Audits are typically conducted at specified frequencies to establish the consistency of data collection activities and comparability of data across a monitoring network.

7.1.3.4 Data Summarization and Characterization

Air quality monitoring programs generate an enormous number of individual pollutant concentration values. After collection, data must be summarized or treated in such a way that their regulatory and environmental significance can be evaluated.

Data summarization and characterization efforts are designed to reduce a large number of measured values to meaningful summary tables, graphical representations, or statistics. Air quality monitoring data are commonly summarized using graphical techniques and statistics.

7.1.3.4.1 Graphical Techniques and Displays

Air quality monitoring data are continuous (i.e., potentially include an infinite number of values), even if they have been integrated over a particular averaging time. Continuous data are often summarized using frequency distributions.

Frequency distributions are prepared by identifying an appropriate number of frequency classes or class intervals. A frequency class is a range of data values that represents a part of the collected data. In standard statistical practice, data are summarized in as few as 10 to as many as 20 or more frequency classes. Frequency class data are graphically presented as histograms. A histogram for TSP data is shown in Figure 7.7. Note that the bar indicating frequency for each class interval touches the adjoining bar; this indicates that data are continuous. Histograms should not be confused with bar graphs used to summarize discreet data. (Note that popular spreadsheet and data manipulation software such as Excel commonly produce bar graphs under the histogram function.)

The histogram is an important tool in evaluating how data are distributed. Data that are normally distributed will be seen as a bell-shaped curve, with an equal number of values to the left and right of center of the data distribution. Symmetry of the distribution is described by its skewness. A normal distribution has a skewness of 0. In Figure 7.7, data appear skewed to the right (positive skewness values). As such, the data are not normally distributed. With some data sets, the frequency curve will be skewed to the left (negative skewness values).

The form of the frequency distribution curve must be known so that appropriate statistical procedures are used in data analysis. If data are not normally distributed, they cannot, in theory, be analyzed by application of parametric statistics (the most widely used statistical procedures). Such procedures can be applied if data, on

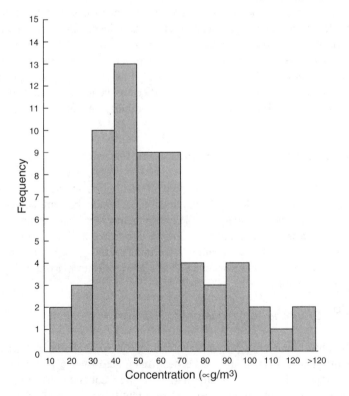

Figure 7.7   Histogram plot of TSP data.

transformation, fit a normal distribution. Not uncommonly, environmental data are lognormally distributed (after log transformation and plotting on semilogarithmic graph paper).

Histograms have been widely used to summarize continuous data (including environmental data). Increasingly, stem and leaf plots and box plots are used to summarize pollution data. Stem and leaf diagrams and box plots show the range of the data (with minima and maxima), as well as medians and quartiles. The median locates the 50th percentile and sorts data into halves. Quartiles divide each half into quarters. The upper quartile locates the 75th percentile, and the lower the 25th percentile.

Frequency distribution data can be summarized using stem and leaf plots. In such distributions, actual values truncated to two or three digits are plotted, with the first value the stem and succeeding numbers the leaves. Stems and leaves are separated by a vertical channel. Numbers in the channel indicate upper and lower quartiles.

A box plot of trends in 8-h average CO concentrations in the U.S. from 1981 to 1990 can be seen in Figure 7.8. The x in the box indicates the location of the 50th percentile, or median. Each box stretches between the 25th and 75th percentiles. Whiskers stretch out from the rest of the box to span the remainder of the data except for outliers. These plots include inner and outer fences. The upper inner fence = [(75th percentile) + (1.5 × midrange)]; the lower inner fence = [(25th percentile) – (1.5 × midrange)]. The upper outer fence = [(75th percentile) + (3.0 × midrange)]; the lower outer fence = [(25th percentile) – (3.0 × midrange)]. Data lying outside the probability distribution described by the box plot are called outliers.

7.1.3.4.2 Statistical Summarization

Data can be summarized using measures of central tendency such as mean, median, and mode. In normally distributed data the mean lies in the center, corresponding to the peak of normal distribution curve. It is calculated simply by summing all data values and dividing the sum by the total number of values (N).

The median is the middle number when data are arranged from the lowest to the highest values. It is the average of the two middle values when N is an even number. The mode is the most commonly observed value in the data set. If data are normally distributed, the mean, median, and mode are all the same value.

The median is the appropriate measure of central tendency when data sets are not normally distributed. In such cases, the arithmetic mean does not accurately reflect where most of the values lie in a frequency distribution. An example is use of the median to report family income by the Bureau of Labor Statistics. Median income is reported since ~5% of Americans earn 40% of total national income. If the mean were reported, 70% or so of American households would appear to have below-average incomes.

In the TSP data summarized in Figure 7.7, median and mean values were 48.8 and 59.9 $\mu g/m^3$, respectively. In this case, a relatively few high values skewed the mean 10 $\mu g/m^3$ higher than the median.

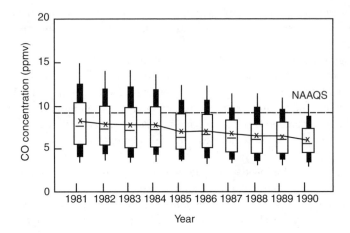

Figure 7.8   Box plot comparisons of trends in 8-h average CO concentrations in the U.S. from 1981 to 1990. (From USEPA, EPA/600/8-90/045f, EPA, Washington, D.C., 1991.)

Though the median is commonly used to describe central values in data sets that are not normally distributed, in most instances it cannot be used to calculate data variation required for statistical analyses. In such cases, it is usually desirable to log-transform data and calculate the geometric mean (GM). The GM can be calculated using the following equation:

$$GM = antilog\ (log\ x_1 + log\ x_2 \ldots + log\ x_n)/N \qquad (7.11)$$

Applying such calculation to the values used in the histogram in Figure 7.7, the GM is found to be 50.8 $\mu g/m^3$, a value relatively close to the median.

The σ describes the distribution of data around the mean. It can be determined by using Equation 7.8. Sixty-seven percent of values in a data set lies within ± 1 σ of the mean. The larger the σ, the more variable the data; the smaller, the less variable.

Air quality standards are based on arithmetic means. When PM concentrations were based on TSP measurements, the TSP standard was based on the GM.

Environmental monitoring data are typically reported as arithmetic numbers. In some cases, e.g., pH values in monitoring acidic deposition and decibel values in monitoring noise (see Chapter 12), data are in geometric form. In both cases, it is inappropriate to sum values and divide by N to calculate the mean.

Since pH is the negative log of H^+ concentration, pH values must be converted to their equivalent molar concentrations. In monitoring acidic deposition weekly over 6 weeks, the following pH values were reported: 4.5, 5.0, 3.8, 4.0, 4.2, and 4.8. Their corresponding molar H^+ concentrations, respectively, are $1 \times 10^{-4.5}$, 1×10^{-5}, $1 \times 10^{-3.8}$, 1×10^{-4}, $1 \times 10^{-4.2}$, and $1 \times 10^{-4.8}\ M$. Summing and dividing equivalent molar concentrations by N, the average molar concentration is 0.000061 M or $1 \times 10^{-4.20}$. By taking the negative log of this value, the average pH is 4.20. Had each of the pH values been summed and divided by N, the average would have been 4.38.

Though these values do not appear to be significantly different from each other, in fact, they are. The actual difference can be determined by dividing the molar concentration of pH 4.20 (0.000061) by the molar concentration of pH 4.38 (0.000042) (the lower H^+ concentration):

$$0.000061 \ M/0.000042 \ M = 1.45$$

A pH value of 4.20 is seen to have a H^+ concentration 1.45 times greater than a pH of 4.38. Similarly, if our average pH value of measured precipitation samples were compared to the theoretical clean air value of 5.65, it would be 28 times more acidic.

7.1.3.5 Air Quality Index

Results of air quality monitoring can be used to provide guidance to public health and environmental officials as well as individuals relative to community air pollution exposures. For this purpose, the USEPA has developed and implemented an Air Quality Index (AQI) program to provide timely and easy-to-understand information on local air quality and related health concerns.

The AQI is used to report daily air quality in metropolitan areas for five pollutants: O_3, PM (PM_{10} and $PM_{2.5}$), CO, SO_2, and NO_2. It uses a scaled approach according to the six levels of health concern in Table 7.2. The index varies from 0 to 500. An AQI value of 100 corresponds to the air quality standard for a single pollutant. AQI values of <100 indicate good to moderate air quality, whereas values of 101 to 500 indicate increased public health risk; values of >300 are considered hazardous. Sensitive groups in the third AQI category (Table 7.2) include children and adults with respiratory disease, adults with cardiovascular disease, and children and adults who are active outdoors.

The AQI is calculated for each of the five pollutants in an AQCR. The reported value is based on the pollutant with the highest concentration relative to its standard. An AQI value can also be reported for a single pollutant, e.g., O_3, when this is the primary pollutant of concern. An AQI value of 100 for O_3 corresponds to an 8-h average concentration of 0.08 ppmv. For PM_{10}, it corresponds to a 24-h average concentration of 150 $\mu g/m^3$; for $PM_{2.5}$, a 24-h average concentration of 65 $\mu g/m^3$. Air quality indices of 100 for CO and SO_2 are based on an 8-h average concentration of 9 ppmv and a 24-h average concentration of 0.14 ppmv, respectively. Since no health effects are known to occur below an AQI of 200 for NO_2, an AQI is not calculated below 201. An NO_2 AQI of 201 corresponds to an NO_2 level of 0.65 ppmv averaged over 24 h.

Relatively good air quality (i.e., reported AQIs of <100) is common in many metropolitan areas in the U.S. However, values between 101 and 200 are reported in many communities several times a year, particularly the summertime high-O_3 season. High AQI values based on CO may occur in some cities during winter. AQI values for O_3 of >101 commonly occur in the Los Angeles Basin and Houston, TX, where smog is a significant problem. Values of >300 in the U.S. are extremely rare. They are likely to be common, however (if the AQI were calculated), in many cities in China and the developing world as well as Mexico City.

The AQI for PM_{10} and $PM_{2.5}$ is based on manual sampling methods that use a 24-h averaging time. Therefore, the AQI for PM is of limited usefulness. As a

Table 7.2   Air Quality Index Values, Levels of Concern, and Color Code

AQI Values	Level of Health Concern	Color Code
0–50	Good	Green
51–100	Moderate	Yellow
101–150	Unhealthy for sensitive groups	Orange
151–200	Unhealthy	Red
201–300	Very unhealthy	Purple
301–500	Hazardous	Maroon

Source: From USEPA, *Air Quality Index: A Guide to Air Quality and Your Health,* http://www.epa.gov/airnow/aqibroch/aqi.html.

consequence, efforts have been made by the USEPA and others to utilize automated equivalent PM_{10} and $PM_{2.5}$ sampling methods to provide hourly values that are strongly correlated (using regression models) with 24-h values.

7.2 EMISSIONS ASSESSMENT

Assessment of emissions of regulated pollutants from stationary and mobile sources is a vital part of federal, state, and local air pollution control efforts. Emissions data are needed to (1) identify sources that are primarily responsible for ambient pollutant concentrations in a region, (2) formulate emissions control policy, (3) comply with permitting requirements, (4) comply with emission reporting requirements for regulated target sources and pollutants, (5) compile national annual emission inventories, and (6) provide a database for ambient air quality modeling.

Emissions assessments of individual sources are conducted by sampling stack gases using manual one-time sampling procedures (stack testing) or continuous emission monitors (CEMs). Such assessments utilize the USEPA reference methods. Though emissions assessment by stack sampling is preferred by air pollution control agencies, most sources of regulated pollutants in the U.S. have never been required or volunteered to conduct stack testing to determine emissions under real-world conditions.

Most emission assessments conducted to meet regulatory reporting requirements have been based on emission factors (see Section 7.2.3).

7.2.1   Source Sampling and Monitoring

Source sampling is typically conducted on waste gas streams that flow to the atmosphere through stacks that elevate emissions above the source. Such stacks may vary from a meter (~3 ft) in diameter and 10 to 20 m (33 to 66 ft) in height to 10 m in diameter and 300 to 400 m (~990 to 1300 ft) in height. For small industrial sources, source sampling can also be conducted on local exhaust ventilation systems.

Manual stack or source sampling is typically conducted by a team of trained technicians using the USEPA reference sampling and analytical methods. Because of the height of some stacks and the height at which representative waste gas flow occurs, sampling teams must often construct scaffolding to collect representative samples. Sample collection in many cases will require several days to a week of on-site activity.

Manual source sampling consists of introducing a probe into the waste gas stream to draw a sample volume into a sampling train. Sampling trains differ somewhat for individual pollutants and sampling methodologies; a sampling train for USEPA Method 5 for PM is illustrated in Figure 7.9. It includes a (1) probe inserted into the stack, (2) temperature sensor, (3) pitot tube, and (4) two-module sampling unit, which includes, in series, a heated compartment, an ice-bath compartment, a series of collecting impingers, and a vacuum pump, dry gas meter, and flow-regulating orifice. This sampling train can be easily modified to sample sulfur oxides (SO_x) and H_2SO_4 mist.

Since particles are subject to inertial forces, samples must be collected at the same rate of flow as stack gases. Such sampling is described as isokinetic.

Stack sampling is conducted in airflows and temperatures that differ considerably from ambient conditions, particularly if the waste stream is produced as a result of combustion or other high-temperature processes. Stack gas temperature and pressure must be measured to standardize the sample volume to reference conditions (i.e., 20°C, 760 mmHg). Gas velocities determined from pressure measurements using a pitot tube are used to calculate stack gas flow rates. A dry test meter is used to determine flow rates (uncorrected to reference conditions) through the sampling train.

Figure 7.9 USEPA reference method for PM stack sampling. (From CFR 40, Part 60, Appendix 5, Method 5, p. 625, July 1, 1989.)

A number of utilities and industrial sources are required to measure stack emissions continuously. The CEM systems must meet USEPA performance specifications for measurement of SO_2, opacity, CO_2, total reduced sulfur (S), and hydrogen sulfide (H_2S). These systems may be *in situ* or extractive. *In situ* systems are, for the most part, no longer used since they cannot be calibrated using standard calibration gases. Extractive systems are configured for wet- or dry-based measurements. Dry extractive systems deliver samples of stack gases to an analyzer cabinet through heated lines. Particles are collected at the sampling probe on a filter. Moisture is removed by a combination of refrigeration and condensation. This keeps the analyzer dry and eliminates interference that may be caused by water (H_2O) vapor. Moisture content is determined so as to convert the volumetric concentration into a mass emission rate.

Most CEM systems utilize dilution extraction. By diluting stack gas samples with clean, dry, instrument air, these systems eliminate the need for heating and conditioning extracted samples. A dilution extraction system allows measurement of the entire sample, including H_2O vapor.

7.2.2 Opacity

New source performance standards (NSPSs) (see Chapter 8) for a number of source categories regulate plume opacity. Plume opacity regulations for PM emission sources have been in place at state and local levels for decades.

Opacity standards for particles require a different form of emission assessment than gas-phase pollutants or conventional emission standards for PM. Opacity describes the relative darkness of a plume and its ability to transmit light. It can be determined by *in situ* CEMs that use open-path optical sensing devices. Opacity can also be determined by trained "smoke readers" who assess plume optical density by comparing the color of a plume to Ringlemann opacity charts (Figure 7.10). Ringlemann measurements have the advantage of simplicity, low cost, and a long history of legal acceptance.

7.2.3 Emission Factors

Emission factors are tools used to estimate source pollutant emissions. They are expressed as pollutant mass divided by weight, volume, heat input, or duration of the activity and are expressed in such units as pounds per ton (lb/ton, kg/metric ton (mton)) of fuel or refuse burned or pounds per million British thermal unit (MBTU) heat input.

Emission factors can be used to estimate source emissions using the following equation:

$$E = [A(EF)(1 - CE/100)] \qquad (7.12)$$

where E is emissions, A is activity rate, EF is uncontrolled emission factor, and CE is control efficiency.

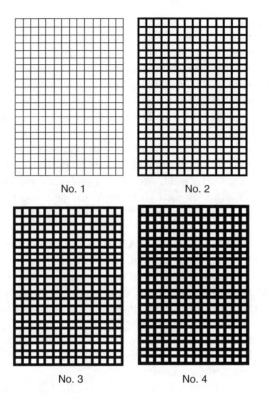

No. 1   No. 2

No. 3   No. 4

Figure 7.10 Ringlemann opacity charts.

Published emission factors represent averages of data of acceptable quality that are assumed to be representative of long-term averages for facilities in a source category. They have been compiled by the USEPA and published in a continuously updated document called AP-42, which is available in hard copy, CD-ROM, and on-line. It contains Volume I, which addresses hundreds of stationary and area sources, and Volume II, which addresses mobile sources. Emission factors in AP-42 are obtained from source tests, material and mass balance evaluations, and engineering estimates. They are rated from A to E to provide an indication of their appropriateness for estimating average emissions for a source activity. An emission rating of A indicates high reliability with comparability equal to a CEM. On the other end, an E rating indicates a highly qualitative "engineering judgment."

Emission factors are particularly appropriate for developing emission inventories. Such inventories may be used in dispersion modeling, control strategy development, and screening for compliance. Because they represent average emissions for a source category, use of emission factors is appropriate when source-specific information is not available. Users are advised to be aware of their limitations in representing facility emissions. The uncertainty of emission estimates based on emission factors is increased when estimates of peak or short-term (hourly, daily) emissions are needed for regulatory purposes. Short-term emissions fluctuations occur as a result of variations in process operating conditions, process materials used, control device operating conditions, ambient conditions, etc. Emission factors are generally based on normal operating conditions and represent long-term average conditions.

Many emission factors are determined from material and mass balance evaluations. For some sources, a material–mass balance approach may provide a better estimate of actual emissions than emission tests, as it would also account for fugitive emissions. Such mass balance estimates are particularly appropriate for processes in which a high percentage of a pollutant is lost (e.g., S from coal and fuel oil, solvent emissions).

Emission factors in AP-42 are given for three pollutant classifications. These are criteria pollutants and their precursors, hazardous air pollutants (HAPs), and greenhouse gases. Ammonia (NH_3) and O_3-destroying pollutants are also included. Since criteria pollutants were the original focus of AP-42 and major regulatory efforts, they are extensively treated. Emission factors for HAPs are increasingly being added to AP-42.

Most emission factors published in AP-42 have been determined from source testing using USEPA-published reference methods. Such reference methods (40 Code of Federal Regulations (CFR) Part 60, Appendix A, and Part 50, Appendix M) exist for PM and PM_{10}, SO_2, NO_2, CO, inorganic Pb, and VOCs. Methods for determining HAP emissions have also been published (40 CFR 61, Appendix, and USEPA document SW-846).

Emission factors can be accessed through the Factor Information and Retrieval of Emissions (FIRE) data system that contains the USEPA's recommended criteria and HAP emission estimation factors.

Emission data from AP-42 for controlled-air medical incinerators for pollutants of both regulatory and special concern are summarized in Table 7.3. Note the relatively high emissions of hydrogen chloride (HCl) and that medical incinerators are a source of dioxins.

For purposes of obtaining a permit under Title V of the Clean Air Act (CAA) Amendments of 1990, let us assume that we have to estimate emissions for a multichamber municipal incinerator that burns 50 tons of waste per day. The published emission factor is 17 lb PM/ton of refuse burned. Assuming an emission factor of 17 lb/ton and an average daily incineration of 50 tons of refuse, the emission rate per day, hour, and year, using Equation 7.13 below, can be calculated:

$$E = A(EF) \qquad\qquad (7.13)$$

where A is the waste-burning rate and EF is the published emission factor. Then

$$E = (50 \text{ tons/day})(17 \text{ lb/ton}) = 850 \text{ lb/day } (386 \text{ kg/day})$$

$$= (850 \text{ lb/day})(1 \text{ day/24 h}) = 35 \text{ lb/h } (15.9 \text{ kg/h})$$

$$= (850 \text{ lb/day})(365 \text{ days/year})(1 \text{ ton/2000 lb}) = 155 \text{ tons/year } (141 \text{ mtons/year})$$

Table 7.3 Emission Factors (No Control Devices) for Selected Pollutants from Controlled-Air Medical Waste Incinerators

Pollutant	Emission Factor Lb/ton	Emission Factor Kg/Mg	Emission Factor Rating
NO_x	3.56	1.78	A
CO	2.95	1.48	A
SO_2	2.17	1.09	B
PM	4.67	2.33	B
HCl	33.50	16.80	C
Mercury	0.107	0.54	C
PCBs	4.65×10^{-5}	2.33×10^{-5}	E
Total chlorinated dibenzyl-p-dioxins	2.13×10^{-5}	1.07×10^{-5}	B

Source: From USEPA, AP-42, EPA, Washington, D.C., 2002.

If the emission limit were 1.5 lb/ton, then a dust cleaning system would have to be installed with a minimum collection efficiency (CE) of

$$CE = [(\text{emission rate} - \text{emission limit})/\text{emission rate}](100) \qquad (7.14)$$

$$= [(17 \text{ lb/ton} - 1.5 \text{ lb/ton})/17 \text{ lb/ton}](100)$$

$$= 91\%$$

Using Equation 7.12, the emission rate would be

$$E = (50 \text{ tons/day})(17 \text{ lb/ton})(1 - 91/100)$$

$$= 76.5 \text{ lb/day} (34.8 \text{ kg/day})$$

7.3 AIR QUALITY MODELING

Because of resource limitations and practical considerations, it is desirable and even necessary to utilize approaches other than air quality monitoring to determine the impact of pollution sources on air quality in a region (particularly under worst-case conditions). Use of air quality models to predict the impact of new and existing sources on ground-level concentrations of emitted pollutants or pollutants produced as a result of atmospheric reactions addresses, in many cases, these limitations and constraints.

Air quality models are important tools for determining the environmental impact of pollution sources. Their use provides a relatively inexpensive and reliable means of determining compliance with NAAQSs and the extent of emissions reductions necessary to achieve standards in an area not in compliance with a NAAQS. They are widely used by regulatory authorities as surveillance tools to assess the effect of emissions on ambient air quality. Models are also used to evaluate permit applications associated with permissible increments under prevention of significant deterioration (PSD) requirements and new source review (NSR) programs.

Models consist of one or more mathematical formulae that include parameters that affect concentrations of pollutants at various distances downwind of emission sources. Models have been developed for a number of pollutants and time periods. Short-term models are used to calculate concentrations of pollutants over a few hours or days. They require good understanding of chemical reactions and atmospheric processes that occur over a period of minutes to hours. Short-term models can be employed to predict worst-case episode conditions and are used by regulatory agencies as a basis for control strategies. Long-term models are designed to predict seasonal or annual average concentrations, which may prove useful in studying atmospheric deposition as well as potential adverse health effects associated with pollutant exposures.

Models may be described according to chemical reactions. So-called nonreactive models are applied to pollutants such as CO and SO_2 because of the simple manner in which their chemistry can be represented. Reactive models address complex multispecies chemical reactions common to atmospheric photochemistry and are used for pollutants such as NO, NO_2, and O_3.

Models can be classified according to the type of coordinate system used. In grid-based models, the region of interest is subdivided into an array of cells that may range in horizontal size from 1 to 2000 km^2 (0.38 to 769 mi^2), depending on the scale desired. The average concentration of each pollutant is then calculated for each grid cell. This approach is commonly used by regulatory agencies to determine compliance with NAAQSs. Trajectory models, unlike the fixed-grid approach described above, follow parcels of air as they move downwind. These are simpler than fixed-grid models since concentrations need not be calculated for every point in the region.

Models can be described as simple or advanced based on assumptions used and the degree of sophistication with which important variables are treated. Advanced models have been developed for photochemical air pollution, dispersion in complex terrain, and long-range transport. Gaussian models are widely used to predict the impact of emissions of relatively unreactive gas-phase substances such as SO_2 and CO, as well as PM on air quality downwind of point sources.

7.3.1 Gaussian Models

Gaussian models have been used in the U.S. since the mid-1960s. They are based on the assumption that plume spread, and dispersion of pollutants within it, results from molecular diffusion, and because of diffusion, pollutant concentrations in both the horizontal and vertical plume dimensions are distributed normally (bell-shaped curve). These models use the σ values of the Gaussian distribution in horizontal and vertical transects to represent the dispersion characteristics of a plume at various distances downwind of a source. Plume shape, as well as the σ values of pollutant concentrations, varies in response to different atmospheric meteorological conditions. The coordinate system in Gaussian models defines the X axis as downwind of the source, the Y axis as horizontal to the X axis, and the Z axis as the vertical

Figure 7.11 Gaussian plume spread associated with plume movement downwind. (From USEPA, EPA/450/2-81-0113, EPA, Washington, D.C., 1981.)

dimension. Overlapping the two single distributions in each of the coordinate directions provides a double Gaussian distribution. Projecting this downwind gives the volume of space containing the plume (Figure 7.11).

As the plume moves downwind, the lowest edge will come into contact with the ground. Assuming no absorption or deposition, that portion of the plume is reflected upward. This reflection results in an increase in ground-level concentrations that must be factored into the model.

In Figure 7.12 and Figure 7.13, σ_y and σ_z are standard deviations of Gaussian concentration distributions in plumes. They are described as horizontal and vertical dispersion coefficients. Their values increase with time and distance. The rate of increase in σ_y and σ_z depends on atmospheric conditions, particularly atmospheric stability. The rate of increase is greatest for stability class A and lowest for stability class F. In general, σ_y is greater than σ_z since σ_y is not influenced by inversion conditions. As both σ_y and σ_z increase in magnitude, pollutant concentrations along the centerline of the plume decrease.

Dispersion coefficients in Figure 7.12 and Figure 7.13 are given for each of six atmospheric stability classes (Table 7.4). Stability classes are based on solar radiation, surface wind speed, and cloud cover. Stability classes A through C only occur in daylight hours; E and F, only at night. Classes E and F are characterized by stable air that minimizes thermal turbulence. Classes A, B, and C represent unstable conditions, with increasing stability from A to C. Class D is neutral, with moderate winds and uniform atmospheric mixing. Stability class A can result in pollutants being brought to the ground relatively close to the emission source, with resultant high concentrations. Stability class F allows the plume to be carried longer distances before it reaches the ground; therefore, it results in lower ground-level concentrations (except when topography is a factor). Both rough terrain and urban heat islands can change the stability classification under ambient conditions (usually upward by one class, i.e., more stable).

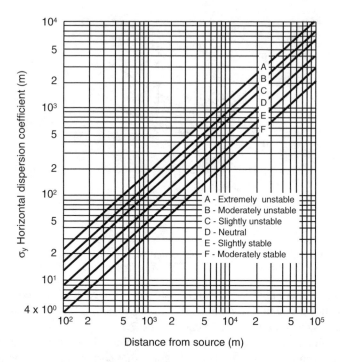

Figure 7.12 Pasquill's horizontal dispersion coefficients. (From Turner, D.B., *Workbook for Atmospheric Dispersion Estimates*, EPA Publication AP-26, EPA, Washington, D.C., 1969.)

Under stable atmospheric conditions or unlimited vertical mixing, ground-level concentrations can be calculated from Equation 7.15:

$$C_x = (Q / \pi \sigma_y \sigma_z \bar{u}) \, e^{-1/2[(H/\sigma z)(H/\sigma z)]} \, e^{-1/2[(y/\sigma y)(y/\sigma y)]} \qquad (7.15)$$

where C_x is the ground-level concentration at some distance x downwind (g/m³), Q is the average emission rate (g/sec), $\bar{u}$ is the mean wind speed (m/sec), H is the effective stack height (m), σ_y is the standard deviation of horizontal distribution of plume concentration (m), σ_z is the standard deviation of vertical distribution of plume concentration (m), y is the off-centerline distance (m), and e is the natural log equal to 2.71828.

Equation 7.15 reduces to Equation 7.16 when y = 0:

$$C_x = (Q / \pi \sigma_y \sigma_z \bar{u}) \, e^{-1/2[(H/\sigma z)(H/\sigma z)]} \qquad (7.16)$$

If the emission source is at ground level and there is no effective plume rise, Equation 7.16 reduces to

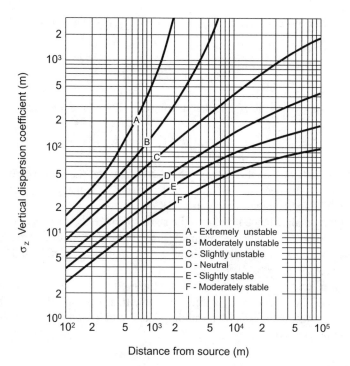

Distance from source (m)

Figure 7.13 Pasquill's vertical dispersion coefficients. (From Turner, D.B., *Workbook for Atmospheric Dispersion Estimates*, EPA Publication AP-26, EPA, Washington, D.C., 1969.)

Table 7.4 Atmospheric Stability Classes

Wind Speed, 10 m (m/sec)	Day Incoming Solar Radiation			Night Thinly Overcast	
	Strong	Moderate	Slight	>4/8 Cloud	<3/8 Cloud
<2	A	A–B	B	E	F
2–3	A–B	B	C	D	E
3–5	B	B–C	C	D	D
>6	C	D	D	D	D

$$C_x = (Q / \pi \sigma_y \sigma_z \bar{u}) \tag{7.17}$$

Plume rise is calculated from equations developed in the early 1960s. They are used to calculate H (effective stack height) in Equation 7.15 and Equation 7.16. The H is the sum of the physical stack height and plume rise. For neutral and unstable atmospheric conditions and for buoyancy fluxes less than 55 m⁴/sec³, H can be estimated from Equation 7.18:

$$H = h + 21.45 \, F^{3/4} / \bar{u} \tag{7.18}$$

where h is the physical stack height (m) and F is the buoyancy flux (m^4/sec^3). When $F \geq 55$ m^4/sec^3,

$$H = h + 38.71 \, F^{3/5} / \bar{u} \qquad (7.19)$$

The buoyancy flux, F, is calculated from the following equation:

$$F = gv_s d^2 \, (T_s - T)/4T_s \qquad (7.20)$$

where g is the gravitational constant (9.8 m/sec^2), v_s is the stack gas velocity (m/sec), d is the internal diameter of the top of the stack (m), T_s is the stack gas temperature ($^{\circ}$K), and T is the ambient temperature ($^{\circ}$K).

The horizontal distance (x_{pr}) from the stack to where the final plume rise occurs is assumed to be $3.5x^*$, where x^* is the distance in kilometers to where atmospheric turbulence begins to dominate entrainment. In Equation 7.18, $x_{pr} = 0.049 \, F^{5/8}$, and in Equation 7.19, $x_{pr} = 0.119 \, F^{2/5}$.

Taking factors into account such as multiple eddy reflections from the ground and bottom of the inversion layer and a σ_z value that is greater than 1.6 the mixing height (m), an equation for determining maximum ground-level concentrations (C_{max}) can be derived:

$$C_{max} = 2 Q\sigma_z / \pi \bar{u} e H^2 \sigma_y \qquad (7.21)$$

This equation is only applicable when the ratio σ_z/σ_y is constant with distance. The distance to the maximum concentration is the distance at which $\sigma_z = (H/2)^{1/2}$.

Ground-level concentration (C_x) at some distance downwind (x) of an industrial boiler burning 12 tons (10.9 mtons) of 2.5% sulfur coal per hour with an emission rate of 151 g/sec can be determined from Equation 7.16. Assume the following: H = 120 m, $\bar{u} = 2$ m/sec, y = 0, it is 1 h before sunrise, and the sky is clear.

From Table 7.4, the stability class for this atmospheric condition is F. From Figure 7.12, the horizontal dispersion coefficient σ_y for a downwind distance of 10 km for F atmospheric stability class is ~270 m; the vertical dispersion coefficient σ_z is 45 m (Figure 7.13).

Therefore, using Equation 7.16,

$$C_x = [151/(3.14)(270)(45)(2)] \, e^{-1/2 \, [(120/45)(120/45)]}$$

$$= 1.98 \times 10^{-3} \, e^{-1/2 \, (7.1)}$$

$$= [1.98 \times 10^{-3}]/e^{1/2 \, (7.1)}$$

$$= [1.98 \times 10^{-3}]/34.8$$

$$= 5.7 \times 10^{-5} \text{ g/m}^3$$

$$= 57 \text{ μg/m}^3$$

If emissions are from a ground-level source with H = 0, $\bar{u}$ = 4 m/sec, Q = 100 g/sec, and a stability class of B, we can use Equation 7.17 to calculate downwind concentrations at various receptor distances (e.g., 200 and 1000 m). For a receptor distance of 200 m,

$$C_x = 100/[3.14 \ (36)(20.5)(4)]$$

$$= 10.8 \times 10^{-3} \text{ g/m}^3$$

$$= 10.8 \text{ mg/m}^3$$

For a receptor distance of 1000 m,

$$C_x = 100/[3.14 \ (155)(110)(4)]$$

$$= 4.67 \times 10^{-4} \text{ g/m}^3$$

$$= 467 \text{ μg/m}^3$$

In the calculations above, H was given. Plume rise must be calculated using the model Equation 7.18 and Equation 7.19. Using these equations, we can calculate H for an 80-m-high source (h) with a stack diameter of 4 m, a stack velocity of 14 m/sec, a stack gas temperature of 90°C (363°K), an ambient temperature of 25°C (298°K), a $\bar{u}$ value at 10 m of 4 m/sec, and a stability class of B.

The buoyancy flux (due to thermal characteristics of the plume) must first be calculated (Equation 7.20) to determine effective stack height:

$$F = 9.8 \ v_s d^2 (T_s - T)(4 \ T_s)$$

$$= 9.8 \ (14)(16)(65)/[4(363)]$$

$$= 98.27$$

Since F ≥ 55 m⁴/sec³, we use Equation 7.19:

$$H = h + (38.71)(98.27)^{3/5}/4$$

$$= 80 + 152$$

$$= 232$$

Using Equation 7.22, the maximum ground-level concentration can be calculated if the ratio σ_z/σ_y is constant with distance. The distance to the maximum concentration is the distance at which

$$\sigma_z = H/(2)^{1/2} \tag{7.22}$$

which, in the case above:

$$\sigma_z = (232/2)^{1/2}$$

$$= 164 \text{ m}$$

From Figure 7.13, the downwind distance for this dispersion coefficient is ~1.4 km (0.87 mi) for stability class B; the corresponding σ (from Figure 7.12) is ~210 m (687 ft). Therefore,

$$C_{max} = [2(100)(164)]/[3.14(4)(2.718)(232)^2(210)]$$

$$= 85 \times 10^{-6} \text{ g/m}^3$$

$$= 85 \text{ μg/m}^3$$

Gaussian dispersion modeling is based on a number of assumptions. These include (1) steady-state conditions (constant source emission strength); (2) wind speed, direction and diffusion characteristics of the plume are constant; (3) no chemical transformations take place; and (4) wind speeds are greater than or equal to 1 m/sec. In addition, Gaussian models are limited to predicting concentrations no greater than 50 km downwind.

7.3.2 Model Options

Dispersion models may be Gaussian, numerical, statistical, or physical; Gaussian-based models are the most widely applied. Their use is based on ease of application and the conservative estimates they provide.

In its efforts to promote consistency in applying models, the USEPA has developed a document entitled "Guidelines on Air Quality Models" in which it lists models based on several comparative analyses and suggests best applications. Those that perform well for a general set of conditions are classified as Appendix A models (Table 7.5). They are classified by whether they are screening or refined models and their application to (1) simple or complex terrains; (2) rural or urban areas, or both; and (3) specific source concerns, pollutant production, and regions.

Table 7.5 Preferred USEPA Air Quality Dispersion Models

Type/Terrain	Mode (Urban/Rural)	Model
Screening		
Simple	Both	SCREEN 3
Simple	Both	ISC 3
Simple	Both	TSCREEN
Simple	Urban	RAM
Complex	Rural	COMPLEX I
Complex	Urban	SHORTZ
Complex	Rural	RTDM 3.2
Complex	Rural	Valley
Complex	Both	CTScreen
Line	Both	BLP
Refined		
Simple	Urban	RAM
Simple	Both	ISC 3
Simple	Urban	CDM 2.0
Complex	Both	CTDMPLUS
Line	Both	BLP
Line	Both	CALINE-3
Ozone	Urban	UAM-V
Coastal	Both	OCD

READINGS

American Conference of Governmental Industrial Hygienists, *Air Sampling for Evaluation of Atmospheric Contaminants*, 8th ed., ACGIH, Cincinnati, OH, 1995.

Axelrod, H.D. and Lodge, J.P., Jr., Sampling and calibration of gaseous pollutants, in *Air Pollution*, Vol. III, *Measuring, Monitoring and Surveillance of Air Pollution*, 3rd ed., Stern, A.C., Ed., Academic Press, New York, 1977, p. 145.

Boubel, R.A., Fox, D.L., Turner, D.B., and Stern, A.C., *Fundamentals of Air Pollution*, 3rd ed., Academic Press, San Diego, CA, 1994.

Briggs, G.A., Plume rise predictions, in *Lectures on Air Pollution and Environmental Impact Analysis*, Haugen, D.A., Ed., American Meteorological Society, Boston, MA, 1975.

Bucholtz, F., Ed., *Environmental Monitoring and Instrumentation*, Optical Society of America, Washington, D.C., 1997.

Couling, S., Ed., *Measurement of Airborne Pollutants*, Butterworth Heineman, Oxford, 1993.

Harrison, R.M., Measurements of concentrations of air pollutants, in *Air Pollution and Health*, Holgate, S.T., Samet, J.M., Koren, H.S., and Maynard, R.L., Eds., Academic Press, San Diego, CA, 1999, p. 63.

Harrison, R.M. and Young, R.J., Eds., *Handbook of Air Pollution Analysis*, 2nd ed., Chapman & Hall, London, 1986.

Jacobson, M.Z., *Fundamentals of Atmospheric Modeling*, Cambridge University Press, Cambridge, U.K., 1998.

Krzyzanowski, M. and Schwela, D., Patterns of air pollution in developing countries, in *Air Pollution and Health*, Holgate, S.T., Samet, J.M., Koren, H.S., and Maynard, R.L., Eds., Academic Press, San Diego, CA, 1999, p. 105.

Lodge, J.P., Jr., Ed., *Methods of Air Sampling and Analysis*, 3rd ed., Intersociety Committee, Lewis Publishers, Chelsea, MI, 1988.

McGregor, G.R., Basic meteorology, in *Air Pollution and Health*, Holgate, S.T., Samet, J.M., Koren, H.S., and Maynard, R.L., Eds., Academic Press, San Diego, CA, 1999, p. 51.

Noll, K.E., *Air Monitoring Survey Design*, Ann Arbor Science Publishers, Ann Arbor, MI, 1976.

Ozkaynak, H., Exposure assessment, in *Air Pollution and Health*, Holgate, S.T., Samet, J.M., Koren, H.S., and Maynard, R.L., Eds., Academic Press, San Diego, CA, 1999, p. 115.

Powals, R.J., Zaner, L.J., and Sporek, K.F., *Handbook of Stack Sampling and Analysis*, Technomic Publishing, Westport, CT, 1978.

Turner, D.B., *Atmospheric Dispersion Estimates*, 2nd ed., Lewis Publishers/CRC Press, Boca Raton, FL, 1990.

USEPA, *Air Quality Index: A Guide to Air Quality and Health*, EPA-454/R-00-005, EPA, Washington, D.C., 2000.

Wight, G.D., *Fundamentals of Air Sampling*, Lewis Publishers/CRC Press, Boca Raton, FL, 1994.

Willeke, K. and Baron, P.A., Eds., *Aerosol Measurement*, Van Nostrand Reinhold, New York, 1993.

Winegar, E.D. and Keith, L.H., Eds., *Sampling and Analysis of Airborne Pollutants*, Lewis Publishers/CRC Press, Boca Raton, FL, 1993.

Zannetti, P., *Air Pollution Modeling: Theories, Computational Methods and Available Software*, Van Nostrand Reinhold, New York, 1990.

ONLINE SOURCES

1. http://www.epa.gov/air/data/index.html — Access to air pollution data
2. http://www.epa.gov/airnow/aqibroch/aqi.html — Air Quality Index
3. www.epa.gov/ttn/chief — Access to AP-42 with Emission Factors and Inventory groups
4. www.epa.gov/ttn/chief/eiip — Emission Inventory Improvement Program
5. www.epa.gov/ttn/chief/nti — Air Toxics Emission Factors and Inventory
6. www.epa.gov/ttn/chief/emch — Emissions Modeling Clearinghouse
7. www.epa.gov/ttn/chief/publications.html — Emissions Factor and Inventory resource material

QUESTIONS

1. Describe differences between qualitative and quantitative assessments of air quality.
2. What is air quality monitoring?
3. Why is air quality monitoring conducted in the U.S.?
4. What is the relationship between sampling and air quality monitoring?
5. Describe passive, grab, intermittent, and continuous sampling and monitoring methods.
6. What is the significance of the following in conducting air sampling?

      a. Sample size.

      b. Sampling rate

      c. Sampling duration

7. What averaging times are used for the following pollutants: O_3, PM_{10}, CO, and SO_2? Why is each averaging time used?

8. Describe the sample collection principles of:

      a. Absorption.

      b. Adsorption

      c. Impaction

      d. Filtration

9. What are grab samples? Why and when are they used?

10. Describe the use of mixing ratios for expressing air pollutant concentrations.

11. The primary 1-h NAAQS for CO is 35 ppmv. What is its SI equivalent concentration in milligrams per cubic meter at reference conditions of 25°C and 760mm Hg?

12. The primary annual average NAAQS for NO_2 in SI units is 100 µg/m³. What is the equivalent concentration in parts per million by volume at reference conditions of 25°C and 760 mmHg? In parts per billion by volume?

13. A PM_{10} sample was collected under ambient average temperature and pressure conditions of 5°C and 740 mmHg at a measured flow rate of 17 ft³/min (0.48 m³/min) over a period of 24 h. What was the volume standardized to reference conditions of 25°C and 760 mmHg?

14. Describe differences between accuracy and precision. How is each determined?

15. What is bias? What is its primary cause?

16. What is calibration? Why must one periodically calibrate sampling and monitoring instruments?

17. Describe differences between primary and secondary standards.

18. What criteria are used to locate individual monitoring sites in SLAM network designs?

19. Describe spatial scales used in identifying monitoring locations.

20. Describe pollutants monitored and the purpose of the following air quality monitoring networks:

      a. NAMS

      b. PAMS

      c. NADP

      d. IMPROVE

21. Describe the concept of federal reference and equivalent methods as it applies to air quality monitoring in the U.S.

22. What are the federal reference methods for SO_2, NO_x, and PM_{10}?

23. Distinguish between quality assurance and quality control as they are applied to air quality monitoring activities.

24. What is the purpose of data summarization procedures and efforts?

25. What information can one obtain from evaluating histogram and stem and leaf plots?

26. What is the usefulness of box plots in summarizing pollutant data?

27. Describe when each of the following measures of central tendency should be used in data analysis:

      a. Mean

      b. Median

      c. Geometric mean

28. Determine the mean, median, and geometric means of the following PM_{10} values: 55.2, 74.5, 102.2, 40.3, 45.6, 20.5, 62.4, 80.7, 35.8, 65.8, and 48.9 $\mu g/m^3$.
29. Weekly samples of wet acid deposition were collected over a 2-month period. Calculate the average pH and determine how many times more acidic it is relative to the theoretical clean air value of 5.65. Measured values were 3.4, 4.2, 5.0, 3.8, 4.0, 4.4, 4.8, and 3.9.
30. Describe how the Air Quality Index is derived. Describe its usefulness.
31. Describe the nature of, and need for, stack sampling.
32. Describe what emission factors are, how they are derived, and their use and limitations.
33. A municipal solid waste incinerator has an emission factor of 20 lbs PM/ton (10 kg/mton) of waste. Assume that the maximum 2-h permissible emission level is 0.18 g/m^3 (corrected to 12% CO_2). If 80 dry standard cubic feet (2.26 m^3) of flue gas are produced per pound (0.45 kg) of waste combusted, determine the control equipment collection efficiency needed to meet the emission limit.
34. Use emission factors from Table 7.3 and assume that a medical waste incinerator burns 10 tons (9 mtons) of mixed hospital wastes per month. What would the daily emission rate (lbs/day, kg/day) of hydrogen chloride be? The annual emission rate of dioxins (lbs/yr, kg/yr)?
35. Describe two procedures that can be used to monitor plume opacity.
36. Why are continuous emission monitors used?
37. Describe how you would determine the emission factor for PM from sewage sludge incinerators.
38. By analyzing/evaluating Equation 7.15, what would be the expected effect of increasing wind speed on downwind pollutant concentrations?
39. What factors affect plume rise and therefore effective stack height?
40. From Figure 7.12 and Figure 7.13, what is the effect of the following on dispersion coefficients?
    a. Stability class
    b. Increasing distance downwind
41. Using Gaussian dispersion equations, the stability class table, and dispersion coefficient data (Figure 7.12 and Figure 7.13) for a source emitting SO_2 at a rate of 100 g/sec on a sunny summer day with a wind speed at anemometer height of 4 m/sec and effective stack height of 36.4 m, determine the stability class and dispersion coefficients for receptor distances of 300 and 1000 m and ground-level concentrations at each distance in parts per billion by volume.
42. Describe model types used in regulatory programs.

CHAPTER 8

Regulation and Public Policy

Nominal regulatory efforts to control air pollution in the U.S. began in the late 19th and early 20th centuries when several large cities approved ordinances to control "smoke" produced by a variety of industrial activities. Though largely ineffectual because they lacked enforcement authority, these smoke ordinances nevertheless served as forerunners of present-day regulatory programs.

In response to the increasingly worsening smog problem in its southern coastal area, California began serious efforts to control air pollution in the 1950s. As scientists unraveled the chemistry of smog, regulatory efforts changed from controlling emissions from stationary sources to those from automobiles. As a consequence, by the late 1960s both California and the federal government began in earnest their significant and continuing regulatory efforts to reduce emissions from motor vehicles.

By the 1960s, there was increasing recognition at all governmental levels that air pollution was a serious environmental problem. Many state and local governments began to form agencies and develop regulatory frameworks to control air pollution. These efforts (in retrospect) were relatively timid and, for the most part, not very effective.

As the modern environmental movement evolved, the premise that polluted cities and countrysides were a price of progress was no longer accepted. A major change occurred in American attitudes. The economic well-being that provided Americans the so-called good life was seen to diminish the quality of the environment in which fruits of economic prosperity were to be enjoyed. The desire of Americans to prevent further deterioration and improve environmental quality gave rise to the significant regulatory efforts to control air pollution and other environmental problems that have taken place since 1970.

8.1 NONREGULATORY LEGAL REMEDIES

Prior to modern regulatory air pollution control programs, individuals or groups of citizens were limited in what they could do to abate pollution adversely affecting them. They could ask corporate officers and others responsible for managing sources to be better neighbors and appeal to their sense of civic responsibility. Such appeals (if many of them were indeed made) would, for the most part, have gone unheeded. Pollution abatement requires allocation of what many economists and corporate financial officers describe as unproductive resources, i.e., money that does not add to product value.

In the absence of strong governmental regulatory programs, a citizen or group of citizens has a right to file suit against a source employing common-law legal principles. Though significant air pollution control regulations have been in place for more than three decades, we continue to have this right.

Individuals claiming injury from emissions produced by a source can file a civil suit against the source in an effort to recover damages and abate the problem. Such suits are called torts. Tort law involves a system of legal principles and remedies that attempts to address harm to individuals or their property. Historically, both nuisance and trespass principles of tort law have been used to address claims of harm associated with air pollution.

8.1.1 Nuisance

A nuisance is an intentional or negligent act that results in an unreasonable interference with the use and enjoyment of one's property. It may be private (affecting some individuals and not others) or public (affecting everyone equally). Under the private nuisance concept, the plaintiff, or injured party, must convince a judge or jury that the action is unreasonable. If the suit is successful, he or she is entitled to recover damages. To abate the problem, however, the plaintiff must seek an injunction against the source.

In deciding whether to grant such an injunction, the court must balance the equities; i.e., it must take into account the following: (1) extent of damage, (2) cost of pollution abatement, (3) whether the pollution source was on the scene first, and (4) the economic position of the polluting source in the community. If the dispute is between a corporation with a large local work force and a single party, or even a group of landowners, courts invariably balance the equity in favor of the polluting source and refuse to grant an injunction. In such cases, courts reason that injunctive relief, if granted, will result in great economic harm to the local community. Although damages are awarded to the affected party or parties, injunctive relief is only rarely provided.

If all members of a community are equally affected, then no individual citizen can sue. In such cases, a public authority, such as a local prosecutor, would have to file a public nuisance suit against the pollution source that is causing "an unreasonable interference with a right common to the general public." As in private suits, courts must assess the seriousness of alleged harm relative to the social utility of the source's conduct. Public authorities have historically been reluctant to file such actions.

8.1.2 Trespass

The trespass concept applies to cases in which there is a physical violation of a landowner's right to exclusive possession of his or her property. Particulate-laden smoke could be considered to violate trespass provisions of common law. Historically, in determining liability, courts have not required that the invasion be proven unreasonable or that damages be proven.

8.2 REGULATORY STRATEGIES AND TACTICS

Successful abatement of air pollution and protection of public health and welfare requires selection of appropriate regulatory strategies and tactics to implement them. A variety of air quality regulatory strategies may be employed. These include air quality management (air quality standards), emission standards, and economics-based approaches. Air quality management and emission standard strategies are the mainstay of pollution control efforts in the U.S. and other developed countries. Several economics-based pollution control approaches have been proposed and adopted. Some have been incorporated into air pollution control laws and regulatory practice.

8.2.1 Strategies

8.2.1.1 Air Quality Management

Air quality management strategies are based on the widely accepted toxicological principle that pollutant exposures below threshold (lowest dose at which toxic effects are observed) values are relatively safe, and therefore some level of atmospheric pollution is acceptable and thus legally permissible. Pollutants identified as posing potentially significant public health risks and welfare concerns may be regulated by the promulgation of ambient air quality standards.

Air quality standards are legal limits on atmospheric concentrations of regulated pollutants. Promulgation of an air quality standard for an ambient air pollutant is typically characterized by intensive review of scientific studies, recommendations on acceptable levels, and ultimate decision making by a regulatory authority.

Setting ambient air quality standards is a difficult and arduous process. In many cases, the relationship between pollutant levels, adverse health effects, and the level of protection necessary is subject to some degree of uncertainty. In the U.S., the Environmental Protection Agency (USEPA) is required to promulgate air quality standards that provide an "adequate margin of safety," with special consideration for those individuals who may be the most sensitive, i.e., asthmatics, children, the aged or infirm, those with existing cardiovascular disease, etc. As a consequence, ambient air quality standards are much more stringent than occupational standards designed to protect nominally healthy working adults.

Since air quality management strategies are based on the premise that a safe level of exposure (a threshold) exists, they are generally not applied to contaminants not known to have threshold values. It has been widely accepted in both the scientific and regulatory communities that there is no generally safe level of exposure to carcinogens.

Once an air quality standard for a pollutant or pollutant category has been promulgated, regulatory authorities at federal, state or provincial, and local levels are responsible for developing and implementing control and management practices (tactics) that will ensure that air quality standards are not exceeded.

Air quality management, though generally assumed to be a cost-effective approach to achieving air quality goals, is a relatively complicated undertaking. Its complexity involves both technical and political dimensions. Regulatory authorities must select and implement tactics that, in their best judgment, will achieve each air quality standard. In the case of ozone (O_3), control efforts have been confounded because of uncertainties associated with the contributions of anthropogenic and biogenic sources of hydrocarbons (HCs) to tropospheric O_3 chemistry, as well as other factors. In efforts to achieve particulate matter (PM) standards, regulatory authorities attempted to employ a control tactic that made both technical and economic sense but was not politically popular. Such was the case with prohibitive bans on leaf burning in autumn. Though relatively cost-effective, it has been politically unacceptable to some communities. On paper, air quality management is a highly practical and cost-effective control strategy. However, as is the case with many things in life, "the devil is in the details."

8.2.1.2 Emission Standards

In an emissions standards strategy, limits are placed on the maximum quantity (typically mass per unit time or thermal input) of a pollutant or pollutants that can be emitted from specific sources. The same emission standard is applied equally to all sources in a source category; i.e., all members of a category must comply with the emission limit without regard to existing air quality. The same emission limits apply to sources in "clean" or "dirty" air regions.

With the exception of hazardous or toxic pollutants in the U.S., emission standards, when used as a strategy, are governed by practicability, i.e., emission standards reflect the highest degree of pollution reduction achievable taking cost of control into consideration.

The "best practicable means" or "good practice" approach is widely used for development of emissions standards in the U.S. and other developed countries. The term *practicable* indicates economic, technical, and political practicality. The "best practicable means" in practice may be the degree of emission reduction achieved by the best industrial plants in a source category, or technology that can be reasonably applied by borrowing from other industries. Implicit in this approach is that emissions standards may become more stringent (at least for new sources) as control technology improves and becomes more affordable. "Best practicable means," the traditional approach to air pollution control in Europe, would be equivalent to what is described in the U.S. as "reasonably available control technology" (RACT). Such requirements are generally applied to new or significantly modified existing sources under new source performance standards (NSPSs), to be discussed later in this chapter, and to achieve air quality standards in some air quality control regions (AQCRs).

An emission standards strategy may also require very stringent emission limits. For pollutants that pose unique health hazards, e.g., mercury (Hg), beryllium (Be), benzene, radioactive isotopes, asbestos, etc., emission reduction requirements reflect "best-available control technology" (BACT). The objective of BACT standards is to achieve the highest degree of emission reduction that technology is capable of, with a more limited consideration of capital and operating costs. New coal-fired utilities in Class I prevention of significant deterioration (PSD) regions (see Section 8.4.6) are required to use BACT.

The 1990 Clean Air Act (CAA) Amendments require setting maximum achievable control technology (MACT) standards for toxic or hazardous air pollutants (HAPs). This concept, in theory, suggests a higher degree of emission control than BACT. This is due to the fact that emissions reductions can be achieved by a variety of control practices other than treating just waste gas streams.

Emissions standards are usually specified in some numerical form. In the case of coal-fired power plants licensed under the 1972 NSPS for sulfur dioxide (SO_2), the emission limit is 1.2 lb of SO_2/MBTU (million British thermal unit). Numerical standards can be based on heat input (lb/MBTU), unit of time (lb/h), weight per unit air volume (grains/standard cubic foot (g/SCF)), or weight of process material (lb/ton). They may also be prescribed in metric units; in some cases, mixed units (U.S. and metric) are used (e.g., g/mi).

Emission standards have the advantage of simplicity. There is no need for extensive and expensive air quality monitoring networks, modeling to determine compliance, and complex technical and public policy decisions. All sources in the source category must meet the same requirements regardless of existing air quality. It is a highly attractive control strategy from an administrative standpoint. As used in the U.S., it has the advantage of being applied to special pollution concerns such as new sources and pollutants believed to pose uniquely hazardous or toxic exposure problems.

Use of emission standards as a control strategy also has limitations or problems. Though costs are taken into consideration in setting emission limits, sources must comply regardless of existing air quality. In areas with relatively good air quality, sources may be required to control emissions beyond that needed to protect public health and possibly welfare as well. In "dirty" air regions, emission limitations influenced by cost considerations may not be sufficiently stringent to protect public health and the environment.

8.2.1.3 Economics-Based Approaches

Several economics-based approaches have been proposed and adopted to achieve air quality objectives. They assume that command-and-control regulations tend to force industry and other sources to implement control measures and practices that do not take cost and economic efficiencies into consideration. They also presume that command-and-control regulation freezes development of new technologies and practices that would result in higher levels of control and provides few or no financial incentives for sources to exceed control targets.

Economics-based approaches proposed or actually adopted include cost–benefit analysis and market-based policies.

8.2.1.3.1 Cost–Benefit Analysis

A control strategy based on cost–benefit analysis has been advocated by free-market economists. It would require quantification of all damage costs associated with atmospheric pollutants and costs of control. Damage costs would be limited to the tangible, i.e., the dollar value of health care and mortality, crop losses, reductions in property values, and effects on materials. Intangible values such as loss of friends and loved ones, changes in ecosystems, and diminished aesthetic quality would not be included. The primary focus of a cost–benefit strategy would be to select one or more pollution control options that minimize damage and control costs.

Theoretical costs of pollution effects on public health and welfare are compared to control costs in Figure 8.1. As damage costs decrease, control costs increase. Note the steep rise in control costs as damage costs decrease toward zero. This illustrates a major problem in pollution control; i.e., there is a disproportionate increase in control costs relative to environmental benefits (decreased damage costs) achieved. In economic theory, it would be desirable to implement the degree of control that provides the greatest reduction of damages (or provides the greatest benefit) per unit cost. Damage and control costs would be "optimized" where the two lines intersect.

Use of cost–benefit analysis is limited (in fact, rendered unusable) by a variety of practical and theoretical problems. First, it assumes that both damage and control costs are easily quantified. Though true for control costs, the uncertainty of the nature and extent of costs and damages associated with the effects of pollutants on humans and the environment makes their quantification very difficult. Second, it assumes that damage costs to public health and welfare above the point of optimization are socially and politically acceptable. A cost–benefit-based control strategy would pose a public policy nightmare (more so than is already the case) for regulatory officials who would have to make decisions on who or what is to be protected.

Figure 8.1 Relationship between control costs and pollutant environmental damage costs.

The cost of control vs. risks to humans and the environment has been a continuous source of contention between the regulated community and regulatory authorities. Under clean air legislation in the U.S., control costs must be taken into consideration when promulgating NSPSs but not air quality standards and standards for HAPs. In 1995, pro-industry legislation that would have required cost–benefit analyses for most new environmental regulations was introduced into the U.S. Congress. Despite considerable congressional support, the measure failed to be enacted. In the late 1990s, the USEPA's revised PM and O_3 standards were challenged (but upheld) in federal courts, in part because the USEPA did not conduct a cost–benefit analysis prior to promulgation of these standards. Control cost considerations are a continual demand of industrial stakeholders and their supporters.

8.2.1.3.2 Economics- and Market-Based Policies

Several economics- and market-based policies have been implemented because of their acceptance by regulators, the regulated community, and environmental groups. These include use of the bubble concept, emission trading, and allowances. Use of allowances and free-market trading of allowances were mandated in the 1990 CAA Amendments.

The bubble concept was the first economics-based regulatory innovation adopted by the USEPA. Under this policy, existing plants and groups of plants may increase (or not decrease) their emissions at one or more sources by decreasing emissions to a more significant degree at other sources within a facility. As a consequence, plant environmental managers have the flexibility to implement more cost-effective means of complying with air quality requirements. To receive regulatory approval, a bubble must result in emissions reductions equivalent to or better than baseline emission limits relative to their environmental impact and enforceability and comply with all relevant provisions of clean air legislation. Excess emission reductions may be banked or traded.

Emission reductions beyond those legally required under the bubble concept, other regulatory requirements, and voluntary emission reduction efforts may be stored in USEPA-approved emission credit banks for later use in emission transactions. These emission reduction credits (ERCs) may be sold or traded to other corporations to meet regulatory requirements. ERCs only apply to the same criteria pollutants (i.e., pollutants for which an air quality standard has been promulgated) and may not be used to meet technology-based emission reduction requirements.

The use, trading, and selling of allowances to achieve objectives of Title IV: Acidic Deposition Control of the 1990 CAA Amendments are excellent examples of a market-based approach to achieve major air quality objectives. Under an allowance system, a source is allowed to emit so many units of a regulated pollutant. This allowance is some fraction of baseline emissions. Sources can "live within their allowance" by employing those emission reduction approaches that they deem to be cost-effective. Emission reductions that exceed allowances may be banked or traded. The allowance system used in the acidic deposition control program facilitated a 9 million metric tons (mtons)/year (10 million tons/year) reduction of SO_2 emissions from utilities without significant increases in the cost of electricity, a remarkable accomplishment.

Pollution charges as a pollution control strategy have received limited attention. Pollution charge systems attempt to reduce emissions by imposing fees or taxes on emissions of regulated pollutants.

Pollution charges are seen as a mechanism for providing incentives to corporations to adopt improved control technologies and operating practices. The imposition of pollution charges based on emissions rather than pollution-generating activities allows regulators the flexibility to encourage sources to reduce emissions when marginal control costs are equal to pollution charge rates. The challenge for regulators is to identify the desirable charge level. If it is too low, insufficient pollution control will result; if it is too high, costs may be unreasonably burdensome.

Pollution charges have had limited use as a means of achieving emission reductions. Examples include charges on (1) SO_2 emissions from large sources in high-pollution areas in Japan, (2) combustion sources with 750 MW or greater electrical generating capacity in France, (3) oil sulfur (S) content in Norway, and (4) chlorofluorocarbons (CFCs) during the CFC phaseout in the U.S.

The pollution charge concept has had a perception problem. Environmental groups have perceived pollution charges as providing industry with the "right to pollute," while conservative groups view them as another tax to oppose.

Despite their limited past use, pollution charges may play a significant role in future international efforts to reduce emissions of carbon dioxide (CO_2) and curb predicted greenhouse gas effects on global climate. The use of broad-based carbon (C) fees to limit emissions of greenhouse gases such as CO_2 has been proposed and is developing increasing support. Their primary effect would be to promote energy conservation and energy-conserving technologies.

8.2.2 Tactics

Once control strategies have been selected, regulatory agencies must develop a plan or plans to achieve control objectives. Specific actions taken to implement a control strategy can be described as tactics.

Tactics employed to reduce emissions from stationary sources have included setting emission standards for specific sources in a control region, restrictions on fuel use (e.g., prohibiting use of coal or oil with a S content of >1%), use of tall stacks, and prohibitive bans on open burning or use of certain types of combustion equipment (e.g., apartment house incinerators).

Tactics used to control emissions from, or reduce air quality problems associated with, motor vehicles include (1) control requirements for blow-by gases, evaporative losses, refueling, and cold starts; (2) exhaust emission standards; (3) inspection and maintenance; (4) fuel-additive standards; (5) special fuel use requirements; (6) use of alternative power systems; and (7) transportation plans.

As can be seen, emission standards can be used as both an overall control strategy and a tactic to achieve air quality standards. Various economics-based approaches may also be used as tactics.

Permit systems are a major component of all air pollution control programs. A permit extends a legal right to a source to emit pollutants into the atmosphere. By requiring emission sources to obtain permits, regulatory authorities can impose emission limits on the source or require other actions that reduce emissions. Permits are important tools for obtaining information about emissions from a source and requiring compliance with emission limits. In the broad context, permitting can be viewed as a major air pollution control tactic.

8.3 FEDERAL LEGISLATIVE HISTORY

At the end of the 19th and through the first half of the 20th century, air pollution in the U.S. was seen as a local problem. This began to change in the 1950s as states, led by California, began a variety of regulatory initiatives. A federal role began to evolve at about the same time. The legislative history of federal air pollution control efforts is described below.

8.3.1   1955 Clean Air Legislation

Legislation enacted in 1955 authorized the Public Health Service, in the Department of Health, Education and Welfare (DHEW), to conduct air pollution research and training programs and provide technical assistance to state and local governments. It affirmed that state and local governments had the fundamental responsibility for air pollution control. Amendments to this law in 1960 and 1962 called for special studies relating to health effects associated with motor vehicle-related pollutants.

8.3.2   1963 Clean Air Act

In response to the deteriorating quality of air in the postwar U.S., Congress passed the CAA of 1963. The CAA broadened the federal role in abating air pollution. Specifically, it provided for (1) grants for program development and improvement of state and local air pollution control efforts; (2) accelerated research, training, and technical assistance; (3) federal enforcement authority to abate interstate air pollution problems; and (4) federal research responsibility for automobile and sulfur oxides (SO_x) pollution. Although the federal role was considerably expanded, state and local governments were primarily responsible for air pollution control.

8.3.3   Motor Vehicle Air Pollution Control Act

In 1965, Congress amended the 1963 CAA. These amendments were described as the Motor Vehicle Air Pollution Control Act. It authorized the secretary of DHEW to promulgate emission standards for motor vehicles. As a consequence, federal emission standards were established for all 1968 model light-duty motor vehicles. The 1965 amendments also established the National Air Pollution Control Administration (NAPCA) within the DHEW to provide regulatory leadership in the nation's efforts to control air pollution.

8.3.4   The Air Quality Act of 1967

Although some progress was made under the CAA of 1963, control programs at all governmental levels remained relatively weak and air pollution problems continued, for the most part, unabated. In 1967, Congress attempted to strengthen this country's air pollution control efforts by enacting the Air Quality Act. Many of the regulatory concepts developed in the 1967 legislation continue to provide the framework for current control efforts. Of special significance were the institution of a regional approach to air pollution control and the development and implementation of air quality standards.

The Air Quality Act of 1967 required NAPCA to designate AQCRs and develop and issue air quality criteria and control technique information. After the federal agency completed these mandated requirements, individual states were to develop state air quality standards and plan for their implementation on a fixed time schedule. air quality criteria documents were to serve as the scientific basis for development of air quality standards.

The Air Quality Act also provided for continuation of the federal role in air pollution research and development, training programs, and grants-in-aid to state and local air pollution control programs. Although federal authorities were given significant new powers to initiate lawsuits whenever interstate violations occurred, states maintained primary responsibility for enforcement.

Progress under the Air Quality Act was slow and limited. States could not promulgate air quality standards until the NAPCA designated AQCRs and issued air quality criteria and control technique documents for specific pollutants. These tasks took considerable time; e.g., air quality criteria documents for SO_x and PM were not issued until 1969; for HCs, carbon monoxide (CO), and oxidants, 1970; and for nitrogen oxides (NO_x), 1971. The process of designating AQCRs was also very slow. Publication of control technique documents was carried out more expeditiously.

8.3.5   1970 Clean Air Act Amendments

In 1970, public concern for environmental quality had reached its zenith. In response to limited progress under the Air Quality Act of 1967 and public support for strong legislation, Congress enacted tough new CAA Amendments. The NAPCA was eliminated and its air pollution control functions transferred to the USEPA, an independent federal agency created by executive order of the president.

The CAA Amendments of 1970 sought to remedy deficiencies of earlier legislation. Most importantly, they completed federalization of air pollution control efforts, provided significant federal enforcement authority, and placed air pollution control on a rigorous timetable.

Specific provisions of the 1970 CAA Amendments included (1) setting uniform national ambient air quality standards (NAAQSs), (2) immediately designating AQCRs, (3) requiring state implementation plans (SIPs) to achieve NAAQSs, (4) setting stringent new automobile exhaust emission standards and standards for fuel additives, (5) setting emission standards for new or significantly modified sources and HAPs, (6) allowing the right of citizen suits, and (7) providing for federal enforcement authority in air pollution emergencies and interstate and intrastate air pollution violations.

8.3.6 1977 Clean Air Act Amendments

The goal of the 1970 CAA Amendments was to achieve clean air for all regions of the country by July 1, 1975. Although significant progress for some pollutants was made, many AQCRs did not achieve one or more NAAQSs by the congressionally mandated deadline.

Recognizing that air quality goals set in 1970 could not be met without major economic disruptions, Congress amended the CAA in 1977. These amendments would best be described as a "mid-course correction." Specifically, they (1) postponed compliance deadlines for national primary (health-based) air quality standards; (2) postponed, and in some instances modified, federal automobile emission standards; (3) included a prevention of significant deterioration (PSD) provision into the language of the act; (4) provided a mechanism by which the USEPA could have flexibility in allowing some growth in "dirty" air regions; and (5) gave the USEPA authority to regulate stratospheric O_3-destroying chemicals. Regulation of O_3-destroying chemicals was the major new initiative in the 1977 Amendments.

8.3.7 1990 Clean Air Act Amendments

Though it is the intent of Congress to reauthorize and update by amendment all major federal legislation on a 5-year schedule, disagreements over policies related to control of acidic deposition contributed to an 8-year hiatus in amending the CAA. In 1990, major agreements on CAA Amendments were reached by Congress and the president. The 1990 CAA Amendments represented a significant political achievement. Major new or expanded authorities included (1) changes in timetables for achievement of air quality standards in nonattainment areas, (2) regulation of emissions from motor vehicles, (3) regulation of hazardous and toxic air pollutants, (4) acidic deposition control, (5) stratospheric O_3 protection, (6) permitting requirements, and (7) enforcement.

8.3.8 Federalization

In the evolution of federal legislation, the most significant trend was increasing involvement of the federal government in the problem, and ultimate primacy in its solution. This federalization of air pollution control efforts was not unique; it was consistent with the centralization of responsibility at the federal level for many problems that beset post-World War II American society. During this period the principle of "states' rights" slowly eroded, with recognition that problems such as air pollution could only be effectively dealt with by a coordinated approach at all levels of government.

8.4 AIR POLLUTION CONTROL UNDER THE 1970, 1977, AND 1990 CAA AMENDMENTS

Prior to enactment of the CAA Amendments in 1970, the NAPCA and Congress considered several different control strategies. The outcome of their deliberations was an ambient air pollution control strategy based on setting NAAQSs, with the proviso that states develop implementation plans to achieve them. Additionally, national emission standards would be established for major new sources and for existing and new sources of HAPs. As a consequence, the nation's ambient air pollution control program has been based primarily on strategies employing both air quality management and emission standards concepts.

8.4.1 Air Quality Standards

Under the 1967 Air Quality Act, states were required to set air quality standards for AQCRs within their jurisdiction. These standards were to be consistent with Air Quality Criteria developed and issued by the NAPCA. Development of state air quality standards under the 1967 legislation was slow, due in part to the slowness of the NAPCA to publish Air Quality Criteria and designate AQCRs. In addition, states were concerned that adoption of stringent standards would result in their industries relocating to states with weaker standards. This concern created a great deal of uncertainty, confusion, and ultimately inaction.

Under the CAA Amendments of 1970, the federal government was charged with responsibility for developing uniform NAAQSs. These were to include primary standards designed to protect health and secondary standards to protect public welfare. Primary standards were to be achieved by July 1, 1975, and secondary standards "at a reasonable time thereafter." Implicit in this dual-standard requirement was the premise that pollutant levels protective of public welfare were to be more stringent than those for public health and that achievement of primary health-based standards had immediate priority. In 1971, the USEPA promulgated NAAQSs for six air pollutants. In 1978, an air quality standard was also promulgated for lead (Pb). Also in 1978, the 1971 photochemical oxidant standard was revised to an O_3 standard, with an increase in permissible levels. In 1987, the suspended PM standard was revised and designated a PM_{10} (PM with an aerodynamic diameter of ≤ 10 μm) standard. This revision reflected the need for a particle size-based PM standard that better reflected respiratory health risks. In 1997, the USEPA promulgated a new 8-h standard for O_3 and a $PM_{2.5}$ standard (in addition to the existing PM_{10}). NAAQSs and averaging times are summarized in Table 8.1.

8.4.2 Air Quality Criteria

Promulgation of NAAQSs must be preceded by publication of Air Quality Criteria. These are issued in document form and summarize all relevant scientific information on the health and welfare effects of individual pollutants. The principal function of the Air Quality Criteria process is to determine the relationship between pollutant concentrations and health and welfare effects so that air quality standards are supported by strong scientific evidence.

Table 8.1   National Ambient Air Quality Standards[a]

Pollutant	Averaging Time	Primary Standard	Secondary Standard
Carbon monoxide	8 h	9 ppmv (10 mg/m^3)	Same
	1 h	35 ppmv (40 mg/m^3)	Same
Nitrogen dioxide	365 days	0.05 ppmv (100 μg/m^3)	Same
Sulfur dioxide	365 days	0.03 ppmv (80 μg/m^3)	Same
	24	0.14 ppmv (365 μg/m^3)	Same
	3 h		0.5 ppmv (1300 μg/m^3)
Nonmethane hydrocarbons	3 h (6–9 A.M.)	0.24 ppmv (160 μg/m^3)	Same
Ozone	8 h	0.08 ppmv (157 μg/m^3)	Same
PM$_{10}$	365 days	50 μg/m^3	Same
PM$_{2.5}$	365 days	15 μg/m^3	Same
	24 h	65 μg/m^3	Same
Lead	3 months	1.5 μg/m^3	Same

[a]Standards, other than those based on the annual average, may not be exceeded more than once per year. Special considerations are applied to some pollutants on how pollutants are to be averaged.

8.4.3   Air Quality Control Regions

The concept of a regional approach to air pollution control was first introduced in the Air Quality Act of 1967. This approach recognized that air pollution was a regional problem that did not respect political boundaries. Initially, the regional concept was only applied to interstate problems. With enactment of the 1970 CAA Amendments, Congress mandated that both interstate and intrastate AQCRs be established for the purpose of achieving NAAQSs.

8.4.4   State Implementation Plans

After promulgation of NAAQSs by the USEPA, individual states are required to develop and submit plans for their implementation, enforcement, and maintenance. These SIPs can be approved by the USEPA in whole or in part. If a portion of an SIP is disapproved, the state must revise it and seek approval for revisions. If the plan, in whole or in part, is not acceptable, the USEPA is empowered to develop an SIP for the state, which the state must then enforce.

In developing implementation plans, states formulate policies for each air pollutant for which NAAQSs have been promulgated. In this process they are required to determine and report on the nature and quantity of emissions from existing sources and require, at a minimum, RACT to limit emissions. They negotiate compliance schedules with major sources and indicate effects emission reductions will have on air quality. Implementation plans are directed to pollutants in AQCRs where levels are above the NAAQSs.

8.4.4.1 Episode Plans

State plans must include control strategies for air pollution episodes or emergencies. Episodes occur as a result of stable atmospheric conditions associated with stagnating, migratory, or semipermanent high-pressure systems. The poor dispersion associated with these atmospheric conditions may result in ambient pollutant levels that pose increasingly greater health risks. States are required to develop response criteria and plans similar to those in Table 8.2. Note that each stage reflects a higher degree of pollution severity. Control actions at each stage are to be sufficient to prevent the next stage from occurring. Episode plans must provide for a reduction of emissions on a prearranged schedule and should therefore, in theory, protect public health.

8.4.4.2 Legal Authority

To gain approval for its SIP, a state must have adequate legal authority to (1) adopt and enforce emission regulations; (2) implement episode control plans; (3) regulate new or modified sources; (4) require stationary sources to provide emission data; and (5) gather information to determine compliance with regulations, including the right to inspect and test. Although federal law is "the law of the land," it cannot be enforced by state and local officials. State and local governmental regulatory agencies must enact enabling legislation and ordinances to enforce provisions of federal law. Although these reflect local circumstances and sentiments, they must be consistent with provisions of federal clean air legislation.

8.4.4.3 State Implementation Plan Approval

In practice, the process of SIP development and approval has been a demanding experience for federal, state, and local officials. Plans had to be submitted within 9 months of primary standard promulgation. In many cases, data on emissions (emission inventory) were scant and manpower for plan development insufficient. Many of these early hastily prepared plans were subject to numerous revision requests before federal approval could be granted. Additional revision requirements were necessitated by federal court decisions resulting from citizen suits filed against the USEPA relative to requirements of SIPs. As a result of these decisions, states in the early 1970s had to revise their implementation plans to provide for maintenance of air quality once standards were achieved, and prevention of significant deterioration of air quality in regions where air quality was better than the standards. Since then, SIPs have had to be revised in response to changes in standards and regulatory policies initiated by the USEPA or mandated by the 1977 and 1990 CAA Amendments.

8.4.5 Nonattainment of Air Quality Standards

As the 1975 statutory deadline for attainment of NAAQSs was approaching, it was evident that many AQCRs would not be in compliance with one or more

Table 8.2   Suggested Criteria for Air Pollution Episodes

Alert Stage

An Air Pollution Alert goes into effect if:

SO_2 reaches 0.3 ppmv, 24-h average

or

PM reaches 3.0 COH, 24-h average

or

Combined SO_2 and PM product reaches 0.2

or

CO reaches 15 ppmv, 8-h average

or

O_3 reaches 0.1 ppmv, 1-h average

or

NO_2 reaches 0.6 ppmv, 1-h average, and 0.15 ppm, 24-h average

and

Meteorological conditions are such that pollutant concentrations can be expected to
remain at these levels for 12 h or more, or increase unless control actions are taken.

Warning Stage

An Air Pollution Warning goes into effect if:

SO_2 reaches 0.6 ppmv, 24-h average

or

PM reaches 5.0 COH, 24-h average

or

Combined SO_2 and PM product reaches 0.8

or

CO reaches 30 ppmv, 8-h average

or

O_3 reaches 0.4 ppmv, 1-h average

or

NO_2 reaches 1.2 ppmv, 1-h average, and 0.3 ppmv, 24-h average

and

Meteorological conditions are such that pollutant concentrations can be expected to
remain at these levels for 12 h or more, or increase unless control actions are taken.

Emergency Stage

An Air Pollution Emergency goes into effect if:

SO_2 reaches 0.8 ppmv, 24-h average

or

PM reaches 7.0 COH, 24-h average

or

Combined SO_2 and PM product reaches 1.2

or

CO reaches 40 ppmv, 8-h average

or

O_3 reaches 0.6 ppmv, 1-h average

or

NO_2 reaches 1.6 ppmv, 1-h average, and 0.4 ppmv, 24-h average

and

Meteorological conditions are such that this condition can be expected to continue for
12 h or more.

From USEPA, GPO Publication 1972 0-452-7291, GPO, Washington, D.C., 1972.

standards. Congress recognized that an extension of deadlines was necessary. Under the 1977 CAA Amendments, attainment deadlines for SO_2 and PM were extended to 1982, and those for auto-related pollutants to 1987.

Since many AQCRs were not in compliance with one or more air quality standards, the USEPA classified these as nonattainment areas. Nonattainment status had considerable significance, since the introduction of new major sources in a dirty air region would be expected to aggravate threats to public health and would therefore be prohibited. The prospect of limitations on economic development where economic growth would naturally occur created major political and economic concerns. In response, the USEPA administratively developed a "compromise" emission offset policy that would allow growth as long as new sources did not produce a net increase in emissions. Under the offset policy, a new industrial facility would have to identify and use emission reductions from existing facilities to offset emissions associated with the new source. In cases where there was to be an expansion of an existing facility, sufficient emission reductions had to be made in the existing facility to qualify as offsets.

The 1977 CAA Amendments mandated that the USEPA offset policy be incorporated into the SIP process. The 1977 CAA Amendments also provided an alternative; i.e., a major new source could be approved for a nonattainment region if it would attain the "lowest achievable emission rate" (LAER). To be eligible under this provision, an industrial source having other facilities within a state would have to be in compliance with all relevant emission reduction programs for all its facilities. In practice, LAER has required the use of BACT on such new sources.

The problem of nonattainment was addressed in a major way in the 1990 CAA Amendments. Deadlines were again postponed, and the USEPA was given increased flexibility in setting deadlines in nonattainment areas. In general, attainment dates were those that could be achieved as expeditiously as practicable, but no later than 5 years after designation. This could be further extended for a period no longer than 10 years if the USEPA deemed it appropriate, considering the severity of the nonattainment problem and availability and feasibility of control measures.

Attainment dates for O_3, CO, and PM_{10} were to be determined on the basis of the classifications in Table 8.3. As can be seen, the more severe the problem and the more difficult its control, the more time a state had to bring an AQCR into compliance with primary air quality standards.

8.4.6 New Source Performance Standards

The 1970 CAA Amendments required that new or significantly modified existing sources (that result in increased pollution emissions) comply with NSPSs. Emissions standards established under NSPSs were to be applied to all new sources in designated source categories uniformly, without regard to whether they were being constructed in clean or dirty air regions. These performance or emission standards were to reflect the degree of emission limitation achievable through application of the best system of emissions reduction, taking cost into consideration. The purpose of these emission limitations was to prevent new pollution problems by requiring installation of control measures during construction, when they would be least expensive.

Table 8.3 Classification and Attainment Dates for 1989 Nonattainment Areas under 1990 CAA Amendments

Area Class	Design Value (ppmv)	Attainment Date
Ozone		
Marginal	0.121–0.138	3 years after enactment
Moderate	0.138–0.160	6 years after enactment
Serious	0.160–0.180	9 years after enactment
Severe	0.180–0.280	15 years after enactment
Extreme	≥0.280	20 years after enactment
Carbon Monoxide		
Moderate	9.1–16.4	December 31, 1995
Serious	≥16.5	December 31, 2000
PM$_{10}$		
Moderate		December 31, 1994
Serious		December 31, 2001

NSPSs are applied to specific source categories, the definition of which is the responsibility of the USEPA administrator. A sample of source categories and pollutants regulated under NSPSs is presented in Table 8.4. The application of NSPSs to pollutants is flexible. In addition to criteria pollutants, NSPSs have required emissions limits on fluoride from phosphate and aluminum (Al) plants, acid mist from sulfuric acid (H_2SO_4) plants, and total reduced S from paper mills. NSPSs can be applied to new sources for pollutants not regulated under NAAQSs or HAP standards. Although formulation and promulgation of NSPSs for any category of sources and pollutants is a federal responsibility, the USEPA delegates enforcement authority to states with acceptable SIPs.

8.4.7 Prevention of Significant Deterioration

Although the principle of PSD in areas where air quality was better than the standards was not specifically mentioned in the CAA Amendments of 1970, environmental groups maintained that nondegradation of air in clean air areas was implicit in the wording of the preamble of the act, "to protect and enhance," and reflected congressional and federal agency intent documented in congressional hearings before its passage. As a result of a lawsuit filed by environmental groups, the USEPA was compelled to develop regulations to protect air quality in clean air areas.

In 1975, the USEPA developed a PSD policy with a classification system that would allow some economic development in clean air areas and still protect air quality from significant deterioration. To eliminate legal uncertainty, the classification system and provisions for its implementation were incorporated into the 1977 CAA Amendments. Clean air areas were divided into three classes. Very little deterioration would be allowed in Class I regions, i.e., all international parks, national wilderness areas with more than 5000 acres, national monuments with more than 5000 acres, and national parks with more than 6000 acres. These included 155 areas in 36 states and 1 international park. The USEPA was required to develop and implement regulations that would protect visibility in these pristine air areas. Class II areas would allow moderate air quality deterioration, and Class III areas would allow air quality deterioration up to the secondary standard. Maximum

Table 8.4 Pollutants Regulated under NSPSs in Selected Source Categories

Sources	Pollutants
Fossil fuel-fired steam generators	PM, opacity, SO_2, NO_x
Coal preparation plants	PM, opacity
Incinerators	PM, opacity
Primary lead, zinc, and copper smelters	PM, opacity, SO_2
Secondary lead smelters	PM, opacity
Secondary bronze and brass smelters	PM, opacity
Primary aluminum reduction plants	Fluorides
Iron and steel plants	PM, opacity
Ferroalloy production facilities	PM, opacity, CO
Steel plants — electric arc furnaces	PM, opacity
Sulfuric acid plants	SO_2, acid mist, opacity
Nitric acid plants	NO_x
Portland cement plants	PM, opacity
Asphalt concrete plants	PM, opacity
Sewage treatment plants	PM, opacity
Petroleum refineries	PM, SO_x, CO, reduced sulfur
Kraft pulp mills	PM, opacity, reduced sulfur
Grain elevators	PM, opacity
Lime manufacturing plants	PM
Phosphate fertilizer industry	Fluorides

allowable increases above baseline values for each of the three PSD classes are summarized in Table 8.5.

Under the PSD program, all areas in attainment for SO_x and PM are designated Class I, II, or III for that pollutant. Additional pollution allowed in these areas is an increment beyond a baseline level established at the time of the location of the first major emitting facility. Similar requirements have also been established for NO_x. All areas of an individual state that meet NAAQSs are initially designated as Class II except for those areas specifically designated by the CAA as Class I.

Special permitting requirements must be met by major new facilities in PSD areas. If they have the potential to affect visibility in Class I areas, they are subject to a visibility review. They must use BACT. Major existing stationary sources in federal Class I areas that adversely affect visibility will be required to install best-available retrofit technology (BART) to minimize visibility impairment.

Although there was considerable congressional debate over visibility protection prior to passage of the 1990 CAA Amendments, actual PSD changes were relatively limited. Most importantly, additional emphasis was placed on regional haze issues. The USEPA was required to work with several western states to establish a commission to address visibility in the Grand Canyon. The agency promulgated regional haze rules (1999) subsequent to the Grand Canyon Visibility Transport Commission's report of 1996.

These regulations are designed to improve visibility in 156 Class I areas. Because fine particles (<2.5 μm) are transported hundreds of miles, most states, including those that do not have Class I regions, will have to participate in planning, analyses, and, in many cases, emissions reductions. States are required to develop long-term strategies with periodic goals for improving visibility in Class I areas. They must develop implementation plans that include enforceable measures by 2008 and identify facilities that will have to install BART controls. Haze rules considerably broaden

Table 8.5   Allowable Increments for PM_{10}, SO_x, and NO_x in Prevention of Significant Deterioration-Regulated Regions

Pollutant	Standard	Class	Allowable Increment ($\mu g/m^3$)
PM_{10}	Annual mean	I	4
		II	17
		III	34
	24-h maximum	I	8
		II	30
		III	60
SO_x	Annual mean	I	2
		II	20
		III	40
	24-h maximum	I	5
		II	91
		II	182
	3-h maximum	I	25
		II	512
		III	700
NO_x	Annual mean	I	2.5
		II	25
		III	50

the scope of existing PSD review requirements that primarily focus on new sources and use existing baseline air quality as a reference for making decisions.

8.4.8   Hazardous and Toxic Air Pollutants

Under the 1970 CAA Amendments, Congress authorized the USEPA to regulate pollutants deemed to be more hazardous than those regulated under NAAQSs. As a result, the USEPA initially promulgated National Emissions Standards for Hazardous Air Pollutants (NESHAPs) for Hg, Be, and asbestos. By 1990, NESHAPs had also been promulgated for vinyl chloride, inorganic arsenic (As), benzene, and radioactive isotopes.

Over a period of 20 years, the USEPA designated and regulated only seven HAPs. Many states had gone their own way in regulating what they described as air toxics. By 1990, state air pollution control agencies had established emission standards for approximately 800 "toxic" pollutants.

The relatively slow pace of designating and regulating HAPs reflected policy uncertainties associated with the NESHAP provision. The USEPA interpreted this provision as requiring the setting of emission limits that would provide an ample margin of safety. Since many HAPs are carcinogenic, the USEPA had at various times interpreted the statutory language as requiring an emission standard of zero for carcinogenic substances. It was therefore reluctant to regulate emissions of economically important substances that were potential human carcinogens, as such regulation would have required a total ban on their production.

Regulation of hazardous or toxic air pollutants received special attention under the 1990 CAA Amendments. In Title III, Congress listed 182 HAPs for which the USEPA was to identify source categories and promulgate emission standards by the year 2000. The USEPA was also authorized to add substances to this list if there were adequate data to show that emissions, ambient concentrations, bioaccumulation, or deposition could reasonably be anticipated to cause adverse human health or environmental effects.

One of the unique features of Title III is that hazards are not limited to human health. By reference to adverse environmental effects, Congress indicated its intent to protect wildlife, aquatic life, and other natural resources, including populations of endangered or threatened species, and prevent degradation of environmental quality over broad areas. Also unique is the use of technology-based standards rather than standards based on known health effects.

The USEPA was required to (1) publish a list of categories and subcategories of major sources (>10 tons/year single pollutant, >25 tons/year aggregate) of HAPs and area sources for pollutants listed, and (2) promulgate emission standards for each in accordance with a congressionally mandated schedule. In promulgating technology-based emission standards, the USEPA was given the flexibility to distinguish among classes, types, and sizes of sources. It was to require the maximum degree of reduction in emissions, taking into account cost of achieving such emission reductions, any non-air quality health and environmental impacts, and energy requirements. Maximum achievable emission reduction could be achieved by (1) process changes, substitution of materials, or other modifications; (2) enclosing systems or processes; (3) collecting or treating pollutants released from a process, stack, storage, or fugitive source; or (4) design, equipment, work practice, or operating standards. Maximum achievable control technology (MACT) standards are used to describe these emission reduction requirements for air toxics and HAPs.

8.4.9   Stratospheric Ozone Depletion

The 1977 CAA Amendments provided the USEPA statutory authority to regulate substances, practices, processes, and activities that could reasonably be anticipated to damage the O_3 layer. Exercising this authority, CFC use for aerosol propellants was banned in 1978. U.S. emissions of CFCs declined by 35 to 40% of their pre-ban levels as a result. The 1977 amendments provided the USEPA the authority it used to meet the emission reduction targets of the 1987 Montreal Protocol and subsequent amendments to it.

The 1990 CAA Amendments repealed the O_3 protection section of the 1977 CAA Amendments by adding Title VI: Stratospheric O_3 Protection. These amendments updated USEPA authority and added specific requirements that reflected more recent needs for protecting the O_3 layer and policies to achieve such protection.

Major provisions of Title VI include a listing of Class I and Class II substances (see below), monitoring and reporting requirements, phaseout of Class I and Class II substances, accelerated phaseout authority, a national recycling and emission reduction program, servicing of motor vehicle air conditioners, regulating non-essential products containing CFCs, labeling requirements for Class I and II substances, a policy for development of safe alternatives, and international cooperation.

Classification of chemicals into Class I and II substances reflects differences in O_3 depletion potential and a recognition that Class II substances, used as substitutes for CFCs, can also destroy O_3, albeit this potential is at least an order of magnitude less. Production of Class I substances, including CFCs, three types of bromine (Br)-containing halons, and carbon tetrachloride (CCl_4), was to be banned by the year 2000, and methyl chloroform by 2002. Class II compounds, which include hydro-chlorofluorocarbons, are to be banned in 2030, with phaseout beginning in 2015.

8.4.10 Acidic Deposition

For many years, congressional efforts to amend the CAA were stymied by contentious debate over control of acidic deposition. Congress put that debate to rest and enacted the 1990 CAA Amendments with specific authority and language addressing acidic deposition.

The purpose of Title IV: Acidic Deposition Control was to minimize adverse effects of acidic deposition by reducing annual emissions of SO_x and NO_x by 10 and 2 million tons, respectively, from 1980 emission levels.

Mindful of potentially significant economic costs and dislocations that may have occurred as a result of traditional "command-and-control" regulatory approaches, Congress developed an economics-based approach that it believed to be both more flexible and cost-effective. Emission reductions to meet the goals of the acidic deposition program would be achieved by establishing a program of emission allowances. An allowance under the Act was equal to 1 ton of SO_x.

Emission reduction requirements for SO_x using the allowance provision were applied to large fossil fuel-fired electric boilers specifically identified in the Act. These were to be achieved using a phased approach, with Phase 1 commencing in 1995 and Phase 2 in 2000. These requirements reduced total annual emissions in the 48 contiguous states and the District of Columbia to ~8 million mtons/year (8.9 million tons/year).

In the allowance program, each unit of each affected source was allocated allowances that were determined from total emissions and emission reductions required to achieve Phase 1 and Phase 2 goals. Each source had to develop a plan that described a schedule and methods to achieve compliance.

Unused allowances at one unit or source could be used by utility operators to offset emissions at other units or sources. They could be traded or banked for use in future years.

Title IV of the 1990 CAA Amendments repealed the SO_2 NSPS (requiring use of scrubbers) prescribed for coal-fired power plants under the 1977 CAA Amendments. Congress mandated that the USEPA promulgate a NSPS for coal-fired power plants that required the best system of emission reduction, taking into account the costs of achieving the reduction, any non-air quality health and environmental impact, and energy requirements.

Nitrogen oxide emission reductions were to be implemented in two phases, in 1996 and 2000. Though the NO_x program utilized some of the same emission reduction principles, e.g., emissions trading and flexibility in how emission reductions were to be achieved, it did not cap NO_x emissions or utilize an allowance system. Compliance was to be achieved by using an individual emission rate for each boiler or averaging emission rates over two or more units to meet an overall emission rate reduction requirement.

Title IV allowed regulated sources to determine their own compliance approaches. These have included switching to cleaner-burning fuels, reassigning some energy production from relatively high emission-producing units to cleaner units, and adopting conservation or efficiency measures. Prior to Title IV requirements, most utilities employed what was described as least-cost dispatching. Under such a system, generating units that produced electrical power at the least cost were used more than higher-cost units. Not surprisingly, lower-cost units had the highest SO_x emissions.

8.4.11 Motor Vehicle Emission Control

The pioneering atmospheric chemistry studies of Dr. Arie Hagen-Smit in the 1950s provided fundamental understanding of the role of automobile emissions in photochemical smog development. As seen in Table 2.1 to Table 2.3, automobiles have been the major anthropogenic source of CO, HCs, NO_x, and Pb. They are the major source of precursor molecules for production of photochemical oxidants such as O_3. It is for this reason that automobile emissions have been subject to significant regulatory requirements.

Automobile air pollution has been of greater concern in southern California than any other location in the world. This is due to a culture heavily dependent on automobile usage, abundant sunshine, topographical barriers to air movement, and frequent elevated inversions (Chapter 2) during summer months. Consequently, California has led the nation and world in controlling automobile emissions. California emission requirements have preceded those required nationally and have, in most cases, been more stringent.

8.4.11.1 Vehicle Emissions

Emissions from light-duty motor vehicles have included blow-by gases, exhaust gases, and volatile nonmethane hydrocarbon (NMHC) losses from the carburetor and fuel tank, and Pb. Control of motor vehicle emissions has included regulatory actions to reduce crankcase emissions, exhaust gases, and evaporative losses. Fuel composition standards have also been necessary to facilitate the use of catalytic systems, phase out the use of Pb, and achieve additional emissions reduction.

8.4.11.2 Regulatory History

The regulatory history of control requirements mandated for motor vehicles sold in California and nationwide is summarized in Table 8.6. Because of the relatively simple and low-cost technology (Chapter 9) required, crankcase emissions were effectively controlled in the early 1960s and evaporative emissions in the early 1970s.

Table 8.6   Regulatory History of Control Requirements on New Light-Duty Motor Vehicles

Model Year	Control Actions
1961	California — crankcase emission controls
1963	U.S. — crankcase emission controls
1966	California — exhaust emission controls for CO and NMHCs
1968	U.S. — exhaust emission controls for CO and NMHCs
1970	California — evaporative control systems; new, more stringent emission standards for CO and NMHCs; required reduction of NO_x emissions for the first time U.S. — Congress passes CAA Amendments; mandates stringent new motor vehicle exhaust emission standards for CO and NMHCs to be achieved by 1975 model year; mandates stringent NO_x emission standard to be achieved by 1976 model year; provides for fuel composition standards; USEPA imposes more stringent CO and NMHC standards
1971	U.S. — evaporative control systems
1972	California — emission standards for NO_x are tightened U.S. — USEPA imposes tighter standards for CO and NMHCs; first federal emission standard for NO_x promulgated
1974	California — Stringent emission standards mandated for 1975 model cars are postponed; California imposes interim standards for all new cars sold in California U.S. — Statutory emission standards for CO, NMHCs, and NO_x mandated in 1970 legislation are postponed; USEPA imposes interim standards for CO and NMHCs
1977	California — Tightens NMHC and NO_x emission standards U.S. — USEPA imposes interim emission standards for NO_x mandated by Congress in 1974 legislation; CAA Amendments passed; interim standards for CO and NMHCs continued through 1979 model year; 1975 statutory emission requirements for NMHCs required in 1980, CO in 1981; NO_x standard relaxed with 1981 deadline
1990	CAA Amendments passed; significant additional reduction in NMHCs required; emission limits for CO and NO_x under 1970 CAA Amendments to be met in the period 1993–1995

Control of exhaust gases, which account for the largest percentage of emissions (~90%), has had a difficult history of regulatory requirements, technology development, cost, and public acceptance.

8.4.11.3   Exhaust Gas Standards

Emission standards for exhaust gases were first required on 1966 model light-duty vehicles sold in California and 1968 models nationwide. Emission standards were applied only to NMHCs and CO, as these were more easily controlled than NO_x.

A history of exhaust gas emission reduction requirements is summarized in Table 8.7. All emission standards are given in grams per mile (g/mi). Note that methods of measuring emissions, as well as expressing their concentrations, have changed several times since emission limits were initially mandated.

Table 8.7 History of Light-Duty Motor Vehicle Exhaust Emission Standards

Model Year	Authority	HCs (g/mi)	CO (g/mi)	NO$_x$ (g/mi)	Evap. (g/test)	PM (g/mi)	HCHO (g/mi)
Pre-control		10.6	84.0	4.1	47		
1966	California	6.3	51.0				
1968	U.S.	6.3	51.0				
1970	California	4.1	34.0		6		
	U.S.	4.1	34.0				
1971	California	4.1	34.0	4.0	6		
1972	California	2.9	34.0		2		
	U.S.	3.0	28.0				
1973	U.S.	3.0	28.0	3.0			
1975	California	0.9	9.0	2.0	2		
	U.S.	1.5	15.0	3.1	2		
1977	California	0.41	9.0	1.5	2		
	U.S.	1.5	15.0	2.0	2		
1978	California	0.41	9.0	1.5	6[a]		
	U.S.	1.5	15.0	2.0	6[a]		
1980	California	0.39[b]	9.0	1.0	2		
	U.S.	0.41	7.0	2.0	6		
1981	California	0.39	7.0	0.7	2		
	U.S.	0.41	3.4	1.0	2		
1983	California	0.39	7.0	0.4	2		
1984	California	0.39	7.0	0.4	2	0.60	
1985	California	0.39	7.0	0.4	2	0.40	
1986	California	0.39	7.0	0.4	2	0.20	
1989	California	0.39	7.0	0.4	2	0.08	
1993	California	0.25 (0.31)[c]	3.4 (4.2)	0.4 (0.6)	2	0.08	
1994	California	0.125 (0.156)[d]	3.4 (4.2)	0.4 (0.6)	2	0.08	0.015 (0.018)
	U.S.	0.25 (0.30)[c]	3.4 (4.2)	0.4 (0.6)			
1997	California	0.075 (0.09)[e]	3.4 (4.2)	0.2 (0.3)	2	0.08	0.015 (0.018)
		0.04 (0.055)[f]	1.7 (2.1)	0.2 (0.3)	2	0.08	0.008 (0.011)
2003	U.S.	0.125[g]	1.7	0.2			

[a] Testing procedure change.

[b] NMHC standard or 0.41 g/mi total HCs.

[c] Standards are for nonmethane organic gas for 5 years or 50,000 mi and 10 years or 100,000 mi, respectively.

[d] Transitional low-emission vehicles.

[e] Low-emission vehicles.

[f] Ultra-low-emission vehicles.

[g] Standards are for 10 years or 100,000 mi.

Source: From Calvert, J.G., et al, *Science*, 261, 37, 1993. With permission.

8.4.11.4 Light-Duty Motor Vehicle Emission Control Requirements

The 1970 CAA Amendments required that emissions of NMHCs and CO for 1975 and NO_x for 1976 model light-duty motor vehicles be reduced by 90% of their 1970 levels. Since an approximate 60% reduction in emissions of NMHCs and CO had already been achieved, total emissions reductions from a precontrol base would be ~95 to 96%. Congress provided the USEPA authority to regulate fuel content, anticipating that compliance with these emission standards would require the use of catalytic converters that would be "poisoned" by Pb fuel additives.

In response to energy problems caused by the Arab oil embargo in 1973–1974, Congress passed the 1974 Energy Supply and Coordination Act. It amended the 1970 CAA to extend the motor vehicle emission standards deadline from 1975 to 1977 model cars for NMHCs and CO and relaxed the NO_x standard for 1977 model vehicles from 0.4 to 2.0 g/mi. In lieu of the 1975 statutory requirements for NMHCs and CO, Congress authorized the USEPA to set interim emission standards that would reflect the greatest degree of emission reduction achievable through use of existing technology, taking costs into account. As a consequence, interim emissions standards were required on 1975 to 1977 model vehicles.

Recognizing the difficulties of achieving mandated emission standards, Congress in 1977 extended the original emission deadlines, set for 1975, to 1980 for NMHCs and 1981 for CO. The NO_x standard was relaxed from 0.4 to 1.0 g/mi, with the deadline extended to 1981. Both the CO and NO_x deadlines were again postponed. After nearly two decades of postponements and extensions, Congress in 1990 set the final course for achieving the tough motor vehicle emission requirements it set in 1970. The original emission standards for CO (3.4 g/mi) and NO_x (0.4 g/mi) and a tougher NMHC standard (0.25 g/mi) were to be achieved on all new light-duty cars and trucks by 1996 model vehicles. The original standards were based on a vehicle operating life of 5 years or 50,000 mi. The 1990 Amendments required somewhat less stringent standards for vehicle operating lives of 10 years or 100,000 mi: NMHC, 0.3 g/mi; CO, 4.2 g/mi; and NO_x, 0.6 g/mi. Congress mandated that manufacturers of light-duty motor vehicles and trucks up to 2727 kg (6000 lb) meet a new PM standard based on a 5-year or 50,000-mi (0.09 g/mi) and 10-year or 100,000-mi (0.10 g/mi) operating life.

Recognizing that motor vehicles operating under cold weather conditions posed unique CO emission control concerns, Congress in 1990 mandated cold-start standards to be phased in over model years 1994 to 1996.

Emission standards for light-duty cars and trucks were specified for Phase 1 (beginning in model year 1996) and Phase 2 (beginning in model year 2001). These include both 50,000- and 100,000-mi standards for nonmethane organic gas (NMOG), CO, NO_x, PM, and formaldehyde (HCOH). Phase 2 standards are more stringent than Phase 1; 50,000-mi standards are more stringent than 100,000-mi standards.

Though evaporative emissions were regulated by California for 1970 model motor vehicles and the U.S. for 1971 models, evaporative emissions continued to be a problem. As a consequence, the 1990 Amendments mandated that the USEPA promulgate regulations for evaporative emissions of NMHCs from all gasoline-fueled motor vehicles for O_3-prone summertime conditions.

Throughout the history of federal involvement in regulating emissions from motor vehicles, California alone had been permitted to regulate emissions more stringently than the federal government. This "California exemption" recognized the state's leadership role in motor vehicle emissions control and its unique motor vehicle-related air pollution problems. Under the 1990 Amendments, other states were allowed to adopt emission standards and requirements similar to, but no more stringent than, those of California.

8.4.11.5 Fuel Additives and Cleaner Fuels

Anticipating that catalytic converters were to be used on 1975 and later model vehicles, Congress authorized the regulation of fuel additives. Lead additives, which had been universally used to increase octane ratings of gasolines, were anticipated to coat catalytic materials, rendering them ineffective. Using this authority, the USEPA began a phasedown in leaded gasoline use. Other regulatory actions included prohibition of (1) misfueling, i.e., introduction of leaded gasoline into a vehicle labeled unleaded gasoline only; (2) use of gasoline or diesel fuel with a S concentration of >0.05% by weight; (3) sale and use of high volatility (Reid vapor pressure of >9 lb/in.^{2} (PSI) (465 mmHg)) gasoline during the high-O_3 season; and (4) sale and highway use of Pb additives after 1995.

Specific provisions of the 1990 CAA Amendments addressed the issues of reformulated gasoline and oxygenated fuels. The USEPA was to promulgate regulations establishing requirements for reformulated gasoline to be used in motor vehicles in specified O_3 nonattainment areas. These were to require significant reductions in emissions of O_3-forming volatile organic compounds (VOCs) during the O_3 season as well as toxic air pollutants. The performance standard for reformulated gasoline required a 15 to 17% reduction in aggregate emissions of O_3-forming VOCs from baseline values, with a 25 to 29% reduction by the year 2000; an aggregate emission reduction of toxic air pollutants of 20 to 22% effective in 2000; and a 9 to 10% reduction in NO_x. Toxic pollutants included in these regulations were benzene, 1,3-butadiene, polycyclic organic matter, acetaldehyde, and HCOH. In CO nonattainment areas, oxygenated gasoline containing no less than 2.7% O_2 by weight is to be sold in high-emission seasons.

8.4.11.6 Emission Test Cycles

Exhaust emission standards must reflect actual emissions in normal driving. Emissions vary, however, as automobiles are operated under different conditions, e.g., long trips, short trips, stop-and-go urban traffic, and superhighway travel. In addition, individual drivers operate vehicles differently under similar driving conditions. Since it is not possible to duplicate all of these driving factors in specifying emission requirements, a standard emission assessment procedure is used to represent traffic that is responsible for emissions in areas where the most severe motor vehicle-related air pollution occurs.

Certification of automobile engines for compliance with emission standards requires that emission tests be conducted under standardized conditions. As a result, both California and U.S. test procedures utilize a dynamometer test cycle that simulates early-morning driving conditions in an urban area.

Dynamometer tests are conducted under laboratory conditions. They are lengthy, somewhat complex, and expensive. As a consequence, certifications are limited to engine systems.

8.4.11.7 Inspection and Maintenance

Seventy U.S. cities and several states had auto emission inspections and maintenance (I&M) programs prior to 1990. The 1990 Amendments expanded I&M requirements to 40 additional metropolitan areas, many in the northeastern U.S. Specific enhanced requirements were mandated for new and existing I&M programs.

I&M programs are deemed to be an essential element of pollution control programs designed to achieve NAAQSs. Long-path monitoring studies have shown that in some major metropolitan areas, ~50% of emissions of regulated pollutants from motor vehicles are associated with ~10% of the motor vehicle population. It is these polluting vehicles with poorly functioning or malfunctioning emission control systems that I&M programs are designed to identify and have repaired and properly maintained.

8.4.11.8 Tampering Prohibitions

Tampering (willful disabling of emission control systems) has been a continuing problem since emissions controls were required on motor vehicles. Tampering prohibitions were included in both 1970 and 1977 CAA Amendments, with a focus on motor vehicle dealers, repair shops, etc. In 1990, tampering prohibitions were extended to individuals who manufacture products designed to disable or interfere with the proper functioning of emission control systems. They prohibited everyone, including "backyard mechanics," from tampering with motor vehicle emission control systems.

8.4.11.9 Onboard Vapor Recovery Systems

A major new requirement of the 1990 CAA Amendments was use of onboard vapor recovery systems to control NMHC emissions during vehicle refueling. The USEPA was mandated to promulgate standards that require new light-duty vehicles to be equipped with vapor recovery systems with a minimum evaporative emission capture recovery of 95%.

8.4.11.10 Onboard Diagnostic Systems

Congress, in 1990, mandated that the USEPA promulgate regulations that would require manufacturers to install onboard diagnostic systems on 1994 and later model vehicles. Such systems were to detect emission-related failures and alert drivers to the need for potential repair. They include dashboard warning lights and trouble code storage that helps mechanics identify malfunctions and possible causes.

8.4.11.11 Emission Requirements for Trucks and Other Vehicles

Due to differences in weight, vehicle use and population, and engine types, trucks (except those less than 3750 lb, 1705 kg) are regulated differently than light-duty vehicles. Emission standards for larger trucks (up to 6000 lb, 2727 kg) and heavy-duty trucks are less stringent than those for light-duty motor vehicles. Light-duty trucks with a gross vehicle weight rating of 3750 to 6000 lb are required to meet emission limits of 0.32 g/mi NMHC, 4.4 g/mi CO, and 0.7 g/mi NO_x for a 5-year or 50,000-mi (80,000 km) operating life, and 0.40 g/mi NMHC, 5.5 g/mi CO, and 0.07 g/mi for a 10-year or 100,000-mi (161,290 km) operating life.

In 1985, the USEPA adopted stringent emission standards for NMHCs, CO, NO_x, and PM for heavy-duty (both gasoline- and diesel-fueled) trucks and buses that were to be phased in beginning in 1991. Under the 1990 Amendments, the USEPA was given authority to revise standards promulgated prior to 1990. Beginning with model year 1994, large diesel truck engines were required to reduce PM emissions by 90%, with greater emission limits on buses.

8.4.11.12 Clean-Fuel Vehicles

The USEPA was mandated to promulgate standards for clean-fuel vehicles under the 1990 CAA Amendments. So-called clean fuels include any fuel, reformulated gasoline, natural gas, liquified petroleum gas (LPG), hydrogen (H_2), or power source (including electricity) used in vehicles that complies with clean-fuel vehicle standards.

8.4.11.13 Clean Fleets

The 1990 CAA Amendments placed new requirements on centrally fueled fleet operations in certain O_3 nonattainment areas. Thirty percent of new vehicles purchased were to use clean fuels and meet exhaust emission standards (0.075 g/mi NMHC, 3.4 g/mi CO) that were lower than those required for general passenger vehicles; these went into effect in 1998, with purchase requirements increased to 70% in 2000. This clean fleet program affects 22 metropolitan areas in 19 states.

8.4.12 Enforcement

Early pollution control programs were largely ineffective because agencies had limited enforcement authority. The 1970 and subsequent CAA Amendments provided considerable federal enforcement authority and mandated similar authority for states under SIP provisions. Since 1970, the administrator of the USEPA has had authority to take enforcement actions when provisions of SIPs, NSPSs, NES-HAPs, and other regulatory requirements are violated.

Under current statutory language, the USEPA can, in response to violations of SIPs and other major provisions of the CAA, issue administrative compliance orders. The order specifies a time period by which a source must be in compliance with emission limits. The USEPA may commence civil actions in federal district courts should the affected party not respond to administrative orders. In such actions, the USEPA can seek a permanent or temporary injunction or civil penalties. If violations are willful, criminal actions may be taken. The enforcement value of USEPA administrative actions was significantly enhanced in 1990 when the USEPA was given authority to impose civil penalties (subject to formal adjudicatory hearings) administratively. The USEPA may also issue noncompliance penalties. These are assessed for certain broad categories of violations in an amount no less than the economic value derived from delayed compliance.

The 1990 CAA Amendments provided authority for tough criminal sanctions. These include fines of upward of a million dollars and prison terms of up to 15 years. Such criminal sanctions may be imposed for "knowing" violations of the law.

The USEPA may file suit to abate any air pollution problem that presents "an imminent and substantial endangerment to the health of persons," even if there has been no violation of the law. It may also issue administrative orders to protect public health if it is not practical to initiate a court action.

Although the USEPA has significant enforcement authority, most enforcement occurs at state and local levels. Federal enforcement authority is only used when state and local agencies are reluctant to enforce provisions of their own SIPs and other programs (e.g., NSPSs). It is applied to violations of regulatory requirements for which the USEPA has direct responsibility, i.e., motor vehicle emissions, O_3-destroying chemicals, and acidic deposition.

8.4.13 Permit Requirements

Permit systems are effective tools for achieving compliance with emission reduction requirements. Permit requirements put the burden of proof of compliance on emission sources. Permits may be required for existing sources to discharge any pollutants into the atmosphere and to install and operate pollution control equipment. They are required for new or significantly modified existing sources before construction commences. Legally, a source cannot operate, i.e., emit pollutants, without a permit. To be granted a permit, a source must comply with all existing emission reduction requirements.

It is not possible in many cases for a source to immediately comply with permit conditions due to time requirements for designing, ordering, and installing emission control systems. For a source to operate in violation of emission regulations, a legal exception or variance is granted for a specified period, e.g., 6 months. These variances provide a reasonable period of time for a source to install control equipment necessary to comply with emission limitations.

The role of permitting in achieving air quality goals was given significant new attention in the 1990 CAA Amendments. Under Title V, the USEPA promulgated regulations that established minimum requirements for air pollution permit programs. Among these were (1) permit applications, including a standard application form and criteria for determining completeness of an application; (2) monitoring and reporting; (3) an annual fee assessed to permittees; (4) personnel and funding to administer the permits program; and (5) adequate authority to issue permits and assure compliance.

Major stationary sources, sources subject to NSPSs, regulated hazardous pollutant sources, and sources subject to PSD requirements must have permits. Permits contain enforceable emission limitations, monitoring and reporting requirements, and in some cases a compliance plan that includes an enforceable schedule.

8.4.14 Citizen Suits

The 1970 CAA Amendments gave citizens the right to sue both private and governmental entities to enforce air pollution requirements. Prior to 1970, citizens did not, in most cases, have legal standing to seek court review of federal agency decisions. The premise of the citizen's suits provision is to make the federal enforcement agency (USEPA) accountable when its actions are deemed to conflict with clean air legislation and regulatory requirements. These provisions have been used by environmental groups, e.g., the Natural Resources Defense Council, Environmental Defense Fund, and Sierra Club, and others when they believed that the USEPA was not complying with its statutory obligations. Successful citizen suits have required the USEPA to promulgate regulations for air quality maintenance, indirect source review, PSD, and an air quality standard for Pb.

One of the principal effects of citizen suits has been to require the USEPA to implement provisions or concepts embodied in clean air legislation when it was reluctant to do so. The reasons for such reluctance include practical and economic considerations, availability of manpower, administrative priorities, and bureaucratic inertia. An attendant result of citizen suits has been legal clarification of the language of clean air legislation.

Although citizen suits have, for the most part, been used to require the USEPA to develop and implement new policies, citizen legal action can also be taken against a pollution source to force compliance with emission reduction requirements. Under the 1990 CAA Amendments, citizens were given a legal right to file lawsuits to seek civil penalties against parties in violation of the CAA.

In the early years of air pollution control, citizen suits were commonly brought against the USEPA by environmental groups. Over time, a number of citizen groups sympathetic to industry interests were formed. Such groups commonly file suits in response to USEPA regulatory actions, often contending that they are arbitrary, capricious, and not supported by "good science."

8.4.15 Pollution Prevention

Congress passed the Pollution Prevention Act (PPA) in 1990. Under it, the scope of federal policy to cover releases of pollutants to all media was expanded.

The focus of the Act was to establish new policies for waste reduction (interpreted broadly). The guiding principle is that source reduction is fundamentally different and more desirable than waste management and pollution control. It focuses on source reduction as the primary means of reducing waste production. A source reduction strategy deemphasizes command-and-control regulation and emphasizes voluntary cross-media pollution prevention activities. The PPA mandates the USEPA to promote governmental involvement in overcoming institutional barriers by providing grants to state technical assistance programs under a new federal pollution prevention program. Pollution prevention is given a new emphasis under the 1990 CAA Amendments. The Amendments required that the USEPA conduct an engineering research and development program to develop, evaluate, and demonstrate nonregulatory technologies for air pollution prevention. In establishing emission standards for HAPs, the USEPA is required to consider cross-media impacts, substitution of materials or other modifications to reduce the volume of HAPs, the potential for process changes, and eliminating emissions entirely.

8.5 STATE AND LOCAL AIR POLLUTION CONTROL FUNCTIONS

All states and many municipalities had some kind of air pollution legislation or ordinances prior to enactment of the 1970 CAA Amendments. These state and local programs varied in scope, enforcement authority, administration, and public support. All were limited in their ability to deal adequately with the complex problems of achieving control objectives.

Passage of the 1970 CAA Amendments, with their attendant requirements for federal approval of SIPs, imposed significant homogeneity on state and local programs. As a consequence, many state and local control programs strongly resemble each other. Despite this homogeneity, differences exist since they reflect unique state and local control concerns.

Most stationary source control functions are carried out at the state or local level. Federal clean air legislation provides the fundamental framework for effective pollution control at these levels. This framework is provided in the USEPA regulations promulgation process, the SIP approval process, grants-in-aid to state and local agencies, and backup federal enforcement authority should state and local agencies fail to carry out their responsibilities.

The federal posture in working with state and local agencies is a combination of carrot and stick. State and local programs that meet federal requirements receive limited monies to maintain air pollution control programs. Noncompliance with federal requirements can result in loss of federal support for pollution control activities as well as highway trust funds.

Most state and local programs operate under the philosophy that local authorities best know how to deal with state and local problems, i.e., which control tactics should be used and how they are to be applied. In theory, this is true. It is also true that political pressure to hamper or slow pollution control efforts is the most intense at state and local levels. As a result, a federal presence is necessary to enforce, if necessary, implementation of air pollution control regulations should state or local agencies choose not to enforce them. In most instances, state and local agencies prefer not to invite "federal interference."

8.5.1 Boards and Agencies

Regulation of air pollution is carried out by state and local authorities through the "police powers" delegated to states under the U.S. Constitution. Utilization of these police powers requires enactment of state legislation and local ordinances to provide a legal framework for air pollution control. Since legislation and ordinances cannot deal with or anticipate all the individual circumstances of source control, they must provide for establishment of an administrative agency to carry out the provisions embodied in them. These agencies may have quasi-legislative powers to adopt rules and regulations and recommend appropriate legal action. In some instances, these powers are vested in separate pollution control boards or commissions.

Air pollution control agencies have the day-to-day executive responsibilities of planning to achieve NAAQSs, source data collection, air quality monitoring, development of emission standards, development of source compliance schedules and permit review, surveillance of sources to determine compliance status, and recommendations for enforcement actions in cases of noncompliance. As previously discussed, states are required to submit SIPs for USEPA approval that outline in detail how NAAQSs are going to be met in each AQCR under their jurisdiction. States may delegate responsibility to local agencies to develop local control plans. These local plans must be compatible with the SIP since they will be incorporated into it.

Once it has been determined that ambient levels of a specific pollutant exceed one or more NAAQSs, it may be necessary to promulgate emission standards for sources in the AQCRs. These standards may be based on computer models that use emission source data and dispersion characteristics of the atmosphere to predict ambient concentrations. Once needed emission reductions are determined from these models, compliance schedules are negotiated with industrial sources.

8.5.2 Effectiveness of State and Local Programs

The effectiveness of state and local air pollution control programs varies from state to state, and from one municipality to another. Among the many factors determining the effectiveness of air pollution control efforts are (1) a legal framework for regulation promulgation and enforcement, (2) availability of economic resources, (3) sufficient trained manpower, (4) commitment of pollution control officials to enforcement of control requirements, (5) cooperation of public officials, and (6) public acceptance of needed pollution control measures. Though each of these is important, the most vital factor is acceptance of pollution control requirements by private citizens, public officials, and sources. Without this support, air pollution control objectives may not be achieved.

8.6 PUBLIC POLICY ISSUES

Clean air legislation has provided a regulatory framework for controlling ambient air pollution. Within that framework, the USEPA is required by law to implement provisions of the CAA in a manner protective of public health and the environment.

Regulatory needs change as new problems are identified, the advancement of science indicates a need for changes in existing standards and requirements, and court decisions mandate new actions or changes in proposed or existing regulatory requirements. Therefore air pollution regulations, and policies that they embody, are continuously evolving to reflect current realities. Policy considerations may involve using existing regulatory authority or seeking expanded authority through congressional action. Policy issues may also involve negotiation, ratification, and implementation of international treaties.

Environmental policy initiatives, or implementation of new regulatory initiatives, commonly result in a measure of controversy. This is particularly the case when proposed policies and regulations have real or perceived major economic implications for those to be regulated. On the other hand, environmental organizations may not accept that regulatory policies are sufficiently stringent.

The nature of air pollution control engenders a continual public policy development effort that involves proposal of new regulations, participation of the regulated community and other publics, and regulation promulgation and subsequent implementation. Major contemporary public policy issues include global warming and the Kyoto Protocol, revised air quality standards for PM and O_3, new source review (NSR), and motor vehicle-related air pollution.

8.6.1 Global Warming and the Kyoto Protocol

As indicated in Chapter 4, an increasing consensus has developed among atmospheric scientists and world leaders that greenhouse gas emissions are having a significant effect on global climate, with even more significant warming effects to come. As a result of concerns expressed at the 1992 Earth Summit in Rio de Janeiro, more than 100 countries (including most industrialized nations) negotiated a treaty in Kyoto, Japan, to reduce greenhouse gas emissions to target levels.

The Kyoto Protocol, signed in December 1997, calls for a worldwide reduction of C-based gas emissions by an average of 5.2% below 1990 levels by 2012 (an actual 33% reduction based on projected increases in emissions). Different targets were adopted by different countries. Members of the European Union committed themselves to an 8% reduction; the U.S., 7%; Japan, 6%; and Russia and the Ukraine agreed to stabilize at 1990 levels. Sensitive to their economic development needs, no C emission reductions were required of developing countries.

The Kyoto Protocol provides considerable flexibility in how emission targets can be achieved. Policy options include emission trading, enhancement of energy efficiency in relevant economic sectors, protection and enhancement of sinks and reservoirs, sustainable agriculture promotion, increased use of renewable energy sources, removal of subsidies and exemptions that serve to increase greenhouse gas emissions, and limitation and reductions in methane (CH_4) emissions.

As soon as the U.S. signed the Kyoto Protocol, the Senate signaled, by an overwhelming vote (on a resolution), that it would not ratify it. Many political leaders claimed the treaty was flawed since it exempted large C emitters such as China. Because of its perceived potential adverse impact on the American economy, the U.S., by an act of the president, withdrew from the treaty. The treaty was subsequently renegotiated in 2001 (without U.S. participation) and went into effect after being ratified by countries whose total C-based emissions exceed 55% of the emission reductions mandated by the treaty.

Economic growth is a powerful factor in American politics. As such, requirements of the Kyoto Protocol are viewed as much, if not more, in political as economic terms. The political hurdles are considerable. Most, but not all, American business leaders believe that emission reduction requirements under the Kyoto Protocol are unrealistic.

Since the U.S. consumes about 25% of the world's C-based fuels, its participation in C emission reductions is vital to any international effort to reduce greenhouse gas emissions, atmospheric accumulation, and related global warming.

The U.S. has the largest per capita consumption of energy in the world. As such, it has the potential to significantly reduce C emissions by moving toward a more energy-efficient economy. A particularly significant sector is transportation, where use of fuel-efficient vehicles and technologies has been notably lagging. Fuel efficiency has not been a priority of automobile manufacturers or consumers. Relatively low gasoline prices in recent years have provided a market incentive to produce, sell, and buy larger and less fuel-efficient vehicles.

Pollution charges on C-based gases have been one of the more significant, and possibly more promising, policy proposals to reduce C emissions to the atmosphere. Pollution charges have the potential to penalize users of fuel-inefficient vehicles and promote fuel efficiency in this sector of the economy as well as others.

Global warming represents one of the most important public policy issues that the U.S. and other countries will deal with in the next half century. Reduction of greenhouse gas emissions will be a continuing public policy issue as a consequence.

8.6.2   Revised Ozone and Particulate Matter NAAQSs

In 1997, the USEPA promulgated new NAAQSs for O_3 and PM. Based on a review of hundreds of epidemiological and toxicological studies, it determined that the existing O_3 standard (0.12 ppmv, 235 $\mu g/m^3$, averaged over 1 h) was not adequate to protect public health. As a consequence, the previous NAAQSs were replaced with an 8-h standard set at 0.08 ppmv (157 $\mu g/m^3$). Prior to setting the new standard, it was clear there was no apparent threshold for adverse effects of ambient O_3 exposures.

As with O_3, considerable scientific evidence (primarily epidemiological) began to accumulate relative to adverse health effects associated with PM exposures below the 24-h PM_{10} standard (see Chapter 5). The evidence indicated a need for a new PM standard based on fine particles. As a consequence, the USEPA promulgated a 24-h $PM_{2.5}$ standard (65 $\mu g/m^3$) and an annual standard (15 $\mu g/m^3$) and retained the annual PM_{10} standard of 50 $\mu g/m^3$.

These new standards had broad support in public health and scientific communities. They were, however, strongly opposed by many industries, business groups such as the Chamber of Commerce, conservative organizations, and congressmen. Opponents attempted to prevent the USEPA from adopting them, then attempted to block them in Congress, and then sued to overturn them in federal court. Initially the standards were overturned by an appeals court decision, but they were subsequently upheld by the U.S. Supreme Court.

From a public policy perspective, challenges to the O_3 and $PM_{2.5}$ standards have been of extraordinary intensity. These reflect (1) the fact that the standards are likely to impose significant new costs on American industry, and (2) the political power of very conservative, antienvironmental protection forces. The public policy debate on the new $PM_{2.5}$ and O_3 standards portends a significantly intensified adversarial relationship among stakeholders in future regulatory initiatives and actions.

8.6.3 New Source Review

New or significantly modified existing sources are required under Title I of the CAA to comply with all applicable NSPSs before they can be granted a permit to operate. They may also be subject to (1) additional emission limitation requirements if they are to be located in nonattainment areas, and (2) PSD requirements in attainment areas. These requirements can be described as new source review (NSR).

An NSR is triggered when a new source's emissions exceed a threshold range of 10 to 100 tons/year (9 to 90 mtons/year) in a nonattainment area and 250 tons/year (225 mtons/year) for a PSD area. An NSR can also be triggered by major modification of an existing source, defined by the CAA as a physical change or change in the method of operation that results in an increase in emissions, or emissions of a new pollutant. Existing regulations exempt routine maintenance, repair or replacement of equipment, increases in hours of operation or production rate, or changes in ownership. If there is no physical change or change in the method of operation that results in significant net increased emissions of a regulated pollutant, then an NSR is not required.

NSR regulations establish significance levels that define major modifications that vary by pollutant and the attainment status of the AQCRs. These regulations allow for net emissions increases if the source offers past or future emissions decreases at its other units to counterbalance an increase from the proposed change. However, NSR only applies if net emissions increase above the significance level.

In 1998, the USEPA launched a major new NSR enforcement initiative in which it notified a number of coal-fired utilities that they were in violation of CAA provisions as a result of what the agency described as "maintenance" activities that were significantly greater than what would be described as "routine." In the utilities' view, the USEPA significantly narrowed its historical interpretation of what constituted routine maintenance and what triggers NSR.

Enforcement actions were taken by the USEPA and several states against the owners and operators of coal-fired power plants, alleging that their facilities had been modified without NSR permits. Modifications in individual cases included construction of entire coal-fired steam-generating units without incorporating modern pollution control equipment, increases in hourly capacity, redesign and modification of original design defects, and life extension projects. In all such cases, the USEPA maintained that net emission increases occurred above significance levels and should have been subject to NSR. Settlements were reached with several power companies and others put on hold as a new presidential administration took office.

Electric utilities cited for violations under NSR claimed their projects were limited to routine maintenance and thus exempt from regulatory requirements. They and the rest of their industry maintained that the Clinton administration erroneously interpreted NSR requirements. They also maintained that the new NSR interpretation was preventing them from expanding production to meet energy supply needs.

In June 2002, a new, more industry-friendly presidential administration announced major proposed changes in enforcement rules governing older coal-fired power plants and refineries. These included (1) promotion of pollution prevention projects, (2) expanded limits on the applicability of NSR designed to provide facilities greater flexibility to modernize without increasing emissions, (3) a clean unit provision that encourages installation of state-of-the-art pollution controls, (4) changes in calculating emissions increases and establishing emissions baselines, (5) clarification of the definition of routine maintenance, and (6) clarification of how NSR applies to multiple projects implemented in a short period of time when a company modifies one part of a facility in such a way that output in other parts of a facility increases. Changes in NSR proposed by the USEPA in 2002 were claimed by proponents as efforts to increase energy efficiency and encourage emission reductions. To opponents, it was a major relaxation of NSR rules that would preclude future federal legal action except in the most flagrant cases.

8.6.4   Motor Vehicle Emissions

As seen in previous sections, significant regulatory efforts have been directed at reducing emissions from motor vehicles. These efforts span more than three and a half decades. Significant progress in reducing emissions from motor vehicles has been achieved. Motor vehicle emissions are, on average, ~80% less than they were in the mid-1960s. Despite these regulatory and technological achievements (see Chapter 9), motor vehicles are still responsible for ~50% of national emissions of HCs and NO_x, 90% of CO in urban air, and more than 50% of HAPs.

Total annual emissions for NMHCs, CO, and NO_x for properly maintained passenger cars and light trucks on the road in July 2000, operating using a typical gasoline on a warm summer day (22 to 35.5°C, 72 to 96°F), are summarized in Table 8.8. These emissions are based on an annual mileage of 12,500 (cars) and 14,000 (light trucks) and assumed average fuel consumption of 21.5 mi/gal (9.1 km/l) and 17.2 mi/gal (7.3 km/l) for cars and trucks, respectively.

Data presented in Table 8.8 are indicative of the relative progress the U.S. has made in controlling motor vehicle emissions, as well as emission control problems that both the U.S. and world still face.

Table 8.8   Average Annual (2000) Emissions from Passenger Cars and Light-Duty
            Trucks

Pollutant	Passenger Cars Emission Rate (g/mi)	Annual Emissions (kg)
Hydrocarbons	2.8	35.0
Carbon monoxide	20.9	261.4
Nitrogen oxides	1.4	17.3
Carbon dioxide	415.9	5204.5
Light Duty Trucks		
Hydrocarbons	3.5	49.1
Carbon monoxide	27.7	388.2
Nitrogen oxides	1.8	25.4
Carbon dioxide	522.1	7288.6

Source: From USEPA, http://epa.gov/otaq/consumer/f00013.htm.

Average emission rates are significantly above statutory requirements for new passenger cars (i.e., 0.25 g/mi NMHC, 3.5 g/mi CO, and 0.4 g/mi NO_x) and the less stringent requirements for light-duty trucks, including sport utility vehicles, i.e., SUVs, whose numbers and percentage of the motor vehicle population have dramatically increased in the past decade.

The regulatory and technological progress that has been achieved in the U.S. in the past three and a half decades has been obviated to some degree by other factors. Most notable of these has been a quadrupling of miles (from 1 to 4 trillion) driven by Americans since 1970. This increase has been due, in part, to an increase in population and continuing urban sprawl that requires increased driving distances, often by single-occupant vehicles. This phenomenon can be expected to continue as urban highways are expanded to relieve the traffic congestion common to urban workplace travel. Paradoxically, roadway improvements facilitate even longer commuting distances.

Motor vehicle emission reductions on a total, national mass emissions basis have been affected by less stringent emission requirements for trucks and buses, changes in motor vehicle fuels that have increased their volatility, and emission control devices whose service life of 50,000 mi is a fraction of the 100,000-mi, and often more, life expectancy of motor vehicles on U.S. highways. In the latter case, the 1990 CAA Amendments require that emission control systems (primarily catalytic converters) must have an effective service life of 100,000 mi. Fuel volatility and more stringent emission reduction requirements on light- and heavy-duty trucks are also addressed under the 1990 Amendments.

Though emission reduction requirements on motor vehicles have resulted in notable progress, the public policy question to be addressed is whether air pollution control efforts that focus on emission reductions achieved by application of control technology are sufficient. Transportation controls, with few exceptions (e.g., fast lanes for multioccupant vehicles), have not been politically popular. Proposals to increase fuel economy requirements have faced stiff opposition from motor vehicle manufacturers. Nevertheless, a significant increase in fuel economy represents the single best approach to reduce overall criteria pollutant and CO_2 emissions in North America and across the planet. The technology to achieve significant fuel economy appears to be available (see "Hybrid Electric Vehicles," Section 9.3.2). The political will to achieve air quality objectives by significantly increasing motor vehicle fuel economy appears not to be.

READINGS

Ackerman, B.A. and Stewart, R.B., Reforming environmental law: the democratic case for economic incentives, *Columbia J. Environ. Litigation*, 13, 171, 1988.

Beaton, S.B., Bishop, G.A., Zhang, Y., Ashbaugh, L.L., Lawson, D.R., and Stedman, D.H., On-road vehicle emissions: regulations, costs, and benefits, *Science*, 268, 991, 1995.

Bedrock, N.R., Enforcing the CAA, *Environ. Protection,* November 10, 1993.

Burke, R.L., *Permitting for Clean Air: A Guide to Permitting under Title V of the CAA Amendments of 1990*, Air and Waste Management Association, Pittsburgh, 1992.

Calvert, J.G., Heywood, J.B., Sawyer, R.F., and Seinfeld, J.H., Achieving acceptable air quality: some reflections on controlling vehicle emissions, *Science*, 261, 37, 1993.

Cannon, J.A., The regulation of toxic air pollutants: a critical review, *JAPCA*, 36, 562, 1986.

Clairborne, M.L., The new air toxics program, *Nat. Res. Environ.*, Fall 1992, p. 21.

Committee for Economic Development, *What Price Clean Air? A Market Approach to Energy and Environmental Policy*, Committee for Economic Development, Washington, D.C., 1993.

DeNevers, N.H., Neligan, R.E., and Stater, H.H., Air quality management, pollution control strategies, modeling, evaluation, in *Air Pollution*, Vol. V, *Air Quality Management*, 3rd ed., Stern, A.C., Ed., Academic Press, New York, 1977, p. 4.

Gibson, J.A., The roads less traveled? Motoring and the Clean Air Act Amendments of 1990, *Nat. Res. Environ.*, Fall 1992, p. 13.

Grant, W., *Autos, Smog and Pollution Control: The Politics of Air Quality Management in California*, Edward Elgar, Brookfield, VT, 1995.

Henz, D.J., Air pollution standards and regulations, in *Handbook of Air Pollution Technology*, Calvert, S. and Englund, H., Eds., John Wiley & Sons, New York, 1984, p. 919.

Lee, B., Highlights of the Clean Air Act Amendments of 1990, *JAWMA*, 41, 48, 1991.

Leonard, L., *Air Quality Permitting*, Lewis Publishers/CRC Press, Boca Raton, FL, 1997.

Lloyd, A.C., Lents, J.M., Green, C., and Nemeth, P., Air quality management in Los Angeles: perspectives on past and future emission control strategies, *JAPCA*, 39, 696, 1989.

Maga, J.A., Emission standards for mobile sources, in *Air Pollution*, Vol. V, *Air Quality Management*, 3rd ed., Stern, A.C., Ed., Academic Press, New York, 1977, p. 506.

Mintzer, I.M., A Matter of Degrees: The Potential for Controlling the Greenhouse Effect, World Resources Institute Research Report 5, Washington, D.C., 1987.

National Research Council, *On Prevention of Significant Deterioration of Air Quality*, National Academy Press, Washington, D.C., 1981.

National Research Council, *Ozone Depletion, Greenhouse Gases and Climatic Change*, National Academy Press, Washington, D.C., 1989.

National Research Council, *Rethinking the Ozone Problem in Urban and Regional Air Pollution*, National Academy Press, Washington, D.C., 1991.

Newell, V.A., Air quality standards, in *Air Pollution*, Vol. V, *Air Quality Management*, 3rd ed., Stern, A.C., Ed., Academic Press, New York, 1977, p. 445.

Office of Technology Assessment, *Acid Rain and Transported Air Pollutants: Implications for Public Policy*, OTA-O-205, OTA, Washington, D.C., 1984.

Padgett, J. and Richmond, H., The process of establishing and revising ambient air quality standards, *JAPCA*, 33, 13, 1983.

P.L. 91-604, *Clean Air Act as Amended*, 42 U.S.C., 1857 et seq., 1970.

P.L. 95-95, *Clean Air Act as Amended*, 42 U.S.C., 1857 et seq., 1977.

P.L.101-549, *Clean Air Act as Amended*, 42 U.S.C., 1857 et seq., 1990.

Roberts, J.J., Air quality management, in *Handbook of Air Pollution Technology*, Calvert, S. and Englund, H., Eds., John Wiley & Sons, New York, 1984, p. 969.

Schneider, S.H., The greenhouse effect: science and policy, *Science*, 243, 771, 1989.

Schulze, R.H., The 20-year history of the evolution of air pollution control legislation in the U.S.A., *Atmos. Environ.*, 27B, 15, 1993.

Squillace, M.S. and Wooley, D.R., *Air Pollution*, 3rd ed., Anderson Publishing Co., Cincinnati, OH, 1999.

Stavins, R.N. and Whitehead, B.W., Market-based incentives for environmental protection, *Environment*, 34, 7, 1992.

Stenvaag, J.M., *Clean Air Act Amendments: Law and Practices*, John Wiley & Sons, New York, 1991.

Stern, A.C., Prevention of significant deterioration: a critical review, *JAPCA*, 27, 440, 1977.

Stern, A.C., Ed., *Air Pollution*, Vol. V, *Air Quality Management*, Academic Press, New York, 1977.

Stern, A.C., History of air pollution legislation in the United States, *JAPCA*, 32, 44, 1982.

Stewart, R., Stratospheric ozone protection: changes over two decades of regulation, *Nat. Res. Environ.*, Fall 1992, p. 24.

Tietenberg, T.H., *Emissions Trading: An Exercise in Reforming Pollution Policy*, Resources for the Future, Washington, D.C., 1985.

USEPA, *Air Pollution Episodes: A Citizen's Handbook*, GPO Publication 1972-0-452-729, GPO, Washington, D.C., 1972.

USEPA, *The CAA Amendments of 1990*, summary materials, EPA, Washington, D.C., November 1990.

USEPA, Implementing the 1990 Clean Air Act: the first two years, in *Report of the Office of Air and Radiation to Administrator William K. Reilly*, EPA 400-R-92-013, EPA, Washington, D.C., 1992.

Watson, A.Y., Bates, R.R., and Kennedy, D., Eds., *Air Pollution, the Automobile, and Public Health*, Health Effects Institute, National Academy Press, Washington, D.C., 1988.

Wolff, G.T., The scientific basis for a new ozone standard, *Environ. Mgr.*, September 27, 1996 (Air & Waste Management Association, Pittsburgh).

QUESTIONS

1. Despite enactment of clean air legislation and regulations, some property owners may wish to sue a source that is alleged to be depositing particulate matter on their property. What would they reasonably expect to gain from filing a civil suit, and what legal principle might they use?

2. What are air pollution control strategies and how do they differ from tactics?

3. What air pollution control strategies are used under federal clean air laws? What is their relative significance in air pollution control?

4. Describe the theoretical and practical uses of cost–benefit analyses, the bubble concept, and pollution charges in achieving pollution control objectives?

5. Describe the principle of using allowances to achieve objectives of acidic deposition control.

6. What is the primary advantage of controlling air pollution by using air quality standards? Emission standards?

7. What tactics are used to achieve air quality standards?

8. Describe differences between reasonably available and best-available control technologies. When is each used?

9. Describe a brief history of local, state, and federal efforts in controlling air pollution in the U.S.

10. What enforcement tools does the USEPA have available under the 1970, 1977, and 1990 CAA Amendments?

11. Under the CAA Amendments of 1970, Congress required that air quality standards be set and applied to every AQCR in the country. What agency or authority has responsibility for (a) setting air quality standards, and (b) enforcing air quality standards?

12. What is the importance of the following in achieving air quality standards: (a) AQCRs, (b) state implementation plans, and (c) emission standards?

13. The deadline for achieving air quality standards in all AQCRs in the U.S. was July 1, 1975. It is many years thereafter, and many AQCRs have yet to attain these standards. Why is this the case?

14. What options does an industry have if it wishes to build a new facility or expand an existing one in an AQCR where the NAAQS for a pollutant it emits has not been attained?

15. Both NSPS and NESHAP provisions of clean air legislation employ the "emission standard" concept of controlling air pollution. In practice, what differences are there in the application of NSPSs and NESHAPs?

16. Prior to the 1990 CAA Amendments, seven pollutants were regulated as hazardous pollutants. Under the Air Toxics provision of the 1990 CAA Amendments, an additional 182 substances were to be regulated under this concept. These additional substances are regulated using technology-based MACT standards. Describe what MACT standards are.

17. What are the responsibilities of federal and state agencies relative to air quality standards, NSPSs, NESHAPs, and regulation of motor vehicle emissions?

18. Describe the purpose and statutory requirements of PSD.

19. Describe regulatory authorities and actions used by the USEPA to protect the environment from O_3-destroying chemicals.

20. In the 1990 CAA Amendments, Congress required that a uniform permit system be developed. Describe the use and significance of permits in achieving air quality objectives.

21. Since 1990, a 10-million ton/year reduction in SO_x emissions has been achieved in the U.S. Describe how this was accomplished.

22. Briefly describe the statutory history of air pollution control requirements on light-duty motor vehicles.

23. The USEPA has authority to control fuel additives. Why and how has this authority been used to reduce the lead content of gasoline?

24. In addition to requiring emission reductions from motor vehicles, Congress has required petroleum refiners to produce gasolines that result in emissions reductions. Explain these requirements and their significance.

25. Despite the fact that most new cars and trucks have emissions that are a small fraction of those of motor vehicles built 35 years or more ago, motor vehicle air pollution is still a major problem. Why?

26. Describe how state boards and agencies function in implementing the requirements of federal legislation.
27. What is the Kyoto Protocol? Why is it so controversial in the U.S.?
28. How could provisions of the Kyoto Protocol as they would have applied to the U.S. be achieved without significant economic impact?
29. New air quality standards for $PM_{2.5}$ and O_3 were promulgated in 1997 but were not immediately implemented. Why was this the case?
30. What is new source review and how does it apply to coal-fired power plants?

CHAPTER 9

Control of Motor Vehicle Emissions

As seen in Chapter 2, transportation in the U.S. is the major emission source of three primary pollutants: hydrocarbons (HCs), carbon monoxide (CO), and nitrogen oxides (NO_x). Most of these emissions are associated with light-duty cars and trucks. Heavy-duty trucks, motorcycles, and off-road vehicles are also significant sources. In urban areas, motor vehicles are estimated to be responsible for 40 to 50% of HC, 50% of NO_x, and 80 to 90% of CO emissions. Similar patterns have been reported in Europe and Japan.

Motor vehicle pollutants such as nonmethane hydrocarbons (NMHCs) and NO_x serve as precursor molecules for the production of elevated tropospheric ozone (O_3) levels. Within the context of primary pollutants and subsequent formation of secondary pollutants such as O_3 and other smog and haze constituents, it is not surprising that motor vehicle emission control has been a major regulatory priority in the U.S. for more than three decades and an increasing priority in other developed and developing countries.

Regulatory efforts to control motor vehicle emissions (Chapter 8) have a history dating back to the early 1960s. Emission reduction requirements over that time have become more stringent, requiring motor vehicle manufacturers to continuously develop and employ new emission control techniques.

As a consequence of regulatory and technological efforts, aggregate CO and HC emissions from motor vehicles were reduced by ~40%, and NO_x by 25%, in the period from 1982 to 1997. This reduction occurred despite a substantial increase in vehicle miles driven.

9.1 MOTOR VEHICLE ENGINES

Two types of engine systems are used in motor vehicles: (1) Otto cycle spark ignition (SI) reciprocating internal combustion engines (ICEs), and (2) diesel cycle compression ignition (CI) ICEs. The former are widely used in cars and light-duty trucks, motorcycles, boats, and other small-horsepower consumer products. The latter are used in utility vehicles, e.g., trucks, earth-moving and mining equipment, farm tractors, and ships, where high-torque, low-speed performance is required. Because of their higher fuel economy, CI engines are widely used in light-duty vehicles in Europe.

9.1.1 Otto Cycle Spark Ignition Engines

9.1.1.1 Engine Characteristics

Otto cycle SI engines may be four-stroke or two-stroke systems. The former are used in cars and light-duty trucks; the latter, in motorcycles and lower-horsepower-requiring vehicles.

Operation of a four-stroke SI engine is illustrated in Figure 9.1. The cycle begins as a mixture of air and fuel is drawn or injected into the combustion chamber (i.e., the volume described by the top of the piston at the beginning of the compression stroke, cylinder walls, and intake valves); this is the intake stroke. In the compression stroke, the valves close and the piston moves up through the cylinder to top dead center where ignition takes place. The piston returns to its original position (power stroke) before it moves upward again to purge exhaust gases (exhaust stroke). As a consequence, there is one power stroke per two piston revolutions.

Ignition of the fuel–air mixture drives the piston downward, providing power to the drive shaft that propels the vehicle. Residence time (the time it takes to complete the power stroke) in a 3000-rpm engine is on the order of 10 msec.

In a two-stroke engine, the basic processes (intake, compression, heat induction, expansion, heat exhaust, gas exhaust) are essentially the same. They occur in half the cycle period of a four-stroke engine. This has the advantage of increasing the power-to-weight ratio and the disadvantage of allowing a shorter time period for each process. This results in a substantial fuel penalty (up to 25%) and higher HC emissions.

Figure 9.1   Four-stroke spark ignition engine operation.

9.1.1.2 Emissions

Pollutant emissions from motor vehicles powered by SI four-stroke ICEs include combustion by-products (exhaust gases), "blow-by" gases, and evaporative fuel losses.

Exhaust emissions consist mainly of unburned or partially burned HCs, CO, and NO_x. They comprise 90 to 92% of emissions from uncontrolled vehicles.

Some unburned HCs "blow by" piston rings and accumulate in the crankcase. In an uncontrolled vehicle, they would be vented to the atmosphere through a small tubular crankcase exhaust port.

Gasoline evaporative emissions from motor vehicle fuel systems are (in decreasing order of importance) diurnal, hot soak, and operating losses. Diurnal emissions occur when fuel tanks of parked vehicles cool at night, drawing air in and "exhaling" vapor-containing air on heating during the day. This fuel tank "breathing" may produce evaporative HC emissions of 50 g/car on hot days. Hot soak emissions occur just after the engine is turned off and residual fuel evaporates from the warm vehicle. Operating, or running, losses occur as vapors are driven from the fuel tank when the vehicle is being operated and the fuel warmed. Operating losses may be significant at elevated ambient temperatures or when the fuel system becomes hot during vehicle operation. Evaporative losses are also associated with vehicle refueling and may result in HC emissions comparable to those associated with the exhaust system.

9.1.1.3 Formation of Combustion Pollutants

A variety of by-products are produced in ICE gasoline combustion. Because of significant emission concentrations and effects on the environment, CO, HCs, and NO_x are of particular importance.

Gasoline consists of a variety of HC species. It has an average composition that can be described by the formula $(CH_m)_n$, where m has a value of ~1.85 to 2.0 and n is ~7.0. An HC molecule with m = 1.85 and n = 7 (C_7H_{13}) would require 10.24 molecules of oxygen (O_2) to be fully oxidized to carbon dioxide (CO_2) and water (H_2O) vapor. This concentration of gasoline HCs to atmospheric O_2 is the stoichiometric ratio and corresponds to a mass-based air:fuel (A:F) ratio of approximately 14.5:1.

The A:F ratio is often expressed as an equivalence ratio, φ:

$$\varphi = \frac{(F/A) \text{ actual}}{(F/A) \text{ stoichiometric}} \tag{9.1}$$

Note that the equivalence ratio is calculated from a fuel:air rather than an air:fuel ratio.

When an HC fuel is oxidized, CO is initially formed and then oxidized to CO_2 by O_2. If O_2 is present in less than stoichiometric proportions, combustion will be incomplete, with some CO remaining unoxidized. The amount of CO produced in exhaust gas is a function of the A:F or equivalence ratios. This relationship can be seen in Figure 9.2. As the A:F ratio increases (φ decreases), CO production significantly declines. At a slightly lean A:F ratio (i.e., less fuel, more air), with a φ of ~0.9, very little CO is produced. As the A:F mixture becomes richer (i.e., more fuel, less air), CO production rapidly increases.

As can be seen in Figure 9.2, the A:F ratio also affects emissions of HCs. However, significant emissions occur even when sufficient air is present due to quenching (cooling) phenomena within the combustion chamber. A severe temperature gradient forms as a result of the large quantity of heat that is transferred from combustion gases to cylinder walls and then to the engine's cooling system. Near cylinder walls, temperatures may be less than required to maintain a flame. The flame is "quenched" and thus goes out, leaving a small zone of uncombusted fuel. These "quench" zones exist in many locations in combustion chambers. A quench zone of 125 μm would represent 1.5% of chamber volume. "Crevice" zones of unburned HCs are also formed. These are spaces that are confined on several sides. They include the upper ring between the piston and wall, behind the rings, between the upper rings, within the spark plug screw threads, where the gasket has a small mismatch with the engine block and head, and where the different shapes of valves and seats form a small gap. Crevice zones can contain as much as 8% gasoline by mass and 3% of the cylinder clearance volume.

Another major source of unburned HCs is sorption on lubricating oils during intake and desorption on exhaust. It has been estimated that ~25% of unburned HCs come from lubricating oil sorption under stoichiometric conditions, and up to 50% under lean A:F ratios ($\varphi = 0.95$).

Major engine operating parameters that affect HC emissions include equivalence and compression ratios (chamber volume remaining at the end of the compression stroke compared to that at the end of the intake stroke), engine speed, and spark

Figure 9.2   Relationships between air:fuel ratios and pollutant emissions.

timing. Unburned HC emissions increase as the compression ratio increases due to the fact that crevice regions, being of constant volume, make up a larger portion of the compressed volume. Emissions of unburned HCs decrease with increased engine speed as the associated increase in turbulence enhances combustion and postcombustion mixing and oxidation. They also decrease as the spark is retarded as more combustion occurs and exhaust gases are hotter.

Formation of NO_x, which includes nitric oxide (NO), nitrogen dioxide (NO_2), and nitrous oxide (N_2O), is highly dependent on combustion chamber temperatures, which are directly related to A:F ratios. As can be seen in Figure 9.2, peak NO_x emissions occur under stoichiometric conditions when combustion chamber temperatures are high (2000 to 3000 K).

9.1.1.4 Emission Control Techniques and Systems

Both engine modifications and specific control technologies have been used to control motor vehicle emissions. These have included engine operation and design factors, engine-based control systems, and emission exhaust systems. Early control efforts were directed to reducing blow-by gases, evaporative emissions, and HCs and CO from exhaust systems. Control of HC emissions reflected U.S. Environmental Protection Agency (USEPA) policies that focused on reducing tropospheric O_3 production.

9.1.1.4.1 Engine Operation and Design

Significant reductions in emissions of CO, HCs, and NO_x have been achieved in SI light-duty motor vehicles by engine operation and engineering design changes. These include use of lean-burn combustion and electronic ignition; changes in spark timing, compression ratios, combustion chamber shape, and fueling systems; and engine temperature control.

9.1.1.4.1.1 Lean-Burn Combustion — When NMHC and CO exhaust emission standards were first required on 1966 model cars in California and 1968 model cars sold nationwide, most automobile manufacturers complied by employing improved combustion systems. The most important combustion modification was to use lean air:fuel mixtures (above the stoichiometric ratio). Carbon monoxide emissions were reduced significantly as A:F ratios were increased from ~12:1 to 16:1. NMHCs were also reduced, but less effectively than CO. Although lean A:F ratios increase combustion efficiency, they reduce engine performance and vehicle driveability and increase emissions of NMHCs (because of ignition failures).

To achieve optimal engine power production, the A:F or equivalence ratio should be set at ~10% rich. Since vehicle power was a major consumer concern in the 1950s and 1960s, vehicles were designed to operate at rich equivalence, or low A:F, ratios. Lean-burn modifications resulted in significant reductions in engine power. Though lean-burn combustion significantly reduces CO and HC emissions, the associated increase in peak combustion temperatures results in optimum conditions for NO_x production (Figure 9.2).

9.1.1.4.1.2 Electronic Ignition — One of the early operating changes was to use electronic ignition. Previous ignition systems had significant maintenance problems that resulted in excessive HC emissions and poor engine performance.

9.1.1.4.1.3 Spark Timing — Emission reductions can be achieved by changing ignition timing, an important performance variable. Optimally, it is set to maximize efficiency to what is described as MBT (minimum advance for best torque, or maximum brake torque). Optimum efficiency results when peak engine pressure occurs at 10 to 15° of top dead center of the compression stroke. Though actual spark timing varies considerably under different operating conditions, it may be retarded from the optimum to minimize emissions. It was one of the earliest engine operation adjustments used for emission control because of its ease of adjustment.

When the spark is retarded, exhaust temperatures are hotter and both CO and HCs continue to be oxidized in the exhaust system. Once the piston commences its descent (expansion or power stroke), the increased volume in the combustion chamber reduces the proportion of the A:F mixture in quench zones.

Unfortunately, spark retardation reduces engine efficiency. It was a major cause of fuel consumption increases associated with emission controls before the use of catalytic converters. After catalytic converters were introduced, ignition timing was adjusted to optimum settings.

9.1.1.4.1.4 Compression Ratios — In light-duty motor vehicles, compression ratios as high as 9 or 10:1 were once used to maximize power production and performance. However, they were decreased in new motor vehicles to reduce emissions of unburned HCs.

As compression ratios decrease, significant reductions (as much as 50%) in HC emissions are achieved. It is believed that these reductions are in part due to the decrease in the ratio of crevice zone:combustion zone volume. In addition, the lower expansion ratio (expansion volume:total combustion chamber volume) maintains exhaust gases at higher temperatures and thus promotes more complete HC oxidation.

Lower compression ratios decrease engine efficiency and power output. They also increase fuel consumption.

9.1.1.4.1.5 Combustion Chamber Shape — Combustion chamber surface area is a major HC emissions factor. A spherical shape has the smallest surface:volume ratio and would, in theory, be an ideal combustion chamber shape. It is approximated by using a hemispherical head and recessed bowl in the piston. This combustion chamber shape is used to achieve high performance with minimal knock. As a consequence, engine efficiency and burning rate improve, quench zone thickness decreases, and HC emissions decrease.

9.1.1.4.1.6 Fueling Systems — Throughout much of the history of light-duty motor vehicle manufacture, a suction-type carburetor was used to mix air and fuel and distribute the mixture to each cylinder (combustion chamber). This carburetor design has been largely replaced by single (SPI) and multiport (MPI) fuel injection systems. The large majority of automobiles manufactured in the U.S. since 1996 use MPI.

Fuel injection systems are readily adaptable to feedback control from measurements of exhaust parameters. An SPI works relatively well to control emissions but has problems maintaining correct A:F mixtures under transient conditions.

The MPI system improves fuel distribution between cylinders and its timewise distribution to each cylinder. Therefore, three-way catalytic converter systems are a viable option for emissions control. The MPI systems have a number of advantages: (1) they increase engine power and performance; and (2) since the intake manifold can be better designed (because it is for air alone), less heating of induction air is required and less performance loss occurs due to abnormal A:F ratios in cylinder-to-cylinder fuel distribution.

9.1.1.4.1.7 Engine Temperature Control — Engine temperature in the cylinders and induction system is an important parameter since a high proportion of emissions are produced when the engine is cold or only partially warmed. As a consequence, it is desirable to bring engines up to normal operating temperatures as quickly as possible. This is achieved by careful adjustment of cooling system thermostats and directing the flow of intake air across the engine. It is now common to use heat from the exhaust manifold to warm cool or cold intake air using a control valve.

9.1.1.4.2 Engine-Based Control Systems

Motor vehicle emissions can be treated or prevented from being formed by engine-based control systems. These include (1) crankcase ventilation to control HC-rich blow-by gases, (2) evaporative controls, and (3) exhaust gas recirculation (EGR) to control NO_x emissions.

9.1.1.4.2.1 Crankcase Ventilation — Blow-by, or crankcase, emissions are controlled by a relatively simple and inexpensive technology known as positive crankcase ventilation (PCV) (Figure 9.3). Gases slipping by the piston rings into the crankcase are returned to intake airflow by negative pressure in the intake manifold. There they are mixed with the incoming air–fuel charge and drawn into a combustion chamber and combusted. A one-way PCV valve regulates this flow, ensuring that it only moves from the crankcase and is not excessive. In modern emission control systems, ambient air is also routed to the crankcase so that it is continually purged.

9.1.1.4.2.2 Evaporative Emissions — Evaporative emissions from the carburetor and fuel tank are controlled by sorbing fuel vapors onto an activated carbon (C) bed in a small canister connected to the fuel system (Figure 9.4). On engine operation, collected vapors are stripped from the sorbent and directed into the combustion chamber. The efficiency of HC collection varies with fuel components. High-volatility, low-molecular-weight substances are less likely to be sorbed and retained than lower-volatility, higher-molecular-weight substances.

Figure 9.3   Positive crankcase ventilation.

Figure 9.4   Evaporative emissions control system.

9.1.1.4.2.3 Exhaust Gas Recirculation — Emissions of NO_x can be reduced to varying degrees by introducing a diluent that absorbs heat. Diluents must be readily available and neutral relative to other engine emissions; they should not decrease engine performance. Exhaust gases meet these requirements.

Heat sorption by CO_2 and H_2O reduces peak combustion temperatures and, consequently, NO_x production. Recirculation of a portion of exhaust gases by means of a control valve is used to control emissions of NO_x in many motor vehicles. On modern systems, an engine management microprocessor couples the flow of exhaust gases with other variables such as engine speed. An exhaust gas recirculation (EGR) system is illustrated in Figure 9.5.

The effectiveness of EGR in reducing NO_x emissions depends on the quantity of exhaust gases used. An EGR value of 10% may reduce NO_x emissions by 30 to 50%. Since combustion instability occurs above 15 to 20% EGR, this range represents its practical limit.

Recycled exhaust

EGR valve

Exhaust

Figure 9.5 Exhaust gas recirculation.

9.1.1.4.3 Exhaust Gas Control Systems

Exhaust gas emissions can be controlled by devices located downstream of the combustion chamber. These include thermal reactors (afterburners) and catalytic converters.

9.1.1.4.3.1 Thermal Reactors — Thermal reactors were used by some automobile manufacturers in the late 1960s to control emissions of HCs and CO. Thermal reactors are devices that under optimal temperature conditions, oxidize CO and HCs to CO_2 and H_2O in the exhaust system.

Reactors require relatively high temperatures (580°C, 1076°F) to oxidize 80% of HCs; they require 750°C (1382°F) to oxidize 80% of CO. To reach these temperatures, rich A:F mixtures are required so that the extra fuel can be burned in the reactor. Since exhaust gases include only small quantities of O_2, an air pump is needed to supply additional air. Due to the use of fuel-rich mixtures and spark retardation, thermal reactors have decreased thermal efficiency and increased fuel consumption.

9.1.1.4.3.2 Oxidizing Catalytic Systems — Oxidizing catalytic converters were first used on 1975 model cars in the U.S. to control emissions of HC and CO. Catalytic systems have the advantage of being able to operate efficiently at moderately low exhaust temperatures and optimal compression ratios and spark timing. Oxidation efficiencies of >90% can be achieved at temperatures of 300°C (572°F) for CO and 350 to 370°C (662 to 698°F) for HCs. Catalysts promote desired oxidation reactions at temperatures much lower than would otherwise be required.

Noble metals such as platinum (Pt), palladium (Pd), and rhodium (Rh) are preferred catalysts for catalytic converters. Most catalytic converter systems use some combination of these metals applied to the surface of small, 3-mm-diameter pellets, a ceramic honeycomb structure, or a metal matrix. Pellet-type systems

Figure 9.6 Movement of exhaust gases through a catalytic chamber.

have been largely replaced by honeycomb or matrix types (because of high pressure drops). Catalyst thickness on substrate surfaces is on the order of 20 to 60 μm. Gas flow through the converter is usually axial (Figure 9.6), but in some cases it is radial.

Oxidation of CO and HCs occurs best using lean A:F fuel mixtures since more O_2 is available for oxidation. Air pumps must be used to provide additional air with richer A:F mixtures. Control efficiencies on the order of 90% for HCs and 95% for CO can be achieved with φ values that are stoichiometric or on the lean side of stoichiometric.

The same catalytic materials promote the reduction of NO_x to nitrogen (N_2) and O_2. Reduction of NO_x is facilitated by removing O_2 (from NO_x) by reaction with CO and hydrogen (H_2). Rhodium is particularly effective in reducing NO_x, but Pt and Pd also work relatively well. Peak conversion of NO_x (90%) can be achieved at φ values that are ~0.5% richer than stoichiometric. Under leaner conditions, conversion efficiency declines significantly. For A:F mixtures 1.5% lean of stoichiometric, conversion efficiency drops to 20%. Therefore, φ values that are optimal for oxidation reactions are less than optimal for NO_x reduction.

For most of their history of use, catalytic converters have been oxidizing systems. This reflects the fact that (1) two of the three major pollutants in auto exhaust are subject to oxidation, and (2) oxidation catalysts obviate the detrimental effects of emission controls on fuel consumption and vehicle operability. Consequently, NO_x emission reduction had to be achieved by other means. EGR was the control technique of choice when oxidizing catalytic systems were used.

9.1.1.4.3.3 Reducing Catalytic Systems — Separate catalytic systems to reduce NO_x to N_2 and O_2 can be used in conjunction with oxidizing systems. A reducing catalyst must be placed upstream of the oxidation catalyst since exhaust gases passing through it need to be fuel-rich and contain CO as a reductant. Since the system is fuel-rich, additional air must be pumped into the exhaust gas stream. Platinum has been the catalyst of choice for reduction systems. Copper ion-exchange zeolites can be used to reduce NO_x under lean conditions; they have been used by Toyota since 1994. The durability of zeolite catalysts (at the time of this writing) has been a concern.

9.1.1.4.3.4 Three-Way Catalytic Systems — In three-way catalytic systems, oxidation and reduction occur in a single catalytic unit. The three-way catalyst is now the most widely used exhaust gas control system, as all three major pollutants are controlled in one bed.

For oxidizing catalysis, an A:F ratio on the lean side of the stoichiometric ratio provides the highest efficiency, whereas for reduction catalysis of NO_x, an A:F ratio that is stoichiometric to rich provides the highest efficiency. However, at φ values of 0.995 to 1.008, conversion efficiencies of >80% are possible for each of the three major pollutants. The effect of A:F ratios on catalyst efficiency in reducing emissions can be seen in Figure 9.7.

Moderately high temperatures (>300°C, 572°F) must be attained and maintained for efficient catalytic system performance. Such temperatures are only attained several minutes after engine operation begins. Under cold-start conditions, catalytic converter performance is relatively poor, resulting in higher exhaust emissions. Catalytic systems must therefore be constructed of materials that have a low specific heat but high thermal conductivity to quickly warm to operating conditions. Catalytic materials are mixed with aluminum oxide (Al_2O_3) to facilitate rapid warming.

A feedback system is required to maintain φ values within the optimum range. Oxygen sensors are used to monitor residual O_2 in exhaust gases. These sensors detect O_2 partial pressure and thus provide direct feedback to fuel injection processes. These sensors work well when φ ≥ 1.0.

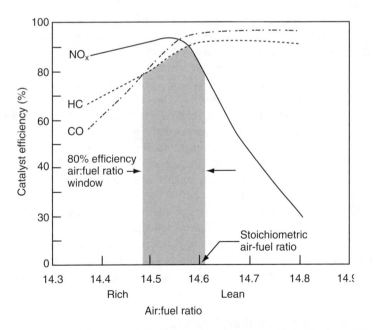

Figure 9.7   Relationship between air:fuel ratios and performance of catalytic converters. (From Sher, E., Ed., *Handbook of Air Pollution from Internal Combustion Engines,* Academic Press, San Diego, 1998. With permission.

9.1.1.4.3.5 Deterioration of Catalyst Performance — Catalytic system performance can be easily degraded. Degradation results from overheating and contamination of catalytic surfaces by substances in fuel (e.g., sulfur (S), phosphorous (P), and lead (Pb)).

Overheating occurs when engine malfunctions allow excessive fuel to pass into the exhaust system. Ignition failure for as little as 20 sec may produce sufficiently elevated temperatures to totally destroy the catalytic system.

Gasoline has to be formulated to minimize concentrations of P, S, and Pb to prevent contamination of catalytic materials. Phosphorous and S levels can be reduced or nearly eliminated at refineries. Lead was a significant concern, as it was added to gasoline to increase octane ratings and engine performance. Motor vehicles equipped with catalytic systems must therefore be operated on Pb-free gasoline.

Lead, S, P, and possibly other substances are said to "poison" catalysts. In fact, they simply coat the surface of catalytic materials, preventing contact with exhaust pollutants.

Catalytic converters can convert S to sulfur dioxide (SO_2) and hydrogen sulfide (H_2S), thereby producing malodors. Sulfur concentrations in year 2001 gasolines were ~300 ppmw/v.

9.1.1.4.4 Emission Controls on Currently Marketed Vehicles

Most, if not all, currently manufactured vehicles are designed with the following systems to control emissions of CO, HC, and NO_x: (1) PCV, (2) evaporative emission control systems, (3) electronic ignition, (4) a compact combustion chamber that minimizes wall effects, (5) a throttle position sensor to allow precise control of idling speeds, and (6) an engine management system.

In addition, special design features are incorporated for SPI and MPI system engines. In SPI engines, a catalytic system is used to oxidize CO and HC. It is placed close to the exhaust manifold. A secondary air supply to the exhaust manifold is provided by an air pump. Intake air is heated by transferring heat from exhaust gases. Intake temperatures are regulated by use of a thermostat and a flap valve. An EGR system is used to limit emissions of NO_x. Temperature control from the engine coolant is used to shut off secondary air to the exhaust during engine warm-up. It also limits purging of fuel tank vent canisters and EGR during cold conditions.

In MPI engines, a microprocessor-type engine management system is utilized. A three-way catalytic converter is placed immediately downstream of the exhaust manifold. The management system controls spark timing as determined by throttle position, engine speed, knock sensor, manifold vacuum, and engine temperature. Purging from the fuel tank vent canister is subject to temperature control from the engine coolant.

9.1.2 Compression Ignition Engines

Emissions reduction concerns associated with CI engines vary across the developed and developing world. In the U.S., they are primarily associated with light-

and heavy-duty trucks and commercial vehicles such as buses. Due to poor consumer acceptance, use of CI engines in automobiles in the U.S. is negligible. However, use of CI engines in automobiles in a number of European countries is significant (average of 20% of all passenger cars sold, with 42% of all passenger cars sold in France). In Europe, diesel cars have a reputation as fuel-efficient, durable vehicles. Lack of consumer acceptance in the U.S. has been due to the poor performance and reliability of early models built by U.S. manufacturers. CI vehicles are noisier and produce malodors and soot particles.

CI engines are naturally aspirated (air is inducted), turbocharged (inlet air is compressed by an exhaust-driven turbine–compressor combination), or super-charged (air is compressed by a mechanically driven pump or blower). Both turbocharging and supercharging increase engine power output by increasing air-flow and therefore fuel flow. Like SI engines, CI engines utilize either four- or two-stroke cycles. Fuel is injected under high pressure into a combustion chamber or bowl in the top of the piston toward the end of the compression stroke. The atomized fuel vaporizes, mixes with high temperature air, and, because of high compression ratios used (12 to 24:1, depending on engine type), spontaneously ignites. Once combustion is initiated, additional fuel mixes with air and is com-busted. Emissions of HC and CO are low since combustion is nearly 100% complete and the engine operates with excess air. Because temperatures produced are high, NO_x emissions are also high. In addition, the fuel-mixing process pro-duces C particles in fuel-rich regions where fuel is sprayed. Some of this elemental C passes through the combustion process unburned and sorbs high-molecular-weight HCs and sulfate S from the fuel and lubricating oil.

Because of lean engine operation, NO_x emissions cannot be controlled using three-way catalysts. Reduction of NO_x has been achieved by careful control of inlet air temperatures and injection retardation to delay most of the combustion process to the early expansion stroke. Nitrogen oxide emission reductions have been on the order of 50 to 65%.

Elemental C, or soot, formation has been reduced by 65 to 80% as a result of modifications to the combustion process. These have included use of fuel injection equipment with very high injection pressures and carefully matching the geometry of the bowl-in-piston combustion chamber, air motion, and spray. Lubricant control has reduced the concentration of high-molecular-weight sub-stances sorbed onto elemental C particles formed during combustion. Oxidation catalysts may also be used to reduce soluble organic components of particulate-phase emissions. Oxidation catalysts reduce emissions of both HC and CO by 80 to 90% and particulate matter (PM) by 50 to 60%. These catalysts are fitted to all European diesel light-duty motor vehicles to meet European stage 2 emis-sion standards.

Particulate-phase emissions can be reduced by filtration systems. A number of design approaches have been utilized in the last decade and a half. Soot filters are located on the tailpipe. They are subject to high cost and technical problems such as filter regeneration and avoidance of excessive backpressure. Soot filters typically work by removing and storing PM until a certain level of resistance is detected. Filters are then regenerated *in situ*.

9.2 AUTOMOTIVE FUELS

Different fuel types are used in SI and CI engines. In the former, HC mixtures called gasoline are used; in the latter, diesel fuel.

9.2.1 Gasolines

Gasoline or gasoline products are produced in crude oil refining. Due to different oil sources and refining procedures, gasoline products are complex mixtures of HC species that include paraffins, cycloparaffins, olefins, and aromatic HCs. Consequently, the composition of HC constituents varies from one refiner to another. It also varies as gasolines are formulated to provide optimum engine performance under different driving and climatic conditions. Individual constituents vary in ignition temperature and other combustion characteristics.

Gasolines may contain small quantities of non-HC substances. These include normal constituents of petroleum such as P and S or contaminants introduced in the gasoline production process (such as H_2O). Phosphorous and S reduce the effectiveness of catalytic converters; as such, additional refining steps are needed to reduce their levels. Water negatively affects engine performance.

Gasolines may contain additives such as Pb, Mn, or oxygenates. The use of Pb in gasolines has been restricted in many countries to minimize damage to catalytic converters and safeguard public health. Nevertheless, it is still used in a number of developing countries. A Mn-containing compound, methylcyclopentadienyl manganese tricarbonyl (MMT), is used in some countries as an octane booster. Refiners have promoted its octane-boosting potential and requested its approval in the U.S. Because of public health concerns, the USEPA has declined to approve its use. Oxygenates are used in the U.S. to boost gasoline octane ratings and decrease emissions of CO.

9.2.1.1 Octane Rating

Gasolines are formulated to prevent knock, the noise transmitted through the engine structure when spontaneous ignition of the end-gas (fuel, air, residual gas) mixture occurs before the propagating flame. Such ignition produces high local pressures and pressure waves with substantial amplitude. Although knock can be severe enough to cause major engine damage, in most cases it is just an objectionable source of noise.

The presence (or absence) of knock is a function of the antiknock quality of gasoline. A gasoline's resistance to knock is defined by its octane number. The higher the octane number, the greater the knock resistance. The octane scale is based on two HCs, heptane and isooctane. The former has an octane rating of zero; the latter, 100. Blends of these define the knock resistance of octane numbers between 0 and 100.

The antiknock quality of gasolines is determined by two different methods. These produce what are called motor octane numbers (MONs) and research octane numbers (RONs). The antiknock quality of a gasoline is determined by averaging its MON and RON values.

From 1923 until the advent of catalyst-equipped motor vehicles, Pb alkyls were added to gasoline to increase octane numbers. Their use allowed an increase in antiknock quality to be achieved at lower cost than modifying fuel composition by additional refining steps. Lead additives used were tetraethyl lead $((C_2H_5)_4Pb)$ and tetramethyl lead $((CH_3)_4Pb)$. As previously indicated, a Mn antiknock agent, MMT, has also been used.

Octane ratings can be increased by adding oxygenates such as ethanol (C_2H_5OH), methanol (CH_3OH), tertiary butyl alcohol, and methyl-tertiary-butyl-ether (MTBE). Octane ratings can also be increased by increasing the aromatic HC content of gasolines. Aromatic HCs such as benzene, toluene, ethylbenzene, and xylene (when combined, commonly referred to as BTEX) significantly enhance antiknock performance. The composition of aromatic HCs in unleaded gasolines has averaged about 30%, but has been as high as 40+%. Because of health concerns (it is regulated as a hazardous pollutant), benzene concentrations in gasoline have recently been limited to 1%.

9.2.1.2 Gasoline Composition and Emissions

Motor vehicle emissions are significantly affected by fuel constituents and formulation. A prime example of this is Pb. Its use was the major source of Pb in the environment and human exposure to it. Emissions are also affected by the presence of low-molecular-weight HCs such as butane that have a high Reid vapor pressure (RVP), or volatility. Fuels with high RVPs have increased evaporative, running, and refueling emissions even in evaporative emissions-controlled systems. Increases in what is described as midrange volatility lead to reduced HC and CO emissions, and increased NO_x emissions.

The olefin content of gasoline is an important emissions concern because of the reactivity of olefins and their role in producing elevated O_3 levels in urban–suburban areas. High olefin contents are therefore undesirable.

The aromatic HC content of gasolines affects emissions of CO, HCs, benzene, polycyclic aromatic hydrocarbons (PAHs), and NO_x. As aromatic HC concentrations in fuels increase, emissions of CO, HCs, benzene, and PAHs also increase; NO_x emissions, on the other hand, decrease.

9.2.1.3 Reformulated Gasolines and Oxygenated Additives

Gasolines described as reformulated have had their compositions changed to reduce the reactivity of exhaust products and decrease emissions of NMHCs, CO, and NO_x. Reformulated gasolines can be used in conventional ICEs with no modification of the propulsion system. They are produced by refining modifications and adding oxygenated compounds. They have lower olefinic and aromatic HC contents and lower RVPs. Oxygenated additives enhance octane ratings decreased when the aromatic and olefinic HC content is reduced. They improve the efficiency of combustion and reduce CO emissions. Commonly used oxygenated HC additives include MTBE, ethyl-t-butyl ether (ETBE), tertiary-amyl-methyl ether (TAME), CH_3OH, and C_2H_5OH. Though C_2H_5OH is widely used in the midwestern U.S., gasoline producers favor the use of ethers. Alcohols increase the vapor pressure of gasoline, and thus increase evaporative emissions. Use of oxygenates in gasoline (which may vary from 2 to 15%) results in increased emission of formaldehyde (HCOH), a potent mucous membrane irritant. Reformulated gasolines have been introduced into fuel markets where significant O_3 or CO problems exist.

Oxygenated compounds such as MTBE, ETBE, TAME, and C_2H_5OH have been added to gasoline for wintertime use in dozens of American cities to achieve emission reductions necessary to comply with the CO air quality standard. The typical oxygenate content is 15%. Introduction of MTBE into gasoline in 1992, and its use to reduce CO emissions during winter months, was followed by scattered health complaints (headaches, dizziness, nausea) from residents in some areas, most notably Fairbanks and Anchorage, AK. As a consequence, a number of epidemiological studies were conducted to determine whether use of MTBE in gasoline may have contributed to adverse health effects in exposed populations. These studies failed to show any significant relationship between MTBE use in gasoline and health complaints. Nevertheless, use of MTBE has been controversial in some areas of the U.S. It is the most widely used oxygenated compound in the national reformulated gasoline program that went into effect in 1995. The USEPA estimates that use of reformulated gasoline in nine targeted metropolitan areas reduced smog-producing emissions by 270,000 metric tons (mtons) (300,000 tons) annually and annual per-car emissions of CO by 13%, NMHC by 15%, NO_x by 3%, and "toxic" air pollutants by 24%.

MTBE use has also been associated with a new environmental issue: contamination of groundwater by leaking underground fuel storage tanks. Storage tank leakage has been a long-standing environmental problem. Since gasoline is insoluble in H_2O, gasoline and H_2O don't mix. As a consequence, gasoline–ground H_2O contamination is not a major "consumer" issue. Since MTBE is H_2O soluble and has a strong ethery odor and taste, contamination of ground H_2O by gasoline containing MTBE is a significant consumer and environmental concern in areas subject to such problems. Since MTBE is a potential carcinogen, several states have taken initial steps to restrict MTBE use in reformulated gasoline.

9.2.2   Alternative Fuels

A variety of fuels have been evaluated or are being used as lower-emission alternatives to conventional gasolines. These include alcohol–gasoline blends, C_2H_5OH, CH_3OH, liquefied natural gas (LNG), and liquefied petroleum gas (LPG). Such fuels have the potential to improve air quality, particularly in urban areas. Reduction in emissions and improvement in air quality depend on the type of fuel used and other factors. Alternative fuels may improve air quality by reducing mass emission rates from motor vehicles or reducing emissions of photochemically reactive HC compounds.

9.2.2.1   Alcohol Fuels

There has been considerable interest in using CH_3OH, C_2H_5OH, and alcohol–gasoline blends as motor vehicle fuels. Methanol is a colorless liquid with a low vapor pressure and high heat of vaporization. Both properties contribute to lower emissions under warm conditions, but tend to make pure CH_3OH-operated vehicles difficult to start when cold (thus resulting in increased emissions). Additives such as gasoline (5 to 15% by volume) are blended with CH_3OH to increase vapor pressure and improve starting under colder weather conditions. The most common blend, containing 85% CH_3OH by volume, is described as M85. Flexibly fueled and dedicated CH_3OH-fueled vehicles have been developed.

The major emission products (on a mass basis) from CH_3OH-fueled vehicles are CH_3OH, HCOH, gasoline-like NMHC components, and NO_x. Of greatest concern is HCOH. Even after catalytic reduction, M85 vehicles may emit ~48 to 64.5 mg HCOH/km (30 to 40 mg/mi), about three to six times more than conventional vehicles. Formaldehyde has high photochemical reactivity. Limited studies of flexibly fueled vehicles indicate that NO_x may be slightly reduced. However, dedicated CH_3OH-fueled vehicles use higher compression ratios and thus produce higher NO_x emissions. Evaporative emissions are lower and CH_3OH is less reactive than many constituents of gasoline.

Flexibly fueled vehicles can also run on C_2H_5OH. Air quality benefits associated with C_2H_5OH are less than those of CH_3OH. Ethanol is more reactive, producing 15% more O_3 on a C atom basis. Little information is available on emission characteristics of well-controlled C_2H_5OH vehicles; limited tests have shown high emissions. Use of C_2H_5OH as a gasoline additive, while simultaneously allowing an increase in vapor pressure, may not have significant air quality benefits. Increased evaporative emissions and increased emissions of C_2H_5OH and acetaldehyde would increase atmospheric levels of peroxyacyl nitrate (PAN).

9.2.2.2 Compressed and Liquified Gases

Both natural gas (which is >90% methane, CH_4) and propane (C_3H_8) can be used to power motor vehicles. Natural gas is compressed and stored at pressures of 4500 PSI (pounds per square inch) (2.32×10^5 mmHg) or liquefied for use as an automotive fuel. Because LNG must be cryogenically cooled and stored, its use has been relatively limited compared to compressed natural gas (CNG). Natural gas vehicles (NGVs) emit primarily CH_4. Because of its low photochemical reactivity, CH_4 has low O_3-forming potential. This benefit is reduced by CNG impurities such as ethane (C_2H_6) and C_3H_8 that may produce up to 25 times more O_3 than CH_4. The presence of olefins may further reduce the benefits of CNG use. Though combustion by-products such as aldehydes are quite reactive, mass emission rates are relatively small. Carbon monoxide emissions from the lean-burn operating condition of NGVs are very low — 90% less than gasoline-powered vehicles. Because of lean-burning conditions, CNG vehicles produce lower engine-out (released prior to the catalytic converter) emissions of NMHCs, NO_x, and CO, with higher fuel economy. Evaporative emissions from CNG vehicles are limited and contribute little to O_3 formation. In the U.S., CNG is used in fleet vehicles by some corporations and institutions. It is not available to consumers at U.S. service stations, but it is available at "petrol" stations in some countries.

LPG consists primarily of C_3H_8, a by-product of petroleum refining. It has many of the attributes of CNG but several disadvantages. These include limited supply and higher exhaust reactivity than CNG. However, it has a higher energy output per unit volume, and thus can be stored in smaller tanks.

9.3 LOW EMISSION AND ZERO EMISSION VEHICLES

The SI ICE has been the dominant propulsion system used in light-duty motor vehicles for a century. It is reliable, economical, and gives excellent performance. Because of pollutant emissions associated with SI engines, a number of alternative propulsion systems have been evaluated by motor vehicle manufacturers and others. These have included the stratified charge, gas turbine, Wankel, and Rankine cycle engines, and electric-powered and hybrid electric vehicles. Most have been shown to have significant limitations both from technical and economic standpoints.

The 1990 Clean Air Act (CAA) Amendments called for development of low emission and zero emission vehicles in air quality control regions where achieving the O_3 air quality standard has been, and will continue to be, difficult (e.g., south coast of California).

9.3.1   Low Emission Vehicles

In the 1990 CAA Amendments, light-duty motor vehicles were to meet more stringent emission limits that were to be phased in starting in 1994. These are referred to as Tier I standards. California was given authority to develop stricter requirements to address the more serious problems that it faces in urban areas. As a consequence, the USEPA approved California's low emission vehicle (LEV) program in 1993. California's LEV program is intended to require development of LEVs that would be phased into the vehicle population over a period of time. These would include transitional LEVs (TLEVs), LEVs, ultra LEVs (ULEVs), and zero emission vehicles (ZEVs). These would reduce, respectively, HC emissions from Tier I levels by 50, 70, 85, and 100%; CO emissions by 0, 0, 50, and 100%; and NO_x emissions by 0, 50, 50, and 100%.

The USEPA established a voluntary national LEV program under which manufacturers agreed to produce cleaner vehicles for sale in eight northeastern states and the District of Columbia beginning in 1999 and nationwide in the year 2000. As of 2003, two manufacturers offered and sold LEVs.

9.3.2   Hybrid Electric Vehicles

Hybrid electric vehicles (HEVs), as implied, combine features of the ICE with the battery and motor of an electric car. This combination approximately doubles the fuel economy of SI engine systems and decreases emissions by a comparable amount. Hybrid systems were developed to compensate for the limitations of battery technology (primarily limited driving range).

Figure 9.8 Hybrid electric vehicle schematic. (From Dept. of Energy, http://www.ott.doe.gov/hev/components.html.)

HEVs combine a power unit, a vehicle propulsion system, and an energy storage unit (Figure 9.8). Power units may include SI engines, CI direct injection engines, gas turbines, or fuel cells. Propulsion can result entirely from an electric motor in a series configuration, or the SI engine and electric motor in a parallel configuration. Energy storage may be achieved by use of batteries, ultracapacitors, or flywheels. Vehicles on the market at the time of this writing use SI engines with parallel configurations and battery storage.

In a series configuration, an SI engine drives a generator that charges a battery that powers an electric motor. In a parallel configuration, the driveshaft is turned by either the SI engine or electric motor. The series hybrid is generally more efficient but less powerful and has a shorter driving range than a parallel HEV. The parallel configuration provides performance that is relatively similar to that of a conventional light-duty vehicle.

An HEV is generally configured with the gasoline engine, electric motor, and electronic circuitry in the front and battery in the rear. As of 2003, three HEVs were marketed, the Toyota Prius, Honda Insight, and Honda Civic (Figure 9.9). These vehicles have been well accepted in Japan. In the U.S., federal and state incentives have been applied to offset some HEV purchase costs.

HEV technology can be applied to a variety of vehicles, including sport utility vehicles (SUVs) and small trucks. The HEV has the potential to receive widespread acceptance and use because it meets stringent emission limits for HCs, CO, and NO_x and has high fuel efficiency. High fuel efficiency would result in a significant decrease in CO_2 emissions, a major greenhouse gas. By increasing fuel economy and thus reducing CO_2 emissions, HEVs have the potential to significantly reduce global emissions of CO_2 in the coming decades.

Figure 9.9   Hybrid electric vehicle.

9.3.3   Electric Vehicles

Electric vehicles have been used for years in a variety of utility applications, ranging from golf carts and forklifts to small automobiles. Unfortunately, they provide only modest acceleration and power and a limited driving range before recharging. The electric car is considered a zero emissions vehicle. Even if power plant emissions are considered, an electric vehicle produces only ~0.5% of the emissions of the cleanest SI-equipped gasoline-powered vehicle.

READINGS

Amman, C.A., Alternative fuels and power systems in the long term, *Int. J. Vehicle Design,* 17, Nos. 5/6 (Special Issue), 1996.

Calvert, J.G., Heywood, J.B., Sawyer, R.F., and Seinfeld, J.H., Achieving acceptable air quality: some reflections on controlling vehicle emissions, *Science*, 261, 37, 1993.

Chang, T.Y., Alternative transportation fuels and air quality, *Environ. Sci. Technol.,* 25, 1190, 1991.

Charlton, S.J., Control technologies in compression-ignition engines, in *Handbook of Air Pollution from Internal Combustion Engines,* Sher, E., Ed., Academic Press, San Diego, 1998, p. 358.

Dekiep, E. and Patterson, D.J., Emission control and internal combustion engines, in *Handbook of Air Pollution Technology,* Calvert, S. and Englund, H., Eds., John Wiley & Sons, New York, 1984, p. 484.

Garret, T.K., *Automotive Fuels and Fuel Systems*, Vol. I, *Gasoline,* Pentach Press and Society of Automotive Engineers, Washington, D.C., 1991.

Harmon, R., Alternative vehicle-propulsion systems, *Mech. Eng.*, 105, 67,1992.

Hawley, J.G., Brace, C.J., and Wallace, F.J., Combustion-related emissions in CI engines, in *Handbook of Air Pollution from Internal Combustion Engines,* Sher, E., Ed., Academic Press, San Diego, 1998, p. 281.

Heck, R.M. and Farranto, R.J., *Catalytic Air Pollution Control,* Van Nostrand Reinhold, New York, 1995.

Heywood, J.B., *Internal Combustion Engine Fundamentals,* McGraw-Hill, New York, 1988.

Heywood, J.B., Motor vehicle emissions control: past achievements, future prospects, in *Handbook of Air Pollution from Internal Combustion Engines,* Sher, E., Ed., Academic Press, San Diego, 1998, p. 1.

Hockgreb, S., Combustion-related emissions in SI engines, in *Handbook of Air Pollution from Internal Combustion Engines,* Sher, E., Ed., Academic Press, San Diego, 1998, p. 119.

Medlen, J., MTBE: the headache of cleaner air, *Environ. Health Perspect.,* 103, 666, 1995.

Milton, B.E., Control technologies in spark ignition engines, in *Handbook of Air Pollution from Internal Combustion Engines,* Sher, E., Ed., Academic Press, San Diego, 1998, p. 189.

National Research Council, *Rethinking the Ozone Problem in Urban and Regional Air Pollution,* National Academy Press, Washington, D.C., 1991.

Olsen, D.R., The control of motor vehicle emissions, in *Air Pollution,* Vol. I, *Engineering Control of Air Pollution,* 3rd ed., Stern, A.C., Ed., Academic Press, New York, 1977, p. 596.

Peyle, R.J., Reformulated gasoline, *Automot. Engin.,* 99, 29,1991.

Watson, A.Y., Bates, R.R., and Kennedy, D., Eds., *Air Pollution, the Automobile and Public Health,* National Academy Press, Washington, D.C., 1988.

QUESTIONS

1. Describe types and sources of pollutant emissions from spark ignition engine-equipped motor vehicles.
2. What is the relationship between air:fuel ratios and emissions of HCs, CO, and NO_x?
3. How is the stoichiometric ratio related to emissions of HCs, CO, and NO_x?
4. How does the compression ratio used in a particular vehicle affect emissions?
5. What is the relationship between spark timing and emissions?
6. How do emissions differ in spark ignition and compression ignition engines?
7. What are blow-by gases? How are they controlled?
8. Describe factors that contribute to evaporative emissions. How are they controlled?
9. What is an equivalence ratio? What is its relationship to pollutant emissions?
10. Since both CO and HCs are products of incomplete combustion, why can't they be controlled by application of the same control principles?
11. Decreased compression ratios result in decreased HC emissions. Why?
12. Even with the best pollution control systems, highest emissions occur under what vehicle operating conditions?
13. Describe the principle of exhaust gas recirculation and the pollutant that it controls.
14. How does a thermal reactor work? What are its limitations in controlling pollutant emissions?
15. Describe how an oxidation catalyst works.
16. What pollutants are controlled using a three-way catalyst?
17. Why are lead, phosphorous, and sulfur levels in gasoline a concern?
18. How are particulate-phase emissions from CI engines reduced?
19. Why do gasolines have a relatively high octane rating?
20. Why was lead used in gasoline? What is being used to replace lead?
21. What are oxygenates? Why are they used in gasoline?
22. From an emissions standpoint, what makes gasoline reformulated?
23. Why use fuels such as CNG, LPG, or methanol in light-duty motor vehicles?
24. Describe how hybrid electric vehicles reduce emissions of HCs, CO, CO_2, and NO_x.

CHAPTER 10

Control of Emissions from Stationary Sources

Stationary sources, as seen in Table 2.1 and Table 2.2, are significant sources of pollutant emissions. These emissions are quite varied in both their nature and quantity, depending on the industrial and commercial activities involved. Pollutants may be emitted to the atmosphere through specially designed stacks and local exhaust ventilation systems or escape as fugitive emissions.

Emissions from stationary sources can be controlled, or their effects reduced, by application of one or more control practices. These include use of tall stacks, changes in fuel use, implementation of pollution prevention programs, fugitive emissions containment, and "end of the pipe" control systems that remove pollutants from waste gas streams before they are emitted into the ambient environment through stacks.

10.1 CONTROL PRACTICES

10.1.1 Tall Stacks

Smokestacks do not reduce source emissions. Rather, they reflect source efforts to reduce potential pollution problems by elevating emissions above the ground where they may be more effectively dispersed. When public concern was limited and control devices not available, a sufficiently sized smokestack was a relatively effective way of reducing ground-level concentrations and human exposures. The effectiveness of a smokestack in dispersing pollutants depends on its height, plume exit velocity and temperature, atmospheric conditions such as wind speed and stability, topography, and proximity to other sources (see Chapter 3 and Chapter 7).

Utility and smelter operations have used tall stacks (200 to 400 m, ~600 to 1200 ft) to reduce ground-level concentrations and thus comply with air quality goals or standards for more than three decades. Such stacks have been necessitated by the high sulfur dioxide (SO_2) emissions of these sources. The higher the stack, the greater the plume height, wind speed, and air volume between the plume and ground available for dilution. Though reducing ground-level concentrations, tall stacks disperse pollutants over large geographical areas.

10.1.2 Fuel Use Policies

Significant reductions in emissions from combustion processes can be achieved by changes in fuel use. Indeed, after passage of the 1970 Clean Air Act (CAA) Amendments, much of the early progress in emissions reductions in the U.S. was achieved by switching from high-sulfur (S) fuel oil and coal to cleaner-burning fuels such as natural gas and lower-S fuel oil and coal. The cleanest-burning fuel is natural gas.

These fuel use changes have not been problem-free. Supplies of low-S fuels such as oil have always been limited, and therefore, these fuels are more expensive. Low-S oil must be imported, increasing this country's reliance on sometimes unreliable foreign suppliers. Use of low-S western coals has decreased markets for high-S eastern coals.

Significant improvements in air quality have been achieved in Chicago and the surrounding area as a result of utility decisions to use nuclear fuels (rather than coal) to generate electricity. More than 40% of the electricity used in the greater Chicago region is produced in nuclear power plants. As a result of accidents at Three Mile Island (Pennsylvania) in the late 1970s and Chernobyl (Ukraine, in the former Soviet Union) in 1986, escalating construction costs, and opposition to nuclear power, no new nuclear power plants have been constructed in the U.S. for nearly 20 years. Nuclear power is receiving some new, albeit limited, attention because of global warming concerns. Reactor safety and radioactive waste disposal remain major environmental concerns. The events of September 11, 2001, suggest that nuclear power plants may be vulnerable to terrorist attacks and thus pose additional safety concerns.

10.1.3 Fugitive Emissions

Significant fugitive emissions of vapor- and particulate-phase materials occur within industrial and commercial facilities and from outdoor materials-handling activities. Fugitive emissions have, for the most part, gone unregulated. This lack of regulation reflected a dearth of information on the nature and extent of such emissions and a regulatory preoccupation with stack emissions, which were amenable to reduction using existing control technology.

Fugitive emissions are increasingly being controlled in industrial facilities. Such control reflects both economic considerations and the fact that fugitive emissions are often a significant problem. Under the bubble concept, which treats a single facility as a source, it may be more economical to control fugitive emissions than apply more expensive technology to control stack gases. A case in point may be a steel mill where fugitive emissions from process and road surface dusts become entrained on disturbance by vehicle movement. Such dusts can be (and are) controlled by periodic roadway surface wetting. Fugitive emissions control can be achieved by implementing a variety of operating practices, materials handling, and process changes, as well as technology.

Considerable potential exists to reduce fugitive emissions by equipment and process changes, equipment maintenance, and containment strategies. Examples of the latter include use of hoods and negative pressure within a facility to minimize the escape of hazardous materials, or construction and operation of negative pressure enclosures. Such enclosures are required under occupational safety and health rules for asbestos and lead (Pb) abatement projects. They have the benefit of reducing emissions to the outdoor environment. Under National Emissions Standards for Hazardous Air Pollutants (NESHAP) asbestos abatement requirements, friable asbestos materials must be wetted prior to and during removal operations to prevent visible emissions.

Fugitive emissions have received considerable attention under Title III, the air toxics provisions of the 1990 CAA Amendments, where maximum achievable control technology (MACT) standards require that plant-wide emissions of hazardous pollutants be controlled.

10.1.4 Pollution Prevention

As indicated in Chapter 8, pollution prevention as a waste control strategy was codified in the Pollution Prevention Act of 1990 and given new attention in achieving air quality goals under the CAA Amendments of 1990. Though pollution prevention has its origin in efforts to reduce solid waste disposal to the land, its application has been broadened to include other media as well.

The goal of all pollution prevention activities is to reduce waste generation at the source. The concept is applied both narrowly (focusing primarily on industrial processes) and broadly (focusing on such things as energy conservation, emissions from finished products, etc.).

At the level of an industrial or commercial facility, pollution prevention may include substitution of process chemicals (materials that are less hazardous to human health or the environment than those presently being used); enhanced maintenance of plant equipment; and changes in process equipment, plant operating practices, plant processes, etc.

10.1.4.1 Substitution

Examples of substitution include use of (1) n-methyl pyrrolidone (NMP) as an alternative for methylene chloride-based paint strippers, (2) reformulated gasoline, and (3) hydrochlorofluorocarbons (HCFCs) and hydrofluorocarbons (HFCs) for chlorofluorocarbons (CFCs). NMP is less toxic and less volatile than methylene chloride. Use of reformulated gasoline to achieve air quality goals was discussed in Chapter 9, and use of HCFCs and HFCs to protect the ozone (O_3) layer in Chapter 8.

10.1.4.2   Process Equipment Changes

A variety of process equipment changes can be used to reduce pollutant emissions. These include the use of completely enclosed vats in place of open vats where solvent emissions occur and, in the case of coke ovens, retrofitting or replacing leaky coke oven doors with state-of-the-art seals, and use of flexible doors and automatic spotting devices to reduce damage to doors and jambs.

10.1.4.3   Plant Operating Practices

Excess production of pollutants and subsequent emissions may occur as a result of poor equipment and plant operation. This is particularly true in the operation of boilers and other combustion equipment. Good boiler and incinerator operation requires adequate supplies of combustion air. Insufficient combustion air results in incomplete combustion, with the production of elemental carbon (C) particulate matter (PM) and a variety of gas-phase substances. Good operating practices are particularly important in achieving emission reductions for which pollution control equipment is designed.

10.1.4.4   Maintenance

The performance of equipment in terms of producing products and releasing contaminants to the environment may be compromised by inadequate maintenance. It is particularly important to maintain combustion equipment in good operating condition. Maintenance is necessary to reduce leakage of solvents and other chemicals from vats, valves, and transmission lines. It is also necessary to reduce spill-related emissions and maintain adequate seals around coke oven doors in coke oven operation.

10.1.4.5   Process Changes

In many cases, pollutant emissions are related to processes used in product manufacturing. Emissions of solvent vapors, for example, can be reduced in painting operations by dry powder painting. In this process, specially formulated, thermoplastic or thermosetting, heat-fusible powders are sprayed on metallic surfaces that are subsequently heat cured. Dry powder painting can also be conducted electrostatically. In either case, significant reductions in hydrocarbon (HC) emission are achieved.

10.1.4.6   Energy Conservation

Reduction in energy use by application of a variety of energy conservation measures can result in significant emission reductions. All fuels, when fully or partially combusted, produce by-products that pollute the atmosphere. Any measure that reduces energy consumption associated with fossil fuel use will result in a reduction in both fuel use and associated emissions.

Energy conservation measures may include the (1) manufacture, sale, and use of fuel-efficient motor vehicles; (2) development and use of energy-efficient combustion and heat-recovery systems in industry and homes; (3) use of mass transit and car pooling; (4) construction of new buildings and retrofitting existing buildings to reduce energy loss; (5) manufacture and use of energy-efficient appliances and equipment; (6) use of secondary (as compared to primary) materials in product manufacturing; and (7) recycling and reuse of materials, etc.

Significant reductions in energy usage have occurred in varied economic sectors in the U.S. since several energy crises were experienced in the 1970s. Most notable have been the improvement in motor vehicle fuel economy and energy-use reductions in manufacturing facilities. Programs with less impact but of some note are USEPA's Energy Star Computer and Green Light Programs.

The Energy Star Program is a voluntary partnership between the USEPA and computer manufacturers. Computers identified with the Energy Star logo are designed to power down automatically when not in use. The Green Light Program encourages voluntary reductions in energy use through more efficient lighting in buildings.

10.2 GAS CLEANING TECHNOLOGY

Compliance with emission limits required under state implementation plans (SIPs) often necessitates the application of pollution control systems. Because of the expense and complexity that may be involved, selection of appropriate equipment requires a careful engineering evaluation of the nature of the emission problem, including physical or chemical characterization of pollutants, process conditions, and determination of gas discharge rates. Selection of a control system will, in good measure, be determined by its performance relative to achieving mandated emission limits. It also involves consideration of capital, operating, and maintenance costs.

10.2.1 System Performance

Gas cleaning systems may be designed to remove PM, individual gases, or both. Control system design and performance is determined by specific control requirements. Performance can be quantified by determining collection efficiency, which is defined as a percentage of influent PM or gas collected or removed from the waste gas stream. Collection efficiency (E) can be calculated from Equation 10.1:

$$E = 100 \ (1 - B/A) \tag{10.1}$$

where A is the concentration of the pollutant entering the control system and B is the emission level achieved or emission reduction required. If the influent to a particle collector were 600 lb PM/h and effluent gas emissions 30 lb PM/h, the calculated collection efficiency would be 95%:

$$E = 100 \ [1 - (30/600)] = 95\%$$

Emission limits or standards generally do not specify the collection efficiency required of control systems. Therefore, performance must be calculated from emission data and applicable emission limits. For example, if process emissions (A) were 600 lb/h and allowable emissions (B) 60 lb/h, the collection efficiency necessary to comply with the emissions standard would be 90%. To minimize costs, industrial sources usually select control equipment that meets only the emission limit mandated.

10.2.2 Particle Collection Efficiency

The major determinants of particle collection efficiency are particle size and control systems employed. Some control systems are inherently more efficient than others. One or several of the following collection principles may be utilized for specific control applications: inertial separation, electrostatic precipitation, filtration, or scrubbing. The overall efficiency and collection techniques for commonly used particle collectors are summarized in Table 10.1 and discussed in the following sections.

Variation in collection efficiency associated with particle size is illustrated for a hypothetical control device in Figure 10.1. Note that fractional collection efficiency decreases sharply for small particles, whereas over a broad range, large particles are collected at efficiencies that approach 100%. Fractional collection efficiencies for commonly used particle control systems are indicated in Figure 10.2.

Although fractional collection efficiency is important in equipment design, performance-based (i.e., overall) collection efficiency is the principal determinant of compliance with emission limits. As shown in Figure 10.1 and Figure 10.2, particle collection efficiency decreases significantly with decreasing particle size, particularly with particles in the submicrometer range. Paradoxically, it is these small particles that have the greatest human health significance.

As shown in Table 10.1, average overall collection efficiencies are dependent on the collection principle used. Under actual operating conditions, performance may vary considerably from the average values of Table 10.1. This variation is due to a number of factors, e.g., gas flow rate, PM loading, and particle properties such as density, shape, viscosity, and resistivity (see Section 10.3.2).

Table 10.1 Collection Efficiencies of Particle Control Devices

Equipment Type	Overall Collection Efficiency, %
Cyclone	
Medium efficiency	68–85
Multitube collector	70–90
Electrostatic precipitator	70–99.5+
Fabric filter	99.5+
Wet scrubber	
Spray tower	90
Baffle plate	95
Venturi	99.5+

Figure 10.1 Generalized fractional collection efficiency of particle collecting devices.

Figure 10.2 Comparative fractional collection efficiencies of particle collecting devices.

10.2.3 Gas-Phase Contaminant Collection Efficiency

The collection efficiency of control systems designed to control gas-phase substances depends on gas treatment techniques employed. Commonly used control principles include thermal oxidation, adsorption, and absorption. In general, the collection efficiency of control systems employing these control principles will be influenced by such factors as gas flow rates and pollutant concentrations. Other factors affecting efficiency may be specific to the treatment process or equipment design.

10.3 CONTROL OF PARTICULATE-PHASE POLLUTANTS

10.3.1 Cyclonic Collectors

Cyclonic systems (i.e., cyclones) are widely used particle collectors. They employ the principle of inertial separation to remove particles from waste streams. Collection efficiencies vary from relatively low for large, simple (i.e., common) cyclones to moderate–high efficiency in multiple-cyclone systems. Cyclonic particle collection systems are characterized by low capital, operating, and maintenance costs.

In a simple, single cyclone, particle-laden air enters an inlet near the top of the collector. Cyclonic flow is induced by the design of the entry, which may be tangential, helical, involute, or vane axial. In simple vane axial systems, air enters the top of the collector and a downward-spiraling vortex is produced whose nature is determined by the shape of the system. This curvilinear flow produces centrifugal forces that cause larger suspended particles (particularly those >10 μm aerodynamic diameter) to move radially outward until they strike the collector wall. They are caught in a thin laminar area of air near the wall and fall downward by gravity into a collection hopper. As pollutants and waste gases continue along their path downward, the narrowing cone-shaped section produces an increase in rotational velocity that confines particles to the wall area and minimizes their reentry. At the bottom of the cone, the vortex reverses direction and cleaned gases move upward and enter the center of the collector. This is a reverse-flow cyclone; other cyclones are designed as straight-through devices.

The particle collection efficiency of cyclones is proportional to the square of the tangential gas velocity and inversely proportional to their radii. Collection efficiency can be optimized by using high inlet velocities, small diameters, and long cylinders. Smaller diameters, however, increase pressure drop and operating costs.

Large individual cyclones such as those serving grain elevators (Figure 10.3) accommodate gas flows up to 18.9 m^3/sec (40,000 actual cubic ft/min (acfm)). Parallel arrays of cyclone tubes are commonly used for higher gas flows and collection efficiencies. These cyclone tubes vary in diameter, with flow capacities up to 2.1 m^3/sec (4500 acfm).

In multiple-cyclone systems, a large number of cyclone tubes (Figure 10.4) have a common gas inlet and outlet in a larger chamber that houses them. Cyclonic motion is imparted by axial inlet vanes. Overall efficiency is determined by the arrangement of individual tubes and their diameters.

The collection efficiency of all cyclones increases significantly with increasing particle size. In common cyclones, collection efficiency is <50% for particles of <5 μm and up to 99% for particles of >40 μm. Efficiency also increases as particle density and loading increase.

Cyclonic collection systems are commonly used to collect PM from grinding, crushing, sanding, conveying, machining, and materials-handling operations. Depending on the volume of waste gas and nature of the particulate dusts produced, cyclones may be used singly (e.g., woodshop), as arrays of simple cyclones (e.g., grain handling), or as arrays of many tubular cyclones in a housing (e.g., boiler soot collection,

Figure 10.3 Cyclone collectors on grain silos in Galveston, TX, harbor.

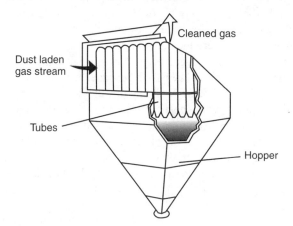

Figure 10.4 Multitube high-efficiency cyclone collector.

precleaners). Multiple-cyclone systems are often used on coal-fired boilers as precleaners for more efficient downstream systems such as fabric filters. They are commonly used downstream of wet scrubbers to collect particle droplets and serve as demisters. In such applications, gas entry occurs near the bottom of the collector.

10.3.2 Electrostatic Precipitation

Electrostatic precipitators are widely used particle collecting systems, particularly where waste gas streams have large, steady volumetric flow rates. They are commonly used to remove fly ash from high-S-coal-using power plants. Historical applications have included collection of sulfuric acid (H_2SO_4) mist (as early as 1907) and collection of metal oxides and particle dusts from a variety of metallurgical operations, including ferrous and nonferrous metal processes.

Electrostatic precipitators remove solid or liquid particles from waste gases passing through an electrical field where negative ions are produced from high-voltage wires or plates and imparted to entrained particles. The negatively charged particles are then collected on positively charged collection plates.

One or more configurations are used in electrostatic precipitator design. These are the (1) plate-wire precipitator, (2) flat-plate precipitator, (3) tubular precipitator, (4) wet precipitator, and (5) two-stage precipitator. The plate-wire precipitator is the most commonly used.

In the plate-wire precipitator (Figure 10.5), effluent gases flow between vertical parallel sheet metal collection plates and high-voltage wire electrodes. Plates are spaced ~30 cm (12 in.) apart, and wire electrodes, spaced ~5 cm (2 in.) apart, are at the center of the channel between collection plates. Each collection plate (present in series) may be up to a meter (3 ft) in length and 3 to 6 m (10 to 20 ft) high. The path length through the collector may be ~6 m (~20 ft) or more.

A high direct current (DC) voltage with negative polarity is applied to the wire electrodes. This causes gas molecules to break down, producing a corona. Ions generated in this corona follow electrical field lines from the electrodes to the collecting plates. Particles intercept these ions and become negatively charged. As they pass each downstream electrode, they are driven closer and closer to the positively charged plates, where they are collected.

With time, the collecting surface becomes coated with particles that reduce the electrical potential of collection plates. As a consequence, collection plates must be periodically "cleaned" by mechanical rapping or wet washing. Particles may become reentrained during rapping, with subsequent loss of collection efficiency.

Although rapping losses are important, collection efficiency is primarily influenced by particle retention time in the electrical field and particle resistivity (resistance to taking a charge). Retention time is determined by gas path length (distance gas travels across the electrical field) and gas velocity. Highest collecting efficiencies are achieved in precipitators employing long gas path lengths and relatively low gas velocity.

When particle resistivity is high, as is the case of fly ash from low-S western coals, the charge is not neutralized at the collection plate; as a result, an electrical potential builds up on the collected particles. As the potential increases, the incoming particles receive less than the maximum charge. As a result, collection efficiency decreases. If this condition continues, a corona discharge may occur at the collection plate, which may cause the collector to malfunction. Ironically, low-S coals, widely used to comply with SO_2 emission standards in many coal-fired utility boilers, produce fly ash of high resistivity. High-S coals produce fly ash with low resistivity; thus, collection efficiency is high. As a result of the resistivity problem, fabric filtration is often used in power plants burning low-S coal.

On the other end of the spectrum are particulate dusts with low resistivity (e.g., carbon black). Collection efficiency is decreased because C particles are highly conductive and lose charges too readily. As collected particles lose their negative charge, they become reentrained in the waste stream. They become negatively

Figure 10.5 Electrostatic precipitator.

charged and are again attracted to collection plates. The process is repeated until particles reach the precipitator outlet where they are discharged to the atmosphere.

Flat-plate precipitators are used in smaller volumetric flow rate applications and for small (1 to 2 μm), high-resistivity particles. Plates instead of wires are used for the high-voltage electrodes. They increase the strength of the electrical field and provide an increased surface area for particle collection. The distance between collection plates is ~16 cm (6.5 in.).

Electrostatic precipitation is a widely used particulate dust collection method because of its high collection efficiency for a range of particle sizes, including submicron particles (Figure 10.2). Collection efficiencies in excess of 99% can be achieved even on very large waste gas flows. Electrostatic precipitators have low operating and power requirements since energy is only needed to impart charges onto particles collected, and not the entire waste stream. Energy costs are therefore low compared to other high-efficiency particle collection systems.

10.3.3 Filtration

In filtration systems, solid particles are physically collected on fibers on the inside of bags (fabric filters) or on filter beds. Particle collection occurs as a result of sieve action, impaction, impingement, and diffusion onto filter fibers. As the filter becomes soiled, a mat of collected particles is produced that increases the filtration system's collection efficiency.

10.3.3.1 Fabric Filters

Fabric filters are employed to control particulate dust emissions from industrial sources such as cement kilns, foundries, oil-fired boilers, carbon black plants, and electric and oxygen furnaces for steel-making operations. They are also used to collect fly ash from electric-generating plants burning low-S coal. Fabric filters are commonly used to control particulate emissions when dust loading is high, particle sizes are small, and high collection efficiencies are required. Fabric filters are designed to handle gas volumes in the range of 4.7 to 23.6 m^3/sec (10,000 to 50,000 acfm).

Fabric filter collection systems consist of multiple tubular collecting bags suspended inside a housing. A single housing, called a baghouse, may contain several hundred to several thousand bag filters. Bags are made from a variety of fibrous materials. Fabric choice depends on temperature, moisture, and chemical composition of the waste gas, as well as the physical and chemical nature of particles collected. Filter bags are made of both woven and nonwoven materials. Glass fiber bags are commonly used for high-temperature or high-corrosion-potential gas cleaning. In the typical baghouse unit (Figure 10.6), waste gas enters from the side and flows downward toward a hopper where the flow is reversed (upward) into the bag array. As the gas stream changes direction, large particles are removed by inertial separation and collected in the hopper below. As gas passes through the tubular bags, particles are collected on the inside of the bag surface and filtered gas is discharged to the atmosphere.

Fabric filtration is similar to the process employed in a home vacuum cleaner except that positive, rather than negative, air pressure causes dirty air to pass into the collection bag. Like a vacuum cleaner, collected dust must be periodically removed. Bags may be cleaned by automated mechanical shaking, reverse air cleaning, or pulse jet cleaners. In each case, cleaning is conducted to dislodge collected particles into a hopper where they are removed from the collection system. Collection efficiencies are lowest when filters are first installed and immediately after cleaning. Maximum efficiency and pressure drop occur when particle mat buildup has occurred.

Although fabric filters provide high overall collection efficiencies (99+%) and represent the best available technology for controlling submicron particles, they have limitations. These include high capital and operating costs, flammability hazards for some particulate dusts, high space requirements, flue gas temperatures limited to 285°C (545°F), and sensitivity to gas moisture content. Operating costs are high because of the high pressure drop that develops as particulate mats form.

Because of the abrasiveness of PM and sensitivity to chemicals, bag wear is a major maintenance concern. Bag life varies with the nature of the application. Bags are commonly changed every 12 to 18 months.

10.3.3.2 Wet Scrubbers

Particulate dust collection by scrubbing involves introduction of a liquid into the effluent stream, with a subsequent transfer of particles to the scrubbing liquid. This transfer increases particle size and traps particles in a liquid film. Both transfer mechanisms are employed in commercially available wet scrubber systems.

Scrubber designs vary from one manufacturer to another. However, all scrubbers have two basic components: (1) a section where liquid–gas contact occurs, and (2) a deentrainment section where wetted particles are removed. Particles in scrubbers come in contact with liquid droplets to form a particle–liquid agglomerate. The contact process is achieved by forcing a collision between liquid droplets and particles. Collisions may be promoted by gravity, impingement, and mechanical impaction. When contact is made, particles significantly increase in size and mass. The resultant particle–gas agglomerates are removed by inertial

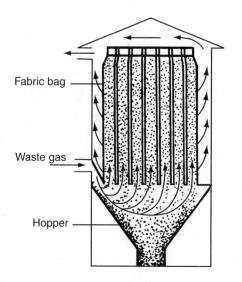

Figure 10.6 Fabric filter (baghouse) particle collection system.

devices. Commonly used deentrainment mechanisms include impaction on extended baffles and centrifugal separation. In most systems the contact liquid is water (H_2O).

10.3.3.2.1 Spray Towers

In the open tower design, a scrubbing liquid is sprayed downward as preformed drops through low-pressure nozzles. The PM-laden waste gas enters from the side and moves downward. Waste gases then change direction and move countercurrent to the flow of the scrubbing liquid and are discharged at the top. Dust particles are captured by the falling droplets. The liquid–particle agglomerate is collected in the liquid pool at the bottom of the tower. Spray towers are limited to low gas flows to prevent liquid drop entrainment in the scrubbed gas stream. These scrubbers achieve relatively low collection efficiencies (80 to 90%) and are usually employed as precleaners to remove particles of >5 µm.

10.3.3.2.2 Venturi Scrubber

The Venturi scrubber is used when high collection efficiencies are required. In the system illustrated in Figure 10.7, effluent gases enter the Venturi section, where they flow against the wetted cone and throat. Waste gas velocity increases as it enters the annular orifice of the Venturi. Resulting high velocities produce a shearing action that atomizes the scrubbing liquid into many fine droplets. The high differential velocity between the gas stream and the liquid promotes impaction of dust particles

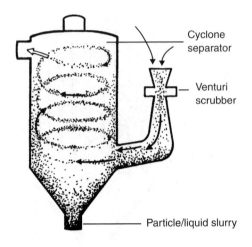

Figure 10.7 Venturi scrubber particle collection system.

on the droplets. As waste gas leaves the Venturi section, it decelerates, and impaction and agglomeration of particles and liquid continue to occur. Particle–liquid droplets are removed from the waste stream by centrifugal forces in a cyclonic deentrainment section.

10.3.3.2.3 Cyclonic Scrubbers

Scrubbing liquids can be injected into cyclonic airflows to increase particle collection efficiencies. The rotating gases cause particles entrained in droplets to migrate to scrubber walls where they are removed by impaction.

10.3.3.2.4 Operating Considerations

Because of their high collection efficiencies, wet scrubbers are used to control fine particles. Venturi scrubbers, for example, are commonly used to collect particles from basic oxygen steelmaking processes.

Although capital costs are relatively low, operating and maintenance costs are often high. These high operating costs are primarily the result of energy requirements needed to provide high collection efficiencies for fine particles and overcome high pressure drops. Additional operating costs are associated with disposal of collected PM and waste liquid. High maintenance requirements often result from the erosion and corrosion of scrubber surfaces produced by abrasion and corrosive acids.

10.4 CONTROL OF GAS-PHASE POLLUTANTS

Public and regulatory concern for the control of PM emissions, for the most part, predated concern for gas-phase pollutants by several decades. In the collection of particulate dusts, specific collection principles can be utilized for a variety of collection problems and sources; the major consideration in selecting control systems is the performance required. Application of control technology for gaseous pollutants cannot be generically applied. Gas cleaning systems must be developed or designed to control specific gas-phase pollutants (or pollutant categories) that vary in their chemical and physical properties. Control of some gas-phase pollutants is a relatively complex activity. It is also an inherently more expensive one, as both capital and operating expenditures for control equipment are often high.

Major physical and chemical principles used to remove gas-phase contaminants from effluent gas streams include thermal oxidation (combustion), adsorption, and absorption. Increasingly, biological systems are being used to control emissions of hazardous air pollutants (HAPs).

10.4.1 Thermal Oxidation

Thermal oxidation is widely used to remove combustible vapors from waste gas streams. The term *incineration* is commonly used to describe such processes. However, its use is problematic since it is also associated with combustion of municipal, medical, and industrial wastes. In the air pollution control sense, incineration is limited to combustion or thermal oxidation of gas- and particulate-phase substances in waste air streams associated with various processes, including the burning of municipal, medical, and industrial waste.

Thermal oxidation as applied to pollution control is a high-temperature or catalytic process in which oxygen (O_2) is combined with organic pollutants to convert them to simpler substances such as carbon dioxide (CO_2) and H_2O vapor, releasing heat in the process.

In most cases, thermal oxidation is applied to dilute streams of waste gases containing HCs or carbon monoxide (CO), such as those associated with paint spraying and baking operations, solvent use, and incompletely combusted gases from other combustion processes (e.g., waste incinerators). It is also used to control organic aerosols, malodorants such as hydrogen sulfide (H_2S) and mercaptans, and combustible gases produced in refineries. Dilute HC airstreams below the lower flammable limit (LFL), better known as the lower explosion limit (LEL), cannot propagate a flame. For many HC–air mixtures, the LEL varies from 1 to 5% HC concentration and >15% O_2. As such, they must be subject to sufficiently high temperatures (800°C, 1500°F), or to catalytic materials at lower temperatures, for complete oxidation to occur.

Three types of systems use the principle of thermal oxidation to control pollutants. These are flare systems, conventional thermal oxidizers, and catalytic systems.

10.4.1.1 Flare

Direct flare systems are used to combust waste gas streams that are HC-rich, with concentrations above the LEL and below the upper explosion limit (UEL). Under these conditions, HCs ignite and propagate a flame. Such waste gas streams are common in chemical processing plants and refineries (Figure 10.8). They pass through stacks that may be 30 to 40 m (100 to 130 ft) or so in height.

Figure 10.8 Refineries in Galveston, TX, using flare incineration.

HCs, released intermittently or as a result of an emergency, are ignited by pilot lights (or some other ignition system). Combustion occurs continuously as long as HCs above the LEL continue to flow through the system. Air jets at the tip of the tower promote turbulent mixing of effluent gases and the immediate atmosphere. Operated properly, flares can achieve HC oxidation efficiencies of 99% or more.

10.4.1.2 Conventional Oxidizers

Conventional oxidizers are used to treat relatively dilute HC effluent gases (often only a few hundred parts per million (ppmv)). Complete oxidation is facilitated by combustion at high temperatures (660 to 1100°C, 1200 to 2000°F), with sufficient residence time (0.5 to 2 sec) and turbulent mixing of effluent gases in air.

Thermal oxidizers are often used as afterburners; i.e., they are used downstream of primary combustion chambers. Afterburners are used on municipal and hazardous waste incinerators.

The use of afterburners as a part of hazardous waste incineration systems is standard practice. They are often required to have removal efficiencies of 99.99% of the principal organic hazardous constituents. U.S. Environmental Protection Agency (USEPA) operating permits often require the use of high temperatures (~1100°C, 2000°F).

10.4.1.3 Catalytic Oxidation

Catalytic oxidizers are used industrially to reduce both the size of control equipment and fuel use. Catalysts reduce temperature requirements for efficient combustion. Catalytic oxidizers operate in the range of 165 to 440°C (300 to 800°F).

Catalysts promote combustion but are not consumed in the process. Catalytic materials are applied to ceramic or metal substrates that have very high surface areas. Catalysts are usually noble metals such as platinum (Pt) or palladium (Pd), but may include chromium (Cr), manganese (Mn), copper (Cu), cobalt (Co), or nickel (Ni).

The length of the catalytic bed varies from ~15 to 60 cm (6 to 12 in.). Despite this short length, the time required to achieve oxidation of HCs is an order of magnitude less for catalytic than conventional thermal oxidizers.

Though catalytic oxidizers are used in many industrial applications, they have disadvantages. These include higher capital costs than thermal oxidizers; a tendency to be "poisoned" by contaminants such as chlorine (Cl), S, and Pb; plugging by PM; and damage by high-temperature excursions associated with transiently higher HC concentrations.

10.4.1.4 Operating Considerations

Conventional thermal oxidation systems have high fuel requirements to maintain system temperatures in the optimum range. To reduce fuel usage costs, most systems are equipped with heat recovery units. Such units have thermal extraction efficiencies on the order of 70%. Recovered heat is transferred to influent gases to increase their temperature and reduce energy needs.

Thermal oxidation in both primary combustion chambers and afterburners can produce unwanted by-products. Most notable are Cl compounds such as hydrochloric acid (HCl) and possibly dioxins and furans associated with combustion of mixed wastes containing chlorinated plastics. Production of such contaminants is common in medical and municipal incinerators.

Any primary or secondary thermal oxidation process will produce nitrogen oxides (NO_x) as an emission product. The higher operating temperatures used in afterburners provide optimum conditions for NO_x production. Lower NO_x production and emissions are associated with catalytic systems because of lower operating temperatures. Nitrogen oxide emission control practices are described in detail at the end of this chapter.

10.4.2 Adsorption

Adsorption is a process in which pollutants are removed from a waste gas stream by the physical attraction of gas and vapors to a solid sorbent as the gas stream moves through a sorbent bed. This adherence is similar to that of van der Waals forces. Physically adsorbed molecules retain their chemical identity and properties.

Adsorption is primarily used to collect organic vapors from solvent-using operations, including degreasing, dry cleaning, surface coating, rubber processing, and several different printing processes. It is also used to control toxic and malodorous substances emitted from food processing, rendering, and chemical and pharmaceutical manufacturing processes.

Sorbents are characterized by their porosity, high surface area, and ability to collect different substances on their surfaces. Commonly used sorbents include activated C, silica gel, activated alumina, and zeolites.

Surface polarity is a major distinguishing feature of sorbents and determines the type of vapors collected. Nonpolar sorbents collect nonpolar vapors; polar sorbents collect polar vapors. Because they sorb H_2O vapor, polar sorbent use in pollution control is very limited.

Activated C is the most widely used sorbent, as it has high affinity for nonpolar compounds that have molecular weights of >45 mass units. Activated C is primarily produced from wood or coal by carbonization at elevated temperatures in the absence of O_2. Such heating drives off volatile compounds. Postcarbonization activation at high temperatures by using steam, air, or CO_2 produces a material of high internal porosity and surface area.

Sorbents are prepared for special applications by impregnating them with reactant substances or catalysts. Examples of impregnants are bromine (Br) to collect ethylene (C_2H_4), iodine (I) for mercury (Hg), lead acetate for H_2S, and sodium silicate for hydrogen fluoride (HF). Catalytic impregnants include metal salts containing Cr, Cu, silver (Ag), Pd, or Pt. These substances catalyze oxidation reactions. In such chemical adsorption (chemisorption), a chemical reaction occurs that converts the sorbate to a new chemical species. As such, the process is not reversible.

Sorbents are produced as granular materials of appropriately small size to increase collection efficiency without excessive resistance to gas flow. They may be used in disposable or rechargeable canisters, fixed regenerable beds, traveling beds, and fluid beds. Disposable and rechargeable canisters and regenerable beds have received the greatest industrial and commercial applications.

Disposable and rechargeable canisters are commonly used for low-volume exhaust flows and waste streams with low contaminant concentrations; a common use is vapor recovery from storage tanks.

Fixed regenerable beds are typically used when the quantity of gas treated or adsorbate concentration is high. They are particularly attractive for solvent recovery when the cost of regeneration is less than the cost of the sorbent.

Both thin-bed (2.5 cm, 1 in. thick) and thick-bed (0.3 to 1 m, 1 to 3 ft thick) regenerable systems are available. The latter, because of their size and capacity, are most appropriate for industrial applications. They are attractive when effluent concentrations of solvent vapors exceed 100 ppmv and exhaust flows exceed 4.7 m^3/sec (10,000 acfm).

A two-unit bed regenerable adsorption system of the type used in vapor recovery is shown in Figure 10.9. In this adsorber system, one unit can be used in an on-stream adsorbing mode, the other in regeneration. They can also be used simultaneously when vapor concentrations are relatively low and regeneration occurs at the end of the work shift. Parallel operation doubles the air handling capacity of the system. Thick-bed systems can be oriented vertically or horizontally.

In regenerable systems, desorption can be achieved by heating, evacuation, stripping with an inert gas, displacing with a more adsorbable material, or a combination of these. Heating is the most common desorption method; the adsorbate can be subsequently collected and reused. Stripping can be used to concentrate contaminants for incineration. Steam displacement is also common. As H_2O vapor is adsorbed, the adsorbate is displaced; desorbed vapors are condensed along with excess steam. The bed is then regenerated by passing hot air over it to remove H_2O.

Figure 10.9 Dual-bed adsorption system. (From Theodore, L. and Buonicore, A.J., Eds., *Air Pollution Control Equipment: Selection, Design, Operation and Maintenance*, Springer-Verlag, New York, 1994. With permission.)

A variety of factors may affect the efficient operation of adsorption systems. These include adequate sorbent capacity, ample contact time, low resistance to airflow, and uniform airflow distribution. It may also be desirable to pretreat the waste gas stream to remove PM and competing vapors, and to cool it. Ample contact time can be achieved by utilizing relatively low flow velocities and extended bed depths.

Adsorption is not generally deemed advisable when waste gas contains PM that may clog the sorbent bed or substances that cannot be easily desorbed from the sorbent. It is also not desirable to use adsorption to collect flammable vapors. Because the adsorption process is exothermic, heat is released. If the adsorption bed is large, a substantial rise in temperature may occur and the collected vapor and sorbent may catch fire by autoignition.

10.4.3 Absorption

Pollution control systems that utilize liquid or liquid–solid media to remove selected gases from waste gas are called scrubbers. Gaseous pollutants are absorbed when they come into contact with a medium in which they are soluble or with which they can chemically react. Though H_2O has been the most common scrubbing medium, it is almost always amended with substances such as lime, limestone, sodium carbonate, or ethanolamine to increase collection efficiency and absorptive capacity. Collection efficiency is also affected by factors such as gas solubility, gas and liquid flow rates, contact time, mechanism of contact, and type of collector.

The fixed-bed packed tower scrubbing system is commonly used to control gas-phase pollutants in industrial applications. It can achieve removal efficiencies of 90 to 95%.

Packed tower scrubbers can be operated as countercurrent, co-current, and cross-flow systems. Countercurrent systems are the most widely used. The scrubbing medium is typically introduced at the top and flows downward; waste gases flow

Exhaust gas

Absorbent in ⟶

Liquid distributor

Waste gas

Absorbent out

Figure 10.10 Countercurrent fixed-bed scrubbing tower.

upward. Countercurrent systems are used for difficult-to-control gases since the packing height is not restricted, thus allowing for higher removal efficiencies. It also has the highest pressure drop and least capacity to handle solids. A schematic cross-section through a countercurrent fixed-bed packed tower can be seen in Figure 10.10.

Co-current systems have both the scrubbing media and waste gases flowing in the same direction. They have a finite limit on absorption efficiency but are useful when the scrubbing medium has a high solids content. They can operate at relatively high scrubber liquid flow rates without plugging.

Cross-flow designs are used for moderately to highly soluble gases or when chemical reactions are rapid. Waste gas flows horizontally while the scrubbing liquid flows downward. An attractive feature of such systems is that they have lower pressure drops.

Absorption of a gas in a scrubbing medium depends on its solubility or chemical reactivity. Solubility depends on gas type and concentration, system pressure, scrubbing liquid, and temperature.

As with sorbents, a scrubbing medium is limited in the amount of a gas it can absorb. On saturation, it cannot absorb additional gases. In most packed tower scrubbers, the liquid is recirculated from a sump. Fresh scrubbing liquid is added and used liquid is withdrawn continuously to prevent saturation.

Collection efficiency in scrubbers is enhanced by using a packing medium. Packing media include steel, ceramic, or thermoplastic materials. These materials come in a variety of shapes and have a characteristically large surface area. By providing a large surface area over which the scrubbing liquid is spread, a greater contact area is provided, thereby increasing collection efficiency. Thermosetting plastics are the most widely used packing materials. They are less subject to corrosion and pressure drop than metal or ceramic types. A plastic packing material can be seen in the photograph in Figure 10.11.

Figure 10.11   Plastic packing material used in a scrubber. (Courtesy of Lantec Products, Inc.)

Scrubbers experience a number of operating and maintenance problems. These include corrosion of scrubber surfaces and plugging of the packed beds. To reduce corrosion, internal scrubber surfaces need to be constructed with stainless steel, lined steel, exotic metals, polyvinyl chloride, or fiberglass-reinforced plastic. Fiberglass or lined steel is commonly used to reduce costs. Plugging may occur as a result of the deposition of undissolved solids or insoluble dusts, or precipitation of dissolved salts or salts produced on absorption. Scaling often occurs when calcium (Ca) or magnesium (Mg) reacts with CO_2 in the alkaline scrubbing medium to produce carbonate precipitates. Such scaling may require flushing with an acidic solution.

The scrubbing medium must be treated and disposed of after it is removed from the system. Waste scrubbing liquids used for acidic gases may require neutralization. Solids may be removed in settling ponds and disposed of in landfills.

10.4.4   Control of Sulfur Oxides

Sulfur oxides (SO_x) generated in the combustion of coal, high-S fuel oils, and smelting of metallic ores have enormous impacts on air quality and the environment. As a result, considerable regulatory attention and allocation of economic resources have focused on reduction of SO_x emissions to the atmosphere.

When SO_x emission reduction requirements were first promulgated by state and local agencies to achieve air quality standards for SO_x, most sources chose to comply by using low-S fuels. Due to the economic impact of fuel use policies (particularly the use of low-S coal) on the midwestern and eastern high-S coal fields of the U.S., Congress, in 1977, mandated that the USEPA promulgate new source performance standards (NSPSs) for large coal-fired power plants that required the use of flue gas desulfurizaton (FGD) systems. As a consequence, new coal-fired power plants given permits to construct or operate between 1977 and 1990 were required to use FGD.

A variety of SO_x control technologies are now used to reduce SO_x emissions to the atmosphere. Some SO_x technologies, e.g., coal beneficiation, coal gasification, and solvent refining, remove S prior to combustion. Others, e.g., fluidized bed combustion (FBC), remove S during the combustion process; FGD systems remove it after combustion has been completed and before gases are emitted to the atmosphere.

10.4.4.1   Coal Beneficiation

Coal is a very heterogeneous mineral consisting of both combustible organic and noncombustible inorganic matter. Both fractions contain S, with concentrations ranging from 0.5 to >5% by weight. The inorganic fraction that contains pyritic S can be removed by coal washing. In coal washing, inorganic minerals are separated from the organic portion by suspending coal fragments in water; differences in specific gravity result in heavy inorganic mineral matter sinking and lighter organic coal floating. Coal cleaning can, on average, remove 50% of pyritic S and 20 to 30% of total S.

Coal washing is widely used by coal producers to reduce the ash content of coal to both enhance its combustibility and meet ash specifications of coal users. Sulfur reduction is a side benefit associated with the enhanced coal quality produced by washing. Sulfur reduction by coal beneficiation or cleaning is usually not adequate to meet emission reduction requirements. However, it may be useful when combined with other control technologies such as FGD.

10.4.4.2   Solvent Refining

Solvent refining, or chemical cleaning, is a technology that has been undergoing development for decades. It is based on the principle that S and other impurities can be removed by solvent extraction. Such a technology has the potential to obviate the need for FGD systems. However, it is not economically viable at this time.

10.4.4.3   Coal Gasification and Liquefaction

Coal can be precleaned on conversion to gases and liquids. In converting coal to synthetic fuels (synfuels), S and ash are removed. These synfuel processes are significantly less energy-efficient than direct coal combustion.

Considerable development efforts were focused on synfuels in the U.S. in the late 1970s and early 1980s when energy concerns were high. As these concerns eased, work on synfuels development projects waned. Coal gasification received major development attention in South Africa when the country was under international economic sanctions because of its apartheid policies.

10.4.4.4   Flue Gas Desulfurization

Flue gas desulfurization (FGD) systems are widely used to remove SO_x from effluent gases of large, and sometimes small, utility boilers, industrial boilers, and primary and secondary metal smelters. Multiple hundreds are currently operated in the U.S., and thousands around the world.

A major use for FGD systems has been to control sulfur dioxide (SO_2) emissions from large coal-fueled power plants. Depending on S content, the 1979 to 1990 NSPSs for coal-fired power plants required 70 to 90% reduction of SO_2 emissions using FGD systems. The average design efficiency for new and retrofit FGD systems applied to coal-fired power plants was 82 and 76%, respectively.

FGD systems vary in the types of sorbents used and how flue gas is treated. In most cases, flue gases are treated downstream of the combustion chamber; in FBC systems, they are treated within the combustion chamber, with further treatment downstream if needed (see Section 10.4.5).

FGD systems can also be described in terms of sorbent reuse. In regenerable systems, expensive sorbents are recovered by stripping SO_2 from the scrubbing medium. Useful by-products may include elemental S, H_2SO_4, liquefied SO_2, and gypsum. In nonregenerable (throwaway) systems, the scrubbing medium is relatively inexpensive and sorbent recovery is not economical. The three dominant FGD processes used to meet NSPS and SIP SO_2 emissions limits are throwaway systems. Such systems have significant waste treatment and disposal requirements.

FGD systems are of two primary types: wet and dry systems, or wet and dry scrubbers.

10.4.4.4.1 Wet Scrubbing Systems

Wet systems typically use a scrubbing medium that is a mixture of a liquid (typically H_2O) and suspended and partially dissolved solids. Thus, the scrubbing medium is a slurry. Common scrubbing media include lime, limestone, lime–limestone, lime–alkaline flyash, limestone–alkaline flyash, sodium carbonate, and dual alkali. A limestone slurry is used in ~50% of FGD applications, and lime in 16%.

Limestone is considerably less soluble in H_2O than is lime. It is also less expensive. Sulfur dioxide is removed from flue gases in a limestone scrubber as a result of the following chemical reactions:

$$SO_2 + H_2O \rightarrow SO_2 \cdot H_2O \qquad (10.2)$$

$$SO_2 \cdot H_2O \rightarrow H^+ + HSO_3^- \qquad (10.3)$$

$$H^+ + CaCO_3 \rightarrow Ca^{+2} + HCO_3^- \qquad (10.4)$$

$$Ca^{2+} + HSO_3^- + 2\ H_2O \rightarrow CaSO_3 \cdot 2\ H_2O + H^+ \qquad (10.5)$$

$$H^+ + HCO_3^- \rightarrow CO_2 + H_2O \qquad (10.6)$$

The summary reaction for this reaction series is:

$$CaCO_3 + SO_2 + 2\ H_2O \rightarrow CaSO_3 \cdot 2\ H_2O + O_2 \qquad (10.7)$$

Initially, SO_2 reacts with H_2O to produce a weak sulfurous acid (H_2SO_3) solution. Hydrogen ions (H^+) then react with calcium carbonate to form soluble calcium bicarbonate. Calcium ions react with sulfite to produce a calcium sulfite hydrate and H^+, which reacts with bicarbonate ions to produce CO_2 and H_2O.

Calcium sulfite hydrate is converted to calcium sulfate hydrate (gypsum) when excess O_2 is present in the slurry:

$$CaSO_3 \cdot 2\ H_2O + \tfrac{1}{2}\ O_2 \rightarrow CaSO_4 \cdot 2\ H_2O \qquad (10.8)$$

As a consequence of these reactions, the slurry, after it passes through the scrubber, contains $CaCO_3$, $CaSO_3 \cdot 2H_2O$, and $CaSO_4 \cdot 2H_2O$.

In lime slurry systems, sulfurous acid reacts with calcium hydroxide to produce calcium sulfite hydrate:

$$Ca(OH)_2 + H_2SO_3 \rightarrow CaSO_3 \cdot 2\ H_2O \qquad (10.9)$$

In the presence of excess O_2, some of the calcium sulfite hydrate is oxidized to calcium sulfate hydrate.

In limestone and lime slurry FGD systems (Figure 10.12), flue gases flow upward through a packed tower countercurrent to the scrubbing medium. After coming into contact with flue gases, the scrubbing medium is pumped to a holding tank where some of the calcium sulfite and sulfate are precipitated. Since considerable sorption capacity remains, scrubbing fluid is recirculated to the scrubber. Scrubbing medium not recycled is pumped to a thickening tank and then to a stabilizing pond. Though continuously recirculated, the scrubbing medium is continuously recharged with freshly prepared scrubbing fluid.

Both limestone and lime scrubbing are similar in process flow and equipment used. Lime, however, is much more reactive than limestone, with better utilization of the scrubbing medium. Lime scrubbers can routinely achieve 95% removal of SO_2; limestone, 90%.

Though lime and limestone scrubbing systems can achieve high FGD efficiencies, their operation is technically demanding. Several factors reduce performance and reliability. Scaling and plugging of scrubber equipment by insoluble compounds is a common occurrence. Scaling problems can be minimized by proper control of pH.

High desulfurization efficiencies require high scrubbing medium:gas ratios. However, high ratios result in higher operating costs because of the pressure drop in the scrubber and higher energy needs for pumping. Energy consumption by FGD systems is on the order of 3 to 6% of the power generated by a coal-fired facility. In addition to these high operating costs, FGD systems used on coal-fired utility boilers have high capital costs; on a 1000-megawatts electric (MwE) unit, these typically exceed 150 million U.S. dollars (USD).

10.4.4.4.2 Dry Scrubbers

In dry scrubbing, flue gas is treated by direct injection of a dry-state reagent aerosol or spraying with a slurry aerosol.

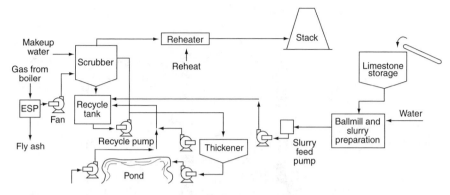

Figure 10.12 Flue gas desulfurization system. (From Theodore, L. and Buonicore, A.J., Eds., *Air Pollution Control Equipment: Selection, Design, Operation and Maintenance,* Springer-Verlag, New York, 1994. With permission.)

In direct injection, the reagents are either trona (naturally occurring sodium carbonate, Na_2CO_3) or nahcolite (naturally occurring sodium bicarbonate, $NaHCO_3$). These dry aerosol particles react with SO_2, and they and fly ash particles are collected on baghouse filters downstream. The PM cake on filters serves as a porous reagent bed that converts SO_2 to sodium sulfite (Na_2SO_3) and sodium sulfate (Na_2SO_4).

In spray dryer or semidry scrubbing, the reagent (typically calcium hydroxide, $Ca(OH)_2$) is mixed with H_2O and atomized into droplets that mix with flue gases in a spray dryer reaction chamber. The hot gases evaporate the H_2O, and SO_2 reacts with the solid suspended reagent. The reagent aerosol is collected on the fabric filter in a large baghouse downstream. Sulfur dioxide not removed in the reaction chamber may be removed by sorption on the filter cake; such removal may be on the order of 15 to 30%.

The baghouse particle collector has to be oversized because of the large PM loading. This cost is more than offset by the larger cost in materials handling, personnel requirements, land, and disposal associated with wet scrubbers.

Dry scrubbing is used in about 10% of coal-fired utility FGD systems. Spray drying with lime is the predominant method employed.

10.4.5 Fluidized Bed Combustion

Because of higher thermal efficiencies, FBC systems are increasingly being used in a variety of applications, including industrial and utility boilers, incineration of solid wastes, and reclamation of coal refuse piles.

FBC units contain a bed of small particles that are brought to a "boil" by air passing through the bed at a sufficient velocity (Figure 10.13). A variety of materials can be introduced into this bed and thermally oxidized. Limestone can be mixed with the fuel to capture SO_2 and convert it to SO_4^{-2}. The ash, SO_2, and SO_4^{-2} can be separated from the circulating mixture as new fuel and limestone are added.

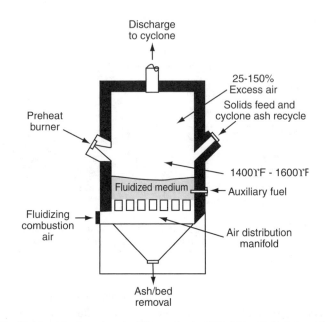

Discharge
to cyclone

25-150%
Excess air

Solids feed and
cyclone ash recycle

Preheat
burner

1400°F - 1600°F

Fluidized medium

Auxiliary fuel

Fluidizing
combustion
air

Air distribution
manifold

Ash/bed
removal

Figure 10.13 Fluidized bed combustion. (From Davis, W.T., Ed., *Air Pollution Engineering Manual*, John Wiley & Sons, New York, 2000. With permission.)

10.4.6 Biological Treatment

Biological treatment systems or bioreactors utilizing microorganisms to metabolize pollutants include three types: bioscrubbers, trickle bed reactors, and biofilters. In addition to wastewater treatment, biological treatment systems are used industrially to control odors associated with food processing and fermentation. They have been used to control alcohols, aldehydes, amines, aromatic HCs, esters, ketones, organic acids, and organic sulfides.

Bioscrubbers consist of a packed bed in which microorganisms are suspended in a scrubbing medium. Inert packing materials serve to increase the contact surface to enhance the transfer of pollutants from the gas to the liquid phase. The scrubbing medium is aerated to provide O_2 and stirred to keep organisms in suspension.

Biofilters consist of one or more beds of biologically active material such as compost, municipal waste, etc., that provides a food base for microorganisms. Bed depths are on the order of 1 m (~3 ft) to minimize pressure drop. Waste gases enter the bottom of the unit and diffuse into the biofilm (biologically active layer) attached to the compost where pollutants are metabolized. If metabolism is complete, end products are CO_2, H_2O, and microbial mass.

Trickle bed reactors consist of a packed bed of compost particles that have a low surface:volume ratio to provide large voids that allow for biological growth. A recirculating fluid of nutrients flows downward, wetting the biofilm attached to compost particles. Waste gases travel upward countercurrent to the flow of liquid. Water-soluble pollutants are absorbed, diffuse to the biofilm, and are metabolized.

10.4.7 Control of Nitrogen Oxides

Because of concerns associated with the role of NO_x in photochemical oxidant formation and acidic deposition, stationary sources, particularly large fossil fuel-fired power plants, are increasingly being required to reduce NO_x emissions.

A variety of control techniques for reducing emissions from utility and industrial boilers have been developed. One of the most widely used is the low NO_x burner, now a standard part of new utility boiler designs.

Low NO_x burners inhibit formation of NO_x by controlling air and fuel mixing. By using these burners, located in pulverized coal boiler walls, the amount of excess air used to ensure good combustion can be reduced. Reducing excess air from 20 to 14%, for example, has been shown to reduce NO_x production by 20%. It also reduces the amount of O_2 available to combine with nitrogen (N_2). Additional NO_x reductions can be achieved by staged combustion (fuel is combusted in a primary zone that is fuel-rich, followed by secondary and following zones that are fuel-lean). Combustion staging methods reduce NO_x formation by either decreasing available O_2 or providing excess O_2 to cool the combustion process.

A variety of flue gas treatment techniques are available for NO_x emission control. One of these, selective catalytic reduction, is widely used in Japan. Catalysts such as mixtures of titanium (Ti) and vanadium (V) can be used in combination with reducing gases such as ammonia (NH_3), hydrogen (H_2), CO, and H_2S to convert NO_x to N_2. The following reactions involving NH_3 are illustrative:

$$4 \, NO + 4 \, NH_3 + O_2 \rightarrow 4 \, N_2 + 6 \, H_2O \tag{10.10}$$

$$2 \, NO_2 + 4 \, NH_3 + O_2 \rightarrow 3 \, N_2 + 6 \, H_2O \tag{10.11}$$

Other NH_3-forming compounds such as urea can also be used. Because of the lower temperatures required for catalysis, catalytic removal efficiencies can exceed 80%.

READINGS

Allen, D.T. and Rosselot, K.S., *Pollution Prevention for Chemical Processes,* Wiley-Interscience, New York, 1997.

Bubenick, D.V., Control of fugitive emissions, in *Handbook of Air Pollution Technology,* Calvert, S. and Englund, H., Eds., John Wiley & Sons, New York, 1984, p. 745.

Buonicore, A.J. and Davis, W.T., Eds., *Air Pollution Engineering Manual,* Van Nostrand Reinhold, New York, 1992.

Cooper, C.D. and Alley, F.C., *Air Pollution Control: A Design Approach,* 2nd ed., PWS Engineering, Boston, 1994.

Croker, B.B. and Schnelle, K.B., Jr., Control of gases by absorption, adsorption and condensation, in *Handbook of Air Pollution Technology,* Calvert, S. and Englund, H., Eds., John Wiley & Sons, New York, 1984, p. 135.

Davis, W.T., *Air Pollution Engineering Manual,* John Wiley & Sons, New York, 2000.

Dempsey, C.R. and Oppelt, E.T., Incineration of hazardous waste: a critical review update, *JAWMA*, 43, 25, 1993.

DeNevers, N., *Air Pollution Control*, McGraw-Hill, New York, 1995.

Devinny, J., Deshusses, M., and Webster, T., *Biofiltration for Air Pollution Control*, Lewis Publishers/CRC Press, Boca Raton, FL, 1999.

Flagan, R.C. and Seinfeld, J.H., *Fundamentals of Air Pollution Engineering*, 2nd ed., Prentice Hall, Saddlebrook, NJ, 1988.

Freeman, H., Harten, T., Springer, J., Randall, P., Curran, M.A., and Stone, K., Industrial pollution prevention: a critical review, *JAWMA*, 42, 618, 1992.

Heck, R.M. and Farrato, R.J., *Catalytic Air Pollution Control: Commercial Technology*, John Wiley & Sons, New York, 1995.

Heinsohn, R.J. and Kabel, R.L, *Sources and Control of Air Pollution*, Prentice Hall, Saddlebrook, NJ, 1999.

Henzel, D.S., Laseke, B.A., Smith, E.O., and Swenson, D.O., *Handbook for Flue Gas Desulfurization: Scrubbing with Limestone*, Noyes Data Corporation, Park Ridge, NJ, 1982.

Hesketh, H.E., *Air Pollution Control: Traditional and Hazardous Pollutants*, Technomics Publishing Co., Lancaster, PA, 1991.

Heumann, W.L., *Industrial Air Pollution Control Systems*, McGraw-Hill, New York, 1996.

Higgins, T.E., Ed., *Pollution Prevention Handbook*, Lewis Publishers/CRC Press, Boca Raton, FL, 1995.

Hinds, W.C., *Aerosol Technology*, Wiley-Interscience, New York, 1982.

Kilgroe, J.D., Coal cleaning, in *Handbook of Air Pollution Technology*, Calvert, S. and Englund, H., Eds., John Wiley & Sons, New York, 1984, p. 419.

Klaassen, G. and Forsund, F.R., Eds., *Economic Instruments for Air Pollution Control*, Kluwer Academic Publishers, Boston, 1994.

Mycak, J.C., McKenna, J.D., and Theodore, L., *Handbook of Air Pollution Control and Engineering Technology*, Lewis Publishers/CRC Press, Boca Raton, FL, 1995.

Noll, K.E. and Anderson, W.C., *Fundamentals of Air Quality Systems: Design of Air Pollution Control Devices*, American Academy of Environmental Engineers, Annapolis, MD, 1999.

Schifftner, K. and Hesketh, H.E., *Wet Scrubbers*, 2nd ed., Technomics Publishing Co., Lancaster, PA, 1992.

Spencer, D.F., Gluckman, M.J., and Alpert, S.B., Coal gasification for electric power generation, *Science*, 215, 1571, 1982.

Strauss, W., *Industrial Gas Cleaning*, Pergamon Press, Tarrytown, NY, 1996.

Turk, A., Adsorption, in *Air Pollution*, Vol. IV, *Engineering Control of Air Pollution*, Stern, A.C., Ed., Academic Press, New York, 1977, p. 329.

Turner, L.H. and McKenna, J.D., Control of particles by filters, in *Handbook of Air Pollution Technology*, Calvert, S. and Englund, H., Eds., John Wiley & Sons, New York, 1984, p. 249.

U.S. DOE, *Clean Coal Technology: The New Coal Era*, DOE/FE-0193P, DOE, Washington, D.C., 1990.

USEPA, *Sulfur Emission: Control Technology and Waste Management*, EPA 600/9-79-019, EPA, Washington, D.C., 1979.

Warner, C.F., Davis, W.T., and Wark, K., *Air Pollution: Its Origin and Control*, 3rd ed., Addison-Wesley-Longman, Boston, MA, 1998.

White, H.J., Control of particulates by electrostatic precipitation, in *Handbook of Air Pollution Control Technology*, Calvert, S. and Englund, H., Eds., John Wiley & Sons, New York, 1984, p. 283.

QUESTIONS

1. What are advantages and disadvantages of using tall-stack technology as a pollution control measure?
2. Why is natural gas such a popular fuel for use in industrial boilers and single-cycle gas turbine electrical power production?
3. Describe several applications of pollution prevention principles for controlling air pollutants.
4. What are fugitive emissions? What is their significance in pollution control?
5. Describe the relationship between pollution prevention principles and control of hazardous air pollutant emissions.
6. The collecting efficiency of particle and dust air cleaners decreases as particle size decreases. What is the environmental significance of this phenomenon?
7. Describe the relative fractional collection efficiencies of the following control devices: (a) simple cyclone, (b) multicyclone, (c) baghouse, and (d) electrostatic precipitator.
8. Describe the physical principles used to collect particles in the following: (a) cyclones, (b) baghouse, and (c) electrostatic precipitator.
9. Why does an electrostatic precipitator experience decreased performance when a coal-fired power plant burns low-sulfur coal?
10. When would thermal oxidization be an appropriate method for controlling pollutants in a waste gas stream?
11. What is an LEL? How does it relate to controlling emissions of hydrocarbon gases and vapors?
12. Under what circumstances would catalytic oxidation be used in an industrial facility to control emissions of pollutants? What are its limitations?
13. What is an afterburner?
14. Describe the process of adsorption and its industrial applications.
15. How do physical and chemical adsorption differ?
16. What physical and chemical processes are utilized to remove SO_2 from a waste gas stream?
17. What is coal beneficiation? What are its advantages and limitations in meeting regulatory SO_2 emission limits?
18. Describe the chemical processes involved in removal of SO_2 from flue gases using a wet scrubber system.
19. What is the role of packing materials in wet scrubbing systems?
20. Describe techniques used to control emissions of SO_2 using dry scrubbing.
21. What is fluidized bed combustion? How can SO_2 emissions be reduced from such a combustion system?
22. Describe methodologies used to reduce emissions of NO_x from industrial and utility fossil fuel-fired systems.

CHAPTER 11

Indoor Air Quality

As we have seen thus far, pollution of the ambient atmosphere is an enormous environmental concern. As a consequence, regulatory measures have been implemented to safeguard public health, protect air quality in pristine air regions, regulate or ban ozone (O_3)-destroying chemicals, and reduce acidic deposition, and measures are being contemplated and debated to reduce emissions of greenhouse gases to mitigate anticipated adverse consequences of global warming. The primary focus of air pollution control activities on a day-to-day basis is to protect public health. In the past three plus decades, these health-based pollution control efforts have required an investment of several trillion dollars.

Exposure to polluted ambient air is but one of several air pollution and public health concerns that humans face. Particularly notable are pollutant exposures in the indoor environments where we live, work, learn, play, etc.

Based on numerous scientific studies and field investigations, the air inside our homes, office buildings, and schools is contaminated by a variety of gas- and particulate-phase substances that may be present in sufficient concentrations to cause acute or chronic symptoms or illness in individuals of a relatively large exposed population. Exposures may vary from ~8 h/day in office, commercial, and institutional buildings to 16 to 24 h/day in homes. This compares, on average, to ~2 h/day in the ambient environment.

Exposure duration, populations at risk, pollutant concentrations, and pollutants themselves are, in many cases, significantly different in indoor and ambient environments. Exposure durations are longer and populations at risk larger in the former. In residential buildings, exposures occur to individuals ranging in age from infants to elderly, and health status from healthy to having a variety of illnesses and infirmities.

Based on studies of illness prevalence, pollutant exposures, and long-term cancer risks associated with exposures to radon, asbestos, pesticides, and combustion by-products, public health concerns associated with indoor air may rival, if not significantly exceed, those associated with the ambient environment.

11.1 PERSONAL, INDOOR, AND OUTDOOR POLLUTANT RELATIONSHIPS

11.1.1 Personal Pollutant Exposures

In National Ambient Air Quality Standards (NAAQSs) promulgated under clean air legislation, there is an implicit (though unrealistic) assumption that exposures occur similarly in both ambient and indoor environments. They do not, as a consequence, take into account actual (personal) exposures. Scientific studies have shown such personal exposures are often considerably different from those implied in NAAQSs. Differences in outdoor, indoor, and personal exposures to respirable particles (RSPs) in a Topeka, KS, study can be seen in Figure 11.1. Note that although all three exposure measures track each other, indoor exposures are greater than outdoor exposures, and for the most part, personal exposures are greater than indoor exposures. Personal exposures include different environments and activities, including motor vehicle travel and hobbies. Research conducted on personal pollution exposures to criteria pollutants indicates that exposures inferred from fixed-site monitoring stations may be significantly less than those experienced by a significant portion of the U.S. population.

11.1.2 Indoor and Outdoor Pollutant Concentrations

Humans are exposed to a variety of pollutants in indoor and ambient air. Though ambient pollutants are present in indoor environments, indoor and outdoor

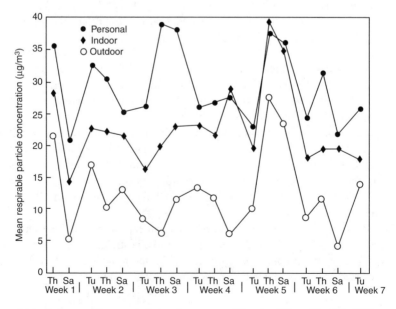

Figure 11.1   Indoor, outdoor, and personal exposure of Topeka, KS, residents to RSPs. (Courtesy of Spengler, J.D., School of Public Health, Harvard University, Boston, MA.)

concentrations often differ, depending on pollutant reactivity and whether indoor sources are present. Indoor levels of reactive chemical species such as O_3 and sulfur dioxide (SO_2) are significantly lower than they are outdoors. In other cases, significantly higher concentrations are found in indoor environments as a result of indoor sources.

The indoor:outdoor ratio (I:O) of O_3 is typically in the range of 0.1 to 0.3, but it may be as high as 0.7 in mechanically ventilated spaces. Typically, ambient air is the major source of O_3 observed indoors. However, elevated localized indoor O_3 concentrations and exposures may result from use of electronic air cleaners or electrostatic photocopy machines.

Sulfur dioxide ratios vary from 0.3 to 0.5 when outdoor levels are moderately high, to 0.7 when they are relatively low. Elevated indoor SO_2 concentrations and exposures may result from kerosene heaters using 0.20% sulfur (S) fuel.

Though ambient air is a source of volatile organic compounds (VOCs), concentrations of individual VOC species indoors are, in most cases, several times to an order of magnitude higher than outdoors. This is due to the fact that many materials used to manufacture building materials, furnishings, etc., emit VOCs for extended periods of time.

Nitrogen dioxide (NO_2) is a relatively reactive gas found in the ambient atmosphere, and as a consequence, I:O ratios of <1.0 should be expected. However, significant sources of NO_2 are often present in residential environments. These include unvented combustion appliances such as gas or kerosene space heaters, gas cooking stoves and ovens, and, increasingly, ventless gas fireplaces.

As carbon monoxide (CO) is a relatively unreactive substance, it passes freely through building envelopes and ventilation systems. In the absence of indoor sources, concentrations in ambient and indoor environments should be the same. However, in many cases, indoor concentrations are higher because of sources such as gas cooking stoves and ovens, ventless gas and kerosene heaters or fireplaces, tobacco smoking, etc.

In a "closed" environment, the building envelope and systems tend to "filter out" ambient particles as they are transported into buildings. In the absence of indoor sources, I:O ratios can be expected to be <0.5. However, the indoor environment is often a major source of particles, particularly when smoking is occurring; in such cases, I:O ratios are >1.0.

Concentrations of asbestos; radon (Rn); formaldehyde (HCHO); tobacco smoke; allergens such as those from dust mites, pets, cockroaches, etc.; and other biological contaminants such as bacterial endotoxins, fungal glucans, certain mold species, and mycotoxins are usually significantly higher indoors than outdoors. Exposure to environmental tobacco smoke (ETS) is, for the most part, unique to indoor environments.

11.2 INDOOR AIR QUALITY CONCERNS

Indoor air quality (IAQ) health and exposure concerns are primarily focused on residential dwellings and nonresidential, nonindustrial buildings such as offices and schools. Though the same pollutants may be found in both building categories, exposures and exposure concerns are often quite different.

Occupants of residences can be exposed to potentially toxic, allergenic, or infectious pollutants upwards of 8 to 24 h/day, with exposed populations including infants/children, healthy adults, the aged, and the infirm. Pollutant exposures that are unique (particularly as a result of elevated concentrations) include Rn, HCHO, ETS, combustion appliance by-products, pesticides, VOCs from personal and home care products and arts-and-crafts activities, and biological contaminants such as bacterial endotoxins and allergens associated with dust mites, cockroaches, pets, and mold. Exposure concerns are increased as a result of limited ventilation during closure conditions in heating and cooling seasons.

Nonresidential, nonindustrial buildings are subject to relatively unique pollutant exposure, health, and comfort concerns. Such buildings are typically mechanically ventilated and may have high occupant densities (e.g., schools). Populations subject to pollutant exposures depend on the building's function, e.g., adults in office buildings; a mixture of children and adults in schools; and the infirm or aged in hospital convalescent, or senior citizen facilities.

Pollutant exposure and health concerns in nonresidential, nonindustrial buildings include:

1. Elevated bioeffluent (pollutants released by humans and other organisms) levels associated with high occupant densities and inadequate ventilation
2. Emissions from office equipment and materials
3. Cross-contamination from contaminant-generating areas
4. Entrainment of contaminants generated outdoors
5. Reentry of building exhaust gases
6. Contamination of air handling units (AHUs) by organisms and biological by-products that can cause illness, e.g., hypersensitivity pneumonitis, humidifier fever, and Legionnaires' disease
7. Transmission of contagious diseases such as flu, colds, and tuberculosis (TB)
8. Exposure to resuspended surface dusts
9. Exposure to ETS where smoking is not restricted

11.3 MAJOR INDOOR POLLUTANTS

A number of pollutants and pollutant categories have been identified as significant exposure and potential health risks in buildings. These include asbestos, Rn, combustion by-products, aldehydes, VOCs, semivolatile organic compounds (SVOCs), pesticides, and a variety of contaminants of biological origin.

11.3.1 Asbestos

Asbestos became a major IAQ concern in the U.S. in the late 1970s when public health and environmental protection authorities concluded that friable (hand-crushable) asbestos-containing building materials (ACBM) in school buildings posed a potential cancer risk to children and other building occupants. Such materials were widely used in construction of schools and other large buildings prior to 1978.

Asbestos is a collective term for a number of fibrous mineral silicates. The most widely used asbestos mineral in the U.S. was chrysotile, or white asbestos. It accounted for >90% of the fibrous mass used in the 3000 or so different asbestos-containing products. Amosite, or brown asbestos, was less commonly used in the U.S. Crocidolite, or blue asbestos, has been widely used in Europe and Southeast Asia.

Asbestos fibers are fire and heat resistant, with high tensile strength. As a consequence, they were used in sprayed-on fireproofing, acoustical plaster, thermal system insulation, friction products, fibrocement board and pipes, roofing products, spackling compounds, mastics, and vinyl floor tile.

Asbestos has been regulated as a hazardous air pollutant by the U.S. Environmental Protection Agency (USEPA) since 1973. Occupational exposures have been causally associated with asbestosis (a fibrogenic, debilitating lung disease), lung cancer, mesothelioma (a cancer of the lining of the chest and abdominal cavities), and scarring of the pleural cavity (found in the chest) called pleural plaques. Asbestosis is caused by heavy, long-term exposures to asbestos fibers and is unlikely to be a health risk to building occupants. Since there is no apparent threshold or safe level of exposure, public health concern has primarily focused on risks of developing lung cancer or mesothelioma. Since children would be exposed earlier in life, providing for a longer latency period, their risk of developing mesothelioma would be increased. Though relatively rare (an estimated 1200 to 1500 cases/year in the U.S.), mesothelioma has a mortality rate of 100%.

Asbestos fibers in many products are held in a matrix that varies in its friability. Fiber release occurs when this matrix is broken by hand pressure (hand friable) or mechanical force (mechanically friable). Since hand-friable ACBM poses the greatest risk of fiber release and building air contamination, the USEPA, in 1973, required friable asbestos removal before building demolition or renovation, and banned sprayed-on fireproofing materials (Figure 11.2). In 1978, the EPA banned sprayed-on and troweled-on acoustical plaster and premolded asbestos products used in thermal insulation (Figure 11.3). These materials are present in hundreds of thousands of buildings in the U.S.

Over time, friable ACBM may become damaged or lose its adhesion to the substrate to which it was applied. Thermal system insulation on steam and hot water lines, boilers, etc., may become friable when damaged by maintenance and service activities. Damage to ACBM increases the likelihood that fibers become airborne, resulting in human exposure. Exposure may be increased by custodial activities such as dusting and cleaning and activities of occupants that may resuspend settled fibers.

The USEPA initiated a guidance and technical assistance program, the Asbestos-in-Schools Program, in 1979 in response to asbestos health concerns in schools. Because of inadequacies in the program, Congress, in 1986, passed the Asbestos Hazard Emergency Response Act (AHERA). This law required all schools to be inspected by accredited inspectors and school corporations and districts to prepare and implement a management plan for each building under their jurisdiction.

Figure 11.2   Sprayed-on asbestos fireproofing.

Figure 11.3   Partially damaged asbestos thermal system insulation.

In the late 1980s, considerable public health attention was given to potential asbestos fiber exposures and health risks to the general school population, i.e., children, faculty, and administrative staff. Subsequently, a consensus evolved in the scientific and regulatory communities that exposure risks to these groups were very small. At the same time, there was increasing scientific evidence that custodial and service staff in buildings containing ACBM were at significant risk of developing asbestos-related disease as a result of asbestos exposures. Custodial workers, sheet metal workers, and carpenters were shown to have significantly elevated prevalence rates of asbestos-related pleural plaques that result in impaired pulmonary function. As a consequence, the Occupational Safety and Health Administration (OSHA) promulgated rules designed to protect custodial and service workers in office and other buildings; materials presumed to contain asbestos are to be identified, and workers notified and trained to reduce exposure risks associated with their disturbance.

11.3.2 Radon

Radon is a noble, nontoxic, radioactive gas produced in the decay of radium-226. Radium is found in relatively high concentrations in uranium (U) and phosphate ores and U mill tailings and, to a lesser extent, in common minerals such as granite, schist, limestone, etc. As a consequence, it is ubiquitous in soil and air near the surface of the Earth. As Rn undergoes radioactive decay, it releases an alpha particle, gamma rays, and progeny that quickly decay (Figure 11.4) to release alpha and beta particles and gamma rays. Since Rn decay products (RDPs) are electrically charged, they readily attach to particles, producing radioactive aerosols. These aerosol particles may be inhaled and deposited in the bifurcations of respiratory airways.

11.3.2.1 Concentrations and Sources

Radon concentrations determined for buildings in the U.S. are calculated and reported as picocuries per liter (pCi/l). A picocurie is equal to a trillionth of a curie, or 2.2 radioactive disintegrations per minute. A working level (WL) is a quantity of Rn that will produce 1.3×10^5 million electron volts (MeV) of potential alpha particle energy per liter of air. One WL is equal to the concentration of RDPs in equilibrium with 100 pCi/l Rn. Since the equilibrium value for RDPs in indoor environments is typically 0.5, Rn concentrations corresponding to 1 WL would be ~200 pCi/l.

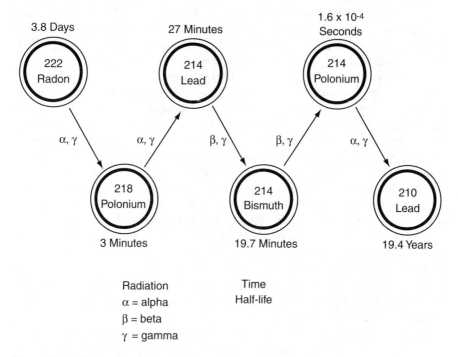

Figure 11.4 Radioactive decay of radon and its progeny.

Because of the relatively common occurrence of radium-226 in the Earth's crust, Rn is present in ambient air with average concentrations in the range of 0.20 to 0.25 pCi/l. Radon levels in buildings may be several times or orders of magnitude higher than ambient levels. This is particularly true in single-family dwellings. In one of the worst cases ever reported, Rn concentrations as high as 2600 pCi/l (13 WL) were measured in a southeastern Pennsylvania home.

Significant geographical differences in Rn levels in the U.S. have been reported. Geometric mean values based on 3-month averages for 30,000 nonrandomly tested houses were reported as 3.43 pCi/l in the Northeast; 2.36, Midwest; 0.64, Northwest; 2.40, mountain states; and 1.43, Southeast. The geometric mean for the entire sample population was 1.74 pCi/l; 0.5% were >100 pCi/l. Other studies have reported a lower geometric mean value of ~1.0 pCi/l.

The major source of indoor Rn is soil gas transported by pressure-induced convective flows. Other sources include well water and masonry materials. Buildings constructed on soils with high Rn release and transport potential typically have the highest Rn levels; these vary in response to temperature- and wind-dependent pressure differentials and changes in barometric pressure. Radon transport is enhanced when the base of a building is under significant negative pressure.

11.3.2.2 Health Effects and Risks

The major health risk definitively linked to Rn exposures is lung cancer. Extrapolations from U miner Rn exposure and health data indicate that humans exposed to the lower concentrations characteristic of buildings may also be at risk. Risk assessments have been made by the National Council on Radiation Protection (NCRP), the National Academy of Sciences Committee on the Biological Effects of Radiation (BEIR IV), and the USEPA. These assessments predict (with some degree of uncertainty) that ~13,000 to 16,000 lung cancer deaths a year can be attributed to Rn exposure, with interaction effects contributing to a higher risk for smokers. For purposes of communicating with the public, the USEPA published a chart (Figure 11.5) wherein lung cancer risks are projected for different average lifetime Rn exposures and compared to equivalent risks from exposures to chest x-rays and tobacco smoking. Note the relatively large lifetime lung cancer risk of 1 to 3 per 1000 population for outdoor levels of 0.2 pCi/l (0.001 WL) and 13 to 50 per 1000 for an average lifetime exposure to 4 pCi/l (0.02 WL), the USEPA action level for remediation. In the latter case, an annual average lifetime exposure to 4 pCi/l would be equivalent to the lung cancer risk associated with smoking five cigarettes per day. Using these risk estimates, or even more conservative ones, Rn appears to be one of the more hazardous substances to which humans are exposed. Because of these risks, the USEPA and the U.S. surgeon general in 1988 issued a public health advisory recommending that all homes be tested and remediation be undertaken when long-term (3-month to 1-year) test results are excessive, i.e., above the action guideline of 4 pCi/l.

There is a limited amount of epidemiological evidence to suggest that Rn exposures may be a risk factor for acute myeloid leukemia, cancers of the kidney, melanoma, and certain childhood cancers. In Great Britain, approximately 6 to 12%

pCi/L	WL	Estimated number of LUNG CANCER DEATHS due to radon exposure (out of 1000)	Comparable exposure levels		Comparable risk
200	1	440-770	1000 times average outdoor level		More than 60 times non-smoker risk
100	0.5	270-630	100 times average indoor level		4 pack-a-day smoker
40	0.2	120-380			20,000 chest x-rays per year
20	0.1	60-210	100 times average outdoor level		2 pack-a-day smoker
10	0.05	30-120	10 times average indoor level		1 pack-a-day smoker
4	0.02	13-50			5 times non-smoker risk
2	0.01	7-30	10 times average outdoor level		200 chest x-rays per year
1	.005	3-13	Average indoor level		Non-smoker risk of dying from lung cancer
0.2	.001	1-3	Average outdoor level		20 chest x-rays per year

Figure 11.5 USEPA radon risk estimates. (From USEPA, EPA-86-004, EPA, Washington, D.C., 1986.)

of acute myeloid leukemia cases are estimated to be associated with Rn exposures; for regions where Rn levels are higher, the predicted range is 23 to 43%. For the world average exposure of 1.3 pCi/l, ~13 to 25% of acute myeloid leukemias are estimated to be associated with Rn.

11.3.3 Combustion By-Products

Combustion by-products are released into indoor air from sources such as (1) open cooking fires fueled by wood, charcoal, dung, etc., in developing countries; (2) unvented gas cooking appliances, space heaters, fireplaces, and kerosene space heaters; and, to a lesser extent, (3) vented combustion systems, e.g., fireplaces, wood-burning appliances, gas and oil furnaces, and hot water heaters. Miscellaneous sources include tobacco smoking (the most ubiquitous source of combustion products in indoor environments), candle and incense burning, propane-fueled burnishers and forklifts, ice resurfacing machines, arena events, entrainment from outdoor sources, and flue gas reentry.

11.3.3.1 Unvented Combustion Systems

11.3.3.1.1 Cooking Stoves in Developing Countries

A high percentage of the world's households use biomass fuels for cooking. Most cooking fires are simple pits and are not vented to the ambient environment (Figure 11.6). As a consequence, indoor concentrations of combustion by-products

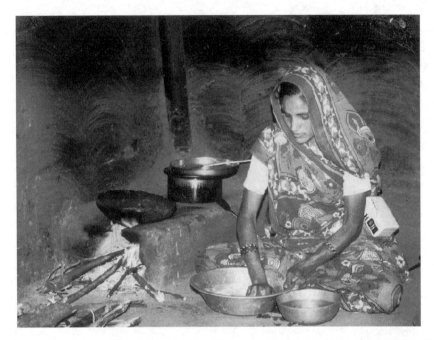

Figure 11.6   Home cooking fire in a dwelling. (Courtesy of Smith, K.R., School of Public Health, University of California–Berkeley.)

are very high. This is evident in particulate matter (PM) concentrations measured in biomass cooking homes in villages in a number of countries (Table 11.1). What is notable about these total suspended particulate (TSP) matter exposure levels (primarily to adult women and young children) is that they are an order of magnitude or more than the old U.S. TSP standard of 260 $\mu g/m^3$ (not to be exceeded more than once per year) and the World Health Organization's (WHO) PM_{10} (PM with an aerodynamic diameter of ≤ 10 μm) guidelines of 100 to 150 $\mu g/m^3$ 24-h average.

High indoor PM concentrations have also been reported in homes in developing countries that use coal for cooking and heating. PM associated with biomass cooking and coal use has high polycyclic aromatic hydrocarbon (PAH) concentrations. Indoor concentrations of these potent carcinogens are also very high.

Indoor levels of gas-phase substances associated with gas cooking are also elevated. These include CO (10 to 50 ppmv, with averages of 30 to 40 ppmv common), NO_2 (70 to 160 ppbv), SO_2 (60 to 100 ppbv), and aldehydes (0.67 to 1.2 ppmv).

The WHO estimates that pollutant exposures associated with biomass cooking are responsible for ~2.5 million premature deaths worldwide each year. Exposures to biomass cooking smoke significantly increase the risk of acute respiratory infections, chronic obstructive lung disease, cor pulmonale, and lung cancer. Untreated acute respiratory infection commonly progresses to pneumonia, a major cause of children's deaths worldwide.

Table 11.1 Indoor Airborne Particulate Matter (TSP) Concentrations Associated with Biomass Cooking in Developing Countries

Location/Report Year	Measurement Conditions	PM Concentration ($\mu g/m^3$)
Papua, New Guinea		
1968	Overnight, floor level	200–4900
1975	Overnight, sitting level	200–9000
India		
1982	Cooking with wood	15,000
	Cooking with dung	18,000
1988	Cooking with charcoal	5500
	Cooking, measured near ceiling	4000–21,000
Nepal, 1986	Cooking with wood	8800
China, 1987	Cooking with wood	2600
Gambia, 1988	24 h	1000–2500
Kenya, 1987	24 h	1200–1900

Source: Courtesy of Smith, K.R., School of Public Health, University of California–Berkeley.

11.3.3.1.2 Gas Cooking Appliances

Gas cooking stoves and ovens can be significant sources of carbon dioxide (CO_2), NO_2, nitrogen oxides (NO_x), aldehydes, RSPs, and a variety of VOCs. Levels of CO in kitchens with operating gas stoves have been reported to be in the range of 10 to 40 ppmv. Nitrogen dioxide levels in homes using gas stoves and ovens have been reported in the range of 18 to 35 ppbv, compared to 5 to 10 ppbv in non-gas-cooking residences.

Exposure to elevated levels of NO_2 associated with gas cooking has been proposed as a potential causal factor of respiratory symptoms and pulmonary function changes in school-age children living in gas-cooking homes. However, a number of epidemiological studies evaluating the relationship between gas cooking (and, to a lesser extent, associated NO_2 levels) and respiratory disease have not shown a consistent pattern between such exposures and respiratory symptoms or lung function changes in either children or adults.

In a recent Australian study, the presence of a gas cooking stove and exposure to NO_2 were observed to be significant risk factors for both asthma (increased threefold) and respiratory symptoms (increased twofold) in children. Though an independent relationship was observed for asthma and NO_2 exposures, statistical analyses revealed that some other aspect of gas stoves posed an additional asthma risk.

11.3.3.1.3 Space Heaters

Unvented gas and kerosene heaters are widely used to heat residential spaces. Gas space heaters are commonly used in southern U.S. states and Australia where winter temperatures are relatively moderate. Kerosene heaters are widely used in Japan and were once popular (late 1970s to early 1980s) in the U.S. as a space heating alternative.

Modern unvented space heaters have low emissions of CO and do not pose the asphyxiation hazard they once did. Nevertheless, they have the potential to significantly contaminate indoor spaces with NO_2, nitric oxide (NO), CO, CO_2, SO_2, aldehydes, VOCs, and RSPs. Emissions depend on heater or burner design, fuel used, and heater maintenance.

Unvented kerosene and gas-fired space heaters are used by homeowners and apartment dwellers under varying home space conditions, ventilation rates, number of heaters used, and daily and seasonal hours of operation. In a large monitoring study, northeastern U.S. homes operating one kerosene heater averaged ~20 ppbv NO_2; with two heaters, 37 ppbv; control homes averaged ~4 ppbv. Approximately 8% of kerosene heater-using homes had peak NO_2 levels of >255 ppbv. Over 20% had average SO_2 levels of >240 ppbv, the 24-h NAAQS. Other studies have reported CO levels of 1 to 5 ppmv in kerosene heater-using homes, with RSP increases above background ranging from 10 to 88 µg/m.[3] Elevated SO_2 levels were associated with use of grade 2-K kerosene, which has a S content of ~0.2%.

Epidemiological studies of households using kerosene heaters for space heating have shown that children under the age of 7 exposed to average NO_2 levels of >16 ppbv had more than a twofold higher risk of lower respiratory symptoms and illness (fever, chest pain, productive cough, wheeze, chest cold, bronchitis, pneumonia, or asthma). They also experienced increased risk of upper respiratory symptoms (sore throat, nasal congestion, dry cough, croup, or head cold).

11.3.3.1.4 Gas Fireplaces

Ventless gas-burning fireplaces have been installed in hundreds of thousands of new homes in North America. Limited studies have indicated that in high-altitude Colorado, such fireplaces could produce CO levels that varied from 40 to >100 ppmv, with elevated levels of NO_x as well. Carbon dioxide levels often exceeded the 5000-ppmv OSHA 8-h permissible exposure limit. At such levels, CO_2 is a respiratory stimulant; i.e., it increases breathing rates.

11.3.3.2 Vented Combustion Systems

In general, vented combustion systems such as gas, oil, and wood furnaces; wood- and coal-burning stoves; and wood-burning fireplaces pose limited indoor contamination concerns. However, flue gas venting systems are not perfect and some flue gas spillage can be expected under normal operation, particularly for systems using natural draft. Wood-burning appliances, because of design deficiencies and a requirement for manual recharging, pose unique indoor contamination concerns.

11.3.3.2.1 Flue Gas Spillage

Flue gas spillage commonly occurs in residences using natural draft gas furnaces, gas water heaters, or wood-burning appliances. In most cases spillage is momentary, resulting in minimal emission of pollutants to the building environment. Serious flue gas spillage does occur, with occasional deaths or, more commonly, sublethal CO poisoning. Such spillage usually occurs in aging or poorly installed or maintained combustion and flue systems.

Flue gas spillage occurs as a result of chimney blockage or backdrafting. Backdrafting results when the chimney is cold, under some weather conditions, and when the house is depressurized by competing exhaust systems (e.g., fireplaces, exhaust fans, and furnaces). Backdrafting is also a concern in energy-efficient houses that do not have sufficient air for combustion or exhaust.

Flue gas spillage also occurs when furnace or flue components malfunction as a result of corrosion, improper installation, or, in the case of flue pipes, accidental disconnection. Perforation of heat exchange elements or flue pipes is more common as systems age.

11.3.3.2.2 Wood-Burning Appliances

During the late 1970s to the mid-1980s, wood-burning appliances (stoves and furnaces) became popular as residential space heating alternatives in the U.S. An estimated 8 to 10 million units were in use. Though no recent surveys have been conducted, it appears that with fuel price stabilization, wood heating appliance use has dramatically decreased.

Combustion by-products may be emitted from wood-burning appliances because of improper installation; negative air pressure and downdrafts; leaks in appliance parts and flue pipes; and in starting, reloading, stoking, and ash removal operations. Wood-burning appliances have been reported to emit CO, NO_x, SO_2, aldehydes, RSPs, and a variety of VOCs. Some of the 200+ VOCs identified in wood smoke are known carcinogens, e.g., PAHs.

Significant contamination of indoor spaces by wood-burning emissions has been reported for conventional stove and furnace units. Airtight appliances, on the other hand, emit only trace quantities of combustion gases and particles into indoor spaces. The effect of heater type on PAH concentrations both indoors and outdoors is illustrated in Figure 11.7.

Because they are open to room air, fireplaces are potentially significant sources of indoor air contamination during use. However, they are employed primarily for their aesthetic appeal, and frequency of use is often limited.

In some parts of the U.S. where wood supplies are abundant and firewood inexpensive, wood-burning appliance use was so extensive that emissions resulted in ambient concentrations that exceeded air quality standards for PM_{10}. This was the case in mountain valley communities such as Missoula, MT, Corvallis, OR, Aspen, CO, and Danbury, VT. As a result, the USEPA promulgated a new source performance standard (NSPS) for wood-burning stoves and furnaces; wood stoves and furnaces subject to the NSPS have significantly decreased emissions of both PM and CO to both ambient and indoor environments.

11.3.3.3 Miscellaneous Sources

A variety of combustion-generated substances cause contamination of indoor spaces as a result of human activities. The most important of these is tobacco

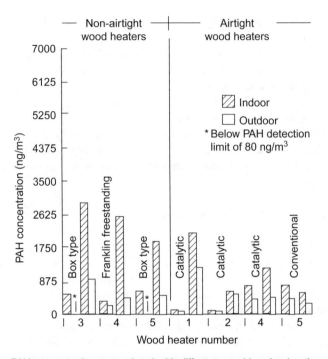

Figure 11.7   PAH concentrations associated with different wood-burning heating appliances. (From Knight, C.V., Humpreys, M.P., and Pennex, J.C., *Proceedings of IAQ'86: Managing Indoor Air for Health and Energy Conservation*, ASHRAE, Atlanta, 1986. With permission.)

smoking. Burning candles and incense in homes has, in the past decade, become an increasingly popular activity. Combustion by-product contamination of indoor spaces may result from use of propane-fueled floor burnishers and forklifts, propane- and gas-fueled ice-resurfacing machines, arena events involving gasoline-powered vehicles, entrainment of motor vehicle emissions, reentry of flue gases, and, surprisingly, deterioration of duct liner materials.

11.3.3.3.1 Tobacco Smoking

Use of tobacco products by approximately 40 million smokers in the U.S. results in significant indoor contamination from combustion by-products that poses significant exposures to millions of others who do not smoke but must breathe contaminated indoor air. Several thousand gas- and particulate-phase compounds have been identified in tobacco smoke. The more significant of these include RSPs, nicotine, nitrosamines, PAHs, CO, CO_2, NO_x, acrolein, HCHO, and hydrogen cyanide (HCN).

ETS is a combination of sidestream (SS) smoke (released between puffs) and exhaled mainstream (MS) smoke. Exposure to ETS depends on the number and type of cigarettes consumed in a space, room volume, ventilation rate, and proximity to the source. Concentrations of major tobacco-related contaminants measured in a

Table 11.2 Tobacco-Related Contaminant Levels in Buildings

Contaminant	Type of Environment	Levels	Nonsmoking Controls
CO	Room (18 smokers)	50 ppmv	0.0 ppmv
	15 restaurants	4 ppmv	2.5 ppmv
	Arena (11,806 people)	9 ppmv	3.0 ppmv
RSPs	Bar and grill	589 µg/m³	63 µg/m³
	Bingo hall	1140 µg/m³	40 µg/m³
	Fast food restaurant	109 µg/m³	24 µg/m³
NO₂	Restaurant	63 ppbv	50 ppbv
	Bar	21 ppbv	48 ppbv
Nicotine	Room (18 smokers)	500 µg/m³	—
	Restaurant	5.2 µg/m³	—
Benzo-α-pyrene	Arena	9.9 ng/m³	0.69 ng/m³
Benzene	Room (18 smokers)	0.11 mg/m³	—

Source: From Godish, T., *Sick Buildings: Definition, Diagnosis & Mitigation*, Lewis Publishers/CRC Press, Boca Raton, FL, 1995.

variety of building environments are summarized in Table 11.2. From these as well as other data, tobacco smoke (where smoking is occurring) appears to be the major source of RSPs in many indoor environments.

Health hazards of MS smoke on smokers are well known. It is the major cause of lung cancer (and causes other cancers as well) and respiratory diseases such as chronic bronchitis and pulmonary emphysema. It is also a major risk factor for cardiovascular disease.

A number of gas- (and presumably particulate-) phase constituents of MS and SS smoke are potent mucous membrane and upper respiratory system irritants in both smokers and nonsmokers. Most notable of these irritants are the aldehydes, acrolein and HCHO. Exposure to acrolein in a smoke-filled room is believed to be the major cause of eye irritation. Nonsmoking office workers exposed to ETS have reductions in small airway function comparable to those in smokers consuming 1 to 10 cigarettes per day. Other workplace studies indicate that nonsmokers exposed to ETS experience more cough, phlegm production, eye irritation, chest colds, and days lost from work than those not exposed.

Respiratory symptoms in infants, especially significant when exposure occurs *in utero,* have been associated with parental smoking. Exposure to ETS from parental smoking is a major risk factor for asthma in children. Asthma attack incidence decreases dramatically when parents quit smoking or smoke outdoors.

Involuntary exposure to ETS appears to be a major risk factor for lung cancer. In a review of a number of epidemiological studies, the USEPA concluded that ETS poses a significant lung cancer risk, with an estimated 3000 deaths per year associated with such exposure.

ETS may also increase the lung cancer risk associated with exposure to Rn. Tobacco smoke particles serve as foci for RDP deposition, suspension in air, and transport and deposition in lung tissue.

Recent epidemiological studies have indicated that the risk of cardiovascular disease mortality associated with ETS exposure constitutes a more serious public health problem than ETS-associated lung cancer. An estimated 35,000 to 40,000 deaths annually in the U.S. have been attributed to ischemic heart disease associated with ETS exposure.

11.3.3.3.2 Candles and Incense

Millions of homeowners and apartment dwellers burn candles daily or several times per week. Because of low flame temperatures, candles can produce significant RSPs. The effect of candle burning and other human activities on indoor PM levels can be seen in Figure 11.8. Soot particles, because of their small size and potential to contain potent carcinogens, may pose a small but unknown health risk.

In epidemiological studies evaluating risk factors for childhood leukemia, a significant increase in childhood leukemia was observed in households where incense was burned once a week or more during pregnancy, with an even greater risk when it was burned regularly. Incense is reported to produce benzo-α-pyrene, several other PAHs, and the nasal carcinogen, sinapaldehyde.

The primary environmental concern allegedly caused by candle burning is the soiling of building surfaces, called ghosting. Though candle burning is known to cause soiling of interior building surfaces, ghosting has, paradoxically, been reported in residences where no candle or incense burning has occurred or, if so, at a minimal level.

11.3.3.3.3 Gas and Gasoline-Using Equipment

A number of propane- or gas-fueled devices and machines used indoors can cause significant indoor contamination. These include floor burnishers, forklifts, ice resurfacers, and motor vehicles operated in arenas.

Propane-fueled burnishers are used to polish vinyl flooring in small to large buildings. They produce significant emissions of CO and other combustion by-products. Employees using such devices have reportedly been exposed to concentrations of several hundred ppmv CO. In a case investigation of illness in an unventilated building (a child care center), CO concentrations of >500 ppmv and carboxyhemoglobin levels as high as 26% were reported.

Propane-fueled forklifts are widely used in industry and warehouse operations. They pose exposure concerns when used in warehouse operations with attached office facilities. Carbon monoxide levels of 50 to 75 ppmv have been observed in office spaces adjoining warehouse operations.

Skating and hockey rinks utilize resurfacing machines that are propane or gasoline fueled. Recent reports indicate CO concentrations in such rinks are in the range of 10 to 30 ppmv. Nitrogen dioxide concentrations are also relatively high. Short-term exposure concentrations of >1 ppmv are not uncommon, with peak levels of ~3 ppmv. Seven-day average NO_2 concentrations of ~225 and 360 ppbv have been reported in rinks in Europe and Japan, and in the northeastern U.S., respectively.

Exposures to relatively high CO and NO_2 in skating rinks may be of particular concern to those who (1) practice for extended hours; (2) exercise significantly, e.g., hockey players; or (3) are full-time employees. For other users, exposures are relatively transient.

Figure 11.8 Effects of various indoor combustion sources and cooking activities on indoor PM$_{2.5}$ concentrations. (Courtesy of Koutraikis, P., School of Public Health, Harvard University, Boston, MA.)

Arena events involving monster trucks, tractor pulls, and motorcross events are popular in North America. Average CO concentrations in the range of 80 to 140 ppmv have been reported for such events, with peak levels of >225 ppmv.

11.3.3.3.4 Entrainment and Reentry

Combustion-generated contaminants from motor vehicles, nearby incinerators and boilers, residential heating appliances (e.g., fireplaces), leaf burning, etc., can enter buildings by infiltration or through open windows or air intakes. Contamination problems arise when major sources are nearby. Motor vehicle pollutants become entrained in building air systems from loading docks and stack effect phenomena.

Indoor air can also be contaminated by combustion by-products as a result of reentry. This may occur in residences when fireplace smoke is drawn indoors after it has been exhausted from a chimney. Factors that cause building depressurization, such as use of exhaust fans and fireplaces themselves, increase reentry risks. Reentry commonly occurs in large buildings when flue gas discharges are located upwind close to outdoor air intakes.

11.3.3.3.5 Duct Liners

Duct liners are fiberglass insulating materials used to reduce (1) loss of heat and cold through supply ducts and AHUs, and (2) sound transmission from supply and return air ducts and AHUs. To reduce friction and dust accumulation, the airstream surface commonly have been coated with acrylic latex polymers.

These polymers are colorless but are both extended (increased bulk) and pigmented by addition of carbon black (~30% by weight). Since they contain unreacted double bonds, they are subject to cracking, even under low ambient O_3 levels.

Over time, the latex fragments may be blown out of the duct system as large visible fragments or smaller particles of liberated carbon black. This deterioration has recently been suggested to be the cause of the ghosting phenomenon reported in some residences in the U.S. Particles are thermophoretically deposited along thermal gradients. Particle deposition is commonly observed to be more intense on cooler wall surfaces. Ghosting can be seen in Figure 11.9.

11.3.4 Aldehydes

Aldehydes are organic substances that belong to a class of compounds called carbonyls, with the carbonyl group in a terminal position. Aldehydes include a number of substances that differ in molecular structure, solubility, reactivity, and toxicity. Only a relatively few have industrial or commercial applications that may cause significant indoor exposures, are by-products of processes, or have biological activities that have the potential to pose major, or even minor, IAQ concerns. Those known to cause significant indoor air contamination or health effects are HCHO, acrolein, and glutaraldehyde. Most aldehydes are sensory (mucous membrane) irritants; some are skin sensitizers; some may be human carcinogens.

11.3.4.1 Formaldehyde

Widespread contamination of residences and, to a much lesser extent, public access buildings, by HCHO came to the attention of public health authorities in the late 1970s and early 1980s as a result of health and odor complaints associated with urea-formaldehyde foam insulation (UFFI) and mobile and conventional homes constructed or furnished with pressed wood products. Like asbestos, HCHO was a widely used industrial and commercial chemical. As a consequence, it was, and continues to be, a ubiquitous contaminant of indoor air. However, changes in product manufacture that significantly lowered emissions, and changes in materials used in construction, have resulted in significantly lower indoor levels than once were common. As such, HCHO is no longer the air quality or public health issue that it once was.

Formaldehyde is a potent mucous membrane irritant causing, on acute exposure, eye and upper respiratory system symptoms. It is also a potent dermal irritant. On chronic exposure, it appears to cause neurological symptoms such as headache, fatigue, disturbed sleep, and depression. It is a sensitizing substance causing allergic reactions of the skin, whole-body reactions from renal dialysis exposures, and asthma. Animal studies, as well as epidemiological studies of exposed workers and residents of mobile homes, indicate that it is a potential human carcinogen. A variety of field investigations, as well as systematic epidemiological studies, have demonstrated strong associations between HCHO exposures and health effects commonly

Figure 11.9 Ghosting in a residential building.

found in residences during the early 1980s. Statistically significant relationships have been reported for residential HCHO exposures and eye irritation, runny nose, sinus congestion, phlegm production, cough, shortness of breath, chest pain, headache, fatigue, unusual thirst, sleeping difficulty, dizziness, diarrhea, rashes, and menstrual irregularities.

11.3.4.1.1   Sources and Levels

Although HCHO has been, and continues to be, used in a variety of products, only a relatively few emitted free HCHO at levels that caused significant indoor air concentrations. Problem products included pressed wood materials such as particleboard (underlayment, paneling, cabinetry, furniture), medium-density fiberboard (furniture, cabinetry), hardwood plywood (paneling, cabinetry), UFFI, and acid-catalyzed finishes (applied to furniture and cabinetry).

Urea-formaldehyde resins are chemically unstable. They release HCHO from the volatilizable, unreacted HCHO in the resin and from hydrolytic decomposition of the resin copolymer itself. Release of the former was responsible for initially high HCHO levels associated with new mobile homes, conventional homes with particleboard underlayment, and homes that had been recently insulated with UFFI.

Formaldehyde-emitting products differ in emission potentials. As a consequence, indoor HCHO levels reflect the nature and potency of sources. Indoor levels are also affected by the quantity of HCHO-emitting material present. Buildings with high load factors (surface area of HCHO-emitting products:air volume ratio), such as mobile homes, experienced the highest reported HCHO levels.

Most research on HCHO levels in residences was conducted in mobile homes, UFFI houses, and conventional houses with a variety of HCHO sources in the late 1970s and early to mid-1980s. As a consequence, data on HCHO levels reflect that period, which can be described as historically high for indoor HCHO.

Highest levels of HCHO were reported for mobile or manufactured homes. Concentrations varied as a function of mobile home age, environmental factors at the time of testing, and geographical area. Concentrations varied from 0.05 to >1 ppmv, with average concentrations in the range of 0.09 to 0.18 ppmv. In general, concentrations of HCHO in mobile homes manufactured today are much lower. Some new mobile homes may have initial HCHO levels as high as 0.15 ppmv, but most are <0.10 ppmv. Residences with particleboard underlayment had HCHO concentrations that, at their extreme, exceeded 0.50 ppmv, but were more typically in the range of 0.05 to 0.20 ppmv. Particleboard underlayment in conventional stick-built houses today is uncommon, as it has been replaced by oriented-strand board, a material with very low HCHO emissions.

Residences with UFFI initially experienced HCHO levels of ≥0.40 ppmv, but these decreased rapidly to a range of 0.03 to 0.12, with an average of 0.06 ppmv 1 or 2 years after application. Though still widely used in Great Britain, very few U.S. houses are newly insulated with UFFI today. Residences with one or more miscellaneous sources (e.g., pressed wood furniture, hardwood plywood, pressed wood cabinets) had HCHO concentrations that ranged from 0.03 to 0.15 ppmv, with average concentrations of 0.03 to 0.05 ppmv.

Formaldehdye concentrations decrease significantly with time. As a consequence, highest concentrations and exposures occur in building environments where HCHO-emitting products are new.

11.3.4.1.2 Guideline Values and Their Adequacy

In response to building-related problems, a number of countries adopted air quality guidelines for HCHO exposures in residences. Countries with guideline values of 0.10 to 0.12 ppmv include Australia, Germany, Denmark, the Netherlands, and Italy. Canada and California have adopted guideline values of 0.10 ppmv as an action level (in need of remediation), with a target level of 0.05 ppmv. The WHO recommends a guideline value of 0.082 ppmv.

Compliance with these guideline values may not be sufficient to protect the health of sensitive populations, e.g., asthmatic children. Studies conducted in Arizona have shown significant pulmonary function deficits in asthmatic children with average household exposures of 0.06 ppmv. Other studies in Canada observed dose–response relationships between HCHO levels and eight symptoms at average exposure concentrations of 0.045 ppmv. Studies in Australia have shown that HCHO exposures of >0.05 ppmv result in increased breath levels of NO, a primary indicator of inflammatory responses.

11.3.4.2 Acrolein

Acrolein is an aldehyde with one carbon–carbon double bond, which makes it highly reactive. It is released into the environment as a combustion oxidation product from oils and fats, wood, tobacco, and automobile or diesel fuels and produced in ambient air as a result of photochemical reactions.

Acrolein is a potent eye irritant causing tearing at relatively low (<0.10 ppmv) exposure concentrations. It is believed to be the primary cause of eye irritation in smoke-filled rooms and one of the major eye irritants in photochemical smog or haze.

11.3.4.3 Glutaraldehyde

Glutaraldehyde is a dialdehyde. It is widely used as a biocidal ingredient in medical and dental disinfectants and not uncommonly in disinfectant applications such as duct cleaning. It is also reported to have been used in carbonless copy paper.

Glutaraldehyde, though relatively nonvolatile, is a potent mucous membrane irritant. Irritant effects include nose, throat, and sinus symptoms. Other symptoms associated with exposures include nausea, headache, chest tightening, and asthma.

11.3.5 Volatile and Semivolatile Organic Compounds

11.3.5.1 Volatile Organic Compounds

A large variety of natural and synthetic organic compounds are found in indoor air. These include compounds that have boiling points ranging from (1) <0 to 50–100°C (<32 to 122–212°F) (very volatile organic compounds (VVOCs)), to (2) 50–100 to 240–260°C (464 to 500°F) (VOCs), to (3) 240–260 to 380–400°C (716 to 752°F) (SVOCs). Because of their variety, relative abundance in indoor air, potential to cause sensory irritation and central nervous system symptoms, and vapor-phase presence, VOCs have received significant attention as potential causal factors in building-related illness symptoms, particularly among office workers. In the office context, VOCs also include the more volatile VVOCs.

VOCs include aliphatic hydrocarbons (HCs), which may be straight, branched chained, or cyclic; aromatic HCs; halogenated HCs (primarily by chlorine); and oxygenated HCs, such as aldehydes, alcohols, ketones, esters, ethers, and acids. They are emitted by a variety of sources, including building materials and furnishings, consumer products, building maintenance materials, humans, office equipment, and tobacco smoking.

Studies in Danish residences and office buildings have reported the presence of 40 different VOCs, consisting mainly of C_2–C_{13} alkanes, C_6–C_{10} alkylbenzenes, and terpenes. In a study of 42 commonly used building materials, Danish investigators identified emissions of, on average, 22 different compounds associated with each material. The 10 compounds with the highest steady-state concentrations were toluene, n-xylene, terpene, n-butylacetate, n-butanol, n-hexane, o-xylene, ethoxyethylacetate, n-heptane, and p-xylene. Organic compounds identified in 40 east Tennessee homes included toluene, ethyl benzene, n-xylene, p-xylene, nonane, cumene, benzaldehyde, mesitylene, decane, limonene, undecane, naphthalene, dodecane, tridecane, tetradecane, pentadecane, hexadecane, and 2,2-methlynaphthalene.

Concentrations of individual VOCs measured in indoor air are typically several orders of magnitude lower than occupational threshold limit values (TLVs) or OSHA permissible exposure levels (PELs). As a consequence, it is not likely that individual VOCs (other than HCHO) are responsible for building-related health complaints.

A potential causal relationship between sick building-type symptoms and exposures to a combination of VOCs at relatively low individual concentrations has been proposed by Danish investigators and supported by human exposure studies. It is described as the total volatile organic compound (TVOC) theory. In the TVOC theory, sensory irritation and possibly neurological symptoms (headache and fatigue) in problem buildings may be due to the combined effect of the many VOCs found in indoor air.

Effects have been reported on exposure to a mixture of 22 different compounds at concentrations as low as 1 ppmv (5 mg/m^3) TVOC (toluene equivalent) as well as submixtures of these compounds. A proposed dose–response relationship between TVOC exposures and discomfort and health effects is indicated in Table 11.3.

A number of investigations have been conducted to determine whether building-related illness symptoms are associated with exposures to TVOCs. Although no such relationships were observed in Danish town hall and California office building studies, significant correlations were observed between the logarithmic value of TVOC concentrations and illness symptoms in problem Swedish school buildings (Figure 11.10). Significant log-linear relationships were also observed between symptom prevalence and concentrations of terpenes, C_8–C_{11} n-alkanes, and butanols in problem Swedish office buildings.

A number of VOCs found in indoor air are mutagenic or carcinogenic (e.g., benzene, styrene, tetrachloroethylene, 1,1,1-trichloroethane, trichloroethylene, dichlorobenzene, methylene chloride, and chloroform). This suggests that VOC exposures in indoor air have the potential (albeit at very low risk) to cause cancer in humans. The nature of potential health risks associated with multiple potential carcinogens at very low exposure levels is not known.

11.3.5.2 Semivolatile Organic Compounds

SVOCs have low vapor pressures (10^{-1} to 10^{-7} mmHg) and exist in gaseous and condensed phases (adsorbed on particles). They include a variety of compound types, including solvents; long-chained alkanes, aldehydes, and acids; pesticides; PAHs; polychlorinated biphenyls (PCBs); and plasticizers. Concentrations of selected SVOCs identified in Danish buildings are summarized in Table 11.4. Many of these are plasticizers, present in relatively low concentrations.

11.3.5.2.1 Plasticizers

Plasticizer compounds such as phthalic acid esters are ubiquitous contaminants of indoor air and dust. They are used in vinyl products to make them soft and relatively flexible. They typically comprise 25 to 50% of the weight of resilient vinyl floor covering. As such materials lose their plasticizers, they become hard and brittle and subsequently crack.

Some scientists have suggested that exposures to phthalic acid ester plasticizers may be partly responsible for the significant increase in asthma observed in devel-

Table 11.3 Dose–Response Model Relationships between TVOC Exposures and Discomfort/Health Effects

TVOC Concentration (mg/m³)	Response	Exposure Range
<0.20	No effects	Comfort range
0.20–3.0	Irritation/discomfort possible	Multifactorial exposure range
3.0–25.0	Irritation and discomfort; headache possible	Discomfort range
>25.0	Neurotoxic effects	Toxic range

Source: From Molhave, L., Proc. 5th Int. Conf. Indoor Air Qual. Climate, Toronto, 5, 15, 1990.

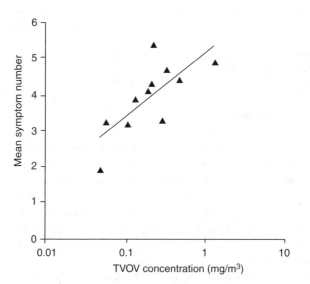

Figure 11.10 Dose–response relationships between building TVOC concentrations and symptom prevalence. (From Norback, D., Michel, I., and Winstrom, J., Scand. J. Work Environ. Health, 16, 121, 1990. With permission.)

oped countries over the past two decades. An association between the presence of plastic interior surfaces and bronchial obstruction in children has been reported. It has been suggested that the structural similarity between plasticizer compounds and prostaglandins (which mediate inflammatory responses) may be a contributing factor to asthma initiation.

11.3.5.2.2 Polychlorinated Biphenyls

Polycholorinated biphenyls are SVOCs that may pose indoor air exposure and health risks. The primary source of PCBs in indoor air and building surfaces appears to be ballasts in old fluorescent lighting fixtures. Such contamination is particularly significant when a building has experienced a structural fire.

Table 11.4 Concentrations of Selected SVOCs Measured in Four Danish Office Buildings, a Day Care Center, and School Classroom (ng/m³)

Compound	Office 1	Office 2	Office 3	Office 4	Day Care Center	School Classroom
Tridecane	1128	890	1070	980	1144	4599
Texanol	2055	4378	1910	3871	5492	1451
Pentadecane	1003	863	1173	1226	1306	1488
Dodecanoic acid	1144	987	2174	2146	3040	2987
TXIB	957	710	658	1282	8341	7803
Heptadecane	993	718	1039	965	952	692
Diisobutyl phthalate	1393	977	891	578	1240	1279
Dibutyl phthalate	789	710	1078	840	1346	1195
Di-2-ethylhexyl phthalate	201	278	241	404	1053	111

Source: From Clausen, P.A., Wolkoff, P., and Svensmark, B., *Proc. 8th Int. Conf. Indoor Air Qual. Climate*, Edinburgh, 2, 434, 1999.
Note: TXIB = 2,2,4-Trimethyl-1,3-pentadioldiisobutyrate

11.3.5.2.3 Pesticides

Pesticides are toxic compounds formulated to control a variety of pest species. Indoor use includes a number of applications, e.g., (1) biocides used by manufacturers to prevent biodeterioration of products and by homeowners or tenants to disinfect surfaces such as toilet bowls, etc.; (2) insecticides to control cockroaches, flies, ants, spiders, and moths; (3) termiticides; (4) flea and tick sprays and pet shampoos; (5) insecticides for indoor and outdoor plants; and (6) fungicides and herbicides used as lawn and garden treatments (the latter two may enter the home on shoes and clothing).

Because of the deliberate use of toxic substances indoors and their passive transport on shoes and subsequent indoor accumulation, pesticides are unique contaminants of indoor environments. Users have only a limited understanding of the potential toxic hazard to which they and other building occupants may be exposed.

Biocides are used to control the growth of microorganisms such as bacteria and fungi. They are used as disinfectants in cleaning agents, on AHU filters and carpeting, and in paints and duct cleaning. Commonly used biocides include hypochlorites, alcohols, o-phenylphenol, and quaternary ammonium compounds. Of these, o-phenylphenol has been one of the most common pesticide compounds observed in indoor air and dust samples. Mercury (Hg) is also commonly found, as it was widely used as a biocide in latex-based paint until its use was discontinued in a voluntary agreement between manufacturers and the USEPA in 1990.

In general, fungicides are not used indoors. However, pentachlorophenol (PCP), a pesticide used to prevent wood decay in outdoor timbers, is one of the most common pesticides found in human blood samples. The source of this widespread exposure is unknown. Exposures to PCP have been associated with PCP-treated logs in log houses. Because it is a teratogen and potential human carcinogen, its use has been reduced.

Insects represent major pest control concerns in buildings. Active ingredients in insecticidal formulations used in or around buildings have included diazinon, p-dichlorobenzene, chlordane, heptachlor, lindane, aldrin, dieldrin, bendiocarb, methoxychlor, propoxur, methyl demeton, naphthalene, and pyrethrins. Para-dichlorobenzene is the active ingredient in mothballs, and naphthalene in moth flakes. Major insecticidal classes include organochlorines, organophosphates, carbamates, and pyrethrins.

Organochlorines, or chlorinated hydrocarbons, such as chlordane, heptachlor, and p-dichlorobenzene, have been used indoors — chlordane and heptachlor as termiticides, and p-dichlorobenzene as a moth control agent. Despite the fact that use of chlordane and heptachlor has been restricted (not used as termiticidal applications in the U.S. for >15 years), these two substances are the most common and abundant pesticidal chemicals found in residential air and dust samples. This is due to their long environmental lifetimes, which are extended indoors because of limited exposure to ultraviolet light. Para-dichlorobenzene is also one of the most common and abundant pesticides found indoors. The primary health concern associated with these compounds, and the reason for restrictions on their use, is their potential human carcinogenicity. Organochlorine exposure risks are increased by the fact that they tend to bioaccumulate in fat tissue.

Organophosphates such as dichlorvos, chlorpyrifos, diazinon, and malathion are characterized by their acute neurotoxicity. Dichlorvos has been used in bug bombs, slow-release pest strips, and dog and cat flea collars. Because of health concerns (animal carcinogenicity), their use has dramatically decreased since 1988. Chlorpyrifos was once widely used indoors for most termiticidal treatments and to control cockroaches and fleas. Because of potential exposures to children, its use was phased out in 2000 in a USEPA–industry agreement. Use of diazinon has been restricted for similar reasons.

Because of their lower toxicity and health risks, pyrethrins and their synthetic cousins, pyrethroids, are increasingly being used to control insects indoors. Though persistent, they are rarely observed in the gas phase.

The herbicide 2,4-D (widely used on lawns to control broad-leaved species) is commonly found in indoor dust samples, apparently due to track-in phenomena. Exposure to 2,4-D has been suggested to be a risk factor for non-Hodgkin's lymphoma.

Pesticides pose a variety of indoor exposure and health concerns. These include acute symptoms associated with misapplication, their persistence in house dust, bioaccumulation of organochlorine compounds in human fat tissue, suspected human carcinogenicity of a number of pesticides, children's exposures to house dust while at play and from pets' flea collars, and potential immunological effects.

11.3.6 Biological Contaminants

It is becoming increasingly evident that exposure to particulate- and, to a lesser extent, gas-phase biological contaminants is responsible for a significant percentage of illness symptoms and disease reported to be associated with building environments. These contaminants include airborne mold (fungi), bacteria, and viruses that may be viable (living) or nonviable; fragments of bacteria and mold that may contain antigens/allergens, endotoxins, glucans, or mycotoxins; microbially produced VOCs (MVOCs); and antigens/allergens produced by insects, arachnids (mites/spiders), and pets.

Biological contaminants such as bacteria, viruses, and fungi can cause infectious disease as a result of airborne transmission. Exposures to mold spores or fragments and other allergenic substances produced by insects, arachnids, and pets can cause immunological sensitization, which may result in allergic rhinitis and, in more severe cases, asthma. Exposures to bacterial cells or mold spores may cause hypersensitivity pneumonitis. Respiratory system inflammatory responses may be produced as a result of exposures to bacterial endotoxins, fungal glucans, or MVOCs. Mycotoxin exposures may, depending on the substance, produce a variety of adverse physiological responses.

Of particular public health concern are those biological contaminants that cause major illness syndromes such as chronic allergic rhinitis, asthma, hypersensitivity lung disease, Legionnaires' disease, and TB.

11.3.6.1   Illness Syndromes

11.3.6.1.1   Chronic Allergic Rhinitis

Chronic allergic rhinitis, or common allergy, is a minor illness characterized by inflammatory symptoms of the upper respiratory system, including runny nose, congestion, phlegm production, cough, and sneezing. These inflammatory responses in sensitized individuals are caused by exposure to substances that have antigenic properties. In indoor environments, major sources of allergens include dust mite and cockroach fecal waste, mold spores and hyphal fragments, and pet danders.

Approximately 40% of the U.S. population has been sensitized to common allergens, with about half this number experiencing clinical symptoms. About 10% of the U.S. population (~25 million people) has experienced chronic allergic rhinitis severe enough to be diagnosed and treated by a physician.

11.3.6.1.2   Asthma

Asthma is an ailment that affects the respiratory airways. It is characterized by episodes of severe bronchial constriction resulting in chest tightness, shortness of breath, cough, and wheeze.

Asthma prevalence in the U.S. has been estimated to be 4.3%; i.e., ~12 million individuals a year experience mild to severe symptoms. Its prevalence is much higher in children, with rates of 8.3% in midwestern children, 11.8% in southern children, and 13.4% in black children. The prevalence rate has increased 50% since 1980, with the annual death rate doubling to ~5000/year.

Asthma is a major medical cause of absenteeism in school-age children and the leading cause of childhood hospital admission and long-term medication use. Approximately 1% of all health care costs are reportedly associated with treatment of asthmatic symptoms.

The primary cause of asthma is exposure to, and sensitization by, inhalant allergens associated with dust mites, mold, pet danders, or cockroaches.

11.3.6.1.3 Inflammatory Lung Disease

Inflammatory lung diseases resulting from indoor exposures include hypersensitivity pneumonitis and humidifier fever.

Hypersensitivity pneumonitis, or extrinsic allergic alveolitis, is a group of immunologically mediated lung ailments associated with repeated exposures to small-diameter biological aerosols that affect alveoli (site of gas exchange).

Classic symptoms occur after immunological sensitization is induced. They may include fever, chills, nonproductive cough, shortness of breath, myalgia (ache all over), and malaise. Onset occurs within hours of exposure, with spontaneous recovery in 18 to 24 h. In office and industrial environments, symptoms diminish as the workweek progresses and increase in severity at the beginning of a new week. Chronic exposures cause irreversible lung tissue damage, progressive shortness of breath, cough, malaise, weakness, and weight loss.

Outbreaks of hypersensitivity pneumonitis in buildings have been associated with contaminated components of heating, ventilating, and air-conditioning (HVAC) systems (including condensate drip pans and ductwork) and a variety of mold-infested materials. Hypersensitivity pneumonitis appears to be caused by exposures to very high air concentrations of bacteria (particularly *Actinomycetes*) or small-diameter mold spores (e.g., those of *Penicillium* and *Aspergillus*).

Humidifier fever is a respiratory ailment not easily distinguished from hypersensitivity pneumonitis. Clinical manifestations include fever, chest tightness, increased respiratory rate, airway hyperreactivity, and bronchitis. Symptoms are produced as a result of inflammatory responses of lung tissue on exposure to bacterial endotoxins. Gram-negative bacterial contamination of cool-mist humidifiers (commonly used in European office buildings) in mechanically ventilated buildings is believed to be the cause of humidifier fever.

11.3.6.1.4 Infectious Diseases

A number of infectious diseases are transmitted in building environments as a result of inhalation exposures. These include viral diseases such as colds and influenza, bacterial diseases such as TB and Legionnaires' disease, and fungal diseases such as aspergillosis. Of major IAQ concern are TB and Legionnaires' disease.

Until the early 20th century, TB was one of the major causes of death in the U.S. and many other countries. As a result of advances in diagnosis, prevention, and treatment, the prevalence of TB dramatically declined. It has experienced a small resurgence as a result of an influx of immigrants from countries where it is endemic, increases in susceptible populations (e.g., those immunocompromised), and development of antibiotic resistance by the causal bacterium.

Recent outbreaks of TB have occurred in hospitals, nursing homes, homeless shelters, correctional facilities, and residential AIDS care centers. The risk of exposure and subsequent infection among health care workers may be high in health care facilities treating patients with undiagnosed TB.

Legionnaires' disease has been reported in a number of indoor environments. The disease syndrome is characterized by cough, fever, shortness of breath, chest pain, myalgia, diarrhea, and confusion. It is a progressive pneumonia caused by infection by *Legionella pneumophila* or similar species, with a 15% fatality rate.

Legionnaires' disease may occur as outbreaks (10% of reported cases) or, more commonly, individual cases.

The causal organism grows well at elevated temperatures (~35°C, 95°F), producing abundant populations in waters of cooling towers and evaporative condensers, hot water heaters, whirlpools, spas, and hot tubs. Exposure results from inhalation of aerosols associated with these sources.

Infections by *Aspergillus fumigatus*, a common mold species found in dust and compost, is a major infection control issue in hospitals where opportunistic infections are a concern in emergency, oncology, and AIDS treatment wards. Aspergillosis is a fungal pneumonia that does not respond to conventional antibiotic therapy.

11.3.6.2 Other Health Concerns

A variety of new health concerns associated with exposures to biological contaminants have developed over the past decade as a result of increased scientific understanding of both exposures and potential health risks. These include exposures and health responses to bacterial endotoxins, fungal glucans, and mycotoxins.

11.3.6.2.1 Endotoxins

Endotoxins are large molecules produced in the outer membrane of gram-negative bacteria and released into the environment on the death of cells.

They are common contaminants of settled dusts in indoor environments. On respiratory tissue exposure, they induce inflammatory responses, producing a number of common symptoms. They are believed to cause some of the symptoms in building occupants reportedly due to "sick building syndrome" (SBS) (see Section 11.4.2) and inflammatory symptoms among children in daycare centers. There is scientific speculation that early childhood endotoxin exposures may subsequently protect individuals from immune system diseases such as chronic allergic rhinitis and asthma. As indicated previously, endotoxin exposure is believed to be the cause of humidifier fever.

11.3.6.2.2 Fungal Glucans

Fungal glucans are large molecules found in the cell walls of most fungal species. They are common in the settled dusts of buildings. Though they can also cause inflammatory responses in human respiratory systems, their potential for causing adverse health effects has not been well documented.

11.3.6.2.3 Mycotoxins

Mycotoxins are large molecules produced by many fungal species when they are subject to nutrient stress. Of particular note are mycotoxins produced by *Aspergil-*

lus species and the black mold, *Stachybotrys chartarum*. *Aspergillus flavus* produces the liver toxins and human carcinogens, aflatoxin A and B; *S. chartarum*, the highly toxic trichothecenes, saratoxin H and G, and verucarin A and B.

Historically, mycotoxin exposures and poisoning have been associated with contaminated livestock feed and human foodstuffs. Recently, however, concerns have been expressed in relation to inhalation exposures to mold spores containing highly toxic mycotoxins. This is particularly the case with *S. chartarum*, widely found in building environments infesting the face paper of gypsum board, ceiling tiles, and other processed wood fiber materials subject to repeated wetting (Figure 11.11).

Building investigations conducted by the U.S. Centers for Disease Control (CDC) in the 1990s linked potential *S. chartarum* mycotoxin exposure to hemosiderosis, a frequently fatal pulmonary hemorrhagic disease of infants. Animal exposure studies show that similar lung hemorrhaging can be induced by exposures to *S. chartarum* spores. However, in 2001, after conducting both internal and external reviews of the data, the CDC concluded that their studies were insufficiently definitive to show a causal relationship between hemosiderosis and exposure to *S. chartarum*. Consequently, the CDC has advised homeowners that for remediation purposes, *S. chartarum* should be treated like other molds. Despite this advice, most *S. chartarum* remediations are conducted using asbestos-type remediation procedures.

11.4 PROBLEM BUILDINGS

Buildings subject to health, comfort, or odor complaints can be best described as problem buildings. They comprise a building population in which some form of

Figure 11.11 *Stachybotrys* growing on water-damaged gypsum board.

dissatisfaction with air quality has been expressed by occupants sufficient to require an IAQ investigation. In many cases, the cause or causes of complaints can be readily identified and remediated. In other cases, there appears to be no single causal factor and mitigation efforts prove ineffective. When a building includes a relatively large percentage of occupants with complaints, it may be described as a sick building.

11.4.1 Building-Related Illness

When causal factors for health complaints have been identified, the air quality problem is described as building-related illness (BRI). BRI is characterized by a unique set of symptoms accompanied by clinical signs, laboratory test results, and specific pollutants. Included in BRI are nosocomial (hospital-acquired) infections, the hypersensitivity diseases (e.g., common allergy, asthma, hypersensitivity pneumonitis, humidifier fever), fiberglass dermatitis, and toxic effects associated with contaminants such as CO, ammonia (NH_3), and HCHO.

11.4.2 Sick Building Syndrome

SBS is used to describe a spectrum of nonspecific symptoms that, on investigation, appear to have no identifiable cause. These commonly include mucous membrane symptoms (irritation of the eyes, nose, throat, sinuses); general symptoms (headache, fatigue, lassitude); skin irritation (dryness, rashes); and, to a lesser extent, respiratory symptoms (cough, shortness of breath).

11.4.3 Field Investigations and Systematic Building Studies

An apparent relationship between illness complaints and building environments has been reported in a large number of problem building investigations conducted by governmental, occupational, and public health agencies, and private consultants responding to occupant or building management requests. In the U.S., more than 1000 such investigations have been conducted by the National Institute of Occupational Health and Safety (NIOSH) health hazard evaluation teams. Symptoms reported have included eye, nose, throat, and skin irritation; headache and fatigue; respiratory symptoms such as sinus congestion, sneeze, cough, and shortness of breath; and, less frequently, nausea and dizziness. Reported potential causal and contributory factors to complaints for 529 investigations conducted between 1971 and 1988 are summarized in Table 11.5.

Field investigations have had an important role in initially identifying and defining problem building phenomena. However, because of inherent bias involved in building studies subject to occupant health and comfort complaints and the relatively unsystematic approaches used in many investigations, they are of limited scientific usefulness. As a consequence, investigators in a number of countries have attempted to conduct systematic studies of noncomplaint office buildings. These have included 47 office buildings in the U.K., 14 town halls in Denmark, 60 office buildings in the Netherlands, a large office building study in northern Sweden, and 12 buildings in California. Significant prevalence rates of mucous membrane, general (headache,

Table 11.5 Problem Types Identified in NIOSH Building Investigations

Problem Type	Number of Buildings Investigated	%
Contamination from indoor sources	80	15
Contamination from outdoor sources	53	10
Building fabric	21	4
Microbial contamination	27	5
Inadequate ventilation	280	53
Unknown	68	13
Total	529	100

Source: From Seitz, T.A., *Proceedings Indoor Air Quality International Symposium: The Practitioner's Approach to Indoor Air Quality Investigations*, American Industrial Hygiene Association, Akron, OH, 1989. With permission.

Table 11.6 Symptom Prevalence Rates (%) among Occupants of 14 Danish Town Hall Buildings

	Males (n = 1093–1115)	Females (n = 2280–2345)
Symptoms		
Eye irritation	8.0	15.1
Nasal irritation	12.0	20.0
Blocked, runny nose	4.7	8.3
Throat irritation	10.9	17.9
Sore throat	1.9	2.5
Dry skin	3.6	7.5
Rash	1.2	1.6
Headache	13.0	22.9
Fatigue	20.9	30.8
Malaise	4.9	9.2
Irritability	5.4	6.3
Lack of concentration	3.7	4.7

Symptom Groups (Males and Females)	
Mucous membrane irritation	20.3
Skin reactions	4.2
General symptoms	26.1
Irritability	7.9

Source: From Skov, P. and Valbjorn, O., *Environ. Int.*, 13, 349, 1987. With permission.

fatigue, lassitude), and skin symptoms were observed in these noncomplaint buildings, with complaint rates varying from building to building. Symptom prevalence rates for both males and females in 14 noncomplaint Danish town halls are summarized in Table 11.6. As a comparison, symptom prevalence rates for 11 Swedish sick (complaint) buildings are summarized in Table 11.7. Note that prevalence rates were relatively high in both cases, with the latter being much higher.

Table 11.7 Symptom Prevalence among Occupants of 11 "Sick" Swedish Office
Buildings

Symptom	Total Mean Prevalence (%)	Range
Eye irritation	36	13–67
Swollen eyelids	13	0–32
Nasal catarrh	21	7–46
Nasal congestion	33	12–54
Throat dryness	38	13–64
Sore throat	18	8–36
Irritative cough	15	6–27
Headache	36	19–60
Abnormal tiredness	49	19–92
Sensation of getting a cold	42	23–77
Nausea	8	0–23
Facial itch	12	0–31
Facial rash	14	0–38
Itching on hands	12	5–31
Rashes on hands	8	0–23
Eczema	15	5–26

Source: From Norback, D., Michel, I., and Winstrom, J., *Scand. J. Work Environ. Health*, 16, 121, 1990. With permission.

11.4.4 Risk Factors

Investigators conducting systematic building studies have attempted to identify potential risk and contributory factors to building- and work-related health complaints. These include people-, work or building environment-, and contaminant-related concerns.

A variety of people-related risk factors are associated with symptom prevalence. These are gender, allergy, a variety of psychosocial factors, and tobacco smoking. All studies have shown that symptom-reporting rates are several times higher among females than males. Though most studies report self-diagnosed allergy as a risk factor for SBS-type symptoms, objective immunological studies have failed to show such a relationship. Symptom prevalence is also significantly associated with psychosocial variables such as work stress and job satisfaction. It increases with increased workplace stress and decreases with increased job satisfaction. Nonsmoker exposure to ETS increases symptom prevalence rates.

A variety of environment factors (temperature, humidity, air movement, etc.) have been investigated as potential risk factors for SBS symptoms. Significant relationships have been reported for elevated temperature (21 to 25°C, 70 to 78°F) and low relative humidity (<30% RH). Low ventilation rates (<10 l/sec/person, <20 cubic feet per minute (CFM)/person) appear to be a risk factor for SBS-type symptoms.

Various office materials and equipment have been implicated as contributory factors to work- or building-related health complaints. These include (1) handling carbonless copy paper and other papers; (2) working with office copy machines such as wet process and electrostatic photocopiers, laser printers, diazo-copiers, microfilm copiers, and spirit duplicators; and (3) using video display terminals.

A variety of gas- and particulate-phase substances have been evaluated as potential risk factors for occupant health complaints. These have included bioeffluents (e.g., CO_2), HCHO, VOCs, and dust. At this time, there is little evidence to implicate bioeffluents as a cause of health complaints. Formaldehyde levels in nonresidential, nonindustrial buildings are relatively low and are not considered sufficient to cause building-related health complaints. TVOCs have been proposed as a causal factor for SBS-type symptoms and airborne and settled dust has been implicated as a risk factor; most notable for the latter has been an association with the macromolecular or organic fraction of floor dust.

11.5 INVESTIGATING INDOOR AIR QUALITY PROBLEMS

As indicated in previous sections, a number of significant IAQ problems exist in buildings. Except in cases of severe, acute illness, most go unrecognized. The need to conduct an investigation to identify the nature of a suspected problem only occurs when occupant complaints are sufficiently intense to convince building management that the air quality needs to be investigated.

Responsibility for investigating building-related health, and sometimes odor, complaints in nonresidential buildings falls to local, state, and federal public or occupational health agencies or private consultants. A variety of protocols are used to conduct such investigations.

The USEPA, in cooperation with NIOSH, has developed a model protocol for conducting health and comfort investigations in public access buildings. Though designed for in-house personnel, it can be readily adapted for use by public agencies and private consultants.

The USEPA/NIOSH investigative protocol is characterized by systematic information gathering and hypothesis testing (Figure 11.12). It includes an initial walk-through of problem areas to gather information on building occupants, ventilation systems, contaminant sources, and pollutant pathways. Additional information gathering is recommended if the walkthrough does not identify the cause or causes of reported problems. Information gathering is intended to provide a basis for developing a hypothesis as to what may be the cause or causes. Since many complaints occur in poorly ventilated buildings and spaces, it is desirable to inspect and evaluate components of ventilation systems that may serve problem areas and other parts of the building.

This investigative protocol gives limited attention to air sampling, suggesting that such sampling may not be required to solve most problems and may even be misleading. Air sampling is only recommended after all other investigative activities have been employed. Exceptions include CO_2 to determine ventilation adequacy and temperature, humidity, and air movement for thermal comfort.

Relatively speaking, the diagnosis of an IAQ problem in residences is a simple undertaking. Residences are small, contain few individuals, and have simple mechanical systems. Additionally, a few contaminants are responsible for most residential health and odor complaints, and causal connections are easily made. Commonly-encountered IAQ problems in residences include exposure to allergens such as mold,

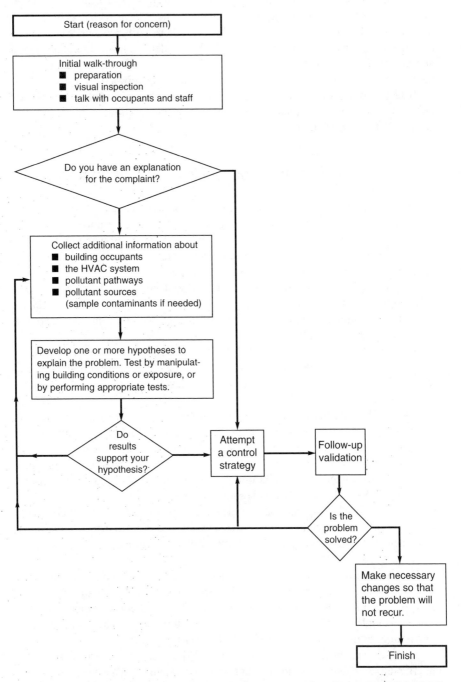

Figure 11.12 USEPA/NIOSH building investigation protocol. (From USEPA/NIOSH, EPA/400/1-91/003, DDHS Publication 91-114, EPA, Washington, D.C, 1991.)

dust mite excreta, cockroach excreta and body parts, and animal dander; HCHO that causes irritant-type symptoms; CO that causes central nervous system symptoms; and a variety of other problems, including sewer gas odors, misapplied pesticides, etc. Diagnosis involves information gathering on occupant symptoms (type and onset patterns) and assessment of the building for possible sources. It may also include air testing or collection of surface samples to confirm a hypothesis of what the potential causal factor or factors may be.

11.6 MEASUREMENT OF INDOOR CONTAMINANTS

Pollutant and contaminant measurements are conducted in most IAQ investigations. In addition to sampling airborne concentrations of gases, vapors, particulate-phase substances, and biological contaminants, other environmental samples may be collected. These include surface dust sampling using wipe or vacuum sampling techniques, surface sampling for mold, and bulk sampling for asbestos or mold.

Commonly measured contaminants include CO_2 (to determine ventilation adequacy), CO, HCHO, VOCs, and mold and its by-products. Measurements are often used as a screening tool to determine whether pollutant levels are within acceptable guideline values.

Though in most cases pollutant measurements are made as a screening tool, they are best conducted to confirm a hypothesis that a substance(s) or organism(s) is causally associated with reported illness symptoms. This is especially the case with exposures to CO and biological contaminants such as mold spores.

11.7 PREVENTION AND CONTROL MEASURES

A variety of approaches are available to prevent IAQ problems before they occur and to control and mitigate them once they are manifested. These can be described as source management, exposure control, and contaminant control. Source management includes source exclusion, removal, and treatment. Exposure control focuses on scheduling activities to reduce occupant exposures. Contaminant control focuses on reducing airborne contaminant levels, usually achieved by ventilation or air cleaning.

11.7.1 Source Management

11.7.1.1 Exclusion

In new building construction and remodeling, IAQ problems can be reduced or minimized by use of low-emission products that are becoming increasingly available. Carpeting and mastic products are being produced and sold that have relatively low emissions of VOCs and the odor-producing substance called 4-phenylcyclohexane (4-PC). Formaldehyde emissions from pressed-wood products are much lower than they once were as lower-emission products have been developed. Low levels of

HCHO can be attained and maintained by avoiding use of HCHO-emitting products such as particleboard, hardwood plywood paneling, medium-density fiberboard, and acid-cured finishes. Alternative products include softwood plywood, oriented-strand board, decorative gypsum board and hardboard panels, and, in the case of finishes, a variety of HCHO-free varnishes and lacquers.

Contamination of indoor air by combustion by-products can be avoided by using electric rather than gas appliances for cooking, using only electric or vented gas or oil space heating appliances, and restricting smoking. Indoor pesticide use can also be largely avoided by implementing integrated pest management practices.

High Rn levels can be avoided by selecting homesites that have soils with a high clay content. However, such soils tend to have moisture problems that increase health risks associated with mold and dust mites.

11.7.1.2 Source Removal

The single most effective measure to reduce contaminant levels is to identify and remove the source. In the case of HCHO control, its effectiveness depends on correct identification of the most potent source or sources in a multiple source environment. Because of interaction effects, HCHO levels are determined by the most potent source or sources present; removal of minor sources would not reduce HCHO levels.

In buildings with mold infestation, mitigation of respiratory health problems, e.g., chronic allergic rhinitis, asthma, and hypersensitivity pneumonitis, will, in many cases, require removal and replacement of infested construction materials and furnishings.

11.7.1.3 Source Treatment

Sources may be treated or modified to reduce contaminant emissions. One can, for example, use encapsulants to prevent release of asbestos fibers from acoustical plaster. Formaldehyde-emitting wood products can be treated with scavenging coatings or encapsulated with vinyl materials. Radon treatment measures include sealing cracks in masonry walls and slabs and gaps around floor-based plumbing. Minor mold infestations can be treated by biocides or encapsulating paints.

11.7.2 Contaminant Control

11.7.2.1 Ventilation

Contaminant levels in building spaces can be reduced by diluting indoor air with less contaminated outdoor air. Such dilution may occur in part or whole by infiltration or exfiltration, natural ventilation, and mechanical ventilation. Ventilation air exchange in residential buildings occurs by natural ventilation, i.e., opening windows and doors during mild to warm seasonal conditions, and by infiltration or exfiltration under closure conditions. Infiltration and exfiltration also play a role in building air exchange in mechanically-ventilated office, commercial, and institutional buildings.

11.7.2.1.1 Infiltration and Exfiltration

All building structures "leak"; i.e., the building envelope contains many avenues whereby air enters and leaves. These include cracks and gaps around windows, doors, electrical and exhaust outlets, ceiling light fixtures, and the building base. Infiltration of outside air and exfiltration of inside air are affected by tightness of the building envelope and environmental factors such as indoor–outdoor temperature differentials and wind speed. High air exchange occurs on cold, windy days, and low air exchange occurs on calm, moderate days when temperature differences are small. During the heating season, indoor–outdoor temperature differences result in pressure differences that draw air into the base area of the building (infiltration) and force air out at the top (exfiltration). This is the so-called stack effect. It occurs in all buildings (including single-story) but is most pronounced in tall buildings.

11.7.2.1.2 Natural Ventilation

A building is "naturally" ventilated when occupants open windows and, in some cases, doors. Air exchange can be considerable, as can contaminant dilution. The degree of air exchange, or ventilation, depends on the magnitude of window and door opening, the position of open windows and doors relative to each other, and the same environmental factors that affect infiltration and exfiltration. In the absence of year-round climate control, natural ventilation is the primary means by which residential and some nonresidential building spaces are made more comfortable during warm weather.

11.7.2.1.3 Mechanical Ventilation

Mechanical ventilation is the most widely used contaminant control measure in large office, commercial, and institutional buildings. Mechanical systems are used to provide general dilution and local exhaust ventilation.

General dilution ventilation is used in most large buildings to dilute and remove human bioeffluents that cause odor and comfort complaints. It is also applied as a generic measure to reduce overall building contaminant levels and mitigate SBS-type symptoms and IAQ complaints.

Use of mechanical ventilation for contaminant control is based on the general dilution theory, wherein contaminant levels are expected to decrease by 50% each time air volume or exchange is doubled. It is likely to be relatively effective in contamination problems that are episodic, e.g., tobacco smoking and emissions from office copiers, or the relatively constant bioeffluent emissions. It is less effective in reducing levels of contaminants (e.g., HCHO and VOCs) emitted by diffusion processes.

There have been attempts to evaluate the effectiveness of general dilution ventilation in reducing both contaminant levels and occupant health complaints, but results have been mixed. In general, it appears that increasing ventilation rates to 10 l/sec/person (20 CFM/person) results in decreased symptom prevalence, with higher ventilation rates being relatively ineffective in achieving further reductions.

The effect of ventilation rate on personal satisfaction or dissatisfaction with air quality can be seen in Figure 11.13.

A consensus has developed that adequate ventilation is essential to maintain comfortable and low-human-odor building environments. As a consequence, guidelines have been developed to help building designers and operators provide adequate ventilation rates to meet the needs of occupants. In North America, ventilation guidelines are developed and recommended by the American Society of Heating, Refrigerating and Air-Conditioning Engineers (ASHRAE). In Europe and East Asia, they have been developed by governmental agencies, regional collaboration, and the WHO.

Ventilation guidelines specify a rate of outdoor airflow that can reasonably be expected to provide acceptable air quality relative to human odor and comfort. Ventilation guidelines specify quantities of outdoor air that will maintain building steady-state CO_2 levels (typically 1000 ppmv (ASHRAE) or 800 ppmv (WHO)) at maximum design capacity. At design occupant capacity, recommended minimum ventilation rates (liters per second per person, CFM per person) would maintain CO_2 levels below guideline values. ASHRAE guideline ventilation values for different building spaces or conditions of use are summarized in Table 11.8.

Local exhaust ventilation is used in special circumstances to minimize levels of contaminants in buildings. It is particularly well suited to situations in which sources are known, emissions are high, and localized and general dilution ventilation would not be acceptable. Its major application is to control odors from lavatories. It is also used to control combustion by-products and odors from cafeteria kitchens, vapors and gases from institutional laboratories, and ammonia from blueprint machines.

Figure 11.13   Relationship between ventilation rate and dissatisfaction with air quality. (From Fanger, P.O., *Proc. 5th Int. Conf. Indoor Air Qual. Climate*, 5, 353, 1990.)

Table 11.8 ASHRAE Ventilation Guidelines

Facility Type	Estimated Occupancy (persons/1000 ft²)	Outdoor Air Requirements (CFM/person)	(L/sec/person)
Food and Beverage			
Dining rooms	70	20	10
Bars, cocktail lounges	100	30	15
Hotels, Motels, Resorts, etc.			
Lobbies	30	15	8
Assembly rooms	120	15	8
Dormitory sleeping areas	20	15	8
Office spaces	7	20	10
Conference rooms	50	20	10
Smoking lounges	70	60	30
Sports and Amusements			
Spectator areas	150	15	8
Game rooms	70	25	13
Ballrooms/discos	100	25	13
Theaters/auditoriums	150	15	8
Education			
Classrooms	50	15	8
Libraries	20	15	8
Hospital			
Patient rooms	10	25	13

Source: Extracted from *ASHRAE Standard 62–1989: Ventilation for Acceptable Air Quality,* ASHRAE, Atlanta, 1989.

11.7.2.2   Air Cleaning

As with ventilation, air cleaning systems are designed to reduce contaminant levels after contaminants become airborne. Air cleaning systems can be designed to control PM (dust), gas-phase contaminants, or both.

Particle or dust cleaning systems have a long history of use. A minimum level of air cleaning (dust-stop filters) is used in most building HVAC systems to protect mechanical equipment from soiling. Filtration systems are less commonly employed to maintain clean building spaces. A variety of filters and filtration systems are available for building dust control. These vary from low-efficiency dust-stop filters, to medium-efficiency pleated panel filters, to high-efficiency pleated panel filters and electronic air cleaners, to very-high-efficiency HEPA (high efficiency particulate absolute) filters, which have the ability to remove 99.97% of particles at the hardestto-control diameter of 0.3 μm. Particle filtration systems can be used relatively effectively to reduce indoor particle and mold spore levels.

Gas-phase contaminant control by air cleaning is more difficult, expensive, and usually less effective than particle control. Such air cleaning typically employs sorbents or specially impregnated sorbents in panel beds or pleated filters. The most widely used sorbent is activated carbon (charcoal). It is effective in removing VOCs that have a relatively high molecular weight (e.g., toluene, benzene, xylene, methyl chloroform, etc.). It is relatively ineffective in removing low-molecular-weight substances such as HCHO, ethylene, acetaldehyde, etc. Such substances may be removed from air by potassium permanganate ($KMnO_4$) on activated alumina and specially impregnated charcoal.

Many air cleaners are commercially available (particularly in the residential market) whose manufacturers claim a high degree of effectiveness in removing a broad range of indoor contaminants. Most claims have not been validated. Based on design features and operating parameters, it is unlikely that such claims are true.

11.8 INDOOR AIR QUALITY AND PUBLIC POLICY

In the U.S., identification of a potential health-threatening environmental problem has historically been followed by enactment of federal legislation that provides a regulatory framework for control or promulgation of specific regulations by the USEPA. In the case of IAQ, a variety of federal agencies have limited authority under existing legislation. These include the Consumer Product Safety Commission (CPSC), which regulates dangerous or defective products; the USEPA, which has specific authority relative to Rn and general authority to conduct IAQ research; the Department of Housing and Urban Development (HUD), which sets standards on the habitability of manufactured houses and federally subsidized housing; OSHA, which sets and enforces workplace health standards; and NIOSH, which has authority to recommend workplace health standards, conduct research, and conduct workplace health hazard evaluations (including IAQ).

Under its statutory authority, the CPSC attempted to ban use of UFFI. The ban was subsequently overturned by a federal appeals court. The CPSC also evaluated emissions of HCHO from wood products and kerosene and gas space heaters for possible regulatory action. Under its statutory authority, HUD issued product standards for emissions of HCHO from particleboard and hardwood plywood used in manufactured housing. It also required manufacturers to prominently post warning signs in or on their products about the health dangers of HCHO.

The USEPA has been the most active federal agency in actions involving IAQ. It has authority to promulgate and enforce regulations under AHERA, which requires schools to use accredited personnel to conduct asbestos inspections, prepare management plans, and conduct any abatement operations.

In 1988, the USEPA, in conjunction with the U.S. Surgeon General, issued a public health advisory recommending that homeowners test their homes for Rn and abate high Rn levels. This advisory was subsequently extended to schools. The

USEPA was authorized by Congress to promulgate rules for Rn testing and abatement activities.

The USEPA has maintained a leadership role in the broad area of IAQ, in part because of its expertise related to air quality and the agency's research and public policy interests. Over the past decade, the USEPA has increasingly emphasized voluntary efforts in controlling indoor air pollution. Since IAQ in many instances (e.g., houses) is not in the public domain, it naturally lends itself to voluntary efforts and the leadership experience of an agency such as the USEPA.

OSHA has attempted to deal with IAQ concerns associated with workplaces. First, it identified ETS as a human carcinogen and subject of workplace exposure concern. Though OSHA proposed rules to regulate smoking and ventilation in nonindustrial workplaces, such rules were never promulgated.

There are inherent limitations in the ability of government to solve IAQ problems through regulatory action. As a consequence, public policy is focused primarily on public education so that individuals can make informed choices about the quality of air they breathe indoors.

READINGS

American Conference of Governmental Industrial Hygienists, Evaluating office environmental problems, *Ann. ACGIH*, 10, 1984.

American Society of Heating, Refrigerating and Air-Conditioning Engineers, *ASHRAE Standard 62-1989: Ventilation for Acceptable Air Quality*, ASHRAE, Atlanta, 1989.

Bearg, D.W., *Indoor Air Quality and HVAC Systems*, Lewis Publishers/CRC Press, Boca Raton, FL, 1993.

Burge, H.A., Ed., *Bioaerosols*, Lewis Publishers/CRC Press, Boca Raton, FL, 1995.

California Environmental Protection Agency, Air Resources Board, *Proposed Identification of Lead as a Toxic Air Contaminant*, California Environmental Protection Agency, Sacramento, CA, 1997.

Cone, J.E. and Hodgson, M.Q., Eds., Problem buildings: building-associated illness and the sick building syndrome, in *State of the Art Reviews: Occupational Medicine*, Hanley and Belfus, Inc., Philadephia, 1989.

Cox, C.S. and Wathes, C.M., Eds., *Bioaerosols Handbook*, Lewis Publishers/CRC Press, Boca Raton, FL, 1995.

Dillon, H.K., Heinsohn, P.A., and Miller, J.D., Eds., *Field Guide for the Determination of Biological Contaminants in Environmental Samples*, American Industrial Hygiene Association, Fairfax, VA, 1996.

Godish, T., *Indoor Air Pollution Control*, Lewis Publishers, Chelsea, MI, 1989.

Godish, T., *Sick Buildings: Definition, Diagnosis and Mitigation*, Lewis Publishers/CRC Press, Boca Raton, FL, 1995.

Godish, T., *Indoor Environmental Quality*, Lewis Publishers/CRC Press, Boca Raton, FL, 2000.

Light, E. and Sundell, J., *General Principles for the Investigation of Complaints, TF11-1998*, International Society of Indoor Air Quality and Climate, Milan, Italy, 1998.

Maroni, M., Siefert, B., and Lindvall, T., *Indoor Air Quality: A Comprehensive Reference Book*, Elsevier, Amsterdam, 1995.

Morey, P.R., Feeley, J.C., and Othen, Q.A., Eds., *Biological Contaminants in Indoor Environments, ASTM STP 1071*, American Society of Testing Materials, Philadelphia, 1990.

National Research Council, Committee on Indoor Pollutants, *Indoor Air Pollutants*, National Academy Press, Washington, D.C., 1981.

National Research Council, *Health Risks of Radon and Other Internally-Dispersed Alpha Emitters*, BEIR IV, National Academy Press, Washington, D.C., 1988.

Norback, D. and Edling, C., Indoor air quality and personal factors related to the sick building syndrome, *Scand. J. Work Environ. Health*, 16, 121, 1990.

Rafferty, P.J., Ed., *The Industrial Hygienists Guide to Indoor Air Quality Investigations*, American Industrial Hygiene Association, Fairfax, VA, 1993.

Salthammer, T., Ed., *Organic Indoor Air Pollution: Occurrence, Measurement, Evaluation*, John Wiley & Sons, New York, 1999.

Samet, J.M. and Spengler, J.D., Eds., *Indoor Air Pollution: A Health Perspective*, Johns Hopkins University Press, Baltimore, MD, 1991.

Seitz, T.A., NIOSH indoor air quality investigations, 1971–1988, in *Proceedings of the International Indoor Air Quality Symposium: The Practitioner's Approach to Indoor Air Quality Investigations*, Weekes, D.M. and Gammage, R.B., Eds., American Industrial Hygiene Association, Akron, OH, 1989.

Smith, K.R., *Biofuels, Air Pollution and Health: A Global Review*, Plenum Publishing Co., New York, 1987.

Spengler, J.D., Samet, J.M., and McCarthy, J.C., Eds., *Indoor Air Quality Handbook*, McGraw-Hill, New York, 2001.

USEPA, *A Citizen's Guide to Radon*, EPA-86-004, EPA, Washington, D.C., 1986.

USEPA/NIOSH, *Building Air Quality: A Guide for Building Owners and Facility Managers*, EPA/400/1-91/1003, DDHS (NIOSH) Publication 91-114, EPA, Washington, D.C., 1991.

USEPA, *Respiratory Health Effects of Passive Smoking: Lung Cancer and Other Disorders*, EPA/600/6-90/006 B, EPA, Washington, D.C., 1993.

U.S. Surgeon General, *The Health Consequences of Involuntary Smoking*, DDHS (CDC) 87-8398, Department of Health & Human Services, Washington, D.C., 1986.

Weekes, D.M. and Gammage, R.R., Eds., *Proceedings of the International Indoor Air Quality Symposium: The Practitioner's Approach to Indoor Air Quality Investigations*, American Industrial Hygiene Association, Akron, OH, 1989.

World Health Organization, Indoor air quality: organic pollutants, in *EURO Reports and Studies*, 171, WHO Regional Office for Europe, Copenhagen, 1988.

Yocum, J.E., Indoor-outdoor relationships: a critical review, *JAPCA*, 32, 500, 1982.

Yocum, J.E. and McCarthy, S.M., *Measuring Indoor Air Quality: A Practical Guide*, John Wiley & Sons, New York, 1991.

QUESTIONS

1. Indoor and outdoor concentrations of pollutants vary considerably. Which pollutants are higher indoors? Outdoors?
2. How do personal exposures to PM differ from outdoor and indoor exposures?
3. Describe the nature of building air quality concerns associated with asbestos.
4. What is radon? Describe its decay.
5. Since radon is inert, why should it be considered a danger to public health?
6. Describe combustion sources that result in elevated NO_x in building environments.

7. What health risks are or may be associated with use of (a) unvented appliances, (b) candle and incense burning, and (c) vented combustion appliances?

8. Describe major health concerns associated with exposures to environmental tobacco smoke.

9. What exposures and health risks are associated with biomass cooking?

10. What exposure and health concerns are associated with aldehydes in buildings?

11. What are VOCs? What exposure and health concerns are associated with them?

12. What are SVOCs? Describe major SVOC contaminants in indoor environments.

13. Why are pesticides an indoor environment concern?

14. Describe illness patterns characteristic of chronic allergic rhinitis, asthma, hypersensitivity pneumonitis, and Legionnaires' disease.

15. Identify primary allergenic biological contaminants in indoor environments.

16. What is the health significance of endotoxins and mycotoxins?

17. What is a problem building? Sick building?

18. Distinguish between building-related illness and sick building syndrome.

19. What are some of the major risk factors for developing SBS-type symptoms?

20. How would one go about conducting an indoor air quality investigation?

21. Why are indoor air quality measurements made in problem buildings?

22. Describe differences between source control and pollutant control measures.

23. What is the relationship between indoor air quality and ventilation? Between ventilation and illness symptoms?

24. How is indoor air quality regulated in the U.S.?

CHAPTER 12

Environmental Noise

Environmental noise associated with human activities is a phenomenon that characterizes the acoustical environment and our perceptions of it in the communities in which we live. Noise is sound energy that is objectionable because of its physiological and psychological effects on humans.

Humans vary in their reception and perception of sound. As such, the concept of noise can and does vary from one individual to another. What is an acceptable or unacceptable sound, or level of it, depends on the individual exposed. Nevertheless, there are sound types and exposure levels that most individuals agree are not acceptable and, therefore, noise. Major noise sources include transportation, industry, construction, buildings and households, and humans and pets. In most cases, noise is produced as a result of human activities.

Noise in industrial, construction, and military environments poses unique exposure concerns because it has the potential to cause hearing loss. Noise exposure in industrial and construction environments is regulated by governmental agencies whose mission it is to protect worker health (in the U.S., the federal Occupational Safety and Health Administration (OSHA) and a number of state agencies). These occupational safety and health concerns differ from noise exposures in the ambient environment, i.e., environmental or community noise.

Noise is treated in this book as a form of ambient or atmospheric pollution since the sound energy that is the cause of noise is primarily transmitted through the air environment.

12.1 SOUND AND ITS MEASUREMENT

Sound is a form of energy produced by vibrating objects or aerodynamic disturbances. The former includes our vocal cords and operating motors, automobiles, and airplanes; the latter, thunder, sonic booms, and air movement.

Disturbance of air molecules by sound energy produces variations in normal atmospheric pressure. As these pressure variations reach our ears, they cause the eardrums to vibrate. Transmission of vibrations to the inner ear, and their interpretation by the brain, results in our sensory perception of sound. Sound can only be transmitted through a medium that contains molecules; i.e., it cannot move through a vacuum. The speed of sound depends on the density of the transmission medium, with increased speed associated with increased density. The speed of sound in air, water, and steel is 343, 1482, and 5030 m/sec (1125, 4860, and 16,500 ft/sec), respectively.

12.1.1 Sound Energy

When objects vibrate, they radiate acoustical energy. The energy produced by a source is described as sound power. In conventional use, sound power (Sp) values are expressed in dimensionless units called decibels (dB), calculated from the following equation:

$$Sp = 10 \log_{10} (W_m/W_r) \text{ dB} \qquad (12.1)$$

where W_m is the measured sound energy in watts and W_r is the reference sound energy, 10^{-12} W. Sound power is the logarithm of the ratio of the measured and reference sound energy multiplied by 10.

When sound energy passes through a given area, its flow is described as sound intensity (SI), expressed as watts per square meter (W/m²). Sound intensity values are also expressed in decibels and are calculated similarly to Sp:

$$SI = 10 \log_{10} (I_m/I_r) \text{ dB} \qquad (12.2)$$

where I_m is the measured sound energy and I_r is the reference sound energy, 10^{-12} W/m².

When sound energy radiating from a source strikes a surface, it induces a pressure. This pressure can be measured by instruments known as sound pressure level meters. Sound pressure values are also expressed in decibels. Calculation of sound pressure differs slightly from those for Sp and SI. Since pressure in sound waves varies sinusoidally, the mean square values of the pressure changes are used to calculate sound pressure levels (Spl):

$$Spl = 10 \log_{10} (P_m^2/P_r^2) \text{ dB} \qquad (12.3)$$

$$= 20 \log_{10} (P_m/P_r) \text{ dB}$$

In calculations of Spl, the reference pressure (P_r) is the threshold of human hearing, 2×10^{-4} μbars or 2×10^{-5} N/m², where N is newtons. P_m is the measured sound pressure. For purposes of illustration, let us calculate the Spl when $P_m = 2.0$ μbars.

$$Spl = 20 \log_{10} (2.0/0.0002) \text{ dB}$$

$$= 20 \log_{10} (10,000) \text{ dB}$$

$$= 20 (4) \text{ dB}$$

$$= 80 \text{ dB}$$

As an object vibrates, sound waves radiate outward in an ever-expanding sphere. At increasing distances, Sp remains the same, but SI declines as the area through which sound energy passes increases. In a free field (space where sound is not reflected), SI and Spl are reduced by 6 dB (factor of 4) each time distance from the source is doubled. This relationship is illustrated in Figure 12.1. At a distance of two radii (2r), the area of the sound sphere is four times as large as the area at a distance of one radius (1r). The change in sound intensity and pressure follows the inverse square law:

$$SI = W/4\pi r^2 \qquad (12.4)$$

where W is watts and $4\pi r^2$ is the area of the sound sphere.

Each time the radius doubles, the area of the sound sphere increases by $4\pi r^2$ of its previous value and SI decreases inversely with this increase in area.

Sound produced by a source is emitted in the form of waves. These waves are characterized by alternating compressions (wave peaks) and rarefactions (wave

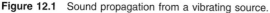

Figure 12.1   Sound propagation from a vibrating source.

troughs) and vary in length, frequency, and amplitude. The shorter the wavelength, the more frequent the waves are per unit time. The higher the wave (increased amplitude), the more energy it has.

Frequency is a major characteristic of sound, and its discrimination by the human auditory organ constitutes the phenomenon of hearing. Frequency is expressed in cycles (waves) per second or hertz. If 1000 complete sound waves pass a point per second, that sound has a frequency of 1000 Hz.

A sound source does not usually produce discrete frequencies. Most sounds include a range of frequencies characterized by dominant frequencies. Male human speech, for example, is characterized by dominant frequencies of <2000 Hz; for female speech, the range of dominant frequencies is somewhat higher.

Humans can hear sounds in the frequency range of 20 to 20,000 Hz. However, frequencies in this range are not equally sensible. As can be seen in Figure 12.2, humans hear those frequencies best that correspond to human speech. Sound frequencies of <16 Hz are characterized as infrasound; those of >20,000 Hz, as ultrasound. Whether humans can hear a sound depends on its spectrum of frequencies and intensity. Decibel readings for a variety of sound sources experienced in the home, work, and ambient environments are summarized in Table 12.1. Note that 0 dBA corresponds to the threshold of hearing. Sound pressure levels of >70 dBA become increasingly objectionable, as they interfere with speech communication. Sounds of <0 dBA occur but cannot be heard. Brief exposures to sound levels of >140 dBA may be painful, and long-term exposures to sound levels of >90 dBA may result in hearing loss.

12.1.2 Sound Measurement

Instruments used to measure sound levels in decibels are described as sound pressure meters (Figure 12.3). Such meters consist of a microphone, attenuator, amplifiers, and a digital or mechanical reading. Sound generates a voltage proportional to the acoustical pressure acting on the microphone. This voltage produces

Figure 12.2 Frequency range of human hearing.

Table 12.1 Sound Pressure Levels, Sources, and Human Responses

Sources	dB	Human Responses
	150	
	140	Painful
	130	
Jet aircraft on ground, 20 ft	120	
Rock and roll band	115	
Loud motorcycle, 20 ft	110	
	100	
Heavy truck, 50 ft	90	Permanent hearing loss
Pneumatic drill, 50 ft	85	
	80	
	75	Annoying
Motor vehicle, 50 ft	70	Interfere with speech communication
	60	
Living room	55	
	50	Quiet
Bedroom	45	
Library	40	
Soft whisper	30	
	20	Very quiet
	10	Just audible
	0	Threshold of hearing

a deflection of the meter needle or is converted to a digitally displayed value. An integral part of sound measuring instruments is weighting scales or networks that emphasize or deemphasize sound in selected frequency ranges.

12.1.2.1 Weighting Scales

Sound level meters are generally equipped with three circuits that correspond to weighting scales. Three weighting scales have been established by the American National Standards Institute (ANSI) for general-purpose use. These are the A, B, and C scales. Other scales (D and E) have been developed and used for special applications. Each scale differs in its ability to measure sound levels across a frequency range (Figure 12.4). The A scale is designed to discriminate against (i.e., not measure) frequencies of <600 Hz. Low-frequency discrimination is more moderate on the B scale. The C scale corresponds more closely to the actual sound energy produced; its response is relatively uniform across the frequency spectrum. Readings on all three scales provide an indication of the frequency distribution of a recorded sound. If the level is greater on the C scale, some of the sound is in frequencies of <600 or >10,000 Hz. If no differences are observed among the three scales, most of the measured sound is in the middle frequencies.

The sound reading should be expressed as a function of the scale used, e.g., 20 dBA, 20 dBB, or 20 dBC. At one time, it was customary to select the weighting scale according to sound level. For levels of <55 dB, the A scale was recommended; from 55 to 85 dB, the B scale; and >85 dB, the C scale. It is now common practice to employ the A scale for most sound measurements. This dominance of A scale

Figure 12.3 Sound pressure level meter.

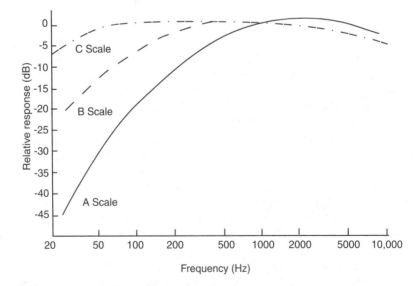

Figure 12.4 Frequency discrimination for A, B, and C weighting scales.

use is due to its application to sound level measurement in occupational environments where compensable hearing impairments are limited to the frequency range of 500 to 2000 Hz. The A scale approximates noise exposure risk to human hearing relatively well. Since humans show limited response to frequencies of <500 Hz, there is little value in measuring such frequencies when controlling worker noise exposure. Similarly, it is the audible middle frequencies that are important in community noise exposures.

Despite its almost universal use for measurements of Spl in industrial and community environments, the A scale has some important limitations. It apparently does not simulate the spectral selectivity of human hearing or its nonlinear relationship to sound energy. If different spectral characteristics are compared, dBA values may not be an accurate indicator of human subjective response. Laboratory and field studies have shown that the A scale predicts perceptions of loudness and annoyance relatively poorly. It is strongly dependent on the exposure pattern with time and underestimates the impact of low-frequency components of sound. It is also unrepresentative of loud sounds that contain a mixture of distinctive tonal components. As such, it is unsuitable for predicting loudness or annoyance.

12.1.2.2 Meter Response and Instrument Accuracy

In addition to weighting scales, most sound pressure level instruments have slow and fast measuring options. The averaging time for the fast response is 0.125 sec; for the slow response, 1 sec. Though the fast response corresponds to the response time of the human auditory system, the slow response averaging time is usually used because it is easier to read.

Commonly available sound pressure level meters are rated as Type I, II, or III. These designations reflect their accuracy. Type I meters have an accuracy of ±1 dB; they are primarily research instruments. Type II meters have an accuracy of ±2 dB; they are general-purpose instruments recommended for use in measuring both industrial and community sound levels. Type III meters have an accuracy of ±3 dB and are not recommended for regulatory or technical use.

12.1.2.3 Measurement of Impulse Sound

General-purpose sound pressure level meters cannot be used to measure impulse noises, i.e., those produced by a variety of industrial processes such as punch presses and pile drivers. These sounds consist of rapidly rising and falling sound pressures. Instruments must be specially designed to measure both peak and average impulse sound levels. An averaging time of 0.035 sec is typically used.

12.1.2.4 Spectrum Analysis

The audible frequency range is covered by 10 octave bands. An octave is a range of frequencies in which the upper frequency is, in most cases, twice the lower frequency (Table 12.2). Each octave band is characterized by a center frequency that is the geometric mean of the upper and lower limits. The octave band sound level

Table 12.2 Octave Band Frequency Ranges and Center Frequencies

Frequency Range, Hz	Center Frequency, Hz
18–45	31.5
45–90	63
90–180	125
180–355	250
355–710	500
710–1400	1000
1400–2800	2000
2800–5600	4000
5600–11,200	8000

is the amount of acoustical energy (measured as sound pressure) associated with sound frequencies in the octave band. Noise assessments are often conducted using one third octave band filters. Each octave is subdivided into thirds to better characterize sound spectra.

12.2 SOUND EXPOSURE DESCRIPTORS

The USEPA uses a system of four sound descriptors to summarize how people hear sound and how environmental noise impacts public health and welfare. These are the A-weighted sound level, A-weighted sound exposure level, equivalent sound level, and day–night sound level.

12.2.1 A-Weighted Sound Level

A weighting is recommended because of its convenience of use, relative accuracy for most purposes, and widespread acceptance around the world. It is used in the other descriptors described below. When used by itself, an A-weighted decibel value describes a sound value at a given instant, a maximum value, or a steady-state value.

12.2.2 A-Weighted Sound Exposure Level

Sound exposure levels take into account the moment-to-moment variations that occur when measuring environmental noise. The changing magnitude of sound levels can be observed on a strip-chart recording. Changes in sound levels associated with different sound sources can be seen in Figure 12.5. Sound levels vary over a range of 30 dBA. They appear to be characterized by relatively steady lower-level (background, ambient) sound values, over which are superimposed sound levels associated with distinct individual sources. These can be characterized by their maximum values and time patterns. By combining the maximum sound level with the length of time that the sound level is within 10 dB of its maximum value, one can determine the single-event noise exposure level (SENEL).

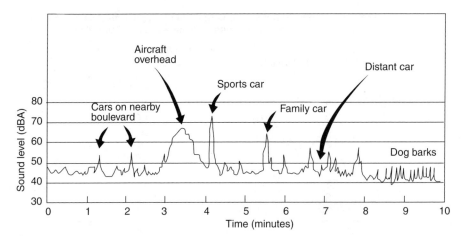

Figure 12.5 Variation in environmental noise levels. (From USEPA, *Protective Noise Levels*, www.nonoise.org/library/levels/levels/htm, 2000.)

12.2.3 Equivalent Sound Level

The equivalent sound level (L_{eq}) is the average of all sounds measured on the A scale over a period of time. It can be used for any exposure duration. The L_{eq} correlates relatively well with effects of noise on humans, even for wide variations in sound level and duration. It is used to report average environmental noise levels.

The L_{eq} can be obtained by using dosimeter-type instruments or calculated from individual sound measurements. In the latter case, we use Equation 12.5:

$$L_{eq} = 10 \log_{10} [\Sigma^n_{i=1} 10^{(L_i/10)}]/N \tag{12.5}$$

where L_i corresponds to each measured value in decibels.

Assume that sound measurements were made in a city every 6 min during an hour, with the following values: 65, 75, 68, 70, 80, 72, 76, 78, 82, and 65 dBA. Using Equation 12.5,

$$L_{eq} = 10 \log_{10} [10^{6.5} + 10^{7.5} + 10^{6.8} + 10^{7.0} + 10^{8.0} + 10^{7.2} + 10^{7.6} + 10^{7.8} + 10^{8.2} + 10^{6.5}]/10$$

$$= 76.3 \text{ dBA}$$

A modification of Equation 12.5 can be used to determine sound levels when one is exposed simultaneously to two sounds of equal or different magnitudes. If an individual is exposed to two different sound sources of 80 dB each, then

$$L_a = 10 \log_{10} [10^8 + 10^8]/2$$

$$= 83 \text{ dBA}$$

Table 12.3 Values Used to Determine Actual Sound Exposures Produced by Two Different Sources and Received at the Same Time

Differences in Sound Values	Value to be Added to Highest Exposure Value
0	3.0
1	2.5
2	2.1
3	1.8
4	1.5
5	1.2
6	1.0
7	0.8
8	0.6
9	0.5
10	0.4

Consequently, a doubling of the sound exposure would be equal to a 3-dBA increase in sound pressure. This relationship is used to construct tables (Table 12.3) to determine equivalent exposures to two or more sounds experienced simultaneously. The exposure value is determined by adding the table value to the highest exposure value. The greatest table value will be 3 dBA. The greater the difference between the two values, the lower the value added to the highest value.

12.2.4 Day–Night Sound Level

The day–night sound level is the equivalent sound level averaged over 24 h, with 10 dB added to the nighttime (10 to 7 P.M.) equivalent sound level. If the daytime L_{eq} were 60 dBA and the nighttime L_{eq} were 50 dBA, the weighted nighttime sound level would be $50 + 10$ dBA = 60 dBA. The day–night sound level (L_{dn}) would therefore be 60 dBA. Table 12.4 indicates L_{dn} values for different outdoor locations. The L_{dn} is commonly used in noise ordinances.

Table 12.4 Outdoor L_{dn} Levels by Location

Location	L_{dn} (dBA)
Airport, near freeway	88
0.75 mi from commercial aircraft landing	86
Urban center construction activity	78.5
Urban high-density apartment	78
Urban row housing — major street	68
Old urban residential area	59
Residential — wooded neighborhood	51
Residential — rural	39
Wilderness	35

Source: From USEPA, *Protective Noise Levels*, www.nonoise.org/library/levels/levels/htm, 2000 (condensed version of EPA Levels document).

12.3 LOUDNESS

The physical magnitude of sound energy can be described as sound intensity or pressure. The subjective, or perceived, magnitude is called loudness. Perception of loudness depends on sound energy, frequency, and duration. Humans do not hear sound frequencies equally well across the frequency spectrum (Figure 12.2). Additionally, physiological processes in the human auditory system protect it from excessive sound stimulation. Consequently, the relationship between sound energy received and our perception of loudness is not linear; it is closer to being logarithmic.

The sone is the basic unit of loudness. It is defined as the loudness of a pure 1000-Hz tone at a sound pressure of 40 dBA under specified listening conditions. A 2-sone sound would be perceived as twice as loud as a 1-sone sound.

Loudness is proportional to sound energy over a significant fraction of audible frequencies. This psychophysical "power law" is referred to as Steven's law. In the middle frequencies, the exponent of the power function is such that a twofold increase in perceived loudness corresponds to a 10-dB increase in sound pressure.

The relationship between loudness and sound pressure can be seen from the suite of equal loudness contours in Figure 12.6. Loudness levels are expressed in units called phons, which at 1000 Hz are directly related to the Spl in decibels. This relationship varies with frequency. Although the 40-phon contour corresponds to 40 dBA at 1000 Hz, it also corresponds to 50 dBA at 100 Hz. Therefore, any tone that is perceived equally well as a 1000-Hz tone assumes the same loudness or phon value as the 1000-Hz tone. At 1000 Hz, phon values are equal to the decibel value.

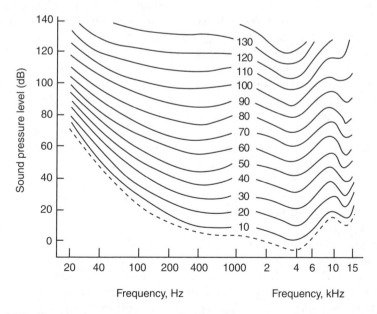

Figure 12.6 Equal loudness contours as a function of frequency and sound pressure level.

The relationship between the scale of loudness, sones (S), and loudness level, phons (P), can be expressed for loudness levels of >40 P:

$$S = 2 \ (P - 40)/10 \qquad\qquad (12.6)$$

If the loudness level increases from 40 to 50 P, sound will be perceived as twice as loud; an increase from 40 to 60 P will be four times as loud.

It would be desirable for measurements of sound pressure to be expressed in loudness readings equal to phons. Due to the complexity of human perceptual processes, this has been difficult to achieve. Such procedures, nevertheless, have been developed and incorporated into modern digital equipment that is not commonly used.

12.4 EFFECTS OF ENVIRONMENTAL NOISE EXPOSURE

Noise exposures experienced daily by an individual depend on how much time one spends in different indoor and outdoor locations and the noise environments present. In the U.S., the noise environment outdoors varies over a range of 50 dBA. Day–night sound levels as low as 30 to 40 dBA occur in wilderness areas; in urban areas, they range up to 80 to 90 dBA. Approximately half of the U.S. population lives in urban areas that have L_{dn} values of 48 dBA or higher due to traffic noise alone. Over 62 million people live in areas with urban or freeway traffic noise of L_{dn} values of 60 dBA or higher; 400,000 people in areas where urban or freeway traffic noise results in L_{dn} values of >80 dBA. Approximately 10 million people live in areas with L_{dn} values of 60 dBA associated with aircraft noise; 200,000 people where aircraft operations result in L_{dn} values of 80 dBA.

Outdoor noise exposures are significantly diminished by building structures when windows are closed. Despite this reduction, indoor noise levels are usually comparable to, or higher than, those outdoors. This is due to indoor sources such as appliances; mechanical equipment; television, radio, and music playing; and people. Indoor sound sources are often used to mask outdoor noise.

Exposure to excessive sound or noise levels may have a variety of adverse effects. These include hearing impairment; adverse effects on speech communication, performance, and behavior; nonauditory physiological effects; sleep disturbance; and annoyance.

12.4.1 Hearing Impairment

Adverse effects of exposure to excessive noise levels in industrial environments are well known. As a consequence, workplace noise exposures in the U.S. are regulated by OSHA. There is very little documentation of hearing impairment resulting from community noise exposures. As a consequence, hearing impairment risks associated with environmental noise are inferred from evaluations of worker exposures.

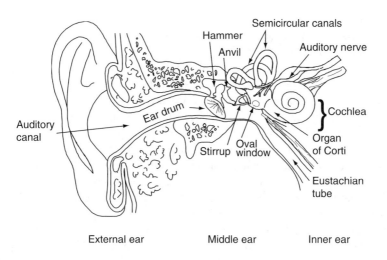

Figure 12.7 Human auditory system.

The human auditory system (Figure 12.7) is comprised of the outer ear, middle ear, cochlea of the inner ear, and auditory nerve–brain connections.

The outer ear collects sound waves and directs them through the ear canal to the eardrum or tympanic membrane. Collected sound causes a resonance vibration of the eardrum. This vibration is transmitted through the three bones (malleus (hammer) , incus (anvil) , stapes (stirrup)) of the middle ear into the inner ear through the oval window in the cochlea; alternatively, sound can pass to the inner ear by bone–bone conduction.

The middle ear plays a major role in protecting the auditory system from excessive noise stimulation. The aural reflex is mediated by two small muscles in the middle ear that contract, pulling the stapes and eardrum toward the middle ear cavity, when intense (>80 dB) sound exposure occurs. This helps protect the middle ear bones from excessive movement and damage and also protects the cochlea. The aural reflex is more responsive to broadband sounds than pure tones, and lower frequencies than higher ones. It is readily activated and maintained by intermittent, intense sound impulses. As middle ear muscles contract, they impede or attenuate sound energy that would have been received by the cochlea.

The organ of Corti is found between two fluid-filled chambers in the cochlea. Pressure waves produced by sound at the oval window displace the organ of Corti. The basilar membrane of the organ of Corti is associated with two groups of hair cells. The inner hair cells serve as sensory receptors. The outer hair cells serve as an amplification system. It is these outer cells that are commonly damaged by noise.

Acoustic information is transmitted from the organ of Corti to the brain by the auditory nerve. At all steps in this pathway, complex processing takes place that enables the brain to interpret sound qualities such as perceived intensity and pitch, speech feature analysis, and noise identification.

As indicated previously, humans can detect sounds in the frequency range of 20 to 20,000 Hz. The wide variation in sensitivity among humans may be affected by different environmental influences.

Hearing sensitivity diminishes significantly with age, particularly at high frequencies (>10,000 Hz). This age-associated decline in hearing sensitivity may be augmented by the cumulative effects of everyday noise exposure.

In evaluating potential hearing impairment, individuals are tested using audiometry. The concept of hearing level is used to determine whether hearing impairment or loss has occurred. Hearing level is the audiometric threshold of an individual or group in relation to an accepted standard. Hearing threshold levels are auditory thresholds expressed in hearing levels. Threshold levels outside the normal range indicate hearing impairment. The term *hearing loss* is used to describe hearing impairment that poses significant difficulties in leading a normal life, particularly in understanding speech, or a hearing threshold that has deteriorated. It is common to use reference levels to determine whether a noise-induced hearing loss is "compensable" under workmen's compensation laws. In the U.S., a hearing loss is compensable only for a 25-dB threshold shift in the frequency range of 500 to 2000 Hz. This is similar to the reference threshold used by the International Standards Organization (ISO). Such a reference standard for hearing disability ignores the fact that sound energy at higher frequencies is important for speech intelligibility and music perception.

Noise may cause a shift in hearing sensitivity that is temporary (temporary threshold shift (TTS)) or permanent (permanent threshold shift (PTS)). One can experience a measurable loss in hearing sensitivity associated with a noisy environment or discrete noise events such as gunshots, explosions, etc. A TTS is reversible and likely to be experienced by a majority of individuals during their lifetime. Recovery, which is exponential, may occur in a matter of minutes, hours, days, or even weeks. Recovery depends on severity of the threshold shift, type of exposure, and individual sensitivity.

Noise-induced PTS commonly occurs in individuals working in very noisy industrial or construction environments. Hearing impairment occurs gradually, becoming progressively worse with time. Maximal loss in hearing acuity occurs at about 4000 Hz. The frequency pattern of hearing impairment and its progress over a time period of exposure can be seen for industrial workers in Figure 12.8. As hearing impairment increases, speech communication becomes increasingly more difficult.

There is limited evidence that environmental noise sources, e.g., activities such as shooting, motorcycling, snowmobiling, fireworks, and even loud music, can cause hearing impairment. The effects of "background" environmental noise in urban areas on hearing impairment cannot be easily determined. As such, hearing impairment risks have to be extrapolated from industrial exposures. From such extrapolations, an 8-h continuous equivalent exposure of 70 dB has been identified by the USEPA as the apparently safe level of protection for significant PTS.

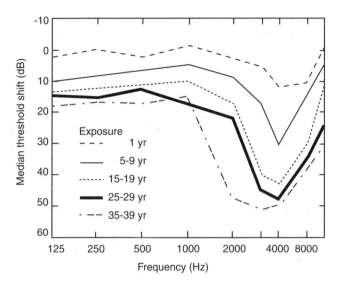

Figure 12.8 Hearing losses in industrial workers. (Adapted from Taylor, W. et al., *J. Acoust. Soc. Am.*, 38, 113, 1965. With permission.)

12.4.2 Speech Communication

One of the major effects of environmental noise is interference with speech communication. As speech is an essential element of human interaction, speech interference can significantly degrade the quality of life. It may disturb normal social and work activities and indirectly cause annoyance and stress. Of primary concern are the effects of noise on face-to-face and telephone conversations and use of audio and video devices, both indoors and outdoors.

Speech is a learned motor behavior. It involves feedback between the act of hearing and speech musculature coordinated by the central nervous system. The acoustical energy of speech falls between 100 and 6000 Hz, with the most important cue-bearing energy between 300 and 3000 Hz. Speech comprehension under normal circumstances involves distinguishing a finite number of mutually exclusive sounds, and temporal features such as variations in loudness, pitch, and rhythm.

Noise-associated interference with speech communication is a masking process in which noise renders speech sounds unintelligible. The more intense the masking sound and the more energy it contains at speech frequencies, the greater the number of speech sounds that cannot be discerned by the listener.

The degree that noise interferes with speech depends on other factors as well. These include vocal effort, distance between speaker and listener, and surrounding acoustics. It also depends on the speaker's enunciation, conversation topic, and the listener's motivation, hearing acuity, and familiarity with the speaker's vocabulary.

In indoor environments, the highest environmental noise level that allows relaxed conversation with 100% intelligibility is about 45 dBA. Conversation is still about 95% intelligible at 65 dBA, but it diminishes dramatically thereafter (Figure 12.9).

Speakers tend to raise their voice level when environmental sounds are ≥60 to 65 dBA. In indoor environments, speech sound levels decrease as an inverse square function of distance up to a distance of 2 m. Further inverse square decreases do not occur because of the effect of reverberations. In outdoor environments, the effect of increasing distance between speakers is a continuum of inverse square decreases.

As can be seen in Table 12.5, if environmental sound levels exceed those at which speech is easily intelligible, speakers must move closer or raise their voices appreciably. For example, if normal voice levels are intelligible at 3m in the presence of 56 dBA environmental sound, 66 dBA of environmental sound requires speakers to be only 1m apart or raise their voices to maintain the same intelligibility.

Figure 12.9 Decrease in speech intelligibility with increasing environmental sound. (From USEPA, *Protective Noise Levels*, www.nonoise.org/library/levels/levels/htm, 2000.)

Table 12.5 Environmental Sound Levels That Allow 95% Speech Intelligibility at Various Outdoor Distances and Two Voice Levels

Distance (m)	Normal Voice (dBA)	Raised Voice (dBA)
0.5	72	78
1	66	72
2	60	66
3	56	62
4	54	60
5	52	58

Source: From USEPA, *Protective Noise Levels*, www.nonoise.org/library/levels/levels/htm, 2000 (condensed version of EPA Levels document).

In a typical residential dwelling with windows closed, sound attenuation is ~10 to 15 dBA. As a consequence, an outdoor L_{dn} value of 60 dBA would (assuming no major indoor sources are present) permit 100% speech intelligibility indoors.

Such speech intelligibility assumes that most individuals have normal speech perception. However, a majority of individuals belong to sensitive populations, e.g., the elderly and those who have hearing impairment. Even individuals as young as 40 years may have difficulty interpreting spoken messages. For sensitive groups or when listening to complex messages (e.g., classrooms, telephone conversations), the signal speech level should be 10 dBA greater than the environmental sound level.

12.4.3 Sleep Disturbance

Sleep disturbance, which includes difficulties falling asleep, alterations in sleep patterns and depth of sleep, and awakenings, is considered to be a major effect of environmental noise. Physiological effects induced by nighttime noise exposure during sleep include increased blood pressure and heart rate, vasoconstriction, change in respiration, and cardiac arrhythmia. Secondary or aftereffects occurring in the morning or day after nighttime exposure may include reduced perceived sleep quality, increased fatigue, and decreased performance and sense of well-being.

The A-weighted equivalent sound level is poorly correlated with sleep disturbance. Better indicators of sleep disturbance risk are the number of sound events exceeding a reference sound level and the difference between maximum and background sound levels.

12.4.4 Psychophysiological Effects

Noise exposure may elicit several kinds of reflex responses, particularly when the noise is sudden, unfamiliar, or of an unwanted nature. These reflect primitive body defense responses and may develop in response to other stimuli as well. Target organs include (1) the heart, blood vessels, intestines, endocrine glands, etc., which are enervated by the autonomic nervous system; and (2) the hypothalamodiencephalic centers that regulate sleep arousal, endocrine secretions, and other functions.

Environmental noise is known to interfere with a number of activities such as recreation, sleep, communication, and concentration. Viewed in this context, it can be considered a stress factor that results in physiological responses that are modified in complex ways by individual psychological processes. Most studies have focused on the possibility that noise may be a risk factor for cardiovascular disease. Of particular note are occupational studies and studies of residents near airports that show a tendency for elevated blood pressure, a major risk factor for heart attack and stroke.

High occupational noise level exposures have been associated with development of neuroses and irritability; high environmental noise exposures have been linked to mental health disorders. In the latter case, noise is not believed to be a direct cause of mental illness; rather, it may accelerate and intensify latent mental health problems.

12.4.5 Performance

Noise exposures may affect human performance. High noise exposures appear to interfere with the conduct of tasks, particularly those (1) that demand attention to multiple cues, (2) for which continuous and detailed attention to frequent signals is required, or (3) that require high levels of working memory. Chronic noise exposures in children appear to impact reading acquisition, which may be related to deficits in auditory discrimination.

12.4.6 Annoyance and Social Behavior

Annoyance resulting from environmental noise exposure is common in urban areas, being experienced in many cases by the majority of inhabitants and visitors. Annoyance can be defined as a feeling of displeasure associated with any agent, condition, or person known or believed to be adversely affecting an individual or group.

Sound-related annoyance may be affected by a number of physical factors. These include the nature of the sound or noise and its generation and temporal patterns. Sound's tonal qualities (low or high frequency, monotonous); its temporal variation (constant, intermittent, random) and intensity (loudness); the highest levels produced by a source as well as the number of sound events; and the time of day affect annoyance. Airport noise is particularly annoying, due in large part to the number of sound events. Because of the intensity and continuous nature of highway traffic, it is a source of considerable individual and community annoyance.

Annoyance is also affected by psychological factors as well as an individual's living conditions, personal attitude toward the noise source, previous noise exposures, and a variety of socioeconomic factors.

Annoyance with noise appears to have significant affects on social behavior. Noise exposures have been reported to reduce helpfulness and promote aggression when combined with provocation or pre-existing anger or hostility. It may affect overt everyday behavior patterns such as opening windows, using balconies, TV and radio use, circulating petitions, and complaining to authorities. It may also be related to increased residential mobility, drug consumption, accident rates, and hospital admissions.

Noise-associated annoyance is, as indicated above, affected by a variety of both physical and psychological factors. As such, it is difficult to quantify. In noise and annoyance reduction efforts, the L_{eq} value is commonly used as a reference, though it is only a fair approximation of the annoyance level.

12.5 NONREGULATORY AND REGULATORY CONTROL APPROACHES

12.5.1 Community Exposure Guidelines

The USEPA has identified a range of L_{eq} and L_{dn} values that would protect the public from effects of environmental noise. The values are meant to be viewed as levels below which the general public would not be at risk for noise exposures. They

Table 12.6 USEPA Community Exposure Guidelines

Effect	Level	Area
Hearing	L_{eq} (24 h) < 70 dBA	All areas
Outdoor activity — Speech interference/annoyance	L_{dn} (24 h) < 55 dBA	Areas where quiet is a basis for use
Outdoor activity — Speech interference/annoyance	L_{eq} (24 h) < 55 dBA	Areas where humans spend limited amounts of time
Indoor activity — Speech interference/annoyance	L_{dn} (24 h) < 45 dBA	Indoor residential areas
Indoor activity — Speech interference/annoyance	L_{eq} (24 h) < 45 dBA	Other indoor areas such as schools

Source: From USEPA, *Information on Levels of Environmental Noise Requisite to Protect Public Health and Welfare with an Adequate Margin of Safety,* EPA/ONAC 550/9-74,004, EPA, Washington, D.C., 1974.

contain a margin of safety to ensure that they are protective. Because the process of identifying noise protection levels did not take into account technical feasibility and economic costs, the USEPA has taken great pains to point out that they should not be viewed as standards, criteria, or goals.

The USEPA's noise protective values are summarized in Table 12.6. Though based on 24-h measurements, they are in fact annual average values. For example, to protect against hearing damage, one's 24-h noise exposure averaged over the whole year should not exceed 70 dBA. Annual yearly outdoor sound levels on the L_{dn} scale of <55 dBA in areas such as residences, schools, and hospitals are deemed sufficiently protective of public health and welfare; for inside buildings, an L_{dn} value of <45 dBA is deemed sufficiently protective.

12.5.2 Community Regulatory Efforts

Abatement of community noise requires identification of noise as a problem of sufficient magnitude to establish a regulatory program for its control. Regulatory programs for noise control exist at local, state, and federal levels. These programs vary in form and scope. Many local noise ordinances are qualitative, prohibiting excessive noise or noise that results in a public nuisance. Because of the subjective nature of such ordinances, they are often difficult to enforce.

Regulation of community noise requires a quantitative approach; i.e., noise regulations should be based on well-defined standards, violations of which can be measured. Standards should be based on scientific criteria as to acceptable noise levels. Noise standards used in Minnesota are summarized in Table 12.7. They are based on Spl values and duration and are defined for four zones: single- and multiple-family residential, commercial, and industrial areas. They are specified for daytime (7 A.M. to 10 P.M.) and nighttime (10 P.M. to 7 A.M.).

The L_{10} value (Table 12.7) indicates that this standard can be exceeded 10% of the time; the L_1 value, 1% of the time. The L_{max} is the maximum acceptable instantaneous noise applied to sudden blaring noises, e.g., a fast motor vehicle. The EPNdB is used for aircraft noise. It is a single-event standard that weighs the

Table 12.7 Minnesota Community Noise Standards

Zone	Daytime (0700–2200 h)					Nighttime (2200–0700 h)				
	$L_{10}{}^a$ (dBA)	L_1 (dBA)	L_{max} (dBA)	EPNdB (dBA)	Impulse (psf)[b]	$L_{10}{}^a$ (dBA)	L_1 (dBA)	L_{max} (dBA)	EPNdB (dBA)	Impulse (psf)[b]
Residential (single-family)	65	70	75	88	1.25	55	60	65	78	0.40
Residential (multifamily)	70	75	80	93	2.22	60	65	70	83	0.70
Commercial	75	80	85	98	3.95	65	70	75	88	1.25
Industrial	85	90	95	105	12.50					

[a]Based on an hourly average.
[b]Pounds per square foot.
Source: From Tyler, J.M. et al., *JAPCA*, 24, 130, 1974. With permission.

presence of very annoying discrete frequencies associated with the whine of jet engines. The impulse standard is designed for noise of very short duration. Note that the most stringent noise standards are applied to residential neighborhoods for nighttime hours. They assume that higher levels are acceptable in commercial, manufacturing, and industrial zones and that sounds of short duration are the most easily controlled.

12.5.3 Federal Noise Control Programs

A variety of federal agencies have authority to regulate and control noise. These include the Federal Aviation Administration (FAA), Federal Highway Administration (FHA), Department of Housing and Urban Development (HUD), OSHA, and the USEPA. The FAA has authority to impose and enforce noise emission standards for commercial aircraft operation. The FHA is responsible for issuing standards for new and improved federal highway construction. OSHA has significant noise control responsibilities that are designed to protect workers from noise-induced hearing losses. HUD's authority involves specification of noise control standards and techniques for HUD-assisted new housing construction.

The principal responsibility for environmental or community noise is vested in the the USEPA. Its role in noise abatement was authorized by Congress in the federal Noise Control Act of 1972 and the Quiet Communities Act of 1978. In the 1972 law, the USEPA was required to develop and publish noise criteria and scientific information on noise effects and identify levels of environmental noise, the attainment and maintenance of which are necessary to protect public health and welfare. The USEPA was also required to identify major noise sources that need regulation. Under this authority, it has promulgated noise emission standards for interstate motor carriers and buses, medium- and heavy-duty trucks, and railroad carriers. Other noise emission standards apply to construction equipment. The Noise Control Act of 1972 also authorized regulation of the manufacture of a variety of equipment and appliances, requiring labeling as to sound levels produced.

The Quiet Communities Act of 1978 was designed to provide financial assistance to state and local governments and regional planning agencies to investigate noise problems, plan and develop a noise control capability, and purchase equipment. It also provides federal assistance to facilitate community development and enforcement of community noise control measures, as well as development of abatement plans for areas around major transportation facilities, including airports, highways, and rail yards.

12.6 CONTROL MEASURES

Achievement of noise abatement objectives requires that noise producers be regulated. Noise control is best achieved at the source. Noise emission standards can be specified and designed into new products such as aircraft engines, motor vehicle tires, industrial machinery, and home appliances. Control of aerodynamically-produced noise may require changing airflow to reduce turbulence or placing sound-absorbent materials along its path. Control of noise from vibrating surfaces may require mounts to limit vibration, use of flexible materials in joints, or materials of sufficient mass that are not easily subject to vibration.

For many existing sources, noise reduction may only be achieved by the retrofit application of engineering modifications, including use of sound-dampening materials. The economics of retrofitting are such that it is far less expensive to initially purchase less-noisy equipment than pay for its control after installation and use. Some existing noise sources can also be isolated, and sound-absorbing enclosures placed around them.

An alternative to noise control at the source is modification of the noise path or its reception. For example, operational changes in commercial airline and flight patterns during takeoff and landing can significantly reduce environmental noise exposures. Such operational requirements are already in effect at the direction of the FAA. At least initially, the location of major highways and airports can minimize noise exposures in residential neighborhoods. Where major highways are in close proximity to residential neighborhoods, sound-absorbing walls, earthen mounds, and vegetation may be utilized to attenuate sound. Inside buildings, sound-absorbing furnishings (carpets, drapery) or construction materials (acoustical tile) can significantly reduce sound.

READINGS

Berglund, B. and Linvall, T., Community noise, *Arch. Center Sensory Res.*, 2, 1, 1995.

Bragdon, C.R., *Municipal Noise Legislation*, Fairmont Press, Atlanta, 1980.

National Bureau of Standards, *Fundamentals of Noise: Measurement, Rating Schemes and Standards*, NTID Publication 300.15. National Bureau of Standards, Gaithersburg, MD, 1971.

Pelton, H.K., *Noise Control Management*, Van Nostrand Reinhold, New York, 1993.

Peterson, A.P., *Handbook of Noise Measurement*, 8th ed., GenRad, Concord, MA, 1978.

Ropes, J.M. and Williamson, D.L., EPA's implementation of the Quiet Communities Act of 1978, *Sound and Vibration*, December 10, 1979.

Tyler, J.M., Hinton, L.V., and Olin, J.G., State standards, regulations and responsibilities in noise pollution control, *JAPCA*, 24, 130, 1974.

USEPA, *Information on Levels of Environmental Noise Requisite to Protect Public Health and Welfare with an Adequate Margin of Safety*, EPA/ONAC 550/9-74,004, EPA, Washington, D.C., 1974.

USEPA, *Noise Effects Handbook: A Desk Reference to Health and Welfare Effects of Noise*, National Association of Noise Control Offices, Fort Walton Beach, FL, 1981.

USEPA, *Protective Noise Levels*, www.nonoise.org/library/levels/levels/htm, 2000 (condensed version of EPA Levels document).

QUESTIONS

1. Describe differences between sound power, sound intensity, and sound pressure.
2. How do sound and noise differ?
3. What range of frequencies can humans hear? What range of frequencies do humans hear best?
4. Identify the primary sources of environmental noise.
5. What reference pressure is used in determining sound pressure levels?
6. Describe the relationship between sound pressure and distance from a source. If the distance is doubled, what is the difference in sound pressure levels in decibels?
7. What are weighting scales? Why are they used?
8. Describe the limitations of the A weighting scale for assessments of loudness and annoyance.
9. Most environmental noise measurements are made using a Type II meter. What is a Type II meter?
10. Why would one measure sound using an octave band analyzer?
11. Describe sound exposure level and equivalent sound level.
12. Calculate the equivalent sound level (L_{eq}) for the following sound values: 65, 70, 60, 55, and 75 dBA.
13. If you are exposed to two sound sources of 83 dBA each, what is your actual exposure in decibels?
14. Describe how the day–night sound level (L_{dn}) is determined and its relevance to community noise exposure.
15. How is loudness related to sound energy?
16. Describe the relationship between sones, phons, and decibels.
17. How does noise exposure affect human health?
18. Describe the relationship between environmental noise and speech communication.
19. How does exposure to noise cause annoyance?
20. The USEPA has identified noise protective levels for community noise exposures. Describe these.
21. If your community wishes to establish an effective noise ordinance, what noise standards would you recommend?
22. Which federal governmental agencies have the authority to regulate noise production and noise exposures? Identify their respective noise-related responsibilities.

23. How could one reduce noise levels in a residential neighborhood adjacent to a freeway?
24. How could exposures to aircraft noise be reduced?

Glossary

Absorption — Process by which pollutants are collected in a sampling or control device by reaction with a collection medium.

Accumulation mode — Particles in the aerodynamic diameter range of 0.1 to 1.0 μm; formed from the agglomeration of nuclei mode particles.

Accuracy — Closeness of a measured value to a true or reference value; measure expressed as ± a percent of the reference value when precision is taken into account.

Acid-neutralizing capacity (ANC) — Measure of the ability of soil and water to neutralize strong acids.

Acid rain — Deposition of H^+ on land or water surfaces by rain droplets.

Acidic deposition — Wet and dry deposition of H^+ on land or water surfaces.

Acidification — Increase of H^+ concentration in water or soil.

Acrolein — Aldehyde produced photochemically in the atmosphere; characterized by its potent eye irritancy.

Activated carbon — Sorbent used to collect nonpolar organic compounds.

Acute exposure — Exposure of humans, plants, and other organisms to high concentrations or doses over a short period of time; physiological responses occur immediately or within hours of exposure.

Adiabatic lapse rate — Theoretical change of temperature with height when a warm, dry parcel of air is released into a cool, dry atmosphere.

Adsorption — Collection of gas-phase pollutants on solid surfaces by physical forces; collection principle used in air sampling and pollution control.

Aerodynamic equivalent diameter — Particle diameter standardized to a sphere with unit density and settling velocity.

Aerometric Information Retrieval System (AIRS) — Repository of pollutant monitoring information maintained by the USEPA.

Aerosol — Solid or liquid particles dispersed in air.

Agglomeration — Process by which atmospheric particles grow in size as a result of collisions and subsequent adherence and associated processes.

Air pollution — Contamination of atmosphere by gas- and particulate-phase substances that results in adverse or undesirable environmental effects.

Air pollution episode — A period of abnormally elevated pollutant levels; typically associated with subsidence inversions.

Air quality control region (AQCR) — Political jurisdiction or geographical area where NAAQSs must be achieved and maintained.

Air quality criteria — Compilation of scientific facts and their evaluation, which are used to support the development and promulgation of NAAQSs.

Air Quality Index — USEPA program that relates measured air pollutants to their health-affecting potential; used to communicate daily air quality risks to the public.

Air quality management — Pollution control strategy based on the use of air quality standards.

Air quality modeling — The use of mathematical models to predict the impact of a pollutant source, atmospheric reactions, etc., on ground-level concentrations.

Air quality standards — Permissible levels of regulated pollutants in a control region.

Air toxics — Noncriteria pollutants regulated under hazardous pollutant standards.

Air:fuel ratio — Mixture of air to fuel used in motor vehicle engines to enhance engine performance and reduce exhaust emissions.

Aitken mode — Particles produced in the atmosphere by gas-to-particle conversion as well as in combustion processes; same as nuclei mode.

Albedo — The ability of a body to reflect light; relative to the Earth, 30% of sunlight is reflected back to space.

Aldehydes — Organic chemical compounds characterized by relatively high chemical reactivities, irritancy to humans, and role as by-products of hydrocarbon oxidation processes.

Allowance system — Pollution control approach that limits sources to specific quantities of emissions to achieve emission reductions economically.

Allowances — Quantity of SO_2 emissions permitted sources under acidic deposition provisions of the 1990 CAA Amendments.

Ambient air — The free-flowing air outside of buildings.

Ammonia — Reduced nitrogen species that serves to neutralize strong acids in the atmosphere.

Annoyance — Unwanted effects of sound or noise that reduce one's sense of personal comfort.

Antagonism — Interaction effect in which one pollutant inhibits the effect of another, thus resulting in reduced physiological harm.

Anticyclones — Clockwise airflow in high-pressure systems in the northern hemisphere.

AP-42 — Emission factor document compiled and continuously updated by the USEPA.

Arctic haze — Reduced visibility in portions of the Arctic in spring months associated with pollutant transport from Europe and Russia.

Area source — Collective emissions from a number of small mobile and stationary sources in an urban area.

Aromatic hydrocarbons — Hydrocarbon species that contain one or more benzene rings; simple aromatic hydrocarbons are characterized by moderate photochemical reactivity.

Asbestos — Fibrous silicate minerals used in fireproofing and insulation products in buildings; hazardous air pollutant.

Asbestos Hazard Emergency Response Act (AHERA) — Law that requires asbestos inspections and management in school buildings.

ASHRAE — American Society of Heating, Refrigerating and Air-Conditioning Engineers; organization that develops and publishes ventilation guidelines.

Asthma — Respiratory disease characterized by severe episodes of shortness of breath.

Atmosphere — Gas-phase environmental medium that surrounds the Earth.

Atmosphere–ocean general circulation models — Models used by atmospheric scientists to predict the impact of changes in greenhouse gases and associated factors on the Earth's climate.

Atmospheric density — Weight of atmospheric molecules per unit volume of air; density decreases exponentially with increasing height.

Atmospheric deposition networks — Monitoring networks created by the USEPA to collect deposition samples and measure concentrations of target pollutants.

Atmospheric lifetime — Average time a pollutant remains in the atmosphere.

Atmospheric pressure — Force of air molecules on a surface; decreases exponentially with height as density decreases.

Atmospheric stability — Relative measure of air motion associated with vertical temperature gradients.

Atmospheric transport — Movement of pollutants over long distances.

Attainment area — An air quality control region or portion thereof that is in compliance with specific National Ambient Air Quality Standards (NAAQSs).

Attenuation — The reduction of image-forming light as it passes from an object to an observer.

Audiometric sound threshold — Lowest sound pressure that an individual can hear at various frequencies in the auditory sound spectrum.

Audit —An evaluation of procedures and activities to ensure high-quality and reliable air quality monitoring data.

Automated surface observing systems (ASOS) — Automated, real-time visibility monitoring systems at U.S. airports.

Averaging time — Time period used to integrate pollution concentrations for data summary, regulatory considerations, and sampling limitations.

A-weighted sound level — Sound pressure levels recorded on the A scale; commonly used measure of sound.

Backscattering coefficient — Measure of light scattering associated with particles and atmospheric gases; major factor contributing to visibility reduction.

Baghouse — System used to control particle emissions from sources by using fabric filtration.

Benzo-α-pyrene — A highly carcinogenic polycyclic aromatic compound commonly found in combustion-derived atmospheric particles.

Best available control technology (BACT) — Control technology demonstrated to achieve the highest level of control; necessary to meet emission reduction requirements in nonattainment areas.

Best available retrofit technology (BART) — Control measures applied to existing sources designed to provide emission limits associated with PSD programs.

Bias — Property of a data set wherein systematic error is occurring; typically associated with calibration errors.

Bioaccumulation — Uptake of pollutants by plants and animals from the environment.

Biocides — Substances used to control pest organisms; commonly used to control microbial growth.

Biogenic pollutants — Pollutants emitted to the atmosphere from biological sources.

Biological contaminants — Indoor pollutants that have a biological origin.

Biomagnification — Increase of pollutant concentrations as pollutants pass upward through food chains.

Biomass cooking — Use of wood, charcoal, or dung for cooking in developing countries; significant source of combustion by-product exposures indoors.

Bioscrubbers/filters — Pollution control devices that use biological processes to remove contaminants from waste gas streams.

Blow-by gases — Uncombusted gasoline vapors that slip by piston rings and accumulate in the crankcase.

Box plot — A graphical representation that shows the distribution of continuous data values around the median.

Brightness — Measure of wavelength-adjusted light received by the human eye from an object.

Bromine monoxide — Chemical species found in the stratosphere that destroys O_3 more effectively than chlorine monoxide.

BTEX — A mixture of aromatic compounds (benzene, toluene, ethylbenzene, and xylene) used as an additive to boost the octane rating of gasoline.

Bubble concept — Source control option wherein emissions are reduced plantwide rather than just from stacks.

Buoyancy flux — Measure of buoyancy of a plume calculated for plume rise equations.

Calibration — Process by which measured values of a pollutant or airflow are compared to standard or reference values; procedure designed to maintain data quality.

Carbon dioxide (CO_2) — Natural constituent of the atmosphere that serves biological processes as a source of carbon and the atmosphere as a greenhouse gas.

Carbon monoxide (CO) — Colorless, odorless, toxic air pollutant produced by incomplete combustion.

Carboxyhemoglobin — Molecular complex in the blood resulting from the reaction of CO with hemoglobin; measure of CO exposure.

Carcinogen — A substance that can cause cancer.

Carcinogenicity — Ability of a substance to cause cancer.

Cascade impactor — Particle-collecting device that utilizes a series of plates to fractionate particles into different size ranges.

Catalyst — Substance that reduces the energy requirement of a chemical reaction.

Catalytic converter — Control device that removes pollutants from automobile exhaust by means of catalytic processes.

Catalytic incinerator — Control device that uses catalytic processes to reduce emissions of oxidizable compounds from an industrial source.

Chapman cycle — Chemical reactions that describe O_3 production and destruction in the stratosphere.

Chemiluminescence — Principle used to measure concentrations of NO_x and O_3 by chemical reactions that release light.

Chlorine monoxide — Primary chemical species involved in the catalytic destruction of stratospheric O_3.

Chlorine nitrate — Chemical species that serves as both a sink and reservoir for stratospheric chlorine.

Chlorofluorocarbons (CFCs) — Once widely used chemicals characterized by their long atmospheric lifetime, O_3-destroying potential, and thermal sorption properties.

Chlorosis — Loss of chlorophyll in plant leaves associated with chronic exposures to phytotoxic pollutants.

Chronic acidification — Long-term acidification of aquatic and terrestrial ecosystems.

Chronic allergic rhinitis — Relatively mild, persistent illness of the upper respiratory system associated with sensitization and exposure to common allergens.

Chronic bronchitis — Disease of respiratory airways caused by exposure to tobacco smoke or ambient air pollution.

Chronic disease — Disease associated with low-level pollutant exposures over an extended time period.

Chronic exposure — Low-level exposure that occurs to humans, plants, or animals over an extended time period (weeks, months, years) resulting in characteristic adverse effects.

Citizen suits — Lawsuits that can be brought by citizens against a source or the EPA; legal standing for citizen suits is provided by the 1970 CAA Amendments.

Class I area — Under the CAA Amendments of 1977, areas around national parks, wilderness areas, etc., that are to receive the greatest degree of protection from new, potential, and existing visibility-reducing sources.

Class II area — Areas that include national forests that are subject to lesser visibility protection limits on emissions than Class I areas.

Clean Air Act (CAA) Amendments — Laws enacted by Congress that grant authority to federal agencies to control air pollution according to specific provisions; basis for all federal air pollution regulation.

Climate change — Change in global surface temperature over time.

Climate models — Models used to describe climatic patterns and variations as a result of changes in Earth, water, and atmospheric systems.

Cloud condensation nuclei — Small particles in the atmosphere that provide surfaces for the condensation of H_2O vapor and formation of cloud droplets.

Coagulation — Growth of particles associated with the collision and subsequent adherance of particles to each other.

Coal beneficiation — Separation of inorganic sulfur from coal by flotation and settling processes.

Coarse particles — Particles with aerodynamic diameters of ≥ 2.5 μm; produced by fragmentation of matter or atomization of liquids.

Collection efficiency — Relative ability of a sampling train or control device to remove or control pollutants from waste gases.

Combustion — Process by which compounds containing hydrogen and carbon are oxidized to CO_2 and H_2O vapor with the release of heat.

Common law — Legal system based on the evolution of accepted principles of human behavior in the matter of disputes concerning property, contracts, and related matters; legal principles under which a property owner can sue a source for damages and injunctive relief.

Compression ignition engines — Engines that utilize high compression ratios and autoignition of air–fuel mixtures; diesel engines.

Compression ratio — Ratio of the air volume at top dead center of the combustion stroke to that at the bottom of the stroke; factor that influences combustion efficiency, power, and emissions.

Concentration — The relative amount of a substance in a medium such as air, water, soil.

Condensation — Gas- to liquid-phase transformation of matter; in the atmosphere, results in the formation and growth of particles.

Coning plume — Behavior of a plume under neutral to isothermal lapse rate conditions.

Continuous emission monitor (CEM) — Monitoring device that provides real-time measurement of stack gases/particles.

Continuous monitor — Sampling device that provides real-time or near-real-time pollutant measurements continuously.

Contrast — Difference in light coming from an object relative to its surroundings; indicator of atmospheric clarity.

Convection — Upward movement of warm air and concomitant subsidence of cooler air to replace it.

Coriolis effect — Apparent deflection of wind associated with the Earth's rotation.

Cost–benefit analysis — A quantitative evaluation of control costs relative to benefits to determine economically efficient emissions reduction.

Criteria pollutant — Pollutant regulated under NAAQS provisions of clean air legislation.

Cross-contamination — Building air quality problem associated with contaminant transport from a source to an exposure location.

Cyclone collector — Control device used to collect large particles by inertial separation.

Cyclones — Airflow associated with low-pressure systems wherein air moves counterclockwise in the northern hemisphere and clockwise in the southern hemisphere.

Day–night sound level — Average 24-h sound level weighted by 10 dB for nighttime hours.

Decibels — Unitless expression of sound power, intensity, and pressure.

Deciview — Measure of visibility that is proportional to the logarithm of atmospheric extinction.

Deciview index — Index that expresses uniform changes in haziness in increments from pristine to very impaired visibility.

Deliquescence — Rapid uptake of water vapor from the atmosphere by particles.

Deposition velocity — Rate at which pollutants are removed from the atmosphere by dry deposition.

Dicarboxylic acids — Organic acids produced in photochemical reactions that condense to produce light-scattering particles.

Dichotomous sampler — Particle sampler that collects and separates atmospheric particles into coarse and fine fractions.

Diesel engines — See *Compression ignition engines*.

Diffraction — Physical process by which light passing near, and parallel to, the surface of a particle is scattered from its optical path.

Diffusion — Random movement of gas-phase substances and very small particles from areas of higher to lower concentration.

Dioxins — Family of long-lived, highly toxic and carcinogenic chlorinated compounds emitted to the environment from the combustion of solid wastes.

Dispersion — Pollutants mixing with air resulting in lower ground-level concentrations.

Dispersion coefficients — Coefficients used in Gaussian models that are based on standard deviations of concentrations of pollutants as a plume spreads under different atmospheric stability conditions.

Dobson units — Measure of column O_3 concentration using a spectrometric technique.

Dose–response — Toxic responses to pollutant exposures that increase with increasing dose; measure of the relationship between adverse health effects and pollutant exposure.

Dry deposition — Deposition of gases and particles on land and water surfaces in the absence of precipitation.

Dry scrubber — Removal of pollutants by injecting dry sorbents into the pollutant-generating process; flue gas desulfurization technique.

Dust dome — A mass of particle-polluted air observed over a city.

Eccentricity — Time-dependent variation of the Earth's orbit around the sun.

Ecosystem changes — The wide-ranging effects on plant and animal communities associated with environmental pollution.

Effective stack height — The combination of both physical stack height and plume rise; used to predict downwind concentrations in air quality models.

Electrostatic precipitator — Control device that collects particles from waste gases by imparting electrical charges, with subsequent collection of charged particles on plates of opposite electrical charge.

Elemental carbon — Inorganic carbon found in particulate matter samples.

Emission factors — Numerical values compiled by the USEPA for pollutants and sources to estimate emissions.

Emission rate — Source pollutant emissions per unit time.

Emission standard — Legal limit on the quantity of a regulated pollutant that can be released from a source; used as a control strategy; used as a control tactic in achieving National Ambient Air Quality Standards.

Emission tax — Pollution control strategy that attempts to reduce emissions from a source by the imposition of a tax; pollution charges.

Emission test cycles — System of testing required of motor vehicle manufacturers to determine compliance with exhaust emission standards under specified driving conditions.

Emission trading — Pollution control tactic in which sources receive credits for control of pollutants in excess of regulatory requirements that may be marketed.

Emissions assessment — Determination of types and quantities of pollutants released from a source per unit time.

Endotoxins — Substances produced in the cell wall of gram-negative bacteria that, on exposure, can cause inflammatory respiratory symptoms; cause of humidifier fever.

Enforcement — Regulatory practice in which legal action is taken against a source for violating air pollution control requirements.

Entrainment — Process whereby outdoor pollutants are drawn into building air.

Environmental noise — Undesirable sound in the ambient environment.

Environmental tobacco smoke (ETS) — By-products of tobacco combustion in indoor air that result from sidestream and exhaled mainstream smoke.

Epidemiology — Study of the statistical relationship between disease in a population and potential causal factors.

Episodes — Periods of high ground-level pollution concentrations commonly associated with subsidence inversions.

Episodic acidification — Short-lived acidification of lakes and rivers associated with spring snowmelt.

Equal loudness contours — Graphic representation of the relationship between sound pressure and loudness.

Equivalence ratio — Ratio of the actual:stochiometric air:fuel ratios; manipulation used to reduce exhaust emissions.

Equivalent method — Any method of sampling and analysis that has been approved by the USEPA as being equivalent to a Federal Reference Method.

Equivalent sound level — Sound averaged over a given time interval.

Ethylene — Pollutant produced in automobile exhaust; natural growth hormone of plants that produces phytotoxic effects on exposure.

Evaporative emissions — NMHC emissions from the fuel system and in refueling.

Exhaust gas recirculation — Technology used in motor vehicles to control emissions of NO_x by using exhaust gases as a heat sink.

Extinction coefficient — Measure of the reduction of light transmission due to scattering and absorption by particles.

Extratropical pump — Mechanism by which air molecules are transported from the troposphere into the stratosphere.

Fabric filtration — Collection of particles by moving polluted air through a fabric medium.

Fanning plume — Plume behavior under inversion lapse rate conditions.

Federal Reference Methods (FRMs) — Primary methods of air sampling and monitoring approved by the USEPA to determine compliance with NAAQSs.

Filtration — Collection of particles on a fibrous medium.

Fine particles — Particles that have an aerodynamic diameter of ≤ 2.5 µm.

Flare oxidation — Pollution control system wherein combustible gases above the lower explosion limit are ignited and burned.

Flecking — Upper surface leaf injury associated with exposure to O_3.

Flue gas desulfurization (FGD) — Control technology designed to remove SO_x from combustion gases.

Flue gas spillage — Combustion by-products that fail to be exhausted by flue gas systems.

Fluidized bed combustion — Combustion process in which a bed of granular material and solid fuel, e.g., coal, are suspended to effect heat transfer and pollutant collection.

Fluorides — Pollutants that can cause injury to plants and animals and etch glass.

Flyash — Uncombusted inorganic solids released in particle form in the combustion of coal.

Forest decline — Progressive death of a forest species initially characterized by chlorosis and dieback of portions of trees; primary cause is reported to be exposure to atmospheric pollutants.

Formaldehyde — Reactive pollutant emitted in motor vehicle exhaust or produced secondarily in the atmosphere; participates in atmospheric reactions and is a potent mucous membrane irritant.

Four-stroke engines — Engines that have a cycle of intake, compression, power, and exhaust strokes.

Frequency — Property of sound characterized by the number of sound waves produced per unit of time.

Friction — Force that reduces wind speed near the Earth's surface and affects macroscale airflow patterns.

FT-IR — Fourier transform infrared spectrometric technique used to measure pollutants in a portion of the atmosphere.

Fuel additives — Substances added to gasoline; provision in clean air legislation that grants the USEPA authority to regulate these substances.

Fuel use policies — Pollution control approach wherein a source uses fuels that release less pollutants.

Fugitive emissions — Emissions not associated with stacks; may be internal or external to a facility.

Fume — An aerosol produced from the heating and vaporization of metals.

Fungal glucans — Substances produced in cell walls of fungi that, on release and exposure, cause inflammatory respiratory responses.

Gas cleaning — Removal of gas- or particulate-phase substances from a waste gas stream.

Gasolines — Liquid hydrocarbon fuels used to power vehicles equipped with spark ignition engines.

Gaussian models — Models use to determine concentrations downwind of a source based on the assumption that plume spread is primarily due to diffusion.

Geometric mean — The mean of lognormally distributed data.

Geostropic wind — Wind flow that is parallel to isobars (lines of equal pressure) at altitudes above 700 m.

Global warming — Climatic phenomenon that may be caused by increases in atmospheric greenhouse gases due to human activities.

Grab sampling — Sample collection over a period of seconds to minutes for purposes of subsequent laboratory analysis.

Gravity — Force of attraction between two bodies; holds atmospheric molecules close to the Earth's surface.

Greenhouse effect — Warming of the Earth's atmosphere resulting from the absorption of thermal energy emitted from the Earth's surface.

Hadley cell — Macroscale circulation pattern that occurs in both tropical and polar regions.

Halogenated hydrocarbons — Widely used industrial and commercial chemicals characterized by their relatively long atmospheric and environmental lifetimes and potential to destroy O_3.

Hazardous air pollutants — Noncriteria pollutants deemed to be especially toxic and regulated under hazardous pollutant provisions of clean air legislation.

Haze — Reduced visibility associated with moderate levels of atmospheric pollutants.

Hearing impairment — Loss of ability to hear sound; usually in the context of exposure to noise.

Heat island — Atmospheric phenomenon of elevated temperatures in urban areas associated with thermal emissions, changes in solar absorption and thermal emission patterns, etc.

Hertz — Quantitative measure of sound frequency; sound waves per second.

High-volume sampler (Hi-Vol) — Particle sampler that collects a broad size range of particles (0.3 to 100 μm) on a glass fiber filter by filtration; concentrations are reported as total suspended particulates (TSPs).

Histogram — Graphical representation of continuous data using frequency classes.

Humidifier fever — Inflammatory lung ailment caused by exposure to bacterial endotoxins.

Humidity — Concentration of water vapor in a volume of air expressed as weight per unit volume (absolute humidity) or percent water vapor that air can hold at a given temperature (relative humidity).

Hybrid electric vehicles — Low-emission vehicles powered by a combination of gasoline and electrical systems.

Hydrocarbons — Substances that contain only hydrogen and carbon in their chemical structure; major pollutants emitted to the atmosphere.

Hydrochloric acid (HCl) — Sink and reservoir chemical species in stratospheric O_3-destroying chemical reactions.

Hydrogen sulfide (H₂S) — Reduced sulfur gas that is a major malodorous pollutant.

Hydroxyl radical (OH·) — Principal atmospheric sink substance that oxidizes pollutants emitted to the atmosphere to more stable chemical forms.

Hygroscopicity — Property of particles that results in their uptake of water vapor from the atmosphere.

Hypersensitivity pneumonitis — Inflammatory lung ailment caused by exposure to high concentrations of bacterial cells or small-spored fungi.

Indirect source — A facility that does not emit pollutants to the atmosphere but attracts sources like motor vehicles that do; may include shopping malls, stadia, arenas, etc.

Indoor air — Air inside nonindustrial buildings.

Indoor air pollution — Contamination of indoor air by gaseous, particulate-phase, and biological pollutants.

Infiltration/exfiltration — Natural movement of air into and out of buildings due to differences in pressure.

Inhalable particulate matter (IPM) — Particles with aerodynamic diameters of ≤30 μm that enter the upper human respiratory airways of the head.

Inspections and maintenance (I&M) — Vehicle inspection and maintenance programs required for motor vehicles in nonattainment areas for specific motor vehicle-related pollutants.

Integrated pest management — Pest control practices that minimize pesticide use.

Interagency Monitoring of Protected Visual Environments (IMPROVE) — Cooperative monitoring program of visibility-impairing chemical species and visibility impairment conducted by the USEPA and the National Park Service.

Intergovernmental Panel on Climate Change (IPCC) — International organization that conducts scientific assessments and policy reviews on climate change.

Intermittent/integrated sampling — Sampling process wherein samples are collected for a given time period (e.g., 24 h) and schedule (e.g., every 6 days); sample concentrations are averaged over the sampling period.

Interveinal necrosis — Death of tissue between the veins of broad-leaved species associated with exposures to SO₂.

Inverse square rule — Decrease in sound intensity or electromagnetic energy as an inverse square of distance from the source.

Inversion — Atmospheric condition in which temperature increases with height; characterized by limited vertical mixing and pollutant dispersion.

Investigative protocols — Methodologies developed to conduct indoor air quality investigations.

Ionizing radiation — Electromagnetic energy and particles emitted by radionuclides that remove electrons from nearby atoms and molecules, thereby causing ionization.

Ionosphere — Region of the lower thermosphere where ionization of air molecules takes place.

Isobars — Lines of equal atmospheric pressure used on weather maps.

Isokinetic sampling — Stack sampling technique for particles wherein the sample flow rate is equal to the stack flow rate.

Jet stream — Relatively fast-moving air in the upper troposphere associated with fronts and discontinuities between macroscale circulation cells.

Knock — Sharp, metallic sounds produced by a motor vehicle engine when fuels do not have desired octane ratings.

Koschmeider equation — Mathematical relationship between visual range and the extinction coefficient.

Land breeze — Lake or seaward movement of air from land areas on calm, clear nights.

Lapse rate — Rate of temperature change with height.

LD_{50}/LC_{50} — Dose of an ingested or injected toxicant, or concentration of an inhaled substance, that will kill 50% of the animals being tested.

Leaching — Loss of plant nutrients from forest trees and soils associated with acidic deposition.

Lead — Criteria pollutant that poses adverse exposure risks to humans and animals.

Lead alkyls — Organic lead additives used in gasolines to reduce knock and increase engine performance.

Lean-burn combustion — Motor vehicle operation that utilizes air:fuel ratios higher than the stochiometric ratio.

Legionnaires' disease — Infectious pneumonia-type disease associated with *Legionella pneumophila*-contaminated cooling towers, whirlpool spas, hot tubs, and hot water heaters.

Limestone scrubbing — Use of a limestone slurry to remove SO_2 from flue gases.

Limit of detection (LOD) — Lowest concentration of a substance that can be accurately quantified.

Lognormal distribution — Environmental data that have been transformed into logarithmic values to form a normal distribution when plotted on semilogarithmic graph paper.

London-type smog — Smog produced when pollutants produced from coal burning are mixed with fog.

Long-range transport — Movement of pollutants tens to hundreds of kilometers downwind from a source.

Looping plume — Behavior of a plume under superadiabatic lapse rate conditions.

Los Angeles-type smog — Smog produced as a result of photochemical reactions in the atmosphere.

Loudness — Human perception of sound intensity.

Low-emission vehicles — Motor vehicles with significantly lower emissions than conventional vehicles; to be phased in under the 1990 CAA Amendments in areas subject to persistently high O_3 levels.

Lower explosion limit (LEL) — Lowest concentration of a flammable gas and oxygen that can propagate a flame.

Lowest achievable emission rate — Most stringent emission rate that can, in practice, be achieved for a new source in nonattainment areas.

Macroscale — Air motions that transport long-lived pollutants planet-wide.

Malodors — Odors that smell "bad"; air pollution problem associated with odor-producing pollutants.

Manual methods — Sampling methods in which air is drawn through a sample medium and later analyzed in a laboratory.

Marginal necrosis — Tissue death observed on the margins of leaves associated with fluoride exposures.

Market strategies — Regulatory approaches that allow sources to choose the most cost-effective control methods to achieve emission limits.

Maximum achievable control technology (MACT) — Maximum reduction requirements for a hazardous pollutant source considering cost and feasibility; standard that is not less than the average of the top 12% performing existing sources within a source category or utility.

Mechanical turbulence — Random motion of air induced by nonthermal processes.

Median — Middle value in a data set.

Melanoma — A serious form of skin cancer associated with severe sunburns.

Mercury — Toxic heavy metal regulated as a hazardous air pollutant; major environmental concern associated with atmospheric deposition.

Mesoscale — Medium-scale air motions associated with coastal areas and river valleys.

Mesosphere — Layer of the atmosphere extending from 55 to 80–90 km characterized by decreasing temperature with height.

Metal corrosion — Metal oxidation caused by exposure to acids or acidic gases.

Methane — Simplest hydrocarbon species characterized by its low chemical reactivity and strong thermal absorption properties.

Methanol — Liquid fuel used as an alternative to gasoline to achieve lower nonmethane hydrocarbon (NMHC) emissions.

Methylmercury — Organic mercury (Hg) compound produced by bacterial action; Hg compound that is bioaccumulated.

Microscale — Air motion that occurs in the near vicinity (meters) of a pollution source.

Mie scattering — Scattering of light by particles in the size range of wavelengths of visible light (0.4 to 0.7 nm).

Milankovitch cycles — Patterns of orbital changes whose periodicity has been linked to major glacial advances.

Mist — An aerosol produced by the atomization of a liquid or condensation of gas-phase substances in the atmosphere.

Mixing height — Vertical volume of air readily available for pollutant dispersion.

Mobile source — Moving pollutant source, e.g., cars, trucks, buses, trains, motorcycles, airplanes, etc.

Mode — Most frequent value in a data set; describes fine particle size range.

Monitoring — Measurement of atmospheric pollutant(s) over time.

Montreal Protocol — International treaty signed in 1987 that began the international effort to phase out and ban the use of O_3-destroying chemicals.

Morbidity — Illness.

Mortality — Death.

Motor vehicle emission standards — Emission limitations placed on major pollutants emitted from motor vehicles.

Mottling — Injury on leaves and conifer needles characterized by large patches of dead cells caused by O_3 exposures.

MTBE — An ether compound (methyl-t-butyl ether) used as an oxygenate additive in reformulated gasoline.

Multiport fuel injection — Fuel injection system used on most late-model light-duty motor vehicles whose accurate control of air:fuel ratios allows for the use of three-way catalytic converters.

Mutagen — Substance that can cause mutations.

Mutagenicity — Ability of a substance to cause heritable changes in a gene.

Mycotoxins — Toxins produced by fungi subject to nutritional stress.

National Acid Precipitation Assessment Program (NAPAP) — Federal interagency program designed to investigate acidic deposition and its effects on the environment.

National Air Monitoring Stations (NAMS) — Monitoring networks administered by the USEPA for purposes of evaluating long-term trends in atmospheric levels of criteria air pollutants in urban areas.

National Ambient Air Quality Standards (NAAQSs) — Permissible levels of criteria pollutants in an air quality control region.

National Council on Radiation Protection (NCRP) — Federal entity that recommends radiation safety guidelines.

National Emissions Standards for Hazardous Air Pollutants (NESHAP) — Provision of the 1970 CAA Amendments that regulated emission of seven hazardous air pollutants.

Necrosis — Plant tissue death caused by acute or chronic exposures to phytotoxic pollutants.

New source performance standards (NSPSs) — Pollutant emission limits on new or significantly modified existing sources.

New source review — Federal requirement that new or significantly modified existing sources undergo a review for compliance with NSPSs, NAAQSs, and PSD requirements before a permit to operate is authorized.

Nitrates — Atmospheric pollutants that serve as plant nutrients and increase nitrogen loadings in aquatic and terrestrial ecosystems.

Nitric acid (HNO$_3$) — Strong acid responsible for ~35% of acidity in acidic deposition in the eastern U.S. and ~90% in the western U.S.

Nitric oxide (NO) — A relatively nontoxic gas produced in high-temperature combustion.

Nitrogen deposition — Deposition of nitrogen to land and water sources associated with atmospheric pollutants.

Nitrogen dioxide (NO$_2$) — A brownish, relatively toxic gas that is a natural constituent of the atmosphere and whose concentrations increase significantly as a result of human activities; gas that absorbs sunlight and initiates photochemical reactions.

Nitrogen oxides (NO$_x$) — Collective term for gas-phase nitrogen compounds that contain one or more oxygen atoms; used to express concentrations of NO + NO$_2$ in the atmosphere.

Nitrogen saturation — Excess nitrogen deposition or use on land that causes nitrate nitrogen to enter standing or moving water.

Nitrous oxide (N$_2$O) — A low-reactivity gas emitted as a result of natural processes and human activities; participates in stratospheric O$_3$ chemistry and serves as a greenhouse gas.

Noise — Unwanted sound; sound that may cause hearing impairment, speech interference, annoyance, and other effects.

Noise Control Act of 1972 — Law that gives the USEPA authority to promulgate noise emission standards for various devices/equipment.

Noise guidelines — Environmental noise exposure guidelines published by USEPA.

Nonattainment area — An air quality control region, or portion thereof, in which one or more air quality standards has not been attained.

Nondispersive infrared (NDIR) photometry — Measurement technique that employs sorption of infrared energy to quantify CO concentrations; FRM for CO.

Nonmethane hydrocarbons (NMHCs) — Collective term for hydrocarbon species regulated for purposes of controlling production of photochemical oxidants; O_3 precursor molecules.

Nonmethane organic gases (NMOGs) — Hydrocarbons and their derivatives, excluding methane, that may serve as precursors for O_3 production.

Normal distribution — Environmental data set that when plotted as a histogram, forms a bell-shaped curve.

Normal lapse rate — Average lapse rate conditions observed in the atmosphere.

Nucleation — Process by which gas-phase substances form liquid particles.

Nuclei mode — Particles in the 0.01- to 0.08-μm range produced from the nucleation of gas-phase substances.

Nuisance suit — Legal action taken against a source by a property owner alleging an unreasonable interference with one's right to the enjoyment of one's property.

Obliquity — Variation in the tilting of the Earth's axis.

Octane rating — Measure of the antiknock quality of gasoline.

Octave band — A range of frequencies in which the upper frequency is twice the lower frequency; characterized by a center frequency.

Odd hydrogen species — Highly reactive oxidized hydrogen species such as hydroxyl radical, hydroperoxy radical, hydrogen peroxide, etc.

Odor — Olfactory sensation produced by an odiferous substance.

Odor threshold — Lowest concentration of an odoriferous substance that can be detected by the human olfactory system.

Offset policy — Regulatory policy that allows new sources to reduce emissions from existing sources to obtain a permit to construct and operate in a nonattainment area.

Olefins — Hydrocarbons characterized by the presence of one or more double bonds and high photochemical reactivity.

Onboard controls — Control systems on motor vehicles that capture gasoline vapors during refueling and route them to the engine during operation.

Opacity — Ability of a plume to absorb light; measure of plume darkness.

Orbital variations — Changes in the Earth's orbit characterized by obliquity, precession, and eccentricity.

Organic carbon — Particulate matter carbon associated with organic compounds.

Organic nitrate compounds — Chemical substances produced in the atmosphere as a result of reactions between hydrocarbon radicals and NO_2.

Organochlorine compounds — Semivolatile chlorinated compounds, e.g., pesticides, PCBs, dioxins, and furans, that move through the environment and are found in atmospheric deposition samples.

Organochlorines — Chlorinated compounds used to control insects; persistent, often carcinogenic compounds that are contaminants of ambient and indoor environments.

Organophosphates — Phosphate hydrocarbons used to control insects; potent, relatively nonpersistent compounds known for their neurotoxicity.

Oxidation — Chemical reactions that often include O_2 and result in the production of heat.

Oxidation catalysts — Catalytic systems on motor vehicles that reduce emissions of CO and NMHCs.

Oxygenates — O_2-containing gasoline additives used to increase both octane ratings and combustion efficiency.

Oxyhydrocarbons — Chemical species produced by the oxidation of hydrocarbons; major atmospheric pollutants.

Ozone (O_3) — Highly reactive gas found in both the troposphere and stratosphere; elevated tropospheric levels occur as a result of photochemical reactions involving anthropogenic pollutants.

Ozone depletion — Destruction of stratospheric O_3 by long-lived chemical species released as a result of human activities.

Ozone hole — Significant decline in column O_3 levels over the Antarctic continent in the austral spring; human-caused stratospheric O_3 depletion over the South Pole.

Packed tower — Pollution control component of a scrubber system.

Packing material — Small plastic or ceramic high-surface-area structures used in scrubbing systems to increase contact time between sorbent and waste gases.

Paraffins — Hydrocarbons characterized by the presence of single covalent bonds and relatively low photochemical reactivity.

Particles — Solid- or liquid-phase substances suspended in the atmosphere or deposited on surfaces.

Particulate matter — Collective term for particles in the atmosphere; more commonly a term used to express atmospheric particle concentrations based on sample collection on a filter.

Passive sampling — Collection of pollutants for quantification using samplers with no moving parts; typically pollutants are collected by diffusion or sedimentation.

Permanent threshold shift (PTS) — Permanent loss in hearing acuity associated with chronic noise exposures.

Permit — Right extended to a source to emit pollutants to the atmosphere at levels that do not violate regulatory requirements; legal right granted by regulatory agency to construct and operate a pollution source; legal right to operate pollution control equipment.

Peroxy radicals — Highly reactive hydrocarbon species produced as a result of photochemical reactions; key chemical species in the production of elevated tropospheric O_3 levels.

Peroxyacyl nitrate (PAN) — An organic nitrate substance produced as a result of photochemical reactions; phytotoxic air pollutant; reservoir chemical for nitrogen transport in the atmosphere.

Personal pollution exposure — Pollutant exposure that includes ambient, indoor, and activity-related human exposures.

Pesticides — Toxic substances used to control a variety of pest species.

pH — A measure of H^+ concentration.

Phons — Measure of sound loudness equal to 1 dB at 1000 Hz.

Photochemical Assessment Monitoring Stations (PAMS) — Federal sampling network in targeted urban locations that measures photochemical pollutants and their precursors.

Photochemical oxidants — Oxidizing chemicals produced in the atmosphere as a result of photochemical reactions; includes such substances as O_3, NO_2, peroxy radicals, etc.

Photochemical smog — Air pollution caused by reactions between primary pollutants and sunlight as well as subsequent secondary pollutants.

Photochemistry — Chemical reactions in the atmosphere that involve the absorption of sunlight and production of new oxidized species.

Planetary boundary layer — Depth of the atmosphere where air motion is affected by the Earth's surface.

Plant injury — Visible phytotoxic pollutant-induced effects.

Plume — Visible manifestation of pollutants being emitted by a source.

Plume blight — Impairment of the visual quality of a scene by the plume of an identifiable source.

Plume rise — Vertical motion of pollutant emissions from a source characterized by the centerline of the plume where it levels off.

PM_{10} — Particulate matter fraction with a 50% cutoff aerodynamic equivalent diameter of ≤ 10 μm.

$PM_{2.5}$ — Particulate matter fraction with a 50% cutoff aerodynamic equivalent diameter of ≤ 2.5 μm.

Point source — Stationary source of pollutant emissions.

Polar stratospheric clouds — Supercold clouds consisting of ice or nitrate hydrates that, through heterogeneous-phase chemistry, participate in formation of the Antarctic O_3 hole.

Polar vortex — Unique atmospheric circulation pattern in the Antarctic that indirectly contributes to the formation of the O_3 hole.

Pollution prevention — Pollution control policies that focus on (1) reducing emissions by changing process materials and operating practices, and (2) increasing energy efficiency.

Polychlorinated biphenyls (PCBs) — Persistent organochlorine compounds that are observed in atmospheric deposition samples, move through food chains, and found in elevated levels in fish.

Polycyclic aromatic hydrocarbons (PAHs) — Hydrocarbon species characterized by multiple benzene rings, presence in the particulate phase, and potent carcinogenicity.

Positive crankcase ventilation — System used on motor vehicles to control emissions of blow-by gases by returning them to the combustion chamber.

Potentiation — Ability of one pollutant to enhance the toxic effect of another.

ppmv — Parts per million by volume; mixing ratio used to express air pollutant concentrations on a volume:volume basis, μl/l.

Precession — Top-like orbital variation of the Earth.

Precision — Reproducibility of a measurement; characterized by the variability of data around the mean of repeated measurements of the same concentration.

Pressure gradient force — Force acting on atmospheric molecules due to differences in pressure; perpendicular to isobars.

Prevention of significant deterioration (PSD) — Regulatory program that limits visibility-impairing emissions in areas of high air quality.

Primary pollutant — Pollutant released into the atmosphere from a mobile or stationary source.

Primary standard — NAAQS designed to protect public health; in the measurement of pollutant concentration and volume flows, directly traceable to a NIST standard.

Problem building — Building in which indoor air quality complaints are reported and requests made to conduct an investigation.

Quality assurance — Overall management program to assure that monitoring data are accurate and reliable.

Quality control — Routine technical actions taken to assure collected data are of high quality; actions include calibration, use of blank and spiked samples, performance checks, etc.

Quenching — Cooling; process by which the flame is extinguishing in motor vehicle combustion chambers due to cooler cylinder walls resulting in NMHC emissions.

Radiational inversion — Ground-based or elevated inversion associated with the radiational cooling of the ground at night.

Radiative forcing — Changes in net radiative energy available to the Earth and its atmosphere.

Radon — Inert gas produced by the decay of radium^{226} that subsequently undergoes radioactive decay; building air contaminant.

Rainout — An in-cloud process in which pollutants participate in rain making and are subsequently removed from the atmosphere.

Rayleigh scattering — Clear-sky light scattering caused by very small particles or clusters of molecules.

Reactive hydrocarbons — Hydrocarbon species with one or more double bonds that react readily with other pollutants in atmospheric photochemistry.

Real-time monitoring — Procedure by which concentration data are recorded instantaneously or near instantaneously.

Reasonably available control technology (RACT) — Pollutant control technology that provides significant pollutant reduction at relatively reasonable capital and operating costs; applications include compliance with NSPSs and SIPs.

Reduced sulfur — Sulfur compounds produced as a result of reduction reactions, e.g., H_2S, CS_2, and COS.

Reduction catalysts — Catalytic systems used on motor vehicles to control emissions of NO_x by reduction reactions.

Reid vapor pressure — Measure of the volatility of gasolines.

Reentry — Return of pollutants that had been exhausted from a building.

Reference conditions — Standard conditions of temperature and pressure used to "correct" air volume in pollutant samples.

Reflection — Backward scattering of light.

Reformulated gasoline — Gasoline specially formulated to reduce emissions of reactive NMHCs, aromatic hydrocarbons, and CO.

Refraction — Scattering of light into different wavelengths by atmospheric particles; change in the direction of a light ray as it passes from one medium to another.

Regulatory strategies — Control approaches based on specific air pollution control principles.

Regulatory tactics — Control approaches used in achieving goals of regulatory strategies.

Resistivity — Relative inability of flyash to accept an electrical charge in an electrostatic precipitator system.

Respirable Particles (RPM) — Particles that can be easily deposited in lung tissue.

Ringlemann chart — A series of shaded charts that are used to visually quantify the opacity of a plume.

Risk — Measure of the probability that exposure to a pollutant or other factor will cause harm.

Risk assessment — Qualitative and quantitative process by which policy makers determine the relative harm associated with a pollutant exposure.

Rowland–Molina hypothesis — Theory that described the potential adverse effects of chlorofluorocarbon pollution of the atmosphere on stratospheric O_3.

Rural ozone — Elevated O_3 levels in rural areas associated with the long-range transport of O_3 and its precursor molecules.

Sampling — The act of collecting an atmospheric or stack pollutant for purposes of determining its concentration.

Saturation vapor pressure — Water vapor concentration at which condensation typically occurs.

Scrubber — A pollution control device used to control particles or soluble gases by absorption in a liquid or slurry.

Sea breeze — Landward movement of air from the oceans during calm, clear days.

Sea ice thinning — Reduced thickness of sea ice believed to be due to global warming.

Secondary pollutant — Pollutant produced in the atmosphere as a result of chemical reactions.

Secondary standard — NAAQS designed to protect public welfare; calibration standard derived from a NIST traceable primary standard.

Settling velocity — The rate (cm/sec) at which particles settle to a surface.

Sick building syndrome — Characteristic illness complex experienced by building occupants wherein a causal factor or factors cannot be identified.

Single-port fuel ignition — System used in early fuel injection systems that limited ability to control air:fuel ratios.

Sinks — Processes by which substances emitted to the atmosphere are removed.

Skin cancer — Malignancies of the skin caused primarily by exposure to UV light.

Slurry — Mixture of water and solid particles used in flue gas desulfurization systems.

Smog — Atmospheric condition characterized by significantly reduced visibility associated with elevated pollutant levels.

Smokestack — Physical structure that elevates pollutants above a source to reduce ground-level concentrations.

Solar constant — Solar energy received by the Earth at an average rate of 1.92 cal/cm^2/sec.

Solar spectrum — Electromagnetic energy emitted by the sun and received on the Earth's atmosphere and at the ground.

Sone — Unit of loudness based on 40 phons at 1000 Hz.

Soot — Elemental carbon particles produced as a result of incomplete combustion of a fuel.

Sorbents — Solid materials that collect contaminants by physical attraction to their surface; liquid substances that absorb pollutants.

Sound — Form of energy produced by a vibrating object or aerodynamic disturbance.

Sound intensity — Measure of sound energy at some distance from a source.

Sound power — Sound energy produced by a source.

Sound pressure level — Sound pressure measured at some distance from a source.

Sound pressure level meter — Instrument used to measure sound to characterize exposures.

Spark ignition engines — Conventional light-duty motor vehicle engines that require an electrically generated spark for ignition.

Spark retardation — Modification of engine operation to maintain high exhaust temperatures to continue CO and HC oxidation.

Spatial scales — Used to locate air monitoring sites in order to better interpret the results of air monitoring in the context of the area affected.

Spectrum analysis — Characterizing sound by its dominant frequencies; conducted with octave band analyzers.

Spray tower — Pollution control system used to control particles by spraying waste gas with water droplets.

Stability classes — Relative atmospheric stability conditions identified by the letters A to F used to calculate and use dispersion coefficients for air quality modeling.

Stable air — Air mass with little vertical mixing.

Stack sampling — Collection and quantification of pollutants from a stack using either manual or automated (continuous) methods.

Standard deviation (σ) — Measure of the variability of measurement data around the mean; mean $\pm$ 1 σ = 67% of measured values.

State and local air monitoring stations (SLAMS) — Network of air monitoring stations that states operate to determine compliance with National Ambient Air Quality Standards.

State implementation plan (SIP) — Specific control actions formulated by a state to meet National Ambient Air Quality Standards in all control regions under its jurisdiction.

Stationary source — Fixed-site source of pollutant emissions.

Stem-and-leaf diagram — Graphical method in which all values are presented to illustrate the distribution of measured data.

Stochiometric ratio — An air:fuel ratio in which there is an exact balance between the amounts of fuel and O_2 present for complete combustion.

Stratosphere — Layer of the atmosphere extending from ~15 to 50 km characterized by increasing temperature with height and elevated O_3 levels.

Stratosphere–troposphere exchange — Mechanisms by which air from the troposphere is transported into the stratosphere and from the stratosphere into the troposphere.

Stratospheric turbidity — Scattering of incoming solar radiation by volcanically derived sulfate aerosols in the stratosphere.

Subsidence inversion — Inverted lapse rate conditions that result from the compression of air aloft; associated with sinking of cool air in a high-pressure system.

Sulfates — Collective term for chemical species that contain SO_4^{-2} as part of the molecule.

Sulfur dioxide (SO_2) — Pollutant produced by oxidation of sulfur-containing fuels and metal ores.

Sulfur oxides (SO_x) — Collective term for sulfur dioxide and sulfur trioxide.

Sulfuric acid (H_2SO_4) — Strong acid responsible for ~60% of the H^+ in acidic deposition in the eastern U.S.

Sunspots — Cool, dark spots on the sun's surface believed to affect solar emissions to the Earth and subsequent weather and climatic conditions.

Superadiabatic lapse rate — Lapse rate conditions in which temperature changes with height more rapidly than that which would occur adiabatically.

SVOCs — Semivolatile organic compounds with boiling points in the range of 240–260 to 380–400°C; chemical compounds characterized by low volatility, low atmospheric concentrations, and long environmental lifetimes and mobility.

Synergism — Pollutant interactions that produce biological responses in exposed organisms that are significantly greater than those resulting from the sum of the individual exposures.

Synoptic scale — Air motions associated with movement of high- and low-pressure systems.

Tall-stack technology — Pollution control principle used to reduce ground-level concentrations by elevating emissions to significant heights above the ground.

Tampering — Illegal practice of disabling motor vehicle emission control systems.

Technology-based standards — Emission limits based on available technology; standards that do not consider health effects directly.

Temporary threshold shift (TTS) — Temporary loss in hearing acuity associated with excessive noise exposure.

TEOM — Tapered element oscillating monitor; continuous particulate air monitoring device that uses a tapered element oscillating balance to determine particle concentrations on a filter.

Thermal emission spectra — Range of infrared energies radiated to space.

Thermal oxidation — Pollution control technique wherein combustible gases below their lower explosion limit are combusted under elevated temperature conditions.

Thermosphere — Outer region of the atmosphere beginning at an elevation of 90 to 100 km where temperature increases significantly with height.

Thoracic Particles (TPM) — Particles 10µm that enter respiratory airways and lung tissue.

Three-way catalysts — Catalytic system used on most late-model motor vehicles that, by precise control of the equivalence ratio, can achieve an 80% reduction in CO, MNHC, and NO_x.

Threshold — Dose or exposure level below which no significant adverse health effects are expected.

Threshold of hearing — The lowest sound pressure a good human ear can hear; 0 dB.

Tip necrosis — Death of the tip of leaves and coniferous needles associated with pollutant exposures.

Topography — Landforms characteristic of an area; topographical features influence air movement and, as a consequence, pollutant dispersion.

Total suspended particulates (TSPs) — Concentration of atmospheric particles determined by using a high-volume sampler.

Toxicant — Substance that may cause biological harm to an exposed organism.

Toxicity — Relative potential of a toxicant to cause biological harm on exposure; toxicity includes the potency of the substance, exposure dose, and genetic susceptibility.

Transmissometer — Instrument that measures light attenuation over a specified path.

Trespass — Common-law principle applied to suits that involve one's exclusive right to the use of property; used when pollutants such as particulate matter "trespass" or violate one's property rights.

Tropopause — Layer of the atmosphere between the troposphere and stratosphere characterized by isothermal lapse rate conditions.

Troposphere — That portion of the atmosphere near the ground where temperature decreases with height.

Turbidity — Reduced ability of air to transmit light; usually considered in the vertical context.

Turbulence — Unstable air movements (eddies) caused by mechanical or thermal factors.

TVOC theory — Total volatile organic compound; theory that building illness symptoms are caused by additive and synergistic effects of low concentrations of multiple VOCs.

Ultrafine particles — Particles with diameters of <0.01 μm produced in the early phases of gas-to-particle conversion processes.

Ultraviolet (UV) light — Region of the solar spectrum that can cause sunburn and skin cancer; light energy below the visible light spectrum and mostly absorbed by stratospheric O_3.

Urban plume — Pollutants transported downwind from an urban area.

UV photometry — Monitoring technique wherein O_3 concentrations are determined by the absorption of UV light by O_3.

UV-A — Region of the UV spectrum (320 to 400 nm) that causes melanin formation and vitamin D production in human skin; regulates plant growth.

UV-B — Region of the UV spectrum (280 to 320 nm) that causes sunburn and skin cancer.

Ventilation — Natural or mechanical process by which outdoor air is brought into buildings to reduce pollutant concentrations.

Ventilation guidelines — Recommended quantity of ventilation air needed to provide acceptable health and comfort conditions in buildings.

Venturi scrubber — Particle control device that achieves high collection efficiency by using large quantities of small water droplets.

Virtual impactor — Sampling device that fractionates particles into coarse and fine modes by impacting them on a "virtual surface" of a slow-moving column of air; see *Dichotomous sampler.*

Visibility — Visual quality of the atmosphere relative to color and contrast; typically described as a function of distance.

Visual range — Distance that a dark object can be clearly seen against its surroundings.

VOCs — Volatile organic compounds with boiling points in the range of 50–100 to 240–260°C.

Washout — Process by which particles and gases are removed from the atmosphere by falling precipitation.

Wet deposition — Removal of pollutants from the atmosphere by precipitation-related processes.

Wet scrubbers — Pollution control systems wherein pollutants are removed by bringing them into contact with water or a watery media.

Wind — Physical characterization of atmospheric motion relative to speed and direction.

Working level — Measure of energy associated with radon decay products.

Index

A

Absorption
  sampling by 221
  control systems 339-342
Accuracy 223, 226-228
Acetaldehyde
  atmospheric lifetime 45
  emissions from motor vehicles 314-315
  filtration, indoor air 390
  oxidation in atmosphere 43, 48
  regulation as toxic pollutant 280
Acetylene 44
Acid
  aerosols, *see* Aerosols, acid
  clouds 198
  effects on humans 159
  fogs, effects on forest trees 192
  formation in atmosphere 38
  mist, new source performance standard 273
  neutralizing capacity (ANC) 109-110, 197
  precipitation 107
  precursors 107
  rain 106
  sulfates, effects on humans 161
Acidic deposition
  control 263, 267
  long-range transport 108
  pH values 238
  phenomenon of 106-110
  requirements for control of
    nitrogen oxides 277
    sulfur oxides 277
Acidification, episodic 109-110
Acid-sensitive ecosystems 109, 197
Acids
  bromic 117-118
  dicarboxylic 46, 49, 62
  hydrochloric, *see* Hydrochloric acid
  hypochlorous, *see* Hypochlorous acid
  nitric, *see* Nitric acid
  nitrous, *see* Nitrous acid

  sulfuric, *see* Sulfuric acid
  sulfurous, *see* Sulfurous acid
Acrolein
  by-product of combustion 45
  effects on humans 149, 166
  in indoor air 364-365, 370-371
Activated carbon use in
  pollutant control 264, 305, 337-339
  sampling 221-222
Additive effects on
  human health 144
  plants 190
Adsorption, use in
  pollutant control 264, 305, 337-339
  sampling 221-222
Aerodynamic equivalent diameter of particles 56,
  162, 218
Aerometric Information Retrieval System (AIRS)
  230
Aerosols
  acid effects on humans 161
  acid neutralization by ammonia 38-39, 42
  dust, 27
  effects on climate 129, 131
  effects on visibility 97, 103
  fume 27
  mist 27
  stratospheric sulfate 121, 124
  sulfate 99, 135
  urban 57
Afterburners 336
Air
  ambient 23
  cleaning, indoor air 389-390
  flows
    convective 15
    cyclonic 15-16
  pollutants, hazardous, *see* Hazardous air
    pollutants
  pristine 95, 273-274
  toxics 156, 167, 231
Air Monitoring Stations

National (NAMS) 230-231
State and Local (SLAMS) 229-230, 235
Air motion
  macroscale 71
  mesoscale 71, 81
  microscale 71
  synoptic scale 71
Air pollution
  episodes 141-142, 144
  odor 211, 213
  personal 175-177
  potential, forecasts 90
  thermal 104
Air quality, see National Ambient Air Quality
      Standards
  Act, 1967 266
  concerns, indoor 353-354
  Control Regions (AQCRs) 218, 260, 266,
      268-270, 288
  criteria
    carbon monoxide 266
    documents 266, 268
    hydrocarbons 266
    nitrogen oxides 266
    photochemical oxidants 266
    sulfur oxides 266
  index 218, 239-240
  investigations, indoor 380-382
  management 259-260
  modeling 90
  monitoring 217-235, 245
Air:fuel ratio 301-303
Albedo 6, 125, 134, 136
Alcohols as fuels 314-315
Aldehydes
  atmospheric formation of 47-49
  chemistry of 45
  control of 347
  in atmospheric photochemistry 53-54
  in indoor air 360-363, 365, 368-371
Allergens 353, 377
Allergic rhinitis, chronic 376, 380, 386
Allowance program 263, 277
Aluminum ions
  effects on ecosystems 197-199
  mobilization 110, 190
Alveoli 152-153
Ammonia
  atmospheric 1-2
  deposition 110
  effects on
    humans 161
    paint 209
    plants 184
  emission factors 244

formation 4, 20
in indoor air 380, 388
neutralization of acid aerosols 38-39, 42, 110
nitrification of 200
use in reducing nitrogen oxides 347
Ammonification 200
Ammonium ions 1, 231
Angiosarcoma 175
Animals, domesticated, pollutant effects on
      194-195
Antagonistic effects on
  humans 144
  plants 190
Antarctic ozone hole 118-121, 124
Anticyclones 15-17
AP-42 243-244
Appliances
  gas cooking 359, 361
  gas/kerosene unvented space heaters 168,
      361-362
  wood-burning 354, 363
Arctic haze 63, 94-95
Aromatic hydrocarbons, see Hydrocarbons,
      aromatic; Polycyclic aromatic
      hydrocarbons
Arsenic, effects on humans 155, 175
Asbestos
  -containing building materials (ACBM)
      354-356
  control of 323
  effects on humans 144, 155, 175
  Hazard Emergency Response Act (AHERA)
      355, 390
  in buildings 351, 353-356, 368
Asbestosis 355
Assessment Monitoring Stations, Photochemical
      230
Assessment, emissions 240-245
Asthma associated with
  community air pollution 153, 155-156, 160,
      164, 168, 170
  indoor air 361, 376-377, 380, 386
Asthmatic attacks 142-143, 156
Asthmatics 163
Atherosclerosis 157, 176
Atmosphere 9-11
Atmospheric
  density 11-12
  deposition
    effects on ecosystems 197-203
    nature of 106-113
  Deposition Program, National/National
      Trends Network 231
  evolution 20-21
  instability 9, 75-76, 83

lifetimes, determination of 87
motion 12-19
particles 55-67
photochemistry 26, 53-55, 99
pressure 12-13
stability 72, 84, 99
turbidity 72, 74
turbulence 72
window 7, 130
Attainment/nonattainment areas 272
Auditory system, human 407
Automated surface observing systems (ASOS) 101
Averaging time 219-220

B

Background concentrations
carbon monoxide 35
nitrogen oxides 39, 41
ozone 54
sulfur dioxide 37
Backscatter by particles 6, 98, 103
Bacteria 375, 377-378
Baghouse 332, 345
Base cations
effects on ecosystems, 190
nature of 110
Benzene
chemistry of 44
effects on humans 175
in gasoline 313
indoor air 372, 390
Benzo-α-pyrene
chemistry of 44
effects on humans 166
Beryllium, effects on humans 175
Best available retrofit technology (BART) 274
Beta particle attenuation 223, 234
Bias, sampling 226, 228
Biocides in indoor air 374, 386
Bioeffluents 354, 383, 387
Biofilters 346
Biological
changes caused by global warming 204-205
contaminants, indoors 375-379
treatment 346
Biomagnification 201-202
Biomass burning 46, 67, 360
Blackbody 6-7
Blow-by gases, control of 264
Boards/agencies, air pollution control 288
Breezes
land 81
sea/lake 81
Bromic acid 117-118

Bromine
atoms 50, 114, 117-118, 122, 124
monoxide 117
nitrate 117-118
Bronchi/bronchial tubes 151-152
Bronchitis, chronic 153, 160, 164, 168, 176
Brownian motion 65
BTEX 44, 313
Bubble policy 263, 322
Building
-related illness 380
sick, syndrome 380
Buildings
non-residential, non-industrial 353-354
problem 379-383
residential 353-354
Buoyancy flux 249-251

C

Calcium sulfate 61, 207, 344
Calibration 228
Cancer
associated with pollutant exposures 142, 351
in risk assessment 177
lung 153-155, 178, 355, 360, 365
skin/melanoma 123-124
Candles
in indoor air 220
lead 359, 363-364, 366
Capture, sampling by 221
Carbon, *see* Activated carbon, use in
activation by carbon dioxide 338
black 368
cycle, global 32-33
disulfide 1-2, 36, 38, 103
in aerosol particles
elemental 44, 60-61, 96, 102, 224, 311, 366
organic 60-61, 102
in motor vehicle emissions 311
oxides 31-36
particles 44, 60-61, 96-97, 102, 311, 324, 366
production in compression engine vehicles 324
Carbon dioxide
absorption of thermal energy 6-8
activation of carbon 338
as cause of scaling 341
as greenhouse gas 128-130, 135-137
as source of carbonic acid 107
atmospheric 1-4
biological treatment 346
budget 33-34
changes in Earth's history 19-21

effects on plants 205
emissions 264, *see* Motor vehicle emissions,
    emissions control
in indoor air 360-367
monitoring of 218
nature and chemistry of 31-34
oxidation product 58
removal from blood stream 150-151
sink processes 33
Carbon monoxide, *see* Motor vehicle emissions,
    emissions control
air quality
    criteria 266
    standard attainment dates 272
as criteria pollutant 183
as oxidation product 335
atmospheric 1-2
atmospheric formation of 48-50
background concentrations 35
effects on humans 153, 156, 159, 163
emission factors 244
federal reference method 233-234
formed in evolution of atmosphere 19-20
in indoor air 360-367, 380, 385
in tobacco smoke 176
indoor:outdoor ratios 353
modeling 246
monitoring of 218, 229
National Ambient Air Quality Standard 35,
    158-159
national emission estimates 28-31
nature and chemistry of 33-36
residence time/lifetime 36
sink processes 35-36
use in air quality index 239
Carbon tetrachloride
as ozone-destroying chemical 117
emissions 50
regulation of 277
Carbonyl sulfide 1-2, 36, 38, 103
Carboxyhemoglobin 157-158
Carburetor 304-305
Carcinogenesis 147
Carcinogenic pollutants in
ambient air 166-167, 173, 259, 275
indoor air 363, 366, 368, 372
Cardiovascular disease 142, 149-150, 365
Catalysts
palladium 307, 337
performance 310
platinum 307, 337
rhodium, 307
zeolite 308
Catalytic converters
hydrogen sulfide emissions from 310

oxidizing 307-308
reducing 308
sulfur dioxide emissions from 310
three-way 308
Catalytic systems 278, 336-337
Centers for Disease Control (CDC) 173-174, 379
Changes, process equipment, for pollution control
    324
Chapman cycle 115
Charcoal canisters 305
Chemiluminescence 232-234
Chlorinated hydrocarbons 51
Chlorine
atoms 21, 50, 103, 114, 116, 119, 122, 124
effects on plants 184
gas 120, 184
monoxide 116-118, 121
monoxide dimers 121
nitrate 86, 116, 120
sink 117
Chlorofluorocarbons
as greenhouse gases 128, 130-131
chemistry and nature of 51
control 323
monitoring of 218
movement into stratosphere 86-87
regulation of 276-277
regulatory control of 264
stratospheric chemistry 116
Chlorosis 185
Chlorotic
dwarf disease 187
mottling 187
Chronic
exposures, human 142, 148
obstructive lung disease 153, 163, 170 176,
    360
Cilia-mucus escalator 152
Class I and II ozone-destroying chemicals 276
Class I, II and III regions 273-274, *see* Prevention
    of significant deterioration
Clean Air Act, 1963 263, 265-266
Clean Air Act Amendments
1970, provisions 174, 183, 266-268, 272-273,
    281, 287, 322
1977, provisions 270, 277
1990, provisions 106, 179, 244, 263, 270, 275,
    278, 281, 283-287, 293, 323
Clean fuel motor vehicles 284
Clearance mechanism, respiratory system 145
Climate change
and human activity 133-134
effect of aerosols 129, 131
effects on animals and plants 205
evidence of 131-133

Intergovernmental Panel on (IPCC) 129, 131, 134-135
Climate forcing 129
Climate, global 124-125
Cloud
   behavior in global warming 135-136
   condensation nuclei 105
Coal
   beneficiation 342
   gasification 342
   high sulfur 322, 329
   low sulfur 322, 330-331
   sulfur dioxide emissions 322, 263-265
   washing 342
Coastal zones 204
Coefficients
   dispersion 247-250
   extinction 97-98, 101
   of variation 227
Collection efficiency
   fractional 326-327
   gas-cleaning systems 325-334
   sampling 219
Collectors
   cyclone 328-329
   multiple cyclone 328-329
   particle 328-334
   simple cyclone 328- 329
Combustion
   by-products, indoors 351, 354, 359-368
   chamber, motor vehicle 301, 303-304
   chemistry, motor vehicle 301-302, *see* Motor
      vehicle emissions, exhaust gases
   flare 335-336
   fluidized bed 342, 345
   sources, indoors, miscellaneous 363-368
   systems, vented 359, 362-363
Condensation
   as sampling method 221-222
   cloud nuclei 105
   of water vapor 5
Consumer Product Safety Commission (CPSC) 390
Contaminants
   control, indoors 386-390
   biological, indoors 375-379
Contrast, visibility-related 94
Control
   measures indoors 385-390
   of indoor gas-phase contaminants 390
   technique documents 266
   technology
      best available (BACT) 261
      maximum achievable (MACT) 261
      reasonably available (RACT) 260, 269

Convection 64
Convergence 16
Cooking
   stoves in developing countries 359-360
   gas appliances for 359, 361
Cooling, radiative 77
Cor pulmonale 149, 360
Corrosion, metal 205-207
Cost-benefit analysis 261-263
Crankcase
   emissions 278
   ventilation 305, 310
Criteria, air quality, *see* Air quality criteria
Cross-contamination 354
Cyclones
   meteorological 15-17
   particle collectors, *see* Collectors; Scrubbers

D

Data summarization 235-239
Data summarization, graphical procedures 235-237
Decibels 396-398
Deciview index 101
Department of Housing & Urban Development (HUD) 390
Deposition
   acidic 106-110, 189, 238, 277-278
   atmospheric 106, 183, 197-203
   control, acidic 263, 267
   dry 42, 86, 107
   Network, *see* Integrated Atmospheric
      Deposition Network; Mercury
      Deposition Network
   processes 86
   Program, National Atmospheric/National
      Trends Network 231
   wet 42, 89
Deposition of
   ammonia 110
   dioxins/furans 106-113
   mercury 106, 196, 201
   nitrates 200
   nitric acid 89, 107, 196
   nitrogen 106, 199-200
   particles in respiratory system 162
   sulfates 110
Detection, limit of 218
Deterioration, paint 205, 208-209
Dew point 5
Dicarboxylic acids 46, 49, 62
Dichlorofluoromethane 51
Diffraction 98-99
Dilution 322

Dimethyl sulfide 1-2, 36
Dinitrogen pentoxide 1, 41, 119
Dioxins
    deposition 106, 113
    emission factor 244
    in ecological systems 202-203
    monitoring 231
Disease
    infectious 145, 153, 156, 377-378
    Legionnaires' 376-378
Dispersion
    coefficients 247-250
    modeling 145
    models
        Gaussian 246
        guidelines 252
        other 252-253
        use of 90
    pollutant 71-85
Divergence 16
Dobson units 3, 118
Dose/dose rate 147
Dose-response relationships 147-148
Downwash 74
Duct liners 367-368
Dust
    dome 103, 105
    mites 353, 385
Dustfall jars 220
Dynamometer test cycles 283

E

Earth's orbital geometry 125-126
Eccentricity 125-127
Economic losses 190-191
Ecosystem changes 195-205
Ecosystems, acid-sensitive 109, 197
Effects,
    Coriolis 14-17
    greenhouse 8, 128-129
Effective stack height 251
Efficiency
    collection, sampling 219
    of collection, control devices 326-327
Electrostatic precipitators 329-331
Emergence tip burn 186
Emission
    allowances 263, 277
    credit banks 263
    estimates, national 27-29, see specific
        pollutants
    limits 264-265, 325
    rate, lowest achievable (LAER) 261, 263, 272
    standards

    as a control policy 259-261
    motor vehicle, see Motor vehicle emission
        standards
    National, for Hazardous Pollutants 267,
        273, 287
    test cycles, motor vehicle 282-283
Emission factors
    ammonia 244
    AP-42 243-244
    carbon monoxide 244
    dioxin 244
    hydrogen chloride 244
    lead 244
    nitrogen dioxide 244
    particulate matter 244
    PM_{10} 244
    sulfur dioxide 244
    use in emissions assessment 242-245
    volatile organic compounds 244
Emissions
    assessment 240-245
    carbon dioxide 264, 361-362, 364, 383, 385,
        387
    continuous monitoring 242
    control, see Catalytic converters
    motor vehicle, see Motor vehicle emissions,
        emissions control, emission
        standards
    sulfur dioxide 322, 263-265
    trading 263
Emphysema 153-154, 167, 176
Endocrine disruption 147, 203
Endotoxins 353, 376, 378
Energy
    balance, global 7-8
    conservation 324-325
    infrared 7-8
    thermal 7-9
    Supply and Coordination Act 281
Enforcement authority 284-285
Engine control, temperature 305
Engines
    compression ignition 299, 310-311
    compression ignition, particulate matter
        emissions 311
    motor vehicle 299-301, 310-312
    motor vehicle, spark ignition 299-301
Entrainment 354, 359
Environmental Protection Agency (USEPA)
    air quality index 239-240
        motor vehicle emissions 284, 292,
            314, 316
    emissions assessment 240-244
    environmental monitoring 101, 229-230, 235
    indoor air 355, 358, 363, 383, 390-391

modeling 252-253
national emissions estimates 28
noise 402, 408, 412, 414
regulatory role 174, 259, 263, 268, 272-275,
     285-287, 289, 336, 341
risk assessments 178
Epidemiological studies 142-146
Episode
   planning 89-90
   plans 270-271
Episodic acidification 109-110
Equal loudness contours 405
Equations
   Koschmeider 98, 101
   radiative transfer 97
Equivalent methods
   nitrogen dioxide 233
   ozone, 234
   sulfur dioxide 230
Erosion, building stone 205, 207-208
Error
   absolute 226-227
   relative 227
Estuaries 200
Ethane 48-49
Ethanol
   as motor vehicle fuel 313-315
   chemistry and nature of 45
Ethylene
   chemistry and nature of 44
   control 338
   effects on plants 184, 189
   filtration, indoor air 390
Ethyl-t-butyl ether (ETBE) 45, 313-314
Eutrophication 200
Evaporative losses/emissions 264, 278, 281-282,
     301
Exclusion as indoor control technique 385-386
Exhaust, see Motor vehicle emissions, exhaust
     gases
Explosion limits 335
Exposures
   assessment 145
   controlled studies 146-148
   indoor/outdoor pollutant 352-353
   occupational 148
   occupational noise 408
   personal pollutant 352
Extinction coefficient 97-98, 101
Extratropical pump 86-87

F

Federal reference methods, see Equivalent
     methods

carbon monoxide 233-234
lead 234-235
nitrogen dioxide 232-233
ozone 234
PM_{10} 234
$PM_{2.5}$ 234
sulfur dioxide 230
Filters
   fabric 331-332
   HEPA 389
   soot 311
Filtration
   fabric 331-332
   for emissions control 222, 326
   indoor air cleaning 389-390
Flammability limits 335
Flecking 186
Flue-gas
   desulfurization 30, 341-342, 344
   spillage 362-363
Fluidized bed combustion 342, 345
Fluorides, effects on
   animals 195
   glass 211
   plants 184, 188
Forest decline 191-194
Formaldehyde
   as a toxic pollutant 282
   atmospheric 1-2
   atmospheric reactions 41
   effects on humans 149, 166
   formation in atmosphere 48, 50
   formation in atmospheric evolution 20
   guidelines 370
   in indoor air 353-354, 364-365, 368-370, 380,
        383, 385-387, 390
   motor vehicles, see Motor vehicle emissions,
        emissions control, emission
        standards
   sampling 220
   urea formaldehyde 368-371
Frequency
   distributions 235-237
   sound 398
Friction 14-17
Fuel
   additives 313-314
   additives, requirements 264, 282
   injection
        multiport 305, 310
        single port 305, 310
   motor vehicle 313-316
   use policies 322
   volatility 313
Fumigation 78

Fungal glucans 353, 376, 378
Furans 113, 202-203

G

Gametotoxicity 147
Gas
    cleaning 325-328
    -cleaning systems, collection efficiency
        325-334
    cooking stoves, effects on humans 168
    exhaust, recirculation 305-307
    fireplaces 359, 362
    -using indoor equipment 361-362, 366-367
Gases
    atmospheric 1-4, 10
    blow-by 301, 305
    compressed/liquefied 315-316
    greenhouse 3-4, 128, 131, 135, 137, 289-290
    motor vehicle exhaust, *see* Motor vehicle
        emissions, exhaust gases
    noble 4, 20
Gasoline
    aromatic hydrocarbons in 312-313
    benzene in 313
    combustion 302
    lead in 282, 310, 313
    olefins in 312
    paraffins in 312
    reformulated 282, 284, 313-314, 323
Gas-phase contaminant control, indoors 390
Gaussian models 246-252
Genotoxic 169, 173
Ghosting 366, 368-369
Glaciations 126
Glass, fluoride etching of 211
Glazing, leaf undersurface 188
Global
    air circulation 17-20
    carbon cycle 32-33
    climate 124-125
    climatic models (GCMs) 134-135
Global warming
    biological effects 133
    description of climate change 124-136
    effects on ecosystems 183, 196, 203-205
    public policy issues 289-290, 351
    sea ice changes 133
    sea level changes 133
    uncertainties 135-136
Glutaraldehyde 371
Gravity 11-12, 64-65, 322
Greenhouse
    effect 8, 128-129
    gases 3-4, 7-8, 128, 130-131, 135, 137, 289-290

Growth abnormalities, plant 185
Guidelines
    community noise exposure 412-413
        formaldehyde 370
    National Research Council Risk Assessment
        177-178
    noise protective values 413
    radon action 358
    ventilation 388-389

H

Hadley cell 17-18
Halocarbons 130-131
Halogenated hydrocarbons
    chemistry of 50-51
    in indoor air 371-372
Halons
    as source of bromine 117
    regulation of 277
Hazardous air pollutants
    arsenic 275
    asbestos 275
    benzene 275
    beryllium 275
    control of 335
    emission factors 244
    mercury 275
    National Emission Standards for 156-157,
        275-276, 285-286, 323, 341, 343
    radioactive isotopes 275
    regulation of 267, 273, 287
    vinyl chloride 275
Haze
    Arctic 63, 94-96
    description of 26-27
    regional 95, 136
    rules 274-275
    South Asian 96-97
Health
    cause-effect relationships 142-145
    interaction effects and 143-145
Hearing impairment 406-409
Heat
    as an air pollutant 23, 26
    island 104
HEPA filters 389
Hertz (Hz) 398
Heterogenous-phase chemistry 119-121
High-pressure systems 25, 79-80, 85, 100
High-volume sampler 60, 65, 234
Histogram 236-237
Humidifier fever 354, 377, 380
Humidity
    absolute 5

air 5, 124, 134
  indoor air 382
  relative 5, 64, 95, 100, 143
Hydrocarbons, *see* Organic compounds
  air quality criteria 266
  atmospheric reactions 94
  biogenic sources of 260
  chemistry and nature of 43-51
  chlorinated 51
  component of Arctic haze 96
  effects on humans 166-167
  halogenated, *see* Halogenated hydrocarbons
  motor vehicles, *see* Motor vehicle emissions,
      emissions control, emission
      standards
  national emission estimates 28
  oxyhydrocarbons 45-46, 56
  thermal oxidation of 335-336
Hydrocarbons, aromatic 46, *see* Polycyclic
      aromatic hydrocarbons; Non-
      methane hydrocarbons
  chemistry of 44
  control of 346
  in gasoline 312-313
  in indoor air 371-372
Hydrochloric acid
  as oxidation by-product 337
  effects on plants 184
  formation in stratosphere 116, 118-120
Hydrochlorofluorocarbons 323
Hydrogen
  cyanide 20, 38, 42, 364
  fluoride, effects on plants 183
  chloride 86
      effects on plants 183
      emission factor 244
  ions
      acidic deposition 106, 110, 197-198
      effects on humans 161
      monitoring 238-239
  odd species 1, 40, 52-53
Hydrogen sulfide
  as malodorant 213
  atmospheric 1-2, 36
  control 338
  effects on paint 209
  emissions from catalytic converters 310
  in atmospheric evolution 20
  sink process 37-38
Hydroperoxy radical 1-2, 36-37, 40, 48
Hydroxyl radical
  atmospheric 1-2
  oxidation reactions 35-37, 40-42, 47, 49-50,
      53, 55, 88

Hypersensitivity pneumonitis 354, 376-377, 380,
      386
Hypochlorous acid 119

I

Ignition, electronic 304, 310
Impaction 222, 331-332, 334
Impactors
  cascade 222
  dichotomous 223
Incense 359, 363-364, 366
Index, deciview 101
Indian Ocean Experiment 96
Indoor air quality
  concerns 143, 145, 153, 353-354
  investigations 380-382
Indoor vs outdoor pollutant exposures 352-353
Indoor:outdoor ratios
  carbon monoxide 353
  nitrogen dioxide 353
  ozone 353
  sulfur dioxide 353
Infections, respiratory system 145, 153, 156
Infiltration/exfiltration 386-387
Inflammation 164, 166, 377
Information, Retrieval System, Aerometric 230
Infrared energy 7-8
Injunctive relief 258
Injury, plant, *see* specific pollutants
  acute 185
  chronic 185
Instability, atmospheric 9, 75-76, 83
Integrated Atmospheric Deposition Network
      (IADN) 231
  organochlorine compounds 201-203
  pesticides, PCBs, dioxins/furans 106, 113
Interagency Monitoring to Protect Visual
      Environments (IMPROVE) 101-102,
      230
Intergovernmental Panel on Climate Change
      (IPCC) 129, 131, 134-135
Inverse square law 397
Inversions
  description of 76-80
  nocturnal, surface 77
  radiational 77
  subsidence 9-10, 25, 78
Investigations, indoor air quality 380-382
Ionosphere 11
Iron
  corrosion of 206
  effects on humans 166, 174
Isobars 13-15
Isokinetic sampling 241

Isoprenes 4, 42, 47

J

Jet streams 17-19, 86

K

Kerosene heaters, unvented, effects on humans
        168, 361-362
Knock, engine 312
Koschmeider equation 98, 101
Kyoto Protocol 289-290

L

Lapse rates
    definition of 75
    dry adiabatic 75
    environmental 75, 79
    isothermal 76
    normal 75
    superadiabatic 76, 81, 83
Lawsuits
    citizen 267, 286
    common law 258-259
        nuisance 258
    trespass 258-259
Lead
    as criteria pollutant 183
    averaging time for 218
    candles 220
    control of 323
    effects on
        humans 149, 156, 170-174
        livestock 195
    emission factor 244
    emissions control progress 30, 231
    emissions from motor vehicles 28-30, 270
    federal reference method 234-235
    in gasoline 282, 310, 313
    in paint 208-209
    National Ambient Air Quality Standard 286
    national emissions estimates 28-30
Leather disintegration 205, 210
Legal remedies 257-259
Leukemia associated with
    benzene 175
    incense burning 360
    radon 358-359
Lifetimes, determination of atmospheric 87
Light scattering 94, 96-99
Limits
    lower explosion (LEL) 335
    lower flammability (LFL) 335

of detection (LOD) 218
    upper explosion (UEL) 335
Little Ice Age 126-127
Loudness 405-406, 409, 412
Lung
    cancer 153-155, 178, 355, 360, 365
    disease
        chronic obstructive 153, 163, 170 176, 360
        inflammatory 377

M

Macrophages 152, 156, 166
Maintenance as a pollution control measure 324
Malodors 38, 211-213
Margin of safety, adequate 259
Materials, effects of air pollutants on 205-211
Maximum ground level concentrations 252
Means
    arithmetic 237
    geometric 207, 238
Measures of central tendency 237
Median 237
Melanoma/skin cancer 123-124
Mercaptans 38
Mercury, see Methylmercury
    biocide in paint 374
    deposition 106, 196, 201, 218
    Deposition Network (MDN) 112, 231
    effects on humans 175
    emissions and environmental fate 112
    in particles 62-63
Mesosphere 10
Mesothelioma 175, 355
Metal corrosion 205-207
Metals 60, 62, 96
Meters, sound pressure level 398, 400-402
Methane
    as greenhouse gas 7, 31, 128, 130-131, 135-
        137
    as sink for atmospheric chlorine 117
    as source of carbon monoxide 33, 36
    atmospheric 1-2
    chemistry and nature of 49-50
    emissions 286
    in atmospheric evolution 20-21
    monitoring 218
    stratospheric 124
    use as motor vehicle fuel 315
Methanol
    as motor vehicle fuel 313-315
    chemistry 45
Methods, see Federal reference methods;
        Equivalent methods
    data summarization 235-239

monitoring 231-235
pararosaniline 232-233
sampling, *see* Sampling methods
Methyl
  bromide 50, 117
  chloride 50, 117
  chloroform 50, 117
Methylcyclopentadienyl manganese tricarbonyl
  (MMT) 312-313
Methylene chloride 323
Methylmercury 112, 200
Methyl-tert-butyl ether (MTBE) 45, 313-314
Milankovitch theory 126
Mixing
  height 80-81, 90
  ratios 11, 224-226
Modeling
  air quality 90, 246-253
  carbon monoxide 246
  dispersion 145
  nitric oxide/nitrogen dioxide 246
  ozone 246
  prevention of significant deterioration 246
  sulfur dioxide 246
Models, *see* Dispersion models
  global climatic (GCMs) 134-135
  options 252-253
  trajectory 246
Mold 375, 377-379, 385
Monitoring, *see* National Air Monitoring Stations;
  State and Local Air Monitoring
  Stations
  air quality 217-235, 245
  continuous 220-221
  continuous emissions 242
  instruments, *see* Nitrogen dioxide; Carbon
    monoxide; Sulfur dioxide;
    Particulate matter
  networks 229-231
  Stations, Photochemical Assessment 230
  to Protect Visual Environments, Interagency
    (IMPROVE) 101-102
Monitors
  fixed site 218
  multigas long-path 232-234
  TEOM 223, 234
Montreal Protocol 276
Morbidity associated with pollutant exposures
  164, 170, 176
Mortality associated with pollutant exposures
  141, 143, 164, 167, 365
Motor vehicle
  combustion chamber 301, 303-304
  Control Act 265

engines, *see* Engines
exhaust gases
  oxygen in 309
  recirculation 305-307
fuels 313-316
inspection and maintenance 264, 283
on-board
  diagnostic systems 284
  recovery systems 283
test cycles 282-283
Motor vehicle emission standards
  carbon monoxide 279-282
  exhaust 264, 267, 279-282
  formaldehyde 281
  nitrogen oxides 279-282
  non-methane hydrocarbons 279-282
Motor vehicle emissions
  acetaldehyde 314-315
  carbon dioxide 301
  carbon monoxide 278, 292-293, 299, 301-304,
    313-314, 316
  crankcase 278
  evaporative losses 278, 281-282, 301
  formaldehyde 314-315
  fugitive 321-333
  hydrocarbons 278, 299, 301-304, 313-314,
    316, 324
  lead 28-30, 270
  nitrogen oxides 278, 292, 299, 301, 303
  non-methane hydrocarbons 278, 292-293,
    299, 313-314
  particulate matter from compression engines
    311
  pollutant exposures due to 142
  sulfur dioxide 322
Motor vehicle emissions control
  carbon monoxide 303, 305, 307-315
  carbon dioxide 264, 317
  evaporative losses 264, 305
  hydrocarbons/non-methane hydrocarbons
    303-305, 307-311, 313-314
  nitrogen oxides 299, 301, 303
  sulfur dioxide 263-265
Motor vehicles
  clean fuel 284
  electric 318
  hybrid electric 316-317
  low-emission 316
  zero-emission 318
Muco-ciliary escalator 152
Mutagenesis 147
Mutagenic pollutants 167, 169, 370
Mycotoxins 353, 376, 378
Myocardial infarction 157, 165

N

Nasopharyngeal region 150
National Air Monitoring Stations (NAMS) 230-231
National Air Pollution Control Administration (NAPCA) 265-266, 268
National Ambient Air Quality Standards (NAAQSs) for
  carbon monoxide 35, 150, 158-159, 183
  lead 156, 183, 286
  nitrogen dioxide 156, 168, 183
  ozone 156, 166, 169, 183
  ozone, revised 263, 289-291
  particulate matter 60, 156, 183
  particulate matter, revised 263, 290-291
  sulfur dioxide 150, 160, 183
National Ambient Air Quality Standards (NAAQSs)
  Clean Air Act 260, 268-269, 352
  compliance with 218-219, 231, 245, 283, 288
National Atmospheric Deposition Program/National Trends Network (NADP/NTN) 112, 231
National Council on Radiation Protection risk estimates for radon 356
National Crop Loss Assessment Network 191
National emission estimates 27-29, see specific pollutants
National Emission Standards for Hazardous Pollutants (NESHAP) 156-157, 275-276, 285-286, 323, 341, 343
National Institutes
  Occupational Safety & Health (NIOSH) 380, 383
  Standards & Technology (NIST) 229
National Research Council Risk Assessment Guidelines 177-178
Necrosis, plants
  interveinal 185
  marginal 188
  tip 187-188
Networks
  Mercury Deposition 112, 231
  National Crop Loss Assessment 191
  National Trends/National Atmospheric Deposition Program 231
New source performance standards
  acid mist 273
  for coal-fired power plants 341
  regulation by 260, 272-274, 285-286, 291
  sulfur dioxide 261
  sulfuric acid 273
New source review (NSR) 246, 289
Nitrate radical 1, 38, 42

Nitrates
  bromine 117-118
  chlorine 116, 120
  deposition of 200
  in particles 60-62, 96-97, 102
  monitoring of 231
  organic compounds 42-43
  peroxyacyl (PAN), see Peroxyacyl nitrate
Nitric acid
  as source of nitrogen dioxide exposure 167
  atmospheric chemistry of 38, 41-42
  deposition 89, 107, 110, 113, 116, 196
  effects on humans 167
  formation in the atmosphere 38, 41-42
  monitoring 231
  neutralization of ammonia 110
  stratospheric 116
Nitric oxide
  atmospheric 1-2
  chemistry and nature of 38-41
  effects on humans 167
  in atmospheric photochemistry 53-55
  motor vehicles, see Motor vehicle emissions, emissions control, emission standards
Nitric oxide/nitrogen dioxide modeling 246
Nitrification 200
Nitrogen
  atmospheric 1-3
  atmospheric evolution 19-20
  compounds 38-40
  deposition 106, 199-200
  from reduction of nitrogen oxides 308, 347
  organic 110
  photoionization of 10-11
  reduced compounds 38, 42, see Ammonia
  saturation 200
Nitrogen dioxide
  atmospheric 1-2
  effects on
    humans 167
    plants 184
  emission factor 244
  equivalent methods 233
  federal reference method 232-233
  formation in atmosphere 36
  in atmospheric chemistry 53-54
  in indoor air 360-362, 365-366
  indoor:outdoor ratios 353
  monitoring 22, 229, 233
  motor vehicle production 303
  standards 168
  use in air quality index 239
Nitrogen oxides (NO_x)
  air quality criteria 266

as criteria pollutants 183
as product of thermal oxidation 337
chemistry and nature 39
concentrations 41
control 347
control under acidic deposition requirements 277
effects on humans 149, 154, 156, 167-168
in indoor air 363-364
in stratospheric chemistry 115-116, 120
national emissions estimates 28
reduction by ammonia 347
sink processes 111
Nitrogen oxides
sink processes 110
emissions, home gas dryers 210
low-emissions burners 357
motor vehicles, *see* Motor vehicle emissions, emissions controls, emission standards
Nitrous acid 38, 53
Nitrous oxide
as greenhouse gas 7, 128, 130, 135
atmospheric 1-2
chemistry and nature 39
environmental formation 200
in stratospheric chemistry 115-116, 120
monitoring 218
motor vehicles, *see* Motor vehicle emissions, emissions controls, emission standards
Noise
as air pollution 23, 26
community regulation of 414-415
control measures 415
effects on humans 406-412
environmental 395
federal programs 415
Non-methane hydrocarbons (NMHCs)
as criteria pollutants 183
effects on humans 156, 166-167
monitoring 230
motor vehicles, *see* Motor vehicle emissions, emissions controls, emission standards

Normal distribution 237, 246-247
Nuclei
cloud condensation 105
freezing 106
giant 106

O

Obliquity 125, 127

Occupational Safety & Health Administrations (OSHA) 356, 371, 390-391
Octane ratings 282, 312-313
Octave band analyzers 402
Odor
as air pollution 211, 213
habituation 213
intensity 212
measurement 212
threshold 212
Offset policy 272
Olefins
as hydrocarbon chemicals 44
emissions of 46
in aerosol particles 62
in gasoline 312
Olfactory epithelium 211
On-board motor vehicle systems
diagnostic 284
recovery 283
Opacity 242-243
Orbital variations 125-126
Organic carbon in particles 60-61, 102
Organic compounds, *see* Hydrocarbons; Motor vehicle emissions, emissions control, emission standards
nitrate compounds 42-43, 110
semivolatile (SVOCs) 44, 113, 231, 372, 375
very volatile (VVOCs) 44-45, 371
volatile (VOCs)
chemistry and nature of 44-46
emission factor 244
in indoor air 354, 362-363, 371-373, 383, 385
national emission estimates 28
sampling 220, 240
total (TVOCs) 372, 383
Organochlorine compounds 201-203, *see* Pesticides
Organophosphate, *see* Pesticides
Oscillation, of ozone layer, quasibiennial 115
Oxidants, photochemical 51-55
Oxidation
catalytic 336-337
thermal 335-337
Oxygen
absorption of sunlight 6, 10-11
and human blood
exchange with carbon dioxide 151
levels with carbon monoxide exposure 159
atmospheric 1-3
from reduction of nitrogen oxides 308
ground state 36, 52-53
in atmospheric evolution 20
in motor vehicle exhaust gases 309

oxidation of carbon 34
oxidation of fuel 301
singlet 39, 115, 116
stratospheric chemistry 115
use in thermal oxidation 335
Oxygenates 313-314
Oxyhydrocarbons 45-46, 56
Ozone
    absorption of solar energy 6-7, 10
    absorption of ultraviolet light 53, 55, 86, 114
    achievement of standard 316
    Antarctic hole 118-121, 124
    as greenhouse gas 7-8, 130-131
    atmospheric 2-4
    attainment dates 272
    calibration 229
    chemiluminescence reactions 233
    control of 260, 303
    cracking of rubber 205, 210-211
    depletion trends 121-122
    effects on
        humans 149, 156
        paint 209
        plants 183-184, 186-187, 190-192, 194,
            196
        rubber 210-211
        textile dyes 210
    equivalent methods 234
    federal reference method 234
    high days 156
    in atmospheric evolution 20
    indoor:outdoor ratios 353
    layer
        characteristic 2-4, 123
        dynamics 114-115
        evolution of 20
        protection of 115, 267, 276, 323
        quasibienniel oscillation 115
        recovery 124
    levels 170, 368
    modeling 246
    monitoring
        ambient 218, 230-231
        instruments 229
    National Ambient Air Quality Standard
        (NAAQS), 166, 169
    National Ambient Air Quality Standard,
        revised 263, 289-291
    protection of stratospheric 267, 323
    rural 54
    sinks 53, 55
    stratosphere-troposphere exchange 87-88
    stratospheric depletion of 103, 113-124, 183,
        196, 203, 276-277

tropospheric 28, 49, 52-55, 128, 131, 135,
    170, 368
Ozone-destroying chemicals 31, 117, 124, 218,
    244, 267, 276-277, 351

P

Packing materials 340-341
Paint deterioration 205, 208-209
Palladium, as catalyst 307, 337
Paper embrittlement 205, 210
Paraffins
    chemistry 43
    emissions of 46
    in gasoline 312
    in particles 62
    in photochemical reactions 49
Particle modes
    accumulation mode 59, 61, 65
    Aiken/nuclei mode 59-60
Particles, see Particulate matter; PM_{10}; $PM_{2.5}$
    aerodynamic equivalent diameter 56, 162, 218
    atmospheric 55-67
    behavior 63-65
    beta attenuation 223, 234
    carbon 44, 60-61, 96, 102, 311, 324, 366
    chemistry 60-63
    coarse 57, 59
    collection 326-334
    condensation 63
    control by
        cyclone systems 328-330
        electrostatic precipitation 324-331
        filtration 222, 326
        wet scrubbers 332-334
    deliquescence 63
    deposition in respiratory system 162
    distributions, multimodal 57-59
    fine 57, 59-60
    hygroscopicity 64
    impaction 331-332, 334
    inhalable 162-163
    long-range transport 66-67
    nucleation 63
    primary 56
    respirable (RSP), 60, 352, 361-365
    samplers 59-60
    secondary 56
    settleable 59
    settling velocity 59
    size 56-60
    soot, indoor air 366
    stratospheric 124
    sulfates 60-61, 102
    suspended 56

thoracic 162-163
transport, long-range 66-67
types 59
ultrafine 59
ultrafine, effects of 166
Particulate matter, *see* Particles; PM_{10}; $PM_{2.5}$
averaging times 219
effects on
    humans 143-144, 153, 156, 163
    paint, 209
    plants 183-184
emission factors 244
emissions from compression ignition engines
    311
in indoor air 360
monitoring 221
National Ambient Air Quality Standard 60
National Ambient Air Quality Standard,
    revised 263, 290-291
nitrate 1
Prevention of Significant Deterioration
    regulation 274
standards 260
total suspended in indoor air 360
Particulates, total suspended (TSP)
in indoor air 360
sampling for 222
PCBs, deposition 106, 113
Perchloroethylene 50
Permits 245, 264-265, 285-286
Peroxyacyl nitrate (PAN)
as component of smog 43, 52, 315
as nitrogen compound 38
effects on
    humans 149
    plants 184, 187-188, 190
interference in nitrogen oxides monitoring
    233
Peroxynitrates 43
Pesticides
deposition 106, 113
in indoor air 351, 354, 372, 374-375
organochlorine 375
organophosphate 375
pH
averaging values 238-239
effects on
    ecosystems 189, 197-198
respiratory system particle deposition 162
of precipitation 106-107
Phagocytes 152
Phons 405-406
Photochemical Assessment Monitoring Stations
    (PAMS) 230
Photochemical oxidants

air quality criteria 266
chemistry and nature of 51-55
Photochemistry, atmospheric 26, 99
Photodecomposition 48-49
Photoionization 11
Phthalic acid esters 372-373
Phytotoxic pollutants 183-189, 196
Picocuries/liter 357-359
Planetary boundary layer 9, 15, 85
Plans
episode 89-90, 270-271
state implementation (SIPs) 266, 270, 285, 287
Plant operating practices 324
Plasticizers 372-373
Platinum, as catalyst 307, 337
Plots
box 237-238
stem and leaf 237
Plume
blight 94-95
rise 82-83, 248-251
Plumes
characteristics 83
coning 83
fanning 83
fumigating 84
lofting 84
looping 83
tall stacks 322
trapping 84
urban 85
PM_{10} (thoracic particles), *see* Particulate matter
concentrations of 65-66
effects on humans 151, 156, 161-165
emission factor 244
federal reference method 234
in indoor air 360
national emission estimates 28-30
samplers 60, 210-211
sampling 222
standard, World Health Organization 360
$PM_{2.5}$ (respirable particles) *see* Particulate matter
concentrations 65-66
effects on humans 150, 155-156, 161-165
standards 289, 290-291
federal reference method 234
samplers 60, 223
Pneumonia 168, 170, 360
Polar
clouds, stratospheric 119, 121
vortex 121
Pollutant exposures
indoor/outdoor 352-353
personal 352

Pollutants, *see* Air quality criteria; Hazardous air
    pollutants; specific pollutants
    concentration expression 224-226
    effects on precipitation 105-106
    irritant 149
    long-range transport 85
    measurements, indoors 385
    non-traditional air 26
    primary 27-29
    secondary 27
Pollution
    anthropogenic 24
    charges 264, 290
    control by
        process changes 324
        process equipment changes 324
    natural 23-24
    prevention 286-287, 322-323
Pollution Prevention Act, 1990 286-287
Polybrominated biphenyls 51
Polychlorinated biphenyls (PCBs) 51, 196, 202,
    231, 372-373
Polycyclic aromatic hydrocarbons (PAHs)
    chemistry of 44
    effects on humans 166, 177
    in indoor air 363-364, 366, 372
    in particles 62, 64
    monitoring 231
    source of 46
Polycyclic organic matter 282
Precession 125, 127
Precipitation, pollutant effects on 105-106
Precipitator, electrostatic 329-331
Precision 223, 226-228
Precursors, acid 107
Pressure
    atmospheric 12-13
    gradient force 13-16
Pressure systems
    high 15-18, 25, 79-80, 85, 100
    low 15-16
Prevention of significant deterioration (PSD)
    Clean Air Act Amendments, 1977 267, 273-
        275
    modeling for 246
    new source review 291
    permit requirements 286
    regulation of
        particulate matter 274
        sulfur oxides 274
    visibility protection 89, 101
Propane, use as motor vehicle fuel 315-316
Protocols
    Kyoto 289-290
    Montreal 276

USEPA/NIOSH, for indoor air investigations
    383-384
Psychrometric chart 5

Q

Quality
    assurance 235
    control 235
Quasibiennial oscillation of ozone layer 115

R

Radiation, *see* Ultraviolet light
    ionizing 23, 30
    solar 6-7, 124, 134
Radiative
    effectiveness 130
    forcing 128-131, 134
    transfer equations 97
Radicals, *see* Hydroxyl radical
    hydroperoxy 1-2, 36-37, 40, 48
    nitrate 1, 38, 42
    peroxy 39
Radioactive isotopes 275
Radionuclides, effects on humans 175
Radon
    action guideline, USEPA 358
    decay products 357, 365
    effects on humans 144, 155, 358-359
    in indoor air 351, 353-354, 391
    risk estimates, National Council on Radiation
        Protection 356
    sampling 220
Rainout 89
Ratios
    air:fuel 301-303
    compression 302-304
    equivalence 301-302, 308-309
    mixing 3, 224-225
    stoichiometric 301-302, 308-309
Reactors
    thermal 307
    trickle-bed 346
Re-entry of pollutants 354, 359
Reformulated gasoline 282, 284, 313-314, 323
Refraction 98-99
Regulations
    community noise 414-415
    federal noise programs 414-415
    for ozone-destroying chemicals 267
Reid vapor pressure 313
Relative humidity 5, 64, 95, 100, 143
Residence time/lifetime
    calculations 87-88

carbon dioxide 33
carbon monoxide 36
sulfur dioxide 37
Resistivity 330-331
Respiratory system
    anatomy and physiology 150-152
    clearance 145, 152
    defense mechanisms 152-153, 161
    infections 145, 153, 156
    particle deposition in 162, *see* $PM_{2.5}$, effects
        on humans
Rhodium, as catalyst 307
Ringlemann charts 242-243
Risk assessment
    management 177-178
    National Research Council Guidelines 177-178
    principles of 177-179
Rowland & Molina hypothesis 114
Rubber, ozone cracking of 205, 210-211

S

Sample size 218
Samplers, particle
    hi-volume 222
    size-selective 222
Sampling
    bias 226, 228
    collection efficiency 219
    duration 218
    gas 221-222
    methods
        automated 218-219
        grab 220
        intermittent 220-221
        manual 218-224
        passive 220
        real-time 219
    particle 222-223
    rate 218-219
    stack 240-241
    techniques 219-220
Scattering
    Mei 96-97
    Rayleigh 94
Scrubbers
    bioscrubber 346
    by-products 343
    cyclone 334
    dry 344-345
    fixed-bed packed towers 339
    lime 343-344
    lime-limestone 343
    limestone 343-344
    Venturi 333-334

wet particle 332-334
Scrubbing media 332, 340
Semivolatile organic compounds (SVOCs) 44,
        113, 231, 372, 375
Senescence 189
Sensitive populations, human 145-146
Sick building syndrome 380
Sink processes
    carbon dioxide 33
    carbon monoxide 35-36
    ozone 55
    sulfur dioxide 37-38
Skin cancer/melanoma 123-124
Sky light 98
Smog
    effects on plants 187-188
    London-type 24-25
    Los Angeles-type 25-26
    nature of 24-26, 94-95
    photochemical 26
Solar
    constant 6, 126
    irradiance 125
    luminosity 126, 128
    radiation 6-7, 124, 134
    spectrum 6-7
Sones 405-406
Soot
    filters 311
    particles, in indoor air 366
Sorbents 337-338, 340, *see* Activated carbon;
        Water vapor
Sound
    energy 395-398
    frequency 398
    impulse 401
    intensity 396-397
    level
        A-weighted 402, 411
        A-weighted, exposure 402
        day-night 404, 412-413
        equivalent 403-404, 412-413
    measurement 398-402
    power 396
    pressure 397-398
Source
    management, indoors 385-386
    removal 386
    treatment 385
Sources
    indirect 27
    mobile 27
    stationary 27
    transportation 27
Spark timing 303-304

Spatial scales 229-230
Spectrometry
    non-dispersive infra-red 233-234
    ultraviolet 234
Spectrum analysis, sound 401-402
Speech communication, interference by noise
        409-411
Spray towers 335
Stability
    atmospheric 75, 321
    classes 90, 247, 249
    neutral 75
Stachybotrys chartarum 379
Stack height
    effective 82-83, 251
    physical 82, 249
Stack sampling 240-242
Stacks, tall 321-322
Standard
    achievement of ozone 316
    deviation 227, 246-247
Standards
    & Technology, National Institute of (NIST)
        229
    motor vehicles
        cold start 281
        emissions, *see* Motor vehicle emission
            standards
    National Ambient Air Quality, *see* National
        Ambient Air Quality Standards;
        National Ambient Air Quality
        Standards for
    National Emission for Hazardous Pollutants,
        see Hazardous pollutants
    new source performance, *see* New source
        performance standards (NSPSs)
    primary 229, 268
    secondary 229, 268
    World Health Organization (WHO), nitrogen
        dioxide 168
State and Local Air Monitoring Stations
        (SLAMS) 229-231, 235
State implementation plans (SIPs) 266, 270, 285,
        287, 325
Stationary sources 321
Steven's law 405
Stipple 186
Stokes number 50
Strategies, economic 259-265
Stratopause 10
Stratosphere 9, 52, 86, 114, 116
Stratosphere-troposphere exchange 86-87
Stratospheric
    circulation 86-87
    cooling 137

nitric acid 116
ozone
    depletion 113-124, 183, 196, 203, 276-277
    protection of 115, 267, 276, 323
    polar clouds (PSCs) 119, 121
    sulfates 103, 121, 124
    turbidity 103
    water vapor 124
Substitution as a pollution control measure 323
Sulfates
    aerosol 99, 135
    ammonium 61
    ammonium acid 61, 64
    atmospheric 1
    calcium 61, 207, 344
    deposition 110
    in Asian haze 96
    in particles 60-61, 102
    monitoring 231
    produced in fluidized bed combustion 345
    stratospheric 103, 121, 124
Sulfide
    carbonyl 36, 38, 103
    dimethyl 36
    hydrogen 36, 38
Sulfur, *see* Coal
    reduced compounds 38
    total reduced 242
    trioxide 36
Sulfur compounds 36-38
Sulfur dioxide, *see* Motor vehicle emissions,
        emissions control
    and visibility reduction 94-96
    atmospheric 1-2
    background concentrations 37
    chemistry and nature of 36-38
    effects on
        building materials 207-208
        humans 143-147
        metal corrosion 206-207
        paper and leather 210
        paints 209-210
        plants 183-187, 190
    emission factor 244
    emissions from catalytic converters 310
    equivalent methods 230
    federal reference method 230
    in air quality index 239
    in indoor air 362-363
    in particle formation 64
    indoor:outdoor ratios 353
    mixing ratios 225-226
    modeling 246
    monitoring 218, 221, 231
    monitoring instruments 229

National Ambient Air Quality Standard 160
new source performance standard 261
residence time/lifetime 37
sink processes 37-38
Sulfur oxides (SO$_x$) 28-30
    air quality criteria 266
    Clean Air Act, 1963 265
    control of 341-345
    control under acidic deposition requirements
        277
    effects on humans 153, 156, 159-160
    Prevention of Significant Deterioration
        regulation of 274
Sulfuric acid, see Scrubbers, by-product
    atmospheric chemistry 37, 42, 110
    control of mist 329
    effects on
        ecosystems 196
        humans 144
        paper and leather 210
        plants 196
    formation 37, 63
    in acidic deposition 107
    in PM$_{2.5}$ samples 61
    mist, control of 329
    neutralization by ammonia 42, 110
    new source performance standard 273
    nucleation of 63
    stratospheric 103
Sulfurous acid 38, 159, see Flue-gas
        desulfurization
Sunspot cycles 126, 128, 134
Supersonic transports 113
Surface roughness 73
Synergism, effects on
    humans 144
    plants 190
System performance 325-326

                            T

Tactics, pollution control 260, 264-265
Tampering prohibitions 283
Target organs 149-150
Temperature
    atmospheric 8-10
    climate average 124
    engine control of 305
    gradient 9-10
    increases, tropospheric 131-132
    pollutant interaction effects on 143-144
TEOM monitor 223, 234
Teratogens 147, 167
Tertiary-amyl-methyl ether (TAME) 45, 313-314
Tetraethyl lead 313

Tetramethyl lead 313
Textiles
    damage 205, 209-210
    dyes, fading of 205
Thermal
    air pollution 104
    emission spectrum 7-8
    energy 7-9
    oxidation 335-337
    reactors 307
    turbulence 74, 79
Thermosphere 10-11
Three-zone model 17-18
Threshold
    shift, hearing 408
    values 259
Tobacco smoke
    carbon monoxide in 176
    environmental (ETS) 353-354, 364, 382
    hydrogen cyanide in 364
Tobacco smoking
    as source of indoor pollutants 353, 359,
        363-366
    effects on humans 143, 145-146, 153-155,
        157, 175-176
Topography 72, 74, 81, 321
Tort law 258
Towers
    fixed-bed packed scrubbers 339
    spray 147, 335
Toxic pollutants, see specific pollutants
Toxicology studies 146-148
Tracheobronchial region 150
Transport
    long-range, pollutant 85
    long-range, particle 66-67
    planetary 85
Transportation plans 263
Trends
    acidic deposition 110
    global warming 131-133
    National Network/Atmospheric Deposition
        Program 231
    ozone depletion 121-122
Trichloroethylene 50
Trichlorofluoromethane 51
Trichlorotrifluoroethane 51
Trigeminal nerve system 211
Tropopause 9-10, 86
Troposphere 9-10, 117, 132
Tropospheric ozone
    as greenhouse gas 128, 131, 135
    control of 303
    production of 28, 49, 52-55
Tropospheric temperature increases 131-133

Tuberculosis 376-377
Turbidity
    atmospheric 72, 74
    stratospheric 103
Turbulence
    atmospheric 72
    mechanical 74
    thermal 74, 79

U

U.S. Environmental Protection Agency (USEPA),
        see Environmental Protection
        Agency
Ultraviolet light
    absorption by
        atmospheric gases 10
        ozone 53, 55, 86, 114
    changes in Earth's evolution 20
    damaging (DUV) 123
    effects on
        ecosystems 209
        humans 123-124
    in solar radiation 2,6
    surface changes 122, 196
    UV-A 122-123
    UV-B 122-123, 203
Ultraviolet spectrometry 234
Unvented gas/kerosene space heaters 361-362
Urea-formadehyde 368-371

V

Vapor pressure, saturation 5
Vapor recovery 305
Variation, coeffient of 227
Vehicles, *see* Motor vehicle(s)
Ventilation
    crankcase 305, 310
    guidelines 388-389
    rates 382, 387
Ventilation systems
    local exhaust 321, 388
    mechanical 387-389
    natural 386-387
Very volatile organic compounds (VVOCs) 44-45,
        371
Vinyl chloride, human health effects of 175
Vinyl chloride
    as hazardous pollutant 275
    effect on humans 175
Visibility 183
    impairment 95-99, 103, 218
    measurements 100-101
    meteorological factors 99-100

trends 101
values, sensor equivalent 101
Visual Environments, Interagency Monitoring to
        Protect 230
Visual range 94, 98, 101
Volatile organic compounds (VOCs), *see* Organic
        compounds, volatile

W

Warming, global
    effects 136, 204-205
    nature 128-136
    treaty 289
    trends 131-133
    uncertainties 135-136
Washout 89
Water
    as oxidation product 335
    biological treatment 346
    for scrubbing 333
Water vapor
    and global warming 136-137
    atmospheric 1-2
    collection by sorbents 338
    concentration 4-5
    condensation 5
    greenhouse gas 7-8, 128, 130
    in atmospheric evolution 19-20
    in stack samples 242
    stratospheric 124
    uptake by particles 56, 64, 105-106
Weighting scales 399-401
Welfare effects 183
Winds
    direction 74
    geostropic 15
    gradient 15
    horizontal 72-74
    in atmospheric motion 13
    in horizontal transport 72, 85
    speed 72-73, 83, 90, 99, 124, 321
Wood-burning appliances 142, 354, 363
Working levels 357-359
World Health Organization (WHO) standards
    nitrogen dioxide 168
    PM_{10} 360

Z

Zeolite, as catalyst 308
Zones
    crevice 302
    quench 302